S0-BNF-757

Reagents for Organic Synthesis

PROOFREADER

ARMIN BURGHART

Fiesers'
Reagents for Organic Synthesis

VOLUME EIGHTEEN

Tse-Lok Ho

National Chiao Tung University
Republic of China

A WILEY-INTERSCIENCE PUBLICATION
JOHN WILEY & SONS, INC.
NEW YORK / CHICHESTER / WEINHEIM / BRISBANE / SINGAPORE / TORONTO

This book is printed on acid-free paper. ∞

Copyright © 1999 by John Wiley & Sons, Inc.

All rights reserved. Published simultaneously in Canada.

No part of this publication may be reproduced, stored in a retrieval system or transmitted in
any form or by any means, electronic, mechanical, photocopying, recording, scanning or
otherwise, except as permitted under Sections 107 or 108 of the 1976 United States
Copyright Act, without either the prior written permission of the Publisher, or authorization
through payment of the appropriate per-copy fee to the Copyright Clearance Center,
222 Rosewood Drive, Danvers, MA 01923, (978) 750-8400, fax (978) 750-4744. Requests
to the Publisher for permission should be addressed to the Permissions Department,
John Wiley & Sons, Inc., 605 Third Avenue, New York, NY 10158-0012, (212) 850-6011,
fax (212) 850-6008, E-Mail: PERMREQ@WILEY.COM.

Library of Congress Cataloging in Publication Data:
ISBN 0-471-24477-5
ISSN 0271-616X
Printed in the United States of America.

10   9   8   7   6   5   4   3   2   1

To the memory of
Professor Louis F. Fieser
and
Mrs. Mary Fieser

PREFACE

My first contact with organic chemistry was through the textbooks and experimental manual written by Professor L. F. Fieser and Mrs. M. Fieser. These works impressed me not only with their organization and lucidity, but the appearance of a compilation of reagents as an appendix really aroused my curiosity about research tools. This feature was not found in any other textbook, and was a harbinger of the famous and successful series, *Reagents for Organic Synthesis.*

The Fiesers will long be remembered for their outstanding contributions to chemical education. Their dedication to providing a quality service during the latter part of their lives played an important role in the rapid advances of organic synthesis during the past thirty years. The quick retrieval of essential information through consultation of the series must have saved uncountable man-hours in research, and in the course of such readings there must also have been occasions that a chemist became inspired to develop improved or new synthetic methods.

The "ROS" reference series is an established institution. That is the main reason Wiley & Sons decided to continue its publication. In accepting an invitation to authorship with great trepidation, I can only hope that my feeble efforts will prove to be more than merely "using a dog's tail to substitute for a mink's." (狗尾續貂)

The previous format is essentially maintained, except that more conventional reference abbreviations are used. Due to space limitations and my attempt to cover many more papers, unnecessary explanations are omitted. With the same considerations, a general transformation that can be described in a short sentence without ambiguity is deemed sufficient, and the corresponding graphics are dispensed with.

TSE-LOK HO

CONTENTS

REFERENCE ABBREVIATIONS

ACR	Acc. Chem. Res.
ACS	Acta Chem. Scand.
ACIEE	Angew. Chem. Int. Ed. Engl.
AJC	Aust. J. Chem.
AOMC	Appl. Organomet. Chem.
BBB	Biosc. Biotech. Biochem.
BCSJ	Bull. Chem. Soc. Jpn.
BSCB	Bull. Soc. Chim. Belg.
BSCF	Bull. Russ. Chim. Fr.
BRAS	Bull. Russ. Acad. Sci.
CB	Chem. Ber.
CC	Chem. Commun.
CCCC	Collect. Czech. Chem. Commun.
CJC	Can. J. Chem.
CL	Chem. Lett.
CPB	Chem. Pharm. Bull.
CR	Carbohydr. Res.
DC	Dokl. Chem. (Engl. Trans.)
G	Gazz. Chim. Ital.
H	Heterocycles
HC	Heteroatom Chem.
HCA	Helv. Chim. Acta
HX	Huaxue Xuebao
IJC(B)	Indian J. Chem., Sect. B
IJS(B)	Int. J. Sulfur Chem., Part B
JACS	J. Am. Chem. Soc.
JCC	J. Carbohydr. Chem.
JCCS(T)	J. Chin. Chem. Soc. (Taipei)
JCR(S)	J. Chem. Res. (Synopsis)
JCS(P1)	J. Chem. Soc. Perkin Trans. 1
JFC	J. Fluorine Chem.
JHC	J. Heterocycl. Chem.
JMC	J. Med. Chem.
JNP	J. Nat. Prod.
JOC	J. Org. Chem.
JOMC	J. Organomet. Chem.

JOCU	J. Org. Chem. USSR (Engl. Trans.)
LA	Liebigs Ann. Chem.
MC	Mendeleev Commun.
NJC	New Journal of Chemistry
NKK	Nippon Kagaku Kaishi
OM	Organometallics
PAC	Pure Appl. Chem.
PSS	Phosphorus Sulfur Silicon
RJOC	Russian J. Org. Chem.
RTC	Recl. Trav. Chim. Pays-Bas
S	Synthesis
SC	Synth. Commun.
SL	Synlett
SOC	Synth. Org. Chem. (Jpn.)
T	Tetrahedron
TA	Tetrahedron: Asymmetry
TL	Tetrahedron Lett.
YH	Youji Huaxue

Reagents for Organic Synthesis

A

4-Acetamido-2,2,6,6-tetramethylpiperidin-1-oxyl (1).

Glycol oxidation.[1] The tosic acid salt of 4-acetamido-TEMPO is a mild oxidizing agent that converts glycols to α-dicarbonyl compounds.

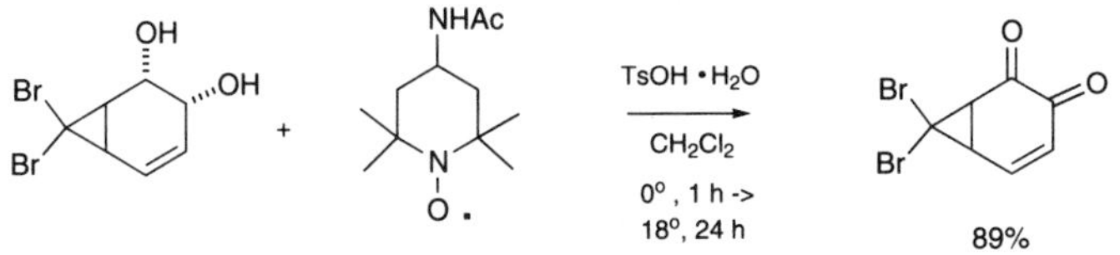

[1]M. G. Banwell, V. S. Bridges, J. R. Dupuche, S. L. Richards, and J. M. Walter, *JOC* **59**, 6338 (1994).

Acetic anhydride.

Acetylation.[1] In the presence of freshly prepared MgI_2 (from Mg and I_2 in dry ether), Ac_2O acetylates primary, secondary, and tertiary alcohols as well as phenols.

Cyclization.[2] Acetic anhydride provides a diketene unit on reaction with ureas in the presence of DMAP.

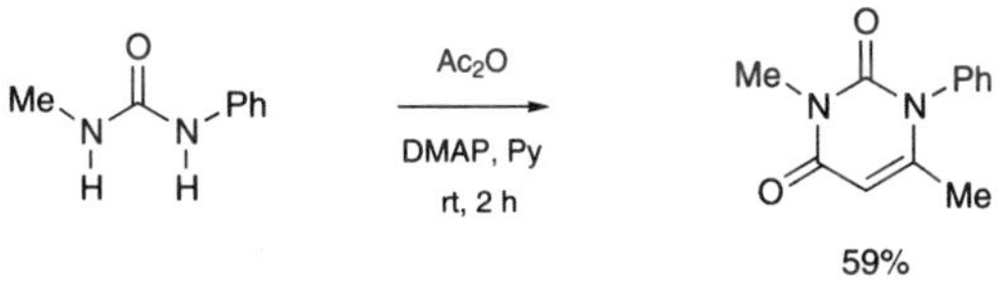

Rearrangement-cyclization.[3] δ-Oximino nitriles form 2-acetamidopyridines on treatment with Ac_2O-AcCl under reflux. The reaction proceeds from rearrangement of *N*-acetoxyenamines to *C*-acetoxy imines, which undergo elimination and cyclization.

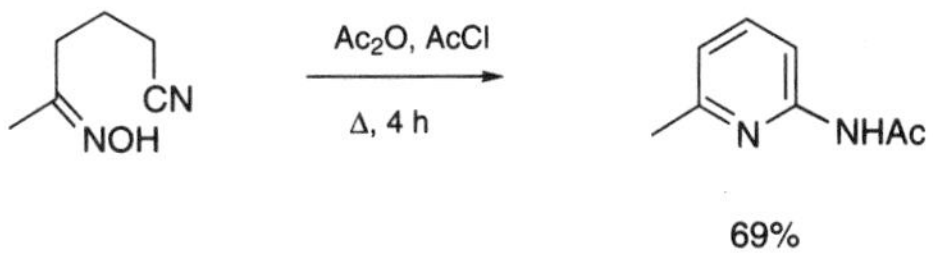

Pummerer rearrangement.[4] γ-Sulfinyl acids give γ-thio-γ-butyrolactones. In a selenium version,[5] the rearrangement product (with a α,α-difluorinated selenoxide) can react with cyclic ethers.

1

[1] P. K. Chowdhury, *JCR(S)* 338 (1993).
[2] J. Aichner, H. Egg, D. Gapp, S. Haller, D. Rakowitz, and U. Ramspacher, *H* **36**, 307 (1993).
[3] R. J. Vijn, H. J. Arts, P. J. Maas, and A. M. Castelijns, *JOC* **58**, 887 (1993).
[4] H. Su, Y. Hiwatari, M. Soenosawa, K. Sasuga, K. Shirai, and T. Kumamoto, *BCSJ* **66**, 2603 (1993).
[5] K. Uneyama, Y. Tokunaga, and K. Maeda, *TL* **34**, 1311 (1993).

Acetone cyanohydrin.

Nitrile synthesis.[1]   As a cyanide ion source in the preparation of alkyl nitriles by displacement of alkyl halides, it permits the reaction in an aprotic solvent.

[1] P. Dowd, B. K. Wilk, and M. Wlostowski, *SC* **23**, 2323 (1993).

Acetonitrile. 15, 1

Ritter reaction.[1]   Benzyl alcohols are converted to amides in acetonitrile in the presence of $BF_3 \cdot OEt_2$.

[1] H. Firouzabadi, A. R. Sardarian, and H. Badparva, *SC* **24**, 601 (1994).

α-Acetoxyisobutyryl chloride. 8, 3

Regioselective chlorination.[1]   Primary alcohols are converted to chlorides; thus unprotected alditols react selectively.

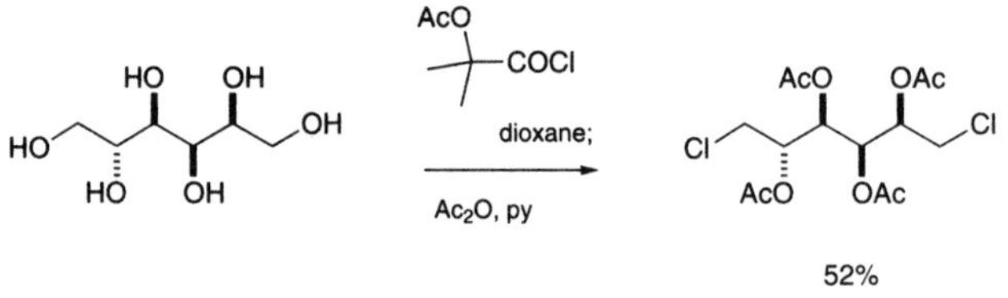

52%

[1] M. Benazza, M. Massoui, R. Uzan, and G. Demailly, *JCC* **13**, 967 (1994).

Acetyl chloride.

Selective acetylation.[1]   Primary alcohols are selectively acetylated in the presence of secondary alcohols with acetyl chloride and varoius bases, such as collidine, diisopropylethylamine, or 1,2,2,6,6-pentamethylpiperidine.

[1] K. Ishihara, H. Kurihara, and H. Yamamoto, *JOC* **58**, 3791 (1993).

Acetylene.

Vinylation.[1]   Pyrrole undergoes *N*-vinylation with acetylene in KOH–DMSO.

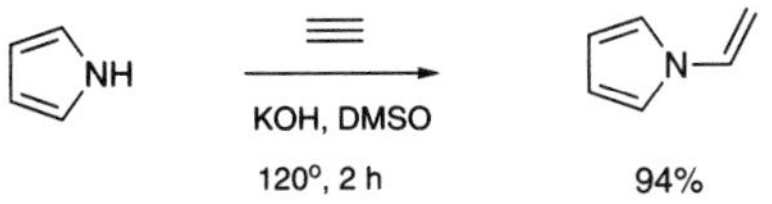

[1] O. A. Tarasova, A. G. Mal'kina, A. I. Mikhaleva, L. Brandsma, and B. A. Trofimov, *SC* **24**, 2035 (1994).

Acetyl hypobromite.

*N***-Bromination.**[1] Amides are brominated with AcOBr in CCl_4 at room temperature.

[1] L. Duhamel, G. Ple, and P. Angibaud, *SC* **23**, 2423 (1993).

2-(4-Acetyl-2-nitrophenyl)ethanol.

Carboxyl group protection.[1] The ω-esters of aspartic and glutamic acids are formed readily using the DCC method, after the geminal functionalities are sequestered (by reaction with Et_3B). The acid can be regenerated under conditions (0.1 M Bu_4NF) that do not affect an *N*-Boc group.

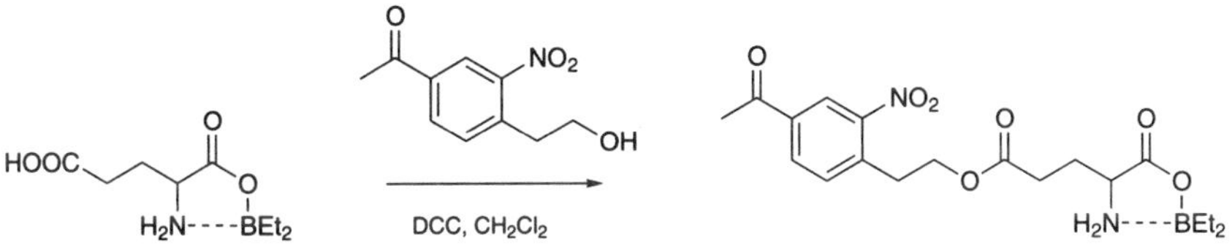

[1] J. Robles, E. Pedroso, and A. Grandas, *S* 1261 (1993).

N-Acylaziridines.

Polyketides.[1] Prepared in a one-pot reaction from RCOOH, $(COCl)_2$, Et_3N, and an aziridine (e.g., 2-methylaziridine), these reagents are suitable acyl donors for chain extension of poly-1,3-dicarbonyl compounds via reaction with their polyanions.

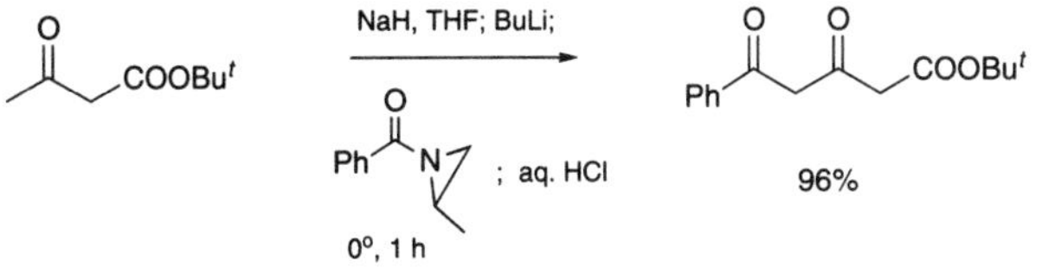

[1] B. Lygo, *TL* **35**, 5073 (1994).

Acylmethylenetriphenylphosphoranes.

Acetylene precursors. Flash vacuum pyrolysis of these stabilized Wittig reagents removes triphenylphosphine oxide to furnish acetylenes. Both terminal[1] and internal acetylenes[2,3] are accessible by this method.

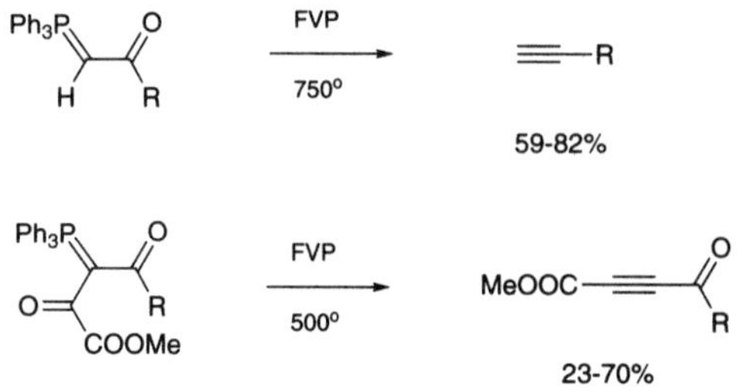

59-82%

23-70%

[1] R. A. Aitken and J. I. Atherton, *JCS(PI)* 1281 (1994).
[2] R. A. Aitken, H. Herion, A. Janosi, S. V. Raut, S. Seth, I. J. Shannon, and F. C. Smith, *TL* **34**, 5621 (1993).
[3] R. A. Aitken, C. E. R. Horsburgh, J. G. McCreadie, and S. Seth, *JCS(PI)* 1727 (1994).

Acylsilanes. 17, 1–2

Cyclopropanation.[1]   Reaction with ketone enolates results in 1,2-cyclopropane-diol derivatives.

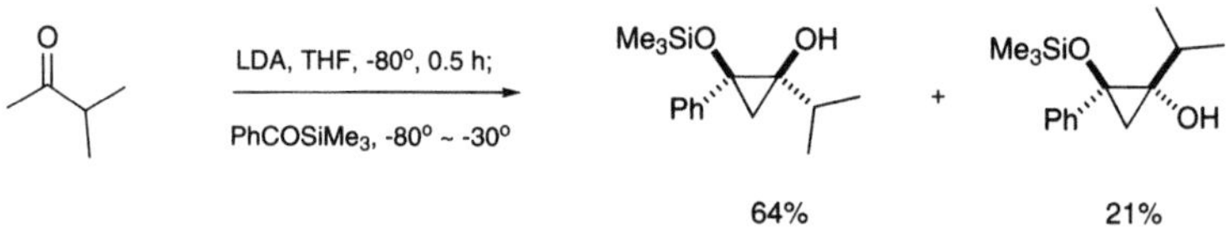

64%   21%

[1] K. Takeda, J. Nakatani, H. Nakamura, K. Sako, E. Yoshii, and K. Yamaguchi, *SL* 841 (1993).

Acyltelluranes.

These reagents are readily obtained[1] from Et₂AlCl-catalyzed reaction of alde-hydes with diisobutylaluminum butyltelluride in THF at room temperature.

Enol silyl ethers of acylsilanes.[2]   On treatment with butyllithium and trimethyl-silyl chloride, these compounds undergo enol silylation, tellurium–lithium exchange, and O → C silyl migration. The lithium enolates are further silylated.

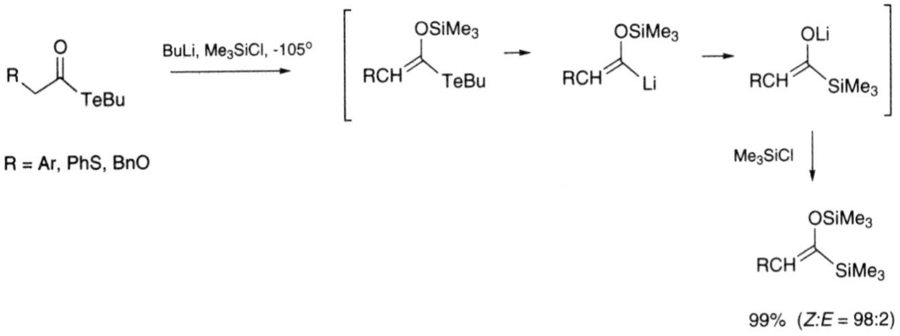

99% (*Z:E* = 98:2)

R = Ar, PhS, BnO

[1] T. Inoue, K. Takeda, N. Kambe, A. Ogawa, I. Ryu, and N. Sonoda, *JOC* **59**, 5824 (1994).
[2] T. Inoue, N. Kambe, I. Ryu, and N. Sonoda, *JOC* **59**, 8209 (1994).

1-Adamantyl fluoroformate (1).

Phenol protection.[1]   The reagent (**1**) transforms phenols carrying strongly electron-withdrawing groups into mixed carbonates. Deprotection is accomplished by use of trifluoroacetic acid at 0–20° for 20–60 min.

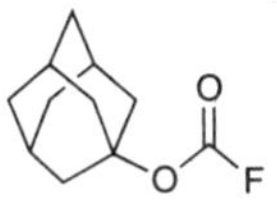

(1)

[1] I. Niculescu-Duvaz and C. J. Springer, *JCR(S)* 242 (1994).

2-Alkanesulfonylbenzothiazole (1).

Olefination.[1]   The anions of (**1**) react with carbonyl compounds to give predominantly (*E*)-alkenes.

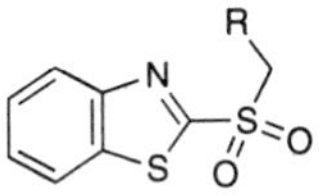

(1)

[1] J.-B. Baudin, G. Hareau, S. A. Julia, R. Lorne, and O. Ruel, *BSCF* 856 (1993).

Alkenylboranes.

Enynes.[1]   The attack of lithioalkynes on alkenyldisiamylboranes followed by oxidation with iodine leads to enynes in which the original configuration of the alkenes is retained.

Carbonylation.[2]   An enone that bears chirality at both α' and β' carbon atoms can be synthesized through successive reactions of a chiral borane with an alkene and an alkyne, followed by exposure to acetaldehyde. Further treatment of the resulting alkenylborane with dichloromethyl methyl ether and Et_3COLi, and finally with H_2O_2 affords the product.

Allylamines.[3]   Alkeneboronic acids are converted to allylamines of defined configuration on treatment with conventional Mannich reaction components (R_2NH, paraformaldehyde). The required vinylboronic acids are available from 1-alkynes by reaction with catecholborane followed by hydrolysis.

γ,δ-Unsaturated ketones.[4,5]   The synthesis from alkenyldiisopropoxyboranes and α,β-unsaturated ketones is catalyzed by $BF_3 \cdot OEt_2$. It is suitable for reaction with substrates that are labile to organometallics (e.g., Cu reagents). The corresponding alkynyl ketones are obtained in an analogous manner.

[1] P. Bovicelli, P. Lupatelli, and E. Mincione, *JNP* **56**, 676 (1993).
[2] H. C. Brown and V. K. Mahindroo, *TA* **4**, 59 (1993).
[3] N. A. Petasis and I. Akritopolou, *TL* **34**, 583 (1993).
[4] E.-i. Takada, S. Hara, and A. Suzuki, *TL* **34**, 7067 (1993).
[5] E.-i. Takada, S. Hara, and A. Suzuki, *HC* **3**, 483 (1992).

Alkenyltriphenylphosphonium tetrafluoroborate, $Ph_3P^+CH=CHR\ BF_4^-$.

1,3-Dienes.[1]   These salts are available from the reaction of epoxides with an acid chloride (AcCl or oxalyl chloride) and $[Ph_3PH]^+BF_4^-$. Diene formation in moderate yields is observed when these salts are treated with ArCHO and DBU.

[1] K. Okuma, M. Ono, and H. Ohta, *BCSJ* **66**, 1308 (1993).

3-Alkoxyacryloyl chlorides.

An improved preparation[1] involves reaction of vinyl ethers with oxalyl chloride followed by heating above 100°C.

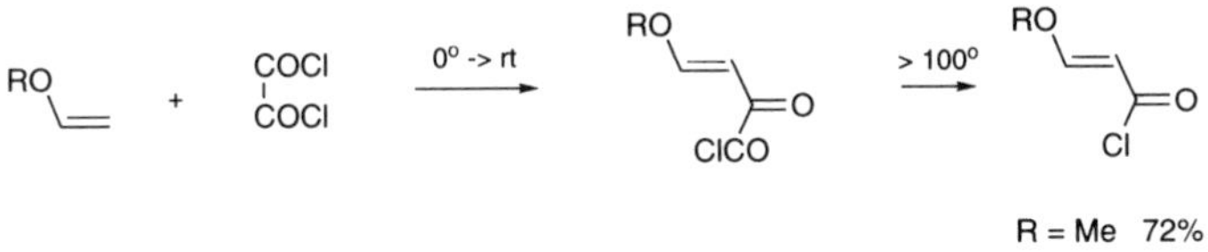

RO‖ + COCl / COCl → (0° -> rt) RO‖ CICO O → (> 100°) RO‖ Cl O

R = Me 72%

[1] L. F. Tietze, C. Schneider, and M. Pretor, S 1079 (1993).

Alkylaluminum chlorides. **13**, 5–8; **14**, 4–5; **15**, 2–5; **16**, 1–3; **17**; 4–7

Mukaiyama aldolization.[1] The intramolecular reaction of a dioxalenium ion with a silyl enol ether side chain gives dioxabicyclic products.

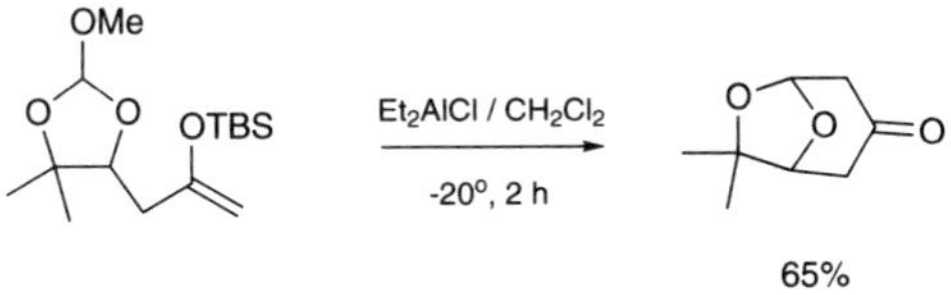

65%

Ene reactions. The intramolecular version is suitable for the construction of 12-, 14-, and 16-membered ring systems.[2] Excellent 1,3-chirality transfer in this process is evident.[3]

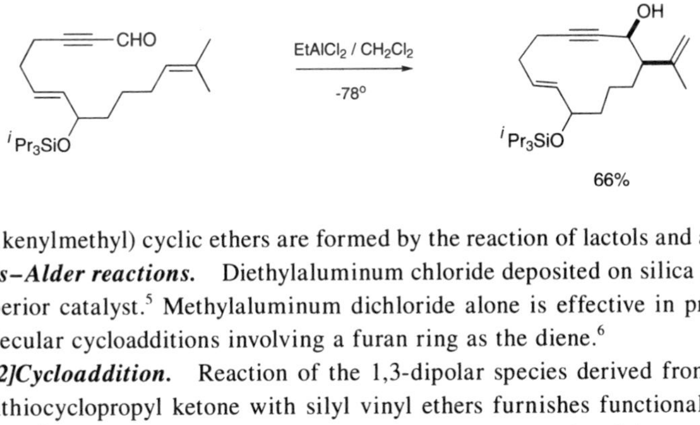

66%

2-(Alkenylmethyl) cyclic ethers are formed by the reaction of lactols and alkenes.[4]

Diels–Alder reactions. Diethylaluminum chloride deposited on silica seems to be a superior catalyst.[5] Methylaluminum dichloride alone is effective in promoting intramolecular cycloadditions involving a furan ring as the diene.[6]

[3+2]Cycloaddition. Reaction of the 1,3-dipolar species derived from methyl 2-phenylthiocyclopropyl ketone with silyl vinyl ethers furnishes functionalized cyclopentanes.[7] A related reaction is the trapping of a fragmented cyclobutane.[8]

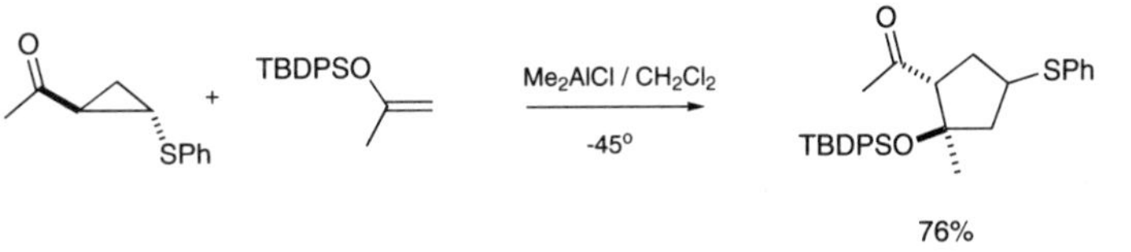

76%

Alkylation of enolizable carbonyl compounds. β-Oxo amides react with organoaluminum chlorides stereoselectively to give alcohol products.[9]

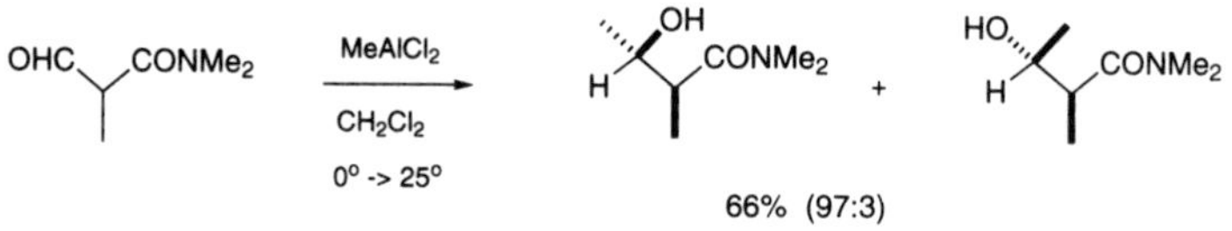

66% (97:3)

γ-Hydroxy esters.[10] The catalyzed epoxide opening by ester enolates is stereoselective, giving predominantly the *syn* products.

Conjugate addition.[11] The addition of diethylaluminum chloride to *N*-crotonyl-4-benzyloxazolidin-2-one shows 1,5-asymmetric induction. The corresponding reaction with Me₂AlCl requires photoactivation.

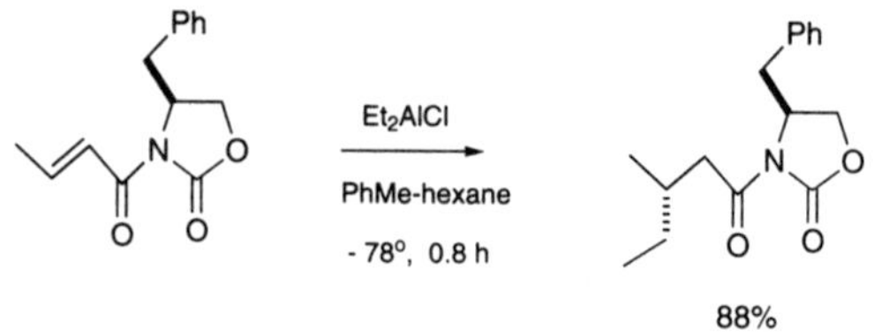

88%

Skeletal rearrangement.[12] The fused ring system (**1**) containing a cyclobutene unit releases its strain by rearrangement to a bridged ring skeleton (**2**). Thus the latter class of compounds are available in two steps, involving the initial TiCl₄-mediated [2+2]cycloaddition of the proper alkynes to cyclic enones.

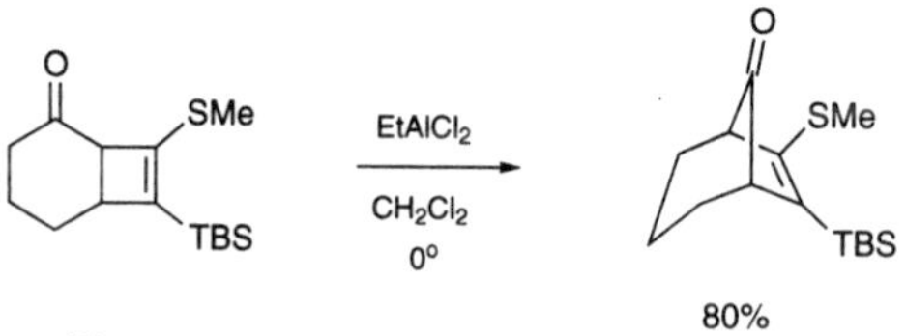

(1)

80%

(2)

[1] S. Leger, J. Omeara, and Z. Wang, *SL* 829 (1994).
[2] J. A. Marshall and M. W. Andersen, *JOC* **58**, 3912 (1993).
[3] K. Masuya, K. Tanino, and I. Kuwajima, *TL* **35**, 7965 (1994).
[4] K. Mikami and H. Kishino, *CC* 1843 (1993).
[5] C. Cativiela, F. Figueras, J. I. Garcia, J. A. Mayoral, E. Pires, and A. J. Royo, *TA* **4**, 621 (1993).
[6] C. Rogers and B. A. Keay, *CJC* **70**, 2929 (1992); **71**, 611 (1993).
[7] Y. Horiguchi, I. Suehiro, A. Sasaki, and I. Kuwajima, *TL* **34**, 6077 (1993).

[8] T. Fujiwara, T. Iwasaki, J. Miyagawa, and T. Takeda, *CL* 343 (1994).

[9] M. Taniguchi, H. Fujii, K. Oshima, and K. Utimoto, *BCSJ* **67**, 2514 (1994).

[10] S. K. Taylor, J. A. Fried, Y. N. Grassl, A. E. Marolewski, E. A. Pelton, T.-J. Poel, D. S. Rezanka, and M. R. Whittaker, *JOC* **58**, 7304 (1993).

[11] K. Ruck and H. Kunz, *S* 1018 (1993).

[12] K. Narasaka, H. Shimadzu, and Y. Hayashi, *CL* 621 (1993).

Alkyl *t*-butyl iminodicarbonates.

Protected amines.[1] Amines can be prepared by Gabriel and Mitsunobu syntheses using the iminodicarbonates (**1**) as nucleophiles. The reagents are available by acylation of 4-methoxybenzylamine with alkyl chloroformates, followed by introduction of the *N*-Boc group and removal of the benzyl moiety with CAN.

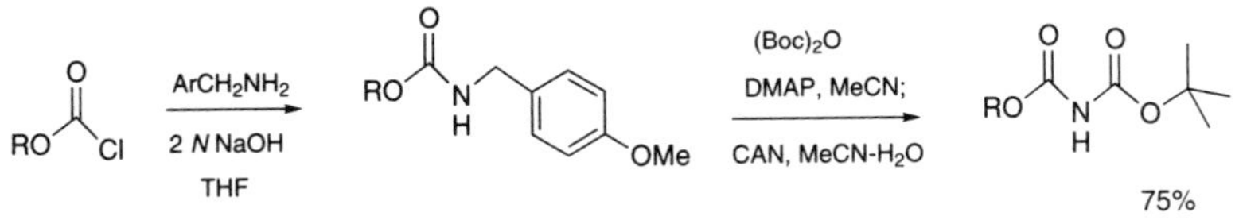

[1] J. M. Chong and S. B. Park, *JOC* **58**, 7300 (1993).

Alkyl phenyl selenides.

Alkylating agents.[1] These selenides, activated through photoinduced electron transfer, react with enol silyl ethers, forming α-alkylated ketones.

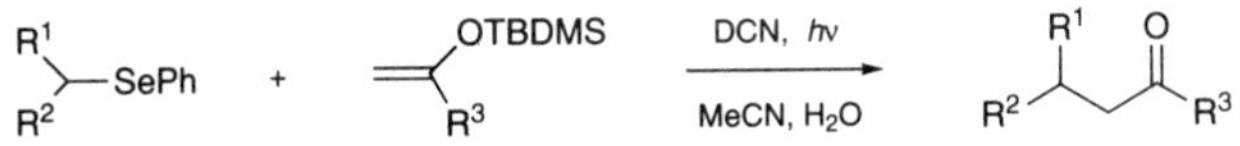

[1] G. Pandey and R. Sochanchingwung, *CC* 1945 (1994).

Alkyl α-phenylthiocrotonates.

Michael reactions. The ester group exerts profound influences on the steric course of the reaction. Thus diastereocontrol[1] is possible by changing solvent, enolate counterion, and activating group at the α-carbon of the acceptor. The phenylthio group increases reactivity, but electron-withdrawing substituents at this position tend to erode the diastereoselectivity. Temperature effects are also dramatic.

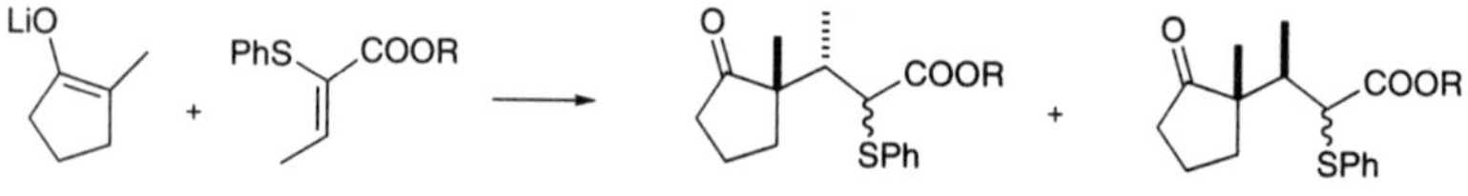

R = Me,   CH$_2$Cl$_2$, -78°, 48 h   81%   (15:1)
R = menthyl,   THF, 0°, 10 min   83%   (1:30)

[1] E. J. Corey and I. N. Houpis, *TL* **34**, 2421 (1993).

Alkyltriphenylarsonium salts.

Unsaturated esters.[1]   Alkylation of the corresponding ylides with a bromoacetic ester generates conjugated esters directly.

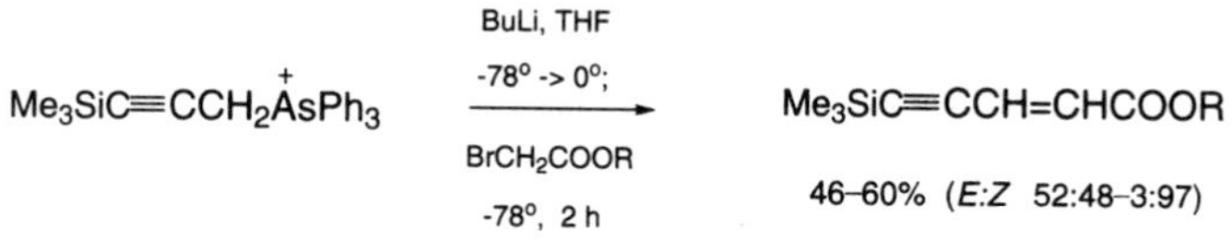

Me$_3$SiC≡CCH$_2$AsPh$_3$

BuLi, THF
-78° -> 0°;

BrCH$_2$COOR
-78°, 2 h

Me$_3$SiC≡CCH=CHCOOR

46–60% (*E:Z* 52:48–3:97)

R = alkyl, propenyl, propynyl

[1] Y. Shen and Y. Xiang, *HC* **3**, 547 (1992).

Allenyl(triphenyl)lead.

1-Alkyn-4-ols.[1]   Propargylation of carbonyl compounds is achieved in a reaction catalyzed by BF$_3$ · OEt$_2$. The reagent is prepared from Ph$_3$PbMgBr and propargyl bromide.

[1] D. Seyferth, D. Y. Son, and S. Shah, *OM* **13**, 2105 (1994).

Allylboranes. **14**, 11–12; 139–141; **16**, 6

Allylation.[1]   2-Allyl-1,2-oxaborolane is an allylating agent prepared from reaction of 2-allyloxy-1,2-oxaborolane with allylmagnesium bromide. Allylation is carried out without solvent, between 0°C to room temperature with aldehydes and at 80–100°C with ketones. Conjugated aldehydes undergo 1,2-addition.

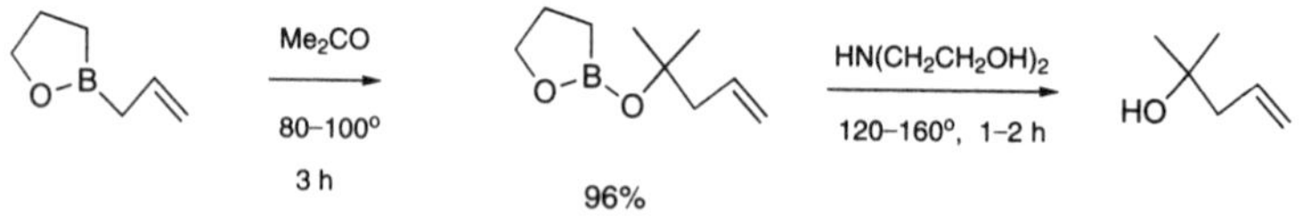

Me$_2$CO
80–100°
3 h

96%

HN(CH$_2$CH$_2$OH)$_2$
120–160°, 1–2 h

B-Allylbis(isocaranyl)boranes (**1**)[2] can deliver the allyl group to carbonyl substrates to give chiral homoallylic alcohols.

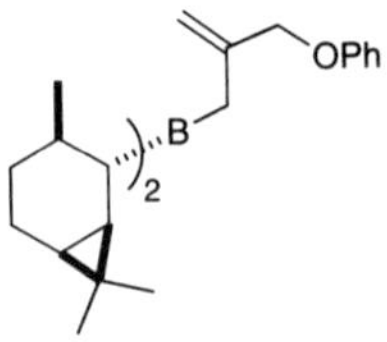

(1)

[1] W. Zhou, S. Liang, S. Yu, and W. Luo, *JOMC* **452**, 13 (1993).
[2] T. A. J. van der Heide, J. van der Baan, E. A. Bijpost, F. J. J. de Kanter, F. Bickelhaupt, and G. W. Klumpp, *TL* **34**, 4655 (1993).

Allyl chloroformate. 13, 9

N ε-Protection of lysine.[1]   Reaction with the copper chelate of lysine furnishes the derivative quantitatively. The method is suitable for large-scale preparation.

[1] A. Crivici and G. Lajoie, *SC* **23**, 49 (1993).

Allyldiisobutyltelluronium bromide.

Heteroatom allylation.   Diisobutyl telluride is a good leaving group; therefore phenols,[1] thiophenols,[2] and anilines[3] are readily allylated by this salt.

Cyclopropanation.[4]   The derived ylide (**1**) is able to participate in cyclopropane ring formation with enones. It is not necessary to use the salt stoichiometically as diisobutyl telluride can be made catalytic.

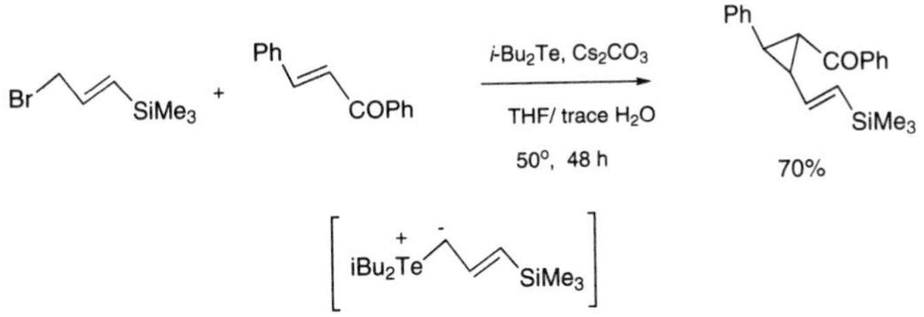

(1)

[1] C. Xu, S. Lu, and X. Huang, *SC* **23**, 2527 (1993).
[2] S.-M. Lu, C.-D. Xu, and X. Huang, *YH* **14**, 545 (1994).
[3] C. Xu, S. Lu, and X. Huang, *HC* **5**, 7 (1994).
[4] Y.-Z. Huang, Y. Tang, Z.-L. Zhou, W. Xia, and L.-P. Shi, *JCS(PI)* 893 (1994).

Allyldicarbonyl(cyclopentadienyl)iron.

Heterocycles. Lewis acid–catalyzed reaction with aromatic aldehydes[1,2] followed by CAN oxidation in methanol leads to methyl 2-aryltetrahydrofuran-4-carboxylates. The reactivity of aliphatic aldehydes and ketones is inferior, and for ketones the reaction proceeds better using $TiCl_4$ as catalyst.

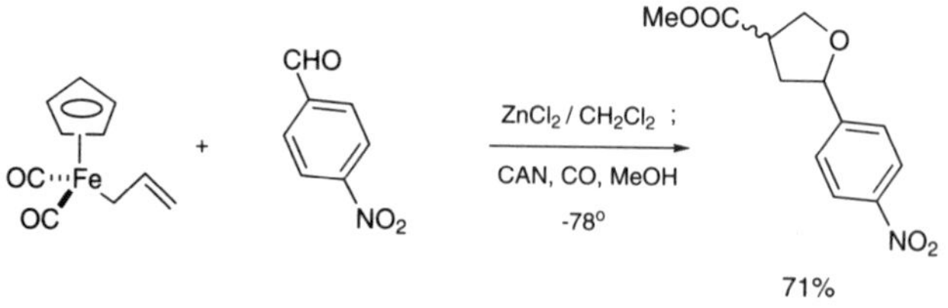

71%

N-Tosylpyrrolidine derivatives[3] are similarly obtained in reactions with tosyl imines.

[1] S. Jiang and E. Turos, *OM* **12**, 4280 (1993); *TL* **35**, 7889 (1994).
[2] S. Jiang and E. Turos, *TL* **35**, 7889 (1994).
[3] T. Chen, S. Jiang, and E. Turos, *TL* **35**, 8325 (1994).

Allyldimethylsilyl triflate.

Allylation of aldehydes.[1] The reagent, prepared by reaction of diallyldimethylsilane with triflic acid, reacts with aromatic aldehydes at low temperatures to afford homoallylic alcohols.

[1] M. A. Brook, G. D. Crowe, and H. Hiemstra, *CJC* **72**, 264 (1994).

Allylidene(triphenyl)phosphoranes.

Cyclopentadienes.[1] One step annulation with α-haoketones involves alkylation and intramolecular Wittig reaction.

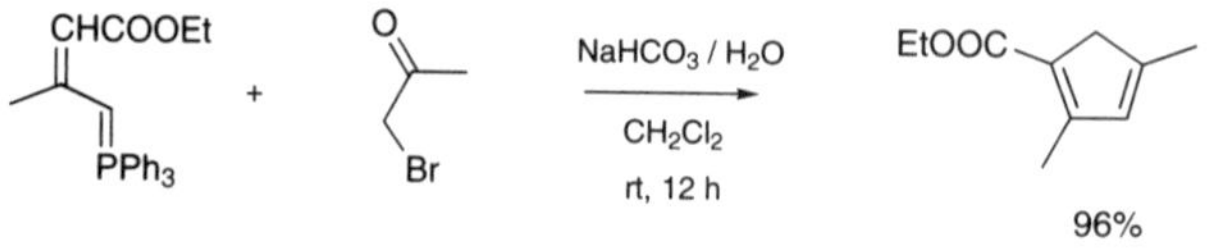

96%

[1] M. Hatanaka, Y. Himeda, and I. Ueda, *JCS(PI)* 2269 (1993).

(Allyloxycarbonylamino)methanol.

Cysteine protection.[1] For masking the thiol group of cysteine, the reagent is applied in trifluoroacetic acid/dichloromethane. Deprotection is achieved by palladium-catalyzed reductive deallylation.

[1] A. M. Kimbonguila, A. Merzouk, F. Guibe, and A. Loffet, *TL* **35**, 9035 (1994).

Allyltributylstannanes. **13**, 10; **14**, 14–17; **16**, 7–9, 342–343; **17**, 12–13

Allylation of carbonyl compounds.[1] The reaction with α-ketols is under chelation control to give *syn*-1,2,diols.

Allylation of β-oxy-o-iodoanilides.[2] Generation of an α-radical is via a radical translocation reaction. The allylation is subject to 1,2-asymmetric induction, giving predominantly the anti products.

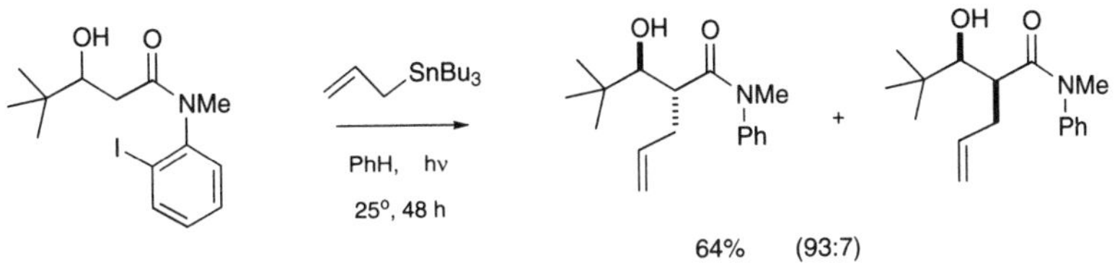

Allylation of heterocycles.[3,4] Upon activation with a chloroformate ester, pyridines and azoles are susceptible to attack by allyltin reagents. Interestingly, the nature of the substituent on C-2 of the reagent has a determining effect on the regio-selectivity in the addition reaction to the pyridine ring. Thus an electron-withdrawing group (cyano and acyl groups) favors addition to C-4, whereas electron-donating substituents (e.g., methyl) direct reaction at C-2.

γ,δ-Unsaturated acids.[5] Free radicals generated from α-iodoacyl derivatives of Oppolzer's camphor sultam are converted to γ,δ-unsaturated amides with defined absolute configuration. The chiral auxiliary can be removed to give the acids.

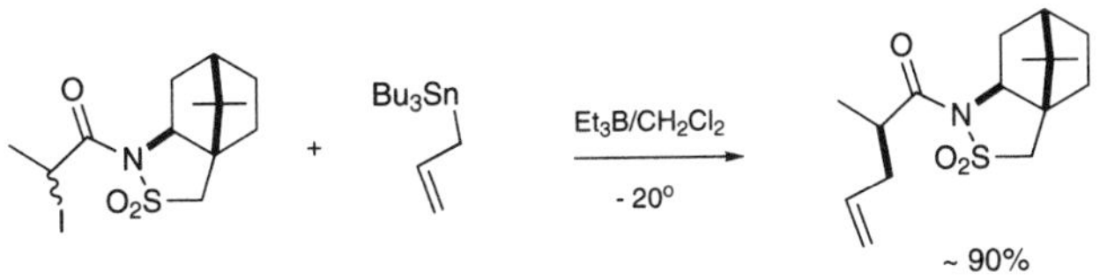

Allyl replacement.[6] The radical reaction is suitable for preparation of homo-allylic fluoro compounds from *gem*-nitroalkyl fluorides.

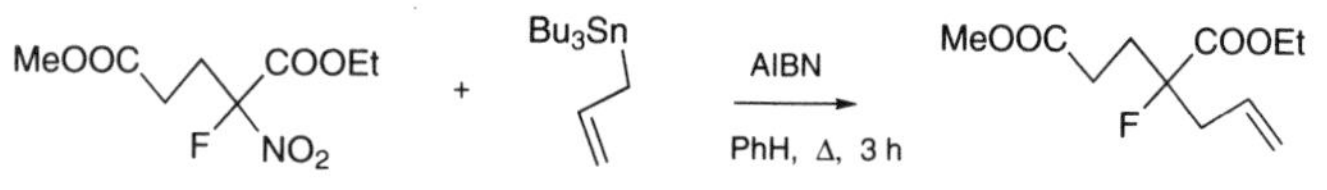

Addition to α,β-epoxy aldehydes.[7] The *syn*-selective reaction catalyzed by LiClO$_4$ is under chelation control.

1-Alken-6-ones.[8] A four-component synthesis from an alkyl halide, carbon monoxide, a Michael acceptor, and the allylstannane in the presence of AIBN starts with generation of an alkyl radical R·, which is subsequently trapped by the three other reagents in a tandem reaction. Yields are excellent.

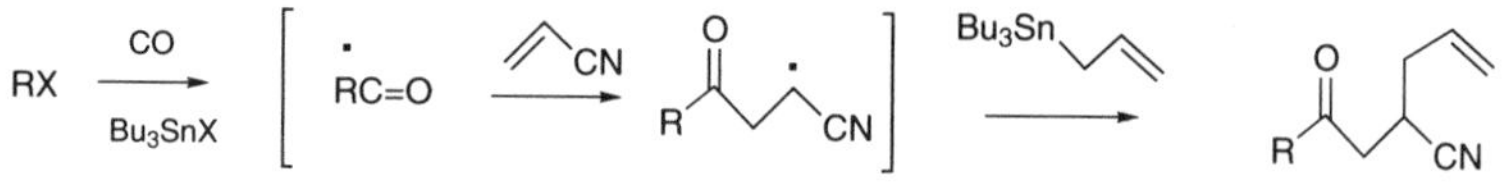

Epoxide opening.[9] Terminal epoxides are opened regioselectively with allyl-lithium reagents derived from the stannanes.

[1] D. J. Hallett and E. J. Thomas, *SL* 87 (1994).
[2] D. P. Curran, A. C. Abraham, and P. S. Ramamoorthy, *T* **49**, 4821 (1993).
[3] R. Yamaguchi, K. Mochizuki, S. Kozima, and H. Takaya, *CC* 981 (1993).
[4] T. Itoh, H. Hasegawa, K. Nagata, and A. Ohsawa, *JOC* **59**, 1319 (1994).
[5] D. P. Curran, W. Shen, J. Zhang, S. J. Gieb, and C.-H. Lin, *H* **37**, 1773 (1994).
[6] Y. Takeuchi, A. Kanada, S. Kawahara, and T. Koizumi, *JOC* **58**, 3483 (1993).
[7] J. Ipaktschi, A. Heydari, and H.-O. Kalinowski, *CB* **127**, 905 (1994).
[8] I. Ryu, H. Yamazaki, A. Ogawa, N. Kambe, and N. Sonoda, *JACS* **115**, 1187 (1993).
[9] L. E. Overman and P. A. Renhowe, *JOC* **59**, 4138 (1994).

Allyltrichlorosilanes.

Allylation of aldehydes. Formation of homoallylic alcohols proceeds in a highly stereoselective manner,[1] depending on the double bond configuration of the allylsilane. The silane reagents can be generated in situ from conjugated dienes by Pd(0)-catalyzed hydrosilylation. In the presence of a chiral phosphoramide,[2] it is possible to achieve asymmetric allylation of aldehydes.

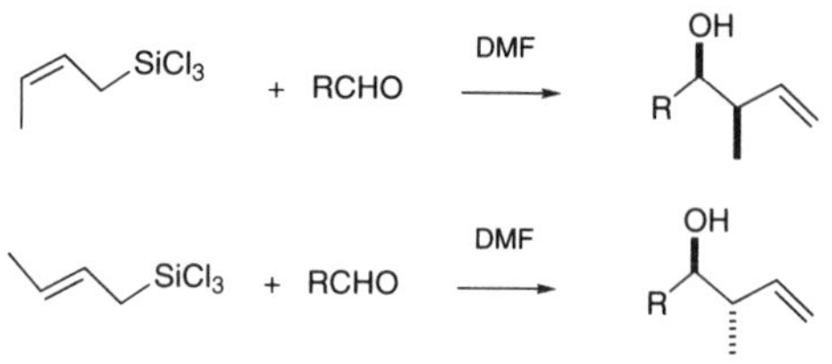

[1] S. Kobayashi and K. Nishio, *JOC* **59**, 6620 (1994).
[2] S. E. Denmark, D. M. Coe, N. E. Pratt, and B. D. Griedel, *JOC* **59**, 6161 (1994).

Allyltrimethylsilane. **13**, 11–13; **14**, 18–19; **15**, 8

Trimethylsilyl fluorosulfonate.[1]   The compound is comparable to trimethylsilyl triflate as a source of trimethylsilyl cation. It is formed by treatment of allyltrimethylsilane (or tetramethylsilane) with fluorosulfonic acid in CH_2Cl_2 at $-78°C$.

α-Allyl amino acids.[2]   α-Acyliminium ions, which are generated in situ from α-alkoxy amino acid derivatives, are efficiently trapped by various allylsilanes.

N-Homoallyl amides.[3]   N-Trifyloxy amides readily give N-acyliminum species on heating in isopropanol. In the presence of allyltrimethylsilane, homoallylic amides are formed as products.

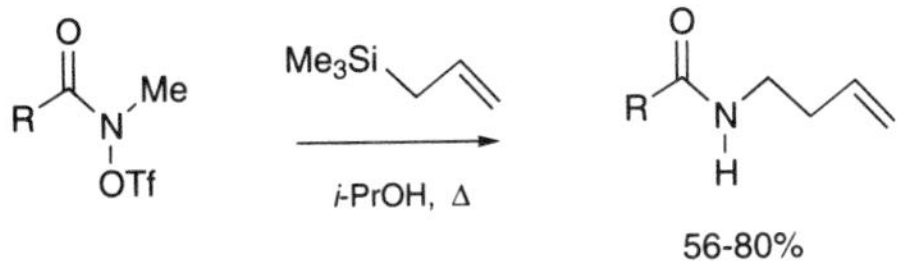

56-80%

Homoallyl selenides.[4]   Selenoacetals are converted to homoallyl selenides in a Lewis acid–promoted reaction.

Pyrrolidines from [3+2]cycloaddition.   Using $BF_3 \cdot OEt_2$ as catalyst, N-Cbz-α-amino aldehydes react with allyltrimethylsilane stereoselectively to give the N-protected *cis, cis*-2-alkyl-3-hydroxy-5-trimethylsilylmethylpyrrolidines.[5] Interestingly, similar reactions of 2-chloromethylallyltrimethylsilane with N-Boc amino aldehydes proceed by the desilylative ene reaction pathway.[6]

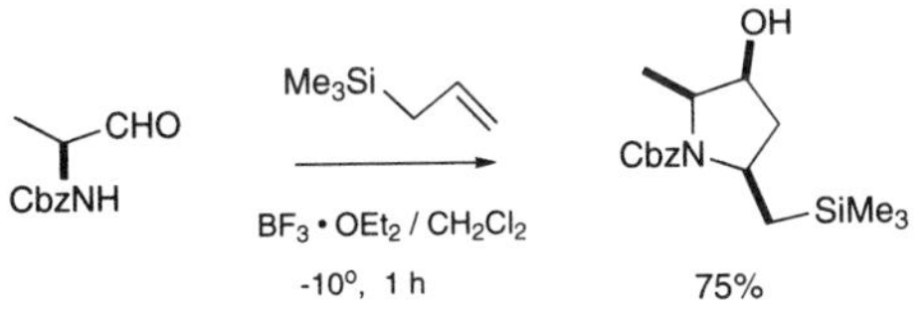

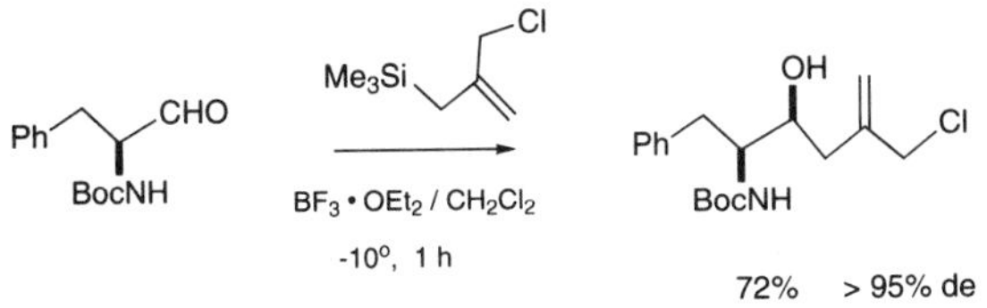

[1] B. H. Lipshutz, J. Burgess-Henry, and G. P. Roth, *TL* **34**, 995 (1993).
[2] E. C. Roos, M. C. Lopez, M. A. Brook, H. Hiemstra, W. N. Speckamp, B. Kaptein, J. Kamphuis, and H. E. Schoemaker, *JOC* **58**, 3259 (1993).

[3] R. V. Hoffman, N. K. Nayyar, J. M. Shankweiler, and B. W. Klinekole, *TL* **35**, 3231 (1994).
[4] B. Hermans and L. Hevesi, *BSCB* **103**, 257 (1994).
[5] S. Kiyooka, Y. Shiomi, H. Kira, Y. Kaneko, and S. Tanimori, *JOC* **59**, 1958 (1994).
[6] F. D'Aniello, A. Mann, D. Mattii, and M. Taddei, *JOC* **59**, 3762 (1994).

Allylzinc reagents.

Homoallylic alcohols.[1] Allylation of carbonyl compounds is straightforward. Under catalyzed conditions enones undergo conjugate addition.

Reductive allylation[2] of perfluorocarboxylic esters involves prior treatment with *i*-Bu$_2$AlH.

$$F_3CCOOEt \xrightarrow[\text{CH}_2\text{Cl}_2, -78°]{\text{i-Bu}_2\text{AlH}} \underset{\substack{| \\ OEt}}{F_3CCHOAlBu_2^i} \xrightarrow[\substack{\text{ZnBr}_2/ \\ \text{CH}_2\text{Cl}_2, 40°}]{\text{Bu}_3\text{Sn}} \underset{\substack{| \\ OH}}{F_3CCHCH_2CH=CH_2}$$

64%

Addition to imines.[3] α-Allylglycine derivatives are readily acquired from α-imino esters.

N-Homoallyl amides.[4] *N*-acyl-α-methoxyamines (available from electrochemical methoxylation of amides and carbamates) are conveniently transformed to homo-allyl amines and β-amino acids by replacement of the methoxy group.

Chlorohydrins.[5] α-Chloroallylzinc reagents are formed by deprotonation of allyl chloride and treatment with ZnCl$_2$. Further reaction with aldehydes and ketones furnishes chlorohydrins.

61-96%

[1] J. J. Eshelby, P. C. Crowley, and P. J. Parsons, *SL* 277 (1993).
[2] T. Ishihara, H. Hayashi, and H. Yamanaka, *TL* **34**, 5777 (1993).
[3] G. Courtois and L. Miginiac, *JOMC* **450**, 33 (1993).
[4] N. Kise, H. Yamazaki, T. Mabuchi, and T. Shono, *TL* **35**, 1561 (1994).
[5] K. Mallaiah, J. Satyanarayana, H. Ila, and H. Junjappa, *TL* **34**, 3145 (1993).

Alumina. 14, 20–21; **16**, 9–10

Ether cleavage.[1] Alumina containing 3% H$_2$O mediates de-*O*-silylation with or without solvent. With microwave irradiation the reaction time shortens dramatically.

Ester hydrolysis.   Efficient cleavage of esters,[2-4] including phenyl pivalate,[5] can be effected in a short time with assistance of microwave. It is possible to cleave a phenyl acetate in the presence of a primary alkyl acetate by shortening the reaction time (e.g., 30 sec vs. 2.5 min).[3]

Aldol condensation.[6] The ultrasound-promoted reaction between silyl enol ethers and aldehydes on alumina surface is *anti*-selective.

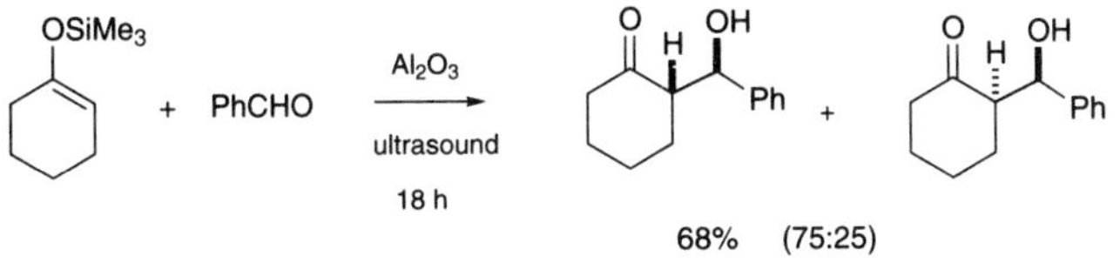

β-Keto esters.[7]   After activation at 200°C under vacuum alumina is used to catalyze the reaction between ethyl diazoacetate and aldehydes, which requires only 10 min at room temperature.

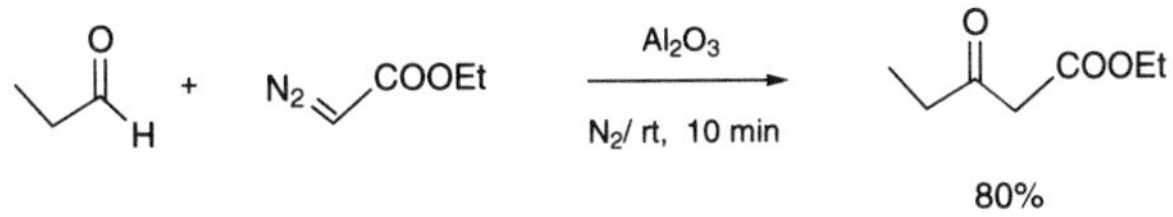

[1] J. Feixas, A. Capdevila, and A. Guerrero, *T* **50**, 8539 (1994).
[2] R. S. Varma, J. B. Lamture, and M. Varma, *TL* **34**, 3029 (1993).
[3] R. S. Varma, M. Varma, and A. K. Chatterjee, *JCS(PI)* 999 (1993).
[4] R. S. Varma, A. K. Chatterjee, and M. Varma, *TL* **34**, 3207; 4603 (1993).
[5] S. V. Ley and D. M. Mynett, *SL* 793 (1993).
[6] B. C. Ranu and R. Chakraborty, *T* **49**, 5333 (1993).
[7] D. D. Dhavale, P. N. Patil, and R. S. Mali, *JCR(S)* 152 (1994).

Aluminum.

Reduction of nitro compounds.   In the amalgamated form aluminum reduces the nitro group to either the hydroxylamine[1] stage or directly to the amine.[2] Ultrasound is used to facilitate the latter reduction.

Pinacol formation.[3]   A combination of KOH and aluminum powder is effective.

Reductive silylation.   *o*-Dichlorobenzene undergoes electrochemical trimethylsilylation.[4] Depending on reaction conditions, various numbers of the silyl groups can be introduced to the ring.

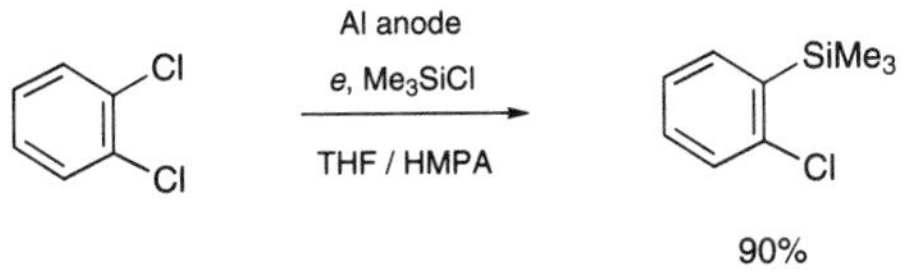

Allylaluminanes.[5]   Formation of the organometallic reagent and in situ reaction with an allenyl β-lactam results in the cephem system. Lead and nickel halides are also involved in the reaction.

85%

Polyfluoroalkyl aldehydes.[6]   Polyfluoroalkyl iodides are metalated by treatment of Al powder and catalytic amounts of PbBr$_2$. The in situ reaction with DMF solvent gives the aldehydes.

[1] C. Yuan, G. Wang, H. Feng, J. Chen, and L. Maier, *PSS* **81**, 149 (1993).
[2] R. W. Fitch and F. A. Luzzio, *TL* **35**, 6013 (1994).
[3] J. M. Khurana and A. Sehgal, *CC* 571 (1994).
[4] D. Deffieux, M. Bordeau, C. Biran, and J. Dunogues, *OM* **13**, 2415 (1994).
[5] H. Tanaka, S. Sumida, K. Sorajo, and S. Torii, *CC* 1461 (1994).
[6] C.-M. Hu and X.-Q. Tang, *JFC* **61**, 217 (1993).

Aluminum chloride. 13, 15–17; 14, 21–22; 15, 10; 16, 10–11; 17, 15

Isomerization.[1]   The (*E*)- to (*Z*)-isomerization of oxoindolin-3-ylidene ketones occurs at room temperature on contact with AlCl$_3$.

Friedel–Crafts reactions.   *N*-Alkylamides, ureas and sulfonamides serve as alkyl donors[2] for benzene in the presence of AlCl$_3$. The selective formation of 2′-hydroxyphenylpyridinemethanols[3] by the reaction of phenols with 2-pyridinealdehyde is due to a metal template effect.

α-Chlorooximes[4] behave normally as electrophiles towards arenes. Salicylaldoximes are formed[5] when phenols and nitromethane are heated with AlCl$_3$.

Chloromethyl methyl sulfide forms a complex with 2 moles of AlCl$_3$, which is very electrophilic towards aromatic compounds.[6] High *p:o* ratios of products are observed in reactions with toluene and related substrates.

Hetero Diels–Alder reactions.   The AlCl$_3$-catalyzed condensation of conjugated carbodiimides with dienophiles gives 2-aminopyridines.[7]

77%

2,3-Dihydro-2,2-dimethylbenzofurans.[8] A one-pot synthesis by treatment of methallyl phenyl ethers with $AlCl_3$ involves Claisen rearrangement and cyclization.

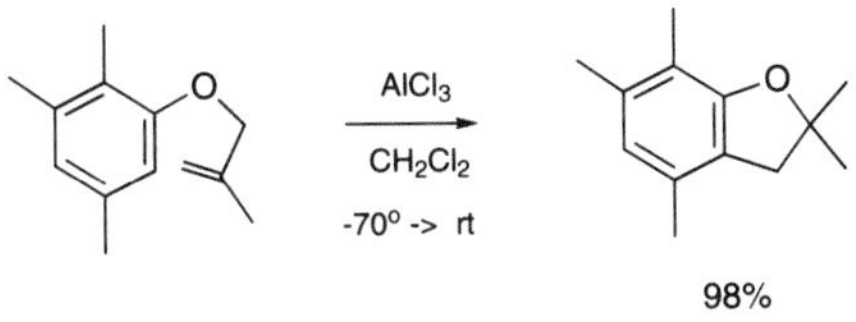

98%

Aliphatic Friedel–Crafts acylations. Reactions involving alkynylsilanes[9] or tricarbonyliron-complexed dienes provide useful synthetic intermediates.[10] A convenient access to conjugated amides and sulfonamides is the dematallative substitution[11] of alkenylstannanes with the proper electrophiles.

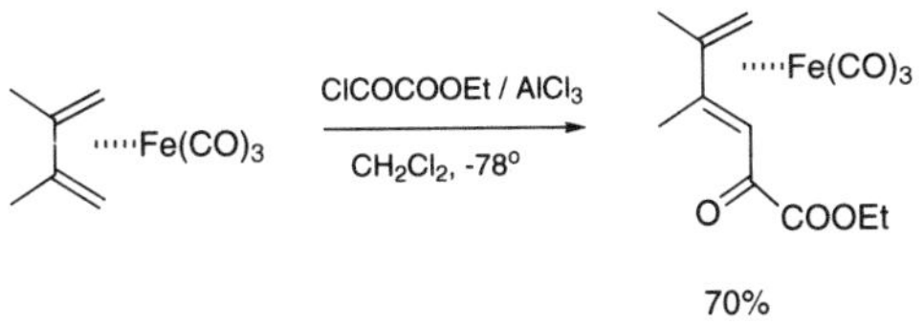

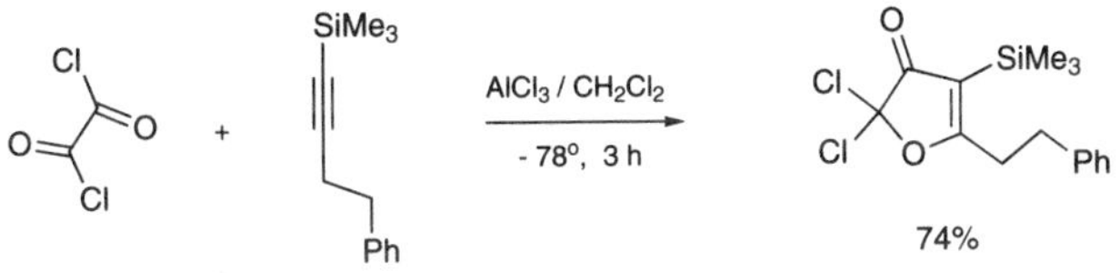

70%

74%

3,5-Dialkyl-6-alkylmethyl-4-hydroxy-2-pyrones.[12] Dehydrochlorinative trimerization of acid chlorides in the presence of $AlCl_3$ is the simplest method for their preparation.

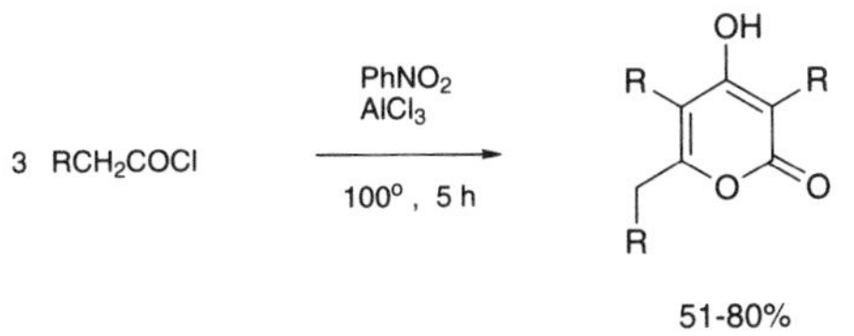

51-80%

Ester cleavage.[13] The $AlCl_3$–$PhNMe_2$ combination is a versatile reagent to cleave methyl, benzyl, methoxymethyl, methylthiomethyl, methoxyethoxymethyl, and β-(trimethylsilyl)ethoxymethyl esters at room temperature.

Transacylation of amines. The effectiveness varies and good yields of products are obtained by using activated amides (imides, N-tosylamides).[14] 4-Tosylaminobutanoylation[15] of amines with N-tosylpyrrolidone proceeds much more smoothly.

[1] G. Faita, M. Mella, P. P. Righetti, and G. Tacconi, *T* **50**, 10955 (1994).
[2] K. H. Chung, J. N. Kim, and E. K. Ryu, *TL* **35**, 2913 (1994).
[3] G. Sartori, R. Maggi, F. Bigi, A. Arienti, C. Porta, and G. Predieri, *T* **50**, 10587 (1994).
[4] J. N. Kim and E. K. Ryu, *TL* **34**, 3567 (1993).
[5] G. Sartori, F. Bigi, R. Maggi, and F. Tomasini, *TL* **35**, 2393 (1994).
[6] G. A. Olah, Q. Wang, and G. Neyer, *S* 276 (1994).
[7] T. Saito, T. Ohkubo, K. Maruyama, H. Kuboki, and S. Motoki, *CL* 1127 (1993).
[8] K. M. Kim, H. R. Kim, and E. K. Ryu, *H* **36**, 497 (1993).
[9] M. Murakami, M. Hayashi, and Y. Ito, *JOC* **59**, 7910 (1994).
[10] M. Franck-Neumann and P. Geoffrey, *TL* **35**, 7027 (1994).
[11] M. Niestroj, W. P. Neumann, and O. Thies, *CB* **127**, 1131 (1994).
[12] G. Sartori, F. Bigi, D. Baraldi, R. Maggi, G. Casnati, and X. Tao, *S* 851 (1993).
[13] T. Akiyama, H. Hirofuji, A. Hirose, and S. Ozaki, *SC* **24**, 2179 (1994).
[14] E. Bon, D. C. H. Bigg, and G. Bertrand, *JOC* **59**, 4035 (1994).
[15] E. Bon, D. C. H. Bigg, and G. Bertrand, *JOC* **59**, 1904 (1994).

Aluminum chloride–carbon disulfide on resin.

Esterification.[1]   The catalyst prepared by treatment of a cross-linked sulfonated polystyrene resin with $AlCl_3$ and CS_2 in ethanol is very effective.

[1] H. Yang, B. Li, and Y. Fang, *SC* **24**, 3269 (1994).

Aluminum hydride.

Reduction of oxime ethers.[1]   The *O*-methyl derivatives of conjugated ketone oximes are reduced to unsaturated amines together with smaller amounts of aziridines and saturated amines.

91% (58:32:10)

[1] M. Zaidlewicz and I. G. Uzarewicz, *HC* **4**, 73 (1993).

Aluminum hydride–triethylamine.

Functional group reduction.[1]   The stable complex is prepared by adding Et_3N to the AlH_3 solution, which is obtained by adding an appropriate quantity of HCl in ether to a solution of $NaAlH_4$ in THF. Its reducing power towards 59 organic compounds has been evaluated. Quantitative reduction using proper amounts of the complex can be achieved.

[1] J. S. Cha and H. C. Brown, *JOC* **58**, 3974 (1993).

Aluminum tris(2,6-diphenylphenoxide).

Cycloaddition.[1]   Selective activation of less hindered aldehydes to participate in the cycloaddition with Danishefsky's diene can be accomplished.

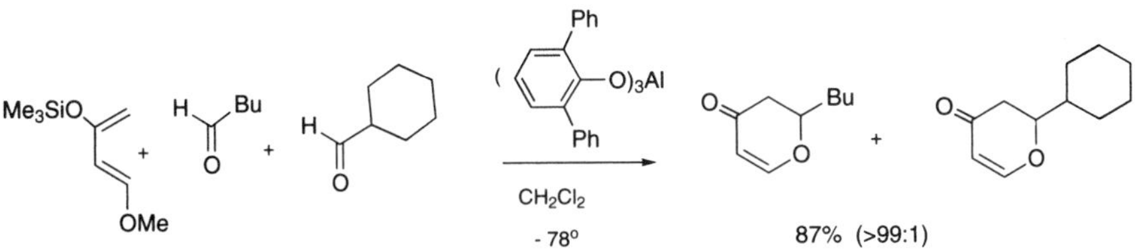

Conjugate addition.[2,3]   Coordination of the very bulky aluminum reagent to the oxygen of a carbonyl disfavors attack of the latter by carbanions.

Homologation of aldehydes.[4]   Reaction with diazomethane gives epoxides. It should be noted that MeAl(OAr)$_2$ promotes formation of ketones.

Organometallic reactions.[5]   Addition of RM to ketones shows chemoselectivity in response to the type of Al-based Lewis acids. 4-Heptanone is attacked exclusively by MeLi at −78°C in the presence of 4-methylcyclohexanone. This result is in complete contrast to the reaction promoted by MeAl(OAr)$_2$.

[1] K. Maruoka, S. Saito, and H. Yamamoto, *SL* 439 (1994).
[2] K. Maruoka, I. Shimada, H. Imoto, and H. Yamamoto, *SL* 519 (1994).
[3] K. Maruoka, I. Shimada, M. Akakura, and H. Yamamoto, *SL* 847 (1994).
[4] K. Maruoka, A. B. Concepcion, and H. Yamamoto, *SL* 521 (1994).
[5] K. Maruoka, H. Imoto, and H. Yamamoto, *SL* 441 (1994).

Aluminum tris(pentafluorophenoxide).

Cleavage of chiral acetals.[1]   *(4R,6R)*-4,6-Dimethyl-1,3-dioxanes undergo reductive ring opening to give mainly the product with *(S)*- configuration at the newly formed stereogenic center.

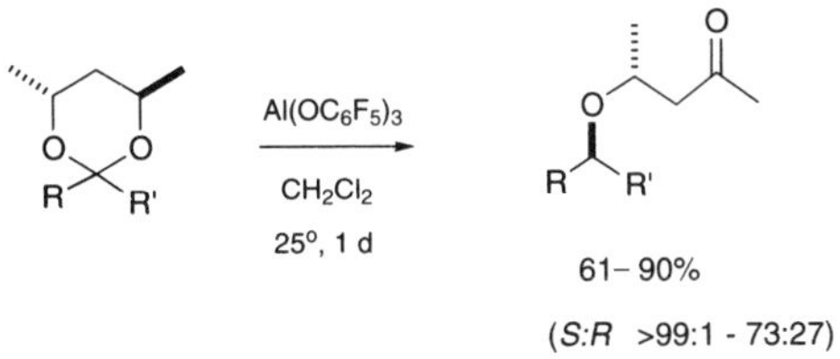

[1] K. Ishihara, N. Hanaki, and H. Yamamoto, *SL* 127 (1993).

N-Amino-2-(1-ethyl-1-methoxy)propyl)pyrrolidine.

Hydrazones. The hydrazone of crotyloxyacetaldehyde undergoes [2,3]Wittig rearrangement resulting in protected γ,δ-unsaturated α-hydroxyaldehyde with stereocenters at C-2 and C-3.[1]

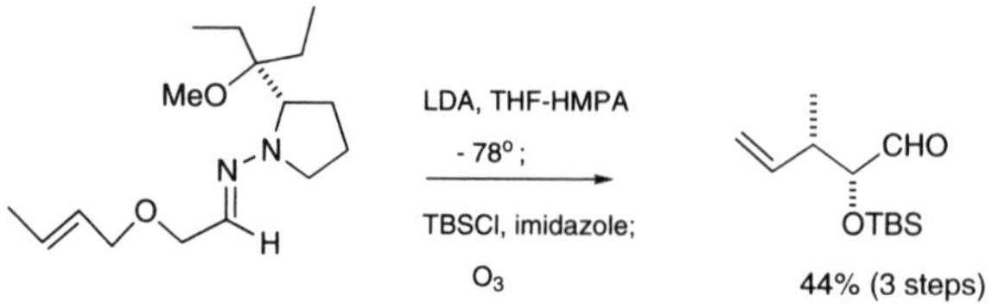

	LDA, THF-HMPA	
	-78°;	
	TBSCl, imidazole;	
	O_3	44% (3 steps)

[1] D. Enders, D. Backhaus, and J. Runsink, *ACIEE* **33**, 2098 (1994).

N-Amino-2-(methoxymethyl)pyrrolidine. 14, 22; 16, 12–13; 17, 15–16

Hydrazones for enantioselective alkylation.[1] The THF solvent can become the electrophile in the presence of an activator such as *t*-BuMe_2SiOTf.

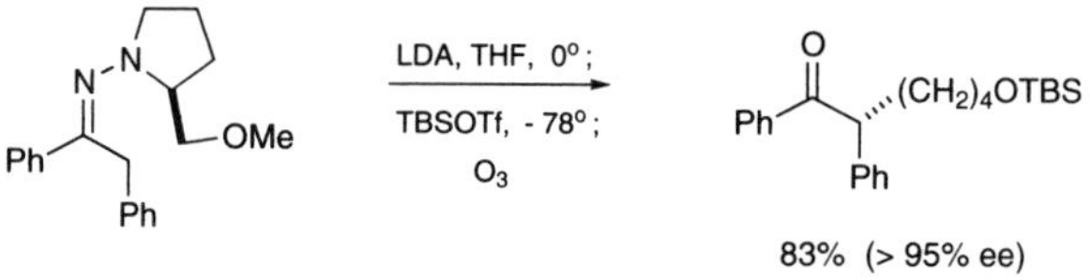

	LDA, THF, 0°;	
	TBSOTf, -78°;	
	O_3	83% (> 95% ee)

Chiral 2-sulfenylated aldehydes.[2] Alkylation of the hydrazone derived from methylthioacetaldehyde, followed by ozonolytic cleavage of the chiral auxiliary constitutes a route to this class of compounds.

[1] B. B. Lohray and D. Enders, *S* 1092 (1993).
[2] D. Enders, T. Schafer, O. Piva, and Z. Zamponi, *T* **50**, 3349 (1994).

endo-3-Amino-5-norbornen-*endo*-2-ol.

Oxazoles.[1] Acylation and pyrolysis of the hydroxy amides in decalin provides 2-substituted oxazoles by way of cyclodehydration and retro-Diels–Alder reaction. Imidazoles are similarly prepared from the *vic*-diamine.

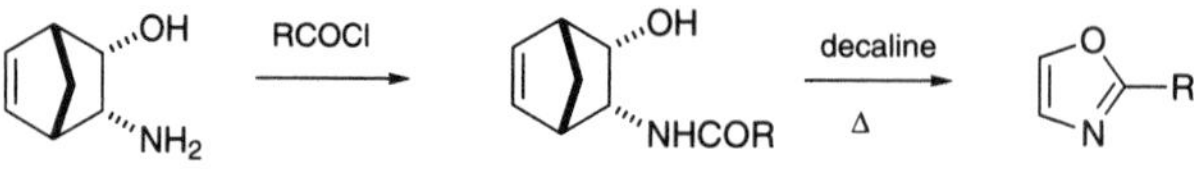

[1] M. A. Eissenstat and J. D. Weaver, *JOC* **58**, 3387 (1993).

Aminosilanes.

Iminium salts.[1] N-Silyl derivatives of secondary amines react with carbonyl compounds afford α-siloxyamines. On further treatment with Me_3SiCl, iminium chlorides are obtained. Trimethylsilyl triflate is superior to Me_3SiCl since it can induce the transformation in the cases of enolizable aldehydes. α-Chloro ethers are more reactive than carbonyl compounds, enabling the preparation of vinylogous Viehe salts.

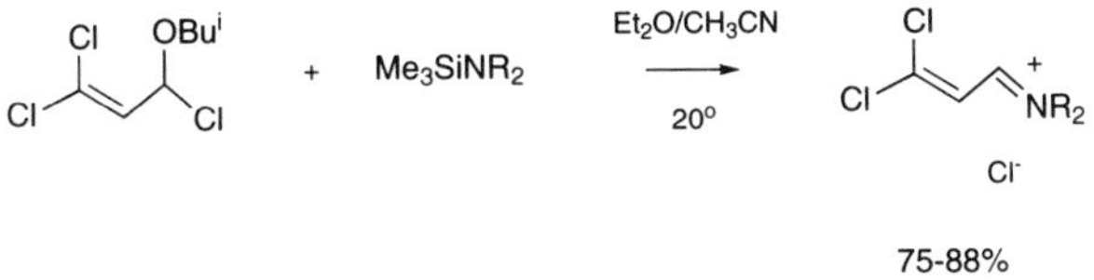

75-88%

[1] W. Schroth, U. Jahn, and D. Strohl, *CB* **127**, 2013 (1994).

Ammonia.

Release of resin-bound peptides.[1]   Bound peptides are quantitatively cleaved from the solid support on exposure to ammonia vapor. The method is general.

Photoinduced amination.   Addition of NH_3 to a styrenic double bond[2] as well as amination of the aromatic ring[3] are possible upon photosensitization.

68%

[1] A. M. Bray, A. G. Jhingran, R. M. Valerio, and N. J. Maeji, *JOC* **59**, 2197 (1994).
[2] T. Yamashita, M. Yasuda, T. Isami, K. Tanabe, and K. Shima, *T* **50**, 9275 (1994).
[3] T. Yamashita, K. Tanabe, K. Tamano, M. Yasuda, and K. Shima, *BCSJ* **67**, 246 (1994).

Ammonium formate.

Transfer hydrogenation.[1,2]   Mediated by Pd/C, alkenes readily accept hydrogen from ammonium formate. Conjugated carbonyl compounds are more readily reduced.[3]

[1] A. K. Bose, B. K. Banik, K. J. Barakat, and M. S. Manhas, *SL* 575 (1993).
[2] H. S. P. Rao and K. S. Reddy, *TL* **35**, 171 (1994).
[3] B. C. Ranu and A. Sarkar, *TL* **35**, 8649 (1994).

Ammonium hypophosphite.

Transfer hydrogenation.[1]   Saturation of double bonds in the presence of Pd/C shows selectivity correlating with steric hindrance. Semihydrogenation of alkynes with this system is reported.

[1] B. T. Khai and A. Arcelli, *CB* **126**, 2265 (1993).

Ammonium persulfate.

Acetoxylactonization.[1]   Unsaturated carboxylic acids undergo oxidative cyclization in the presence of triflic acid in acetic acid.

1,4-Naphthoquinones.[2]   Polymethoxynaphthalenes give quinones in a AgNO$_3$-catalyzed oxidation. The results are similar to those of the Ce(IV) oxidation.

Isocyanate synthesis.[3]   A convenient and general method for isocyanate synthesis is by oxidation of oxalylamines. Silver nitrate and copper(II) sulfate are cocatalysts.

$$\text{BuNHCOCOOH} \xrightarrow[\substack{\text{AgNO}_3,\ \text{CuSO}_4 \\ \text{hexane} \\ 40^\circ,\ 3\,\text{h}}]{(\text{NH}_4)_2\text{S}_2\text{O}_8} \underset{83\%}{\text{BuN=C=O}}$$

Selenosulfonates.[4]   Diaryl diselenides are cleaved, and the products may be intercepted with sodium benzenesulfinate.

[1] M. Tiecco, L. Testaferri, and M. Tingoli, *T* **49**, 5351 (1993).
[2] Y. Tanoue, K. Sakata, M. Hashimoto, S. Morishita, M. Hamada, N. Kai, and T. Nagai, *BCSJ* **67**, 2593 (1994).
[3] F. Minisci, F. Coppa and F. Fontana, *CC* 679 (1994).
[4] L. Wang and X. Huang, *SC* **23**, 2817 (1993).

Ammonium vanadate.

Meyer–Schuster rearrangement.[1]   Ammonium vanadate, diphenylsilanediol, and an alkanedicarboxylic acid constitute an efficient catalytic system for the conversion of propargyl alcohols to α,β-unsaturated aldehydes.

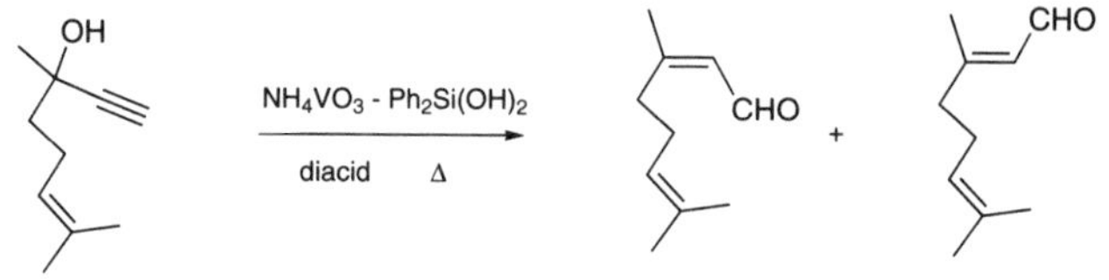

[1] M. B. Erman, S. E. Gulyi, and I. S. Aulchenko, *MC* 89 (1994).

Antimony(III) chloride.

Hydrophenylation of enones.[1]   A remarkable catalytic effect exerted by $SbCl_3$ has been observed in the palladium-mediated transfer of a phenyl group from sodium tetraphenylborate to enones, furnishing β-phenyl-carbonyl compounds.

[1]C. S. Cho, S. Motofusa, and S. Uemura, *TL* **35**, 1739 (1994).

Antimony(V) chloride–silver hexafluoroantimonate.

Aryl sulfides.[1]   Disulfides are activated by this catalytic system to react with arenes.

[1]T. Mukaiyama and K. Suzuki, *CL* **23**, 1 (1993).

Antimony(V) fluoride.

Halogen replacement.[1] α-Halogen atoms (such as chlorine) of ketones can conveniently be replaced with fluorine on reaction with SbF_5, and the extent can be controlled by the amount of the reagent.

4-Chlorocyclohexa-2,5-dienones.[2]   The preparation from 4-alkylphenol ethers probably involves $[Cl_2CX]^+$ $[Sb_2F_{10}Cl]^-$ as the reagent in which the carbocation species are derived from dichloromethane or chloroform, which is used as solvent.

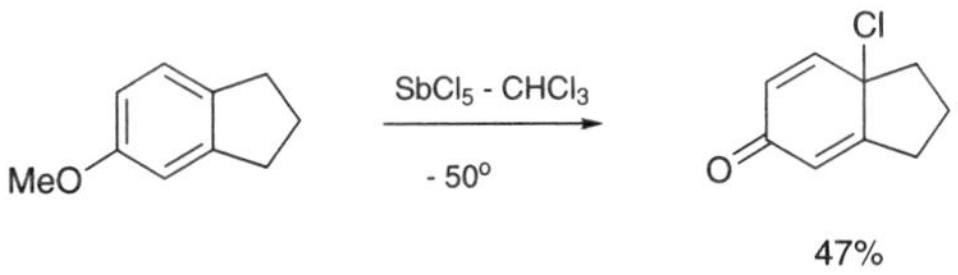

[1]D. J. Burton and I. H. Jeong, *JFC* **65**, 153 (1993).
[2]B. Ferron, J.-C. Jacquesy, M.-P. Jouannetaud, O. Karam, and J.-M. Coustard, *TL* **34**, 2949 (1993).

Arenediazonium salts.

Reaction with silyl enol ethers.[1]   Derivatives from ketones and esters behave towards arenediazonium salts according to their relative nucleophilicities. α-Arylation of ketones by a free radical pathway and nonradical α-amination of esters are observed.

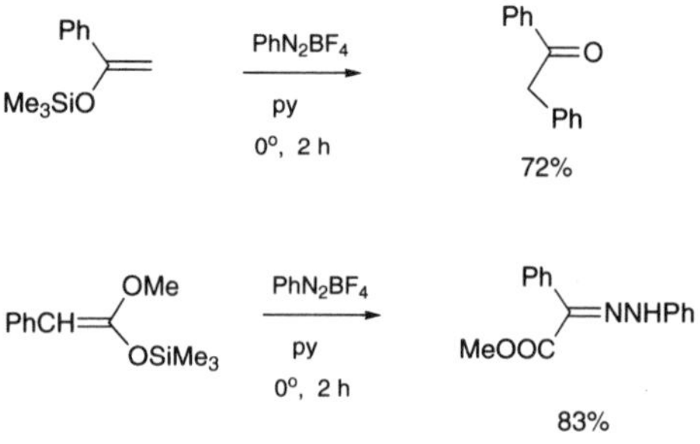

[1] T. Sakakura, M. Hara, and M. Tanaka, *JCS(P1)* 283, 289 (1994).

Areneselenenyl bromides.

Asymmetric methoxyselenenylation.[1]   A chirally constituted Ar*SeBr reagent induces asymmetric addition to double bonds. Chiral allyl ethers are accessible after oxidation and selenoxide elimination.

[1] K. Fujita, M. Iwaoka, and S. Tomoda, *CL* 923 (1994).

N^1-Arenesulfonyloxybenzotriazoles.

Peptide synthesis.[1,2]   The β-Naphthalenesulfonyloxy derivatives of benzotriazole (1) and of 6-nitrobenzotriazole have been evaluated as rapid coupling agents. These can be used with hindered amino acids as well as Gln and Asn residues.

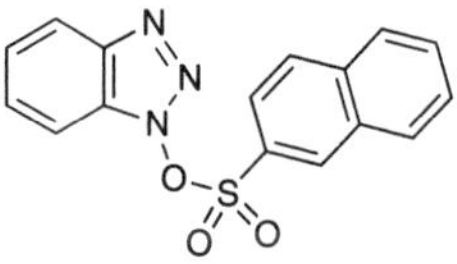

(1)

[1] B. Kundu, S. Shukla, and M. Shukla, *TL* **35**, 9613 (1994).
[2] B. Devadas, B. Kundu, A. Srivastava, and K. B. Mathur, *TL* **34**, 6455 (1993).

Aryl(cyano)iodonium triflates.

Iodonium ion transfer.[1]   These reagents are prepared from aryliodonium bis(trifluoroacetate)s and trimethylsilyl cyanide in the presence of trimethylsilyl triflate. They can be used for the preparation of aryl and alkynyl iodonium salts from the corresponding tributyltin derivatives.

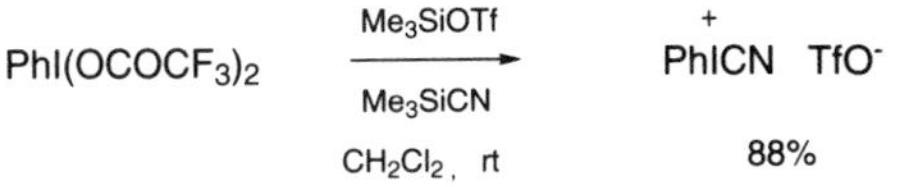

$$PhI(OCOCF_3)_2 \xrightarrow[\substack{Me_3SiCN \\ CH_2Cl_2, \ rt}]{Me_3SiOTf} \overset{+}{PhICN} \ TfO^-$$

88%

[1] V. V. Zhdankin, M. Scheuller, and P. J. Stang, *TL* **34**, 6853 (1993).

Aryllead triacetates.

Arylation of activated ketones.   The reagents are useful for the preparation of isoflavanones[1] and α-arylglycine derivatives.[2]

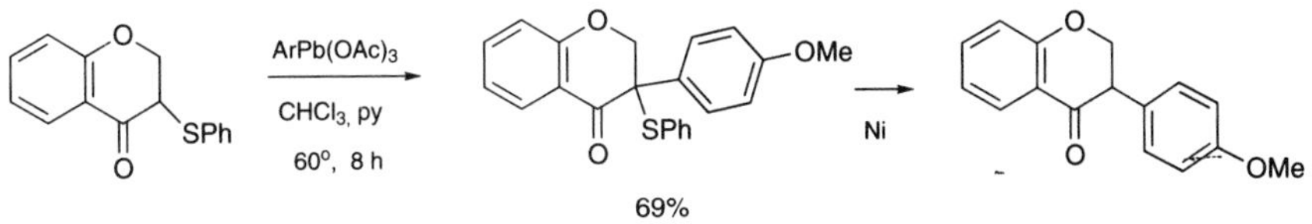

69%

o-Arylation of 3,5-di-t-butylphenol.[3]   Highly hindered 2-hydroxybiphenyls and terphenyls are obtained.

Diorganolead diacetates.[4]   Exchange of one acetoxy group of an aryllead triacetate with a vinyl or another aryl group takes place when it is treated with $RB(OH)_2$.

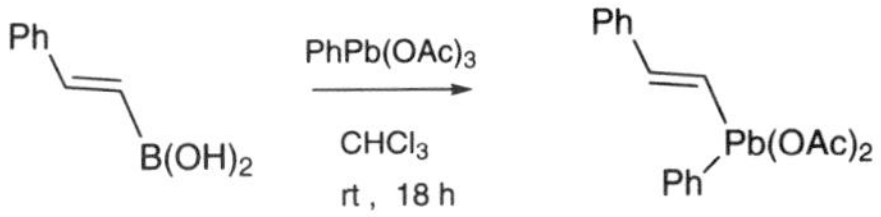

[1] D. M. X. Donnelly, B. M. Fitzpatrick, B. A. O'Reilly, and J.-P. Finet, *T* **49**, 7967 (1993).
[2] J. Morgan and J. T. Pinhey, *TL* **35**, 9625 (1994).
[3] D. H. R. Barton, D. M. X. Donnelly, P. J. Guiry, and J.-P. Finet, *JCS(PI)* 2921 (1994).
[4] J. Morgan, C. J. Parkinson, and J. T. Pinhey, *JCS(PI)* 3361 (1994).

1-Aza-1,3-bis(triphenylphosphoranylidene)propane.

=CHCH₂N=synthon.[1]   This reagent is prepared by reaction of $BtCH_2N{=}PPh_3$ (Bt=1-benzotriazolyl) with $Ph_3P{=}CH_2$ and then BuLi. When used in situ in the condensation with araldehydes, cinnamylamines are produced.

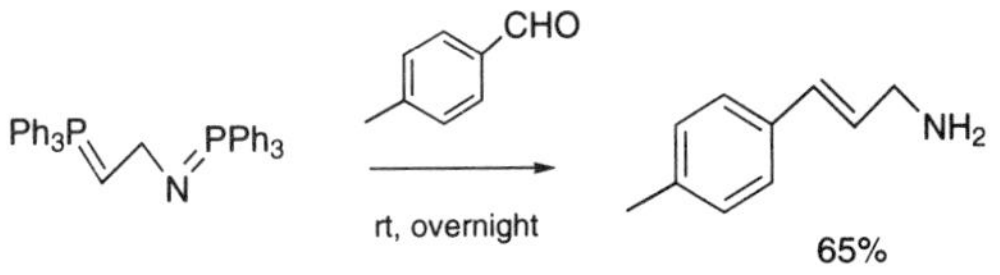

65%

[1] A. R. Katritzky, J. Jiang, and P. J. Steel, *JOC* **59**, 4551 (1994).

B

Baker's yeast.

Carbonyl reduction. Many substrates have been reduced enantioselectively to give alcohols: trifluoromethyl ketones,[1] α-acetoxyketones,[2] pyridinophenones,[3] ethyl α-methylacetoacetate,[4] α-keto acid derivatives,[5] and 3-chloro-2-oxoalkanoic esters.[6] α,β-Epoxy ketones undergo reduction and hydrolytic ring opening.[7]

It is possible to use a cell-free extract and a catalytic amount of NADPH to perform the reduction.[2] In this approach, glucose is used as the hydride source.

Hydrogenation. The double bonds of (E)-β-nitrostyrenes[8] are saturated.

Reduction of other functional groups. Certain thiones are reducible to afford chiral mercaptans.[9] Using nitroaryl ketones[10] as substrates, the nitro group is preferentially reduced.

Kinetic resolution.[11] In the resolution of 1-aryl- and 1-heteroarylethanols by enantioselective oxidation the ee value of the (S)-isomer reaches 86–100% when the ethanol has either a phenyl, 2-furyl, or 2-thienyl group. The value of the 2-pyridyl derivative is lower (40% ee). The method is not applicable to 1-(2-thiazolyl)ethanol (0% ee).

[1] T. Fujisawa, T. Sugimoto, and M. Shimizu, *TA* **5**, 1095 (1994).
[2] K. Ishihara, T. Sakai, S. Tsuboi, and M. Utaka, *TL* **35**, 4569 (1994).
[3] M. Takemoto and K. Achiwa, *CPB* **42**, 802 (1994).
[4] Y. Kawai, M. Tsujimoto, S. Kondo, K. Takanobe, K. Nakamura, and A. Ohno, *BCSJ* **67**, 524 (1994).
[5] G. Pedrocchi-Fantoni, S. Redaelli, and S. Servi, *G* **122**, 499 (1992).
[6] S. Tsuboi, H. Furutani, M. H. Ansari, T. Sakai, M. Utaka, and A. Takeda, *JOC* **58**, 486 (1993).
[7] O. Meth-Cohn, R. M. Horak, and G. Fouche, *JCS(P1)* 1517 (1994).
[8] M. Takeshita, S. Yoshida, and Y. Kohno, *H* **37**, 553 (1994).
[9] J. K. Nielsen and J. O. Madsen, *TA* **5**, 403 (1994).
[10] W. Baik, J. L. Han, K. C. Lee, N. H. Lee, B. H. Kim, and J.-T. Hahn, *TL* **35**, 3965 (1994).
[11] M. Fantin, M. Fogagnolo, A. Medici, P. Pedrini, S. Poli, and M. Sinigaglia, *TL* **34**, 883 (1993).

Barium hydroxide.

Olefination.[1] Although the conventional procedure for the Horner–Emmons–Wadsworth reaction calls for enolate generation with NaH in an aprotic solvent, more complex substrates tend to give unsatisfactory results due to elimination and/or epimerization. The Ba(OH)$_2$/aqueous THF system is uniquely effective at room temperature in such cases. Usually 0.3 equivalent of activated (by heating 100–140°C for

2 h) Ba(OH)$_2$ is sufficient, but more complicated phosphonates may require 0.7–1.0 equivalent of the base.

[1] I. Paterson, K.-S. Yeung, and J. B. Smaill, *SL* 774 (1993).

Barium ruthenate.

Alkane oxidation.[1] Hydrocarbons are oxidized to carbonyl compounds in reasonably good yields. For example, cyclohexane gives cyclohexanone (60%). The reaction is greatly accelerated by the addition of Lewis acids (FeCl$_3$, ZnCl$_2$, etc.).

[1] T.-C. Lau and C.-K. Mak, *CC* 766 (1993).

Benzenediazonium chloride.

Indole synthesis.[1] *m*-Hydroxystyrenes react with the diazonium salt to afford *N*-anilinoindoles. Cleavage of the N–N bond is readily achieved by hydrogenolysis.

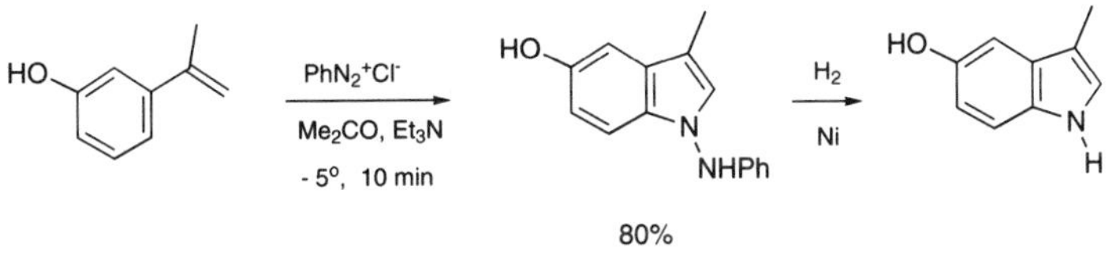

80%

[1] M. Satomura, *JOC* **58**, 3757 (1993).

Benzeneselenenyl bromide. 13, 26–27

Alkynyl phenyl selenides.[1] The selenenylation of 1-alkynes is mediated by CuI.

Heterocyclization.[2,3] Participation of an imino group to form five-membered heterocyclic products or intermediates thereof is observed during PhSeBr-induced functionalization of alkenes. Likewise, the cyclofunctionalization of allylic benzimidates and benzamides can be achieved.[4]

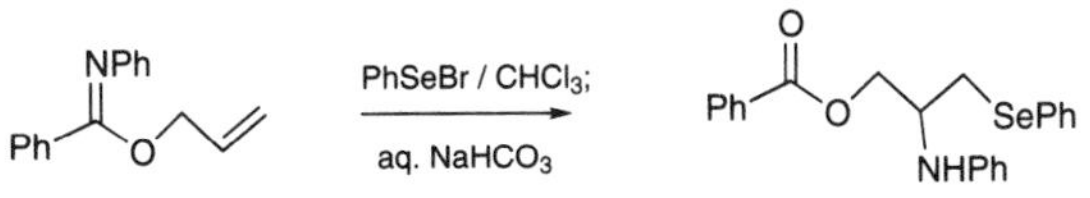

[1] A. L. Braga, C. S. Silveira, A. Reckziegel, and P. H. Menezes, *TL* **34**, 8041 (1993).
[2] N. De Kimpe and M. Boelens, *CC* 916 (1993).
[3] M. Tiecco, L. Testaferri, M. Tingoli, L. Bagnoli, and C. Santi, *T* **51**, 1277 (1995).
[4] L. Engman, *JOC* **58**, 2394 (1993).

Benzeneselenenyl chloride. **13**, 26–27; **14**, 27; **16**, 19–21; **17**, 26

Substitutive deselenenylation.[1]   The reagent and conditions for the substitution are sufficiently mild that functionalities such as the azido group are not affected. *O*-Glycosylation can be effected by treatment of phenyl selenoglycosides with alcohols acting as nucleophiles.

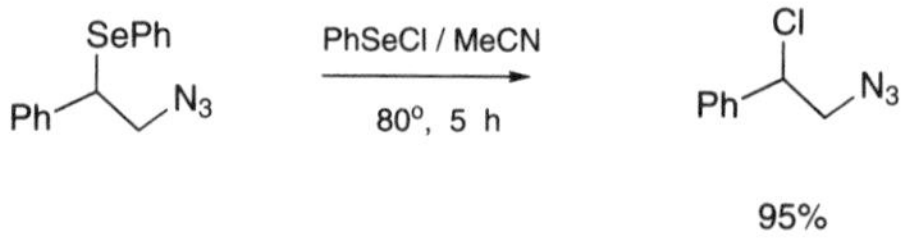

Cyclopropyl ketones.[2]   While normal selenenylchlorination occurs when γ,δ-unsaturated ketones are treated with PhSeCl, subsequent reaction of the adducts with a base (e.g., NaH) effects an intramolecular alkylation to form cyclopropane derivatives.

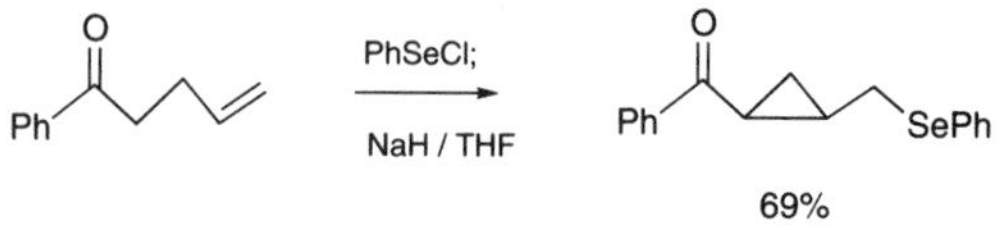

[1] M. Tingoli, M. Tiecco, L. Testaferri, and A. Temperini, *CC* 1883 (1994).
[2] L. Dechoux, L. Jung, and J. F. Stambach, *SL* 965 (1994).

N-Benzeneselenenylphthalimide.

α-Selenenylation of carbonyl compounds.[1]   The reaction conditions are mild, and a separate enolization step is not required.

Glycosylation.[2]   Together with TMSOTf the selenide reagent activates thioglycosides. Disaccharides can be obtained under these conditions in 77–100% yield.

[1] J. Cossy and N. Furet, *TL* **34**, 7755 (1993).
[2] H. Shimizu, Y. Ito, and T. Ogawa, *SL* 535 (1994).

2-Benzeneselenenylethanol.

Vinyl esters.[1]   Esterification followed by oxidation (in situ thermolysis) generates the target esters.

[1] M. I. Weinhouse and K. D. Janda, *S* 81 (1993).

Benzeneseleninic anhydride–trifluoroacetic anhydride.

Trifunctionalization of alkenes.[1]   Addition to a double bond using the combination of anhydrides leads to a thermally unstable selenoxide as the primary product. *syn*-Elimination occurs and the allylic trifluoroacetate is susceptible to a second round of addition by the new reagent PhSeOCOCF₃, which is formed in situ. The process is thus based on an otherwise undesirable side reaction.

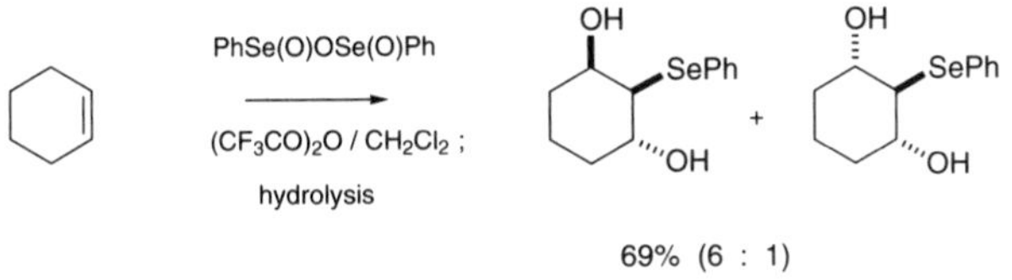

69% (6 : 1)

[1] A. G. Kutateladze, J. L. Kice, T. G. Kutateladze, and N. S. Zefirov, *JOC* **58**, 995 (1993).

Benzenesulfenyl chloride. 14, 27; **15**, 19

2-Indolylmethyl phenyl sulfoxides.[1] α-Ethynyl-*o*-acetamidobenzyl alcohols undergo *O*-sulfenylation, which is immediately followed by a [2,3]sigmatropic rearrangement. The ensuing allenyl sulfoxides are prone to cyclize.

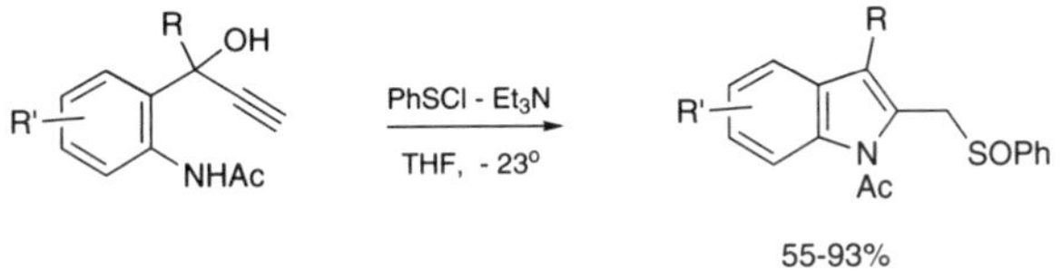

55-93%

o-Quinodimethane generation.[2]   When *o*-stannylmethylstyrenes are exposed to PhSCl, sulfenylation triggers a fragmentation. The *o*-quinodimethanes thus generated can be trapped by dienophiles.

[1] M. Gray, P. J. Parsons, and A. P. Neary, *SL* 281 (1993).
[2] H. Sano, K. Kawata, and M. Kosugi, *SL* 831 (1993).

Benzenesulfonselenyl chloride.

Selenenylation.[1]   PhSO₂SeCl reacts with enolizable carbonyl compounds to give α-seleno ketones. Its reaction with *N*-trimethylsilylanilines results in the very reactive ArN=Se species, which can be trapped as Diels–Alder adducts.

[1] M. R. Bryce and A. Chesney, *CC* 195 (1995).

Benzenesulfonylmethyl *p*-tolyl sulfoxide.

RCH₂CHO → RCH(OH)CH=CHSO₂Ph.[1]   Condensation of aldehydes with TolS(O)-CH₂SO₂Ph is promoted by piperidine. The β-substituent of the carbon chain influences the [2,3]sigmatropic rearrangement of the allylic sulfoxides, which are in equilibrium with the vinylic isomers, such that *anti* products predominate. Enantiopure polypropionate chains can be assembled by an iterative process incorporating this step.

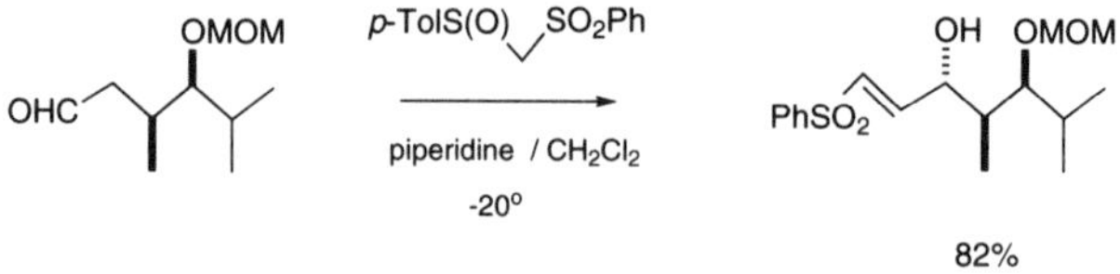

[1] J. C. Carretero and E. Dominguez, *JOC* **58**, 1596 (1993).

2-Benzenesulfonyl-3-phenyloxaziridine.

Alkynyl sulfoxides.[1]   The reagent transfers its oxygen atom to alkynyl sulfides in refluxing chloroform.

[1] S. T. Kabanyane and D. I. MaGee, *CJC* **70**, 2758 (1992).

1-Benzenesulfonyl-2-trimethylsilylacetylene.

Acetylene equivalent.[1]   As a dienophile for the Diels–Alder reaction, it incorporates into the adducts two hetero substituents on trigonal carbons. After conjugate reduction with LiAlH$_4$ the β-silyl sulfone moiety is susceptible to fluoride ion-initiated fragmentation.

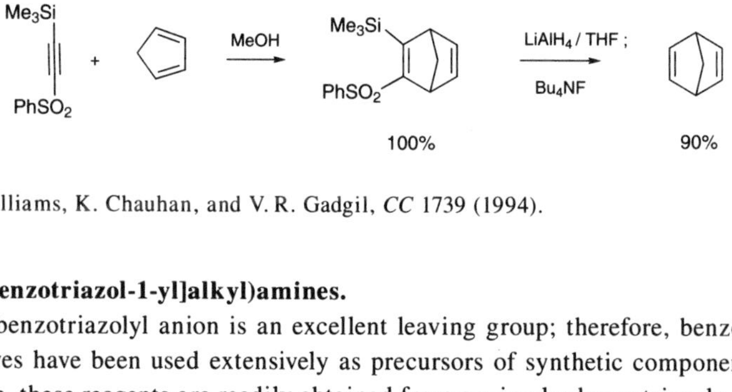

[1] R. V. Williams, K. Chauhan, and V. R. Gadgil, *CC* 1739 (1994).

N-(α-[Benzotriazol-1-yl]alkyl)amines.

   The benzotriazolyl anion is an excellent leaving group; therefore, benzotriazole derivatives have been used extensively as precursors of synthetic components. Furthermore, these reagents are readily obtained from equimolar benzotriazole, an aldehyde, and a secondary amine. A review[1] of synthetic uses of benzotriazole derivatives appeared in 1991.

   Aminal exchange.[2]   Aminals are very easily available by admixture of the reagents with other amines. This replacement of the benzotriazolyl group should be accomplished with the same amine as the extant subunit in order to avoid the generation of a mixture of compounds.

N-(α-Alkoxyalkyl)- and N-(α-[alkylthio]alkyl)amines.[3] The high-yielding reactions with sodium alkoxides and sodium thiolates proceed at room temperature.

Trialkylstannylmethylamines.[4] Products are obtained from reaction with the stannyllithium reagents.

Condensations with alkenes.[5] In the presence of LiBF$_4$ iminium species are generated. Trapping of these species with dienes and alkenes provides tetrahydropyridinium salts or 1,2,3,4-tetrahydroquinolines.

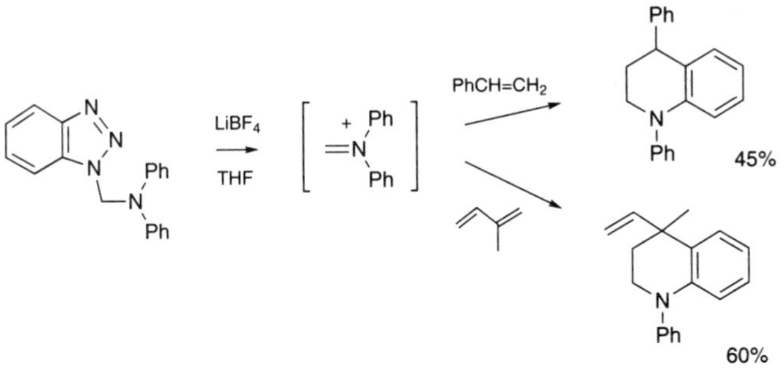

β-Amino ketone synthesis.[6] The benzotriazole derivatives enable 9-vinylcarbazole to become an acetaldehyde 1,2-dianion equivalent. The acid-catalyzed reaction gives rise to 1,1,3-trifunctionalized propanes in which the migrated benzotriazolyl group activates the α-carbon for alkylation. Subsequent hydrolysis of the aminal unit reveals the keto group.

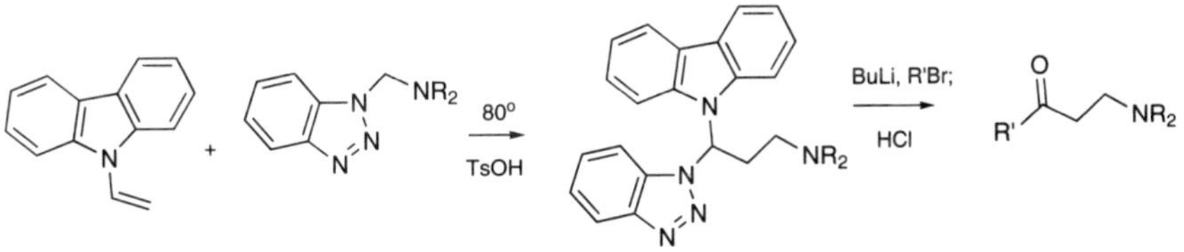

Enaminones.[7] When the amine geminal to the benzotriazolyl group is secondary, elimination promoted by a strong base results in a metalloenamine, which can be condensed with esters. Enaminones are the final products.

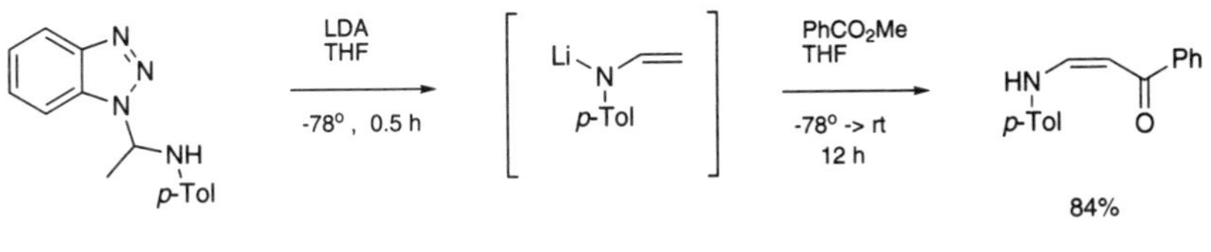

[1] A. R. Katritzky, S. Rachwal, and G. J. Hitchings, *T* **47**, 2683 (1991).
[2] A. R. Katritzky, K. Yannakopoulou, and H. Lang, *JCS(P1)* 1867 (1994).
[3] A. R. Katritzky, W.-Q. Fan, and Q.-H. Long, *S* 229 (1993).
[4] A. R. Katritzky, H.-X. Chang, and J. Wu, *S* 907 (1994).
[5] A. R. Katritzky and M. F. Gordeev, *JOC* **58**, 4049 (1993).
[6] A. R. Katritzky, Z. Yang, and J. N. Lam, *JOC* **58**, 1970 (1993).
[7] A. R. Katritzky, R. A. Barock, Q.-H. Long, M. Balasubramanian, N. Malhotra, and J. V. Greenhill, *S* 233 (1993).

Benzotriazol-1-ylmethanol.

N-Alkylation of amides.[1]   Benzotriazol-1-ylmethylation on the nitrogen atom of an amide is accomplished in refluxing acetic acid. Benzotriazole can then be displaced by reaction with an organometallic reagent (R_2Zn, etc.). On $LiAlH_4$ reduction *N*-methylamines are obtained.

[1] A. R. Katritzky, G. Yao, X. Lan, and X. Zhao, *JOC* **58**, 2086 (1993).

N-(Benzotriazol-1-ylmethyl)carbamates.

1,1-Bis(heteroaryl)alkanes.[1]   The benzotriazole unit has a lower leaving tendency than the carbamoyl group in the presence of a Lewis acid. Accordingly, Friedel–Crafts alkylation of very reactive arenes (heteroarenes) is feasible. However, the benzotriazole can be replaced in the second step.

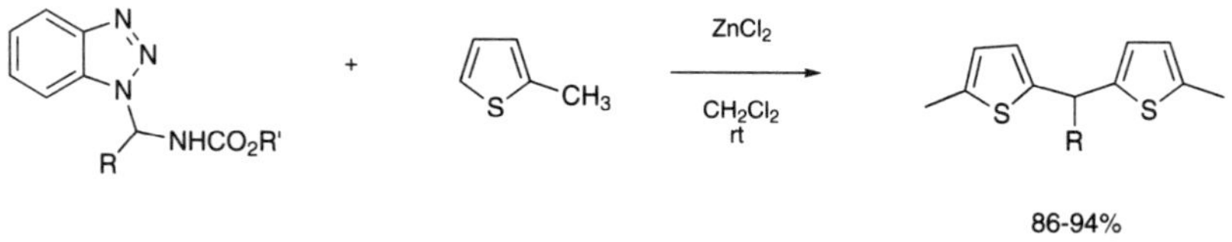

86-94%

[1] A. R. Katritzky, L. Xie, and W.-Q. Fan, *JOC* **58**, 4376 (1993).

Benzotriazol-1-ylmethylimino(triphenyl)phosphorane.

Wittig–Horner reactions.[1]   Phosphonomethyliminophosphoranes are readily prepared. These novel compounds form unsaturated heterocycles on reaction with dialdehydes.

Homologous phosphine imines undergo elimination of benzotriazole to give (*N*-vinylimino)phosphoranes, which are useful for the synthesis of pyridines.[2]

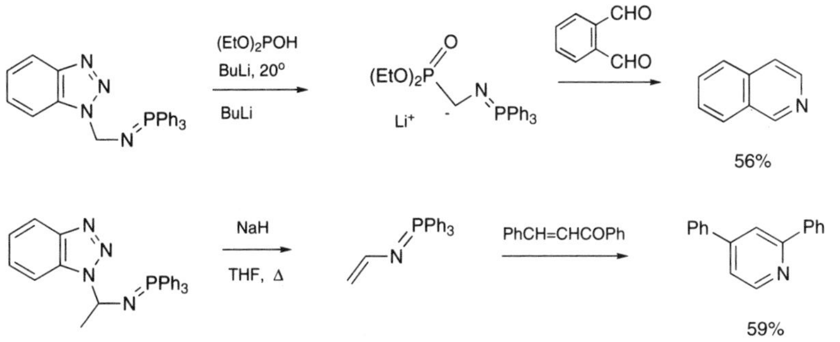

56%

59%

[1] A. R. Katritzky, J. Jiang, and J. V. Greenhill, *JOC* **58**, 1987 (1993).
[2] A. R. Katritzky, R. Mazurkiewicz, C. V. Stevens, and M. F. Gordeev, *JOC* **59**, 2740 (1994).

Benzotriazol-1-ylmethyl methyl ether.

Dimethyl acetals.[1] Deprotonation of the methyl ether with BuLi and subsequent alkylation serve to elongate the side chain. The α-benzotriazolyl ethers give dimethyl acetals on heating with TsOH-MeOH. Thus the reagent is a methylal anion equivalent.

[1] A. R. Katritzky, Z. Yang, and D. J. Cundy, *SC* **23**, 3061 (1993).

Benzotriazol-1-ylmethyl phenyl sulfide.

2,2-Diarylcyclopropyl phenyl sulfides.[1] The cyclopropanation takes place when the reagent is deprotonated with LDA in the presence of 1,1-diarylethene. Benzotriazole acts as a leaving group in the ring-closure step. Simpler benzotriazolyl derivatives also undergo the same reaction.

[1] A. R. Katritzky and M. F. Gordeev, *SL* 213 (1993).

Benzotriazol-1-yloxy tris(dimethylamino)phosphonium hexafluorophosphate (1).

Dipeptide synthesis. Among various coupling reagents the best yields and the least problems due to racemization are observed with this salt (**1**)[1]. Similar reagents are the 6-trifluoromethylbenzotriazole[2] and 7-azabenzotriazole[3] analogs. [*Note:* 1-Hydroxy-7-azabenzotriazole itself is an efficient peptide coupling additive.[4] It has the additional advantage of being a visual indicator (yellow to colorless) of the reaction endpoint.]

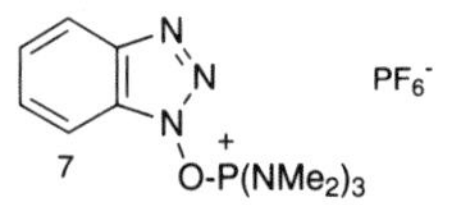

(1)

[1] J. Dudash, Jr., J. Jiang, S. C. Mayer, and M. M. Joullie, *SC* **23**, 349 (1993).

[2] J. C. H. M. Wijkmans, J. A. W. Kruijtzer, G. A. van der Marel, J. H. van Boom, and W. Bloemhoff, *RTC* **113**, 394 (1994).

[3] A. Ehrlich, S. Rothemund, M. Brudel, M. Beyermann, L. A. Carpino, and M. Bienert, *TL* **34**, 4781 (1993).

[4] L. A. Carpino, *JACS* **115**, 4397 (1993).

3-(Benzotriazolyl)propyne.

2-Arylpyrroles.[1]   Reaction with N-tosylimines at the *sp* terminus of the propargylbenzotriazole followed by intramolecular substitution (and aromatization–detosylation) represents a new synthetic protocol. 2-Hetarylpyrroles are also available by this method.

[1] A. R. Katritzky, J. Li, and M. F. Gordeev, *S* 93 (1993).

N-Benzoyltetrazole.

Acylation.[1]   Derivatization of alcohols and amines with the N-acyl-5-phenyltetrazole in the presence of pyridine is uniformly applicable. The mixture in THF is kept at $-10°C$ and then at room temperature to complete the reaction.

[1] B. S. Jursic, *SC* **23**, 361 (1993).

N-Benzoyl-(4S)-t-butyl-2-oxazolidinone.

Benzoylation.[1]   n-Alkyl carbinols show kinetic selectivities in the range of 20–30:1 for reaction of the (R)-enantiomers.

[1] D. A. Evans, J. C. Anderson, and M. K. Taylor, *TL* **34**, 5563 (1993).

Benzylamine.

Fluoroalkyl- and fluoroarylamines.[1] Imines formed (with Dowex-50 H^+-form resin as catalyst) with fluoro carbonyl compounds undergo prototropic shift on treatment with Et_3N. Hydrolysis of the benzylidene isomers gives the fluorinated amines.

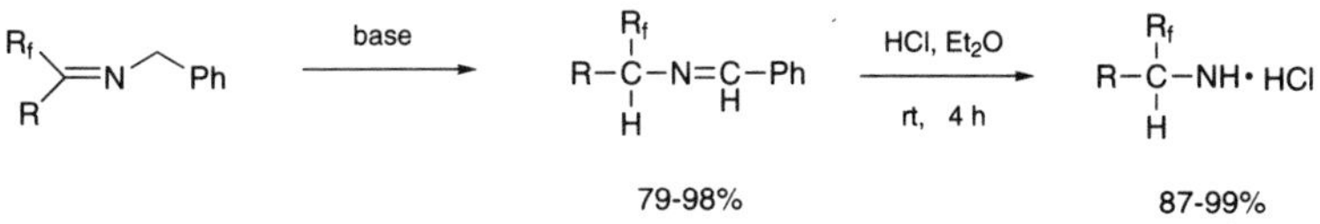

79-98%   87-99%

[1] V. A. Soloshonok, A. G. Kirilenko, V. P. Kukhar, and G. Resnati, *TL* **35**, 3119 (1994).

(*R,R*)-3-Benzyl-1,5-diphenyl-3-azapentane-1,5-dioxysamarium(III) iodide (1).

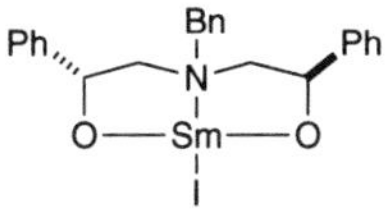

(1)

Meerwein–Ponndorf–Verley reduction.[1] The samarium iodide is a catalyst for asymmetric reduction (36–96% yield, up to 97% ee) of carbonyl compounds by *i*-PrOH.

[1] D. A. Evans, S. G. Nelson, M. R. Gagne, and A. R. Muci, *JACS* **115**, 9800 (1993).

N-Benzylidenebenzenesulfonamide.

Knoevenagel reaction.[1] As surrogate for benzaldehyde in the condensation with active methylene compounds, the benzenesulfonimine is superior. In the presence of Et_3N in chloroform the reaction with, *inter alia*, β-diketones, β-keto esters, α-sulfonyl ketones, α-nitro esters, and malononitrile proceeds at room temperature.

[1] W. W. Zajac, T. R. Walters, J. Buzby, J. Gagnon, and M. Labroli, *SC* **24**, 427 (1994).

Benzyloxyketene.

Iterative chain extension.[1] [2+2]Cycloaddition with imines provides β-lactam derivatives, which can be opened and processed into new imines to renew the reaction sequence. The amino group may be retained or removed as desired. The chain extension method is useful to construct *syn*-poly-1,3-diols and 2-amino-1,3-diols.

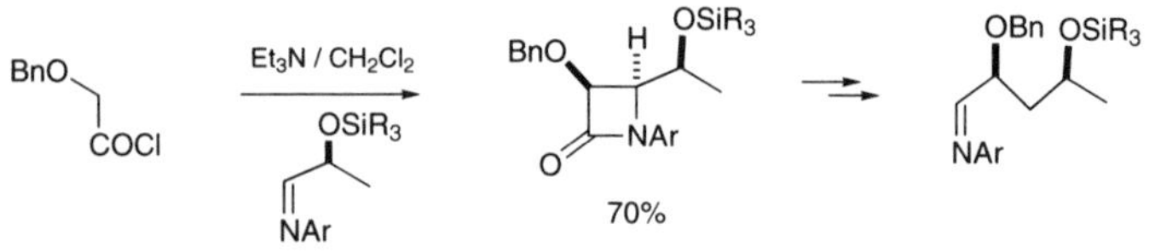

[1] C. Palomo, J. M. Aizpurua, R. Urchegui, and J. M. Garcia, *JOC* **58**, 1646 (1993).

o-(2-Benzyloxyethyl)benzoic acid.

Alcohol protection.[1]   For the esterification of an alcohol with this acid conventional methods (e.g., DCC) are usually suitable. The cleavage is achieved by hydrogenolysis (H$_2$, Pd/C) followed by treatment with *t*-BuOK to induce lactonization, which liberates the alcohol unit.

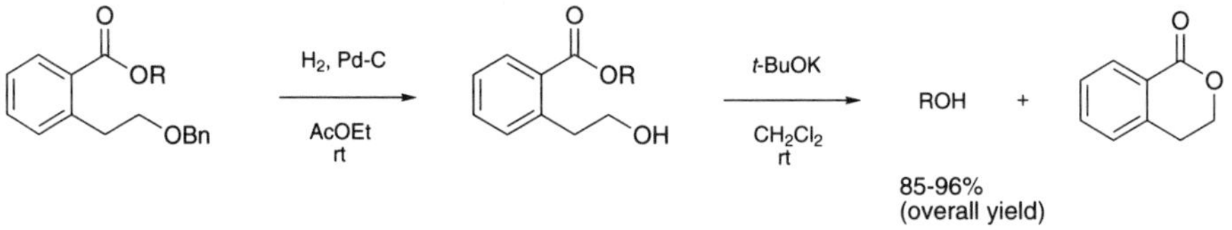

[1] Y. Watanabe, M. Ishimaru, and S. Ozaki, *CL* 2163 (1994).

Benzyl trichloroacetimidate. 13, 32

Alcohol benzylation.[1]   The derivatization is catalyzed by TMSOTf, which does not cause racemization of sensitive chiral compounds.

Lactim benzyl ethers.[2]   Cyclodipeptides are converted to the bislactim ethers, which are valuable intermediates for stereoselective α-amino acid synthesis.

[1] P. Eckenberg, U. Groth, T. Huhn, N. Richter, and C. Schmeck, *T* **49**, 1619 (1993).
[2] U. Groth, C. Schmeck, and U. Schöllkopf, *LA* 321 (1993).

Benzyltriethylammonium tetrathiomolybdate.

Sulfur transfer.[1]   The reaction with arenediazonium salts leads to diaryl disulfides. Interestingly, 1,1'-binaphthalene-2,2'-dithiol is the only product (61% yield) in the reaction with the bis-diazonium salt derived from 1,1'-binaphthyl-2,2'-diamine.

[1] D. Bhar and S. Chandrasekaran, *S* 785 (1994).

Benzyltrimethylammonium dichloroiodate.

Iodination.[1]   Introduction of an iodine atom to the central carbon atom of enamino ketones is readily achieved with the reagent.

[1] K. Matsuo, S. Ishida, and Y. Takuno, *CPB* **42**, 1149 (1994).

2,2'-Bis(diphenylphosphino)-1,1'-binaphthyl (BINAP). 13, 36–37; 14, 38–44, 15, 34; 16, 32–36; 17, 34–38

Palladium complexes.

Asymmetric alkenylation and arylation.[1,2]

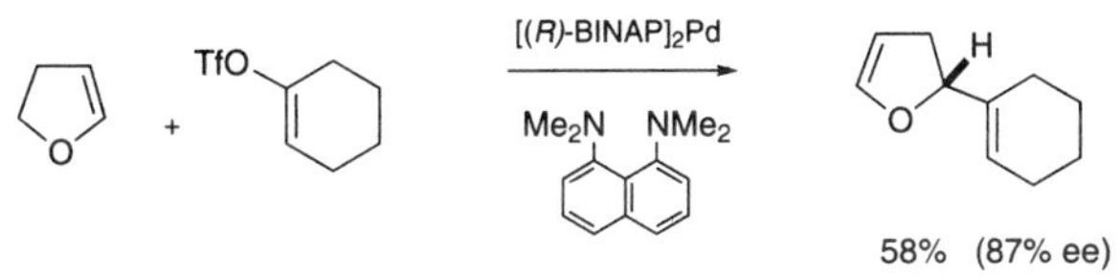

58%   (87% ee)

Chiral morpholines and piperazines.[3]   The ring formation by tandem allylic substitutions of 1,4-diacetoxy-2-butene with 1,2-amino alcohols and 1,2-diamines is promoted by Pd(0) complexes. In the presence of a chiral BINAP ligand this reaction gives optically active products.

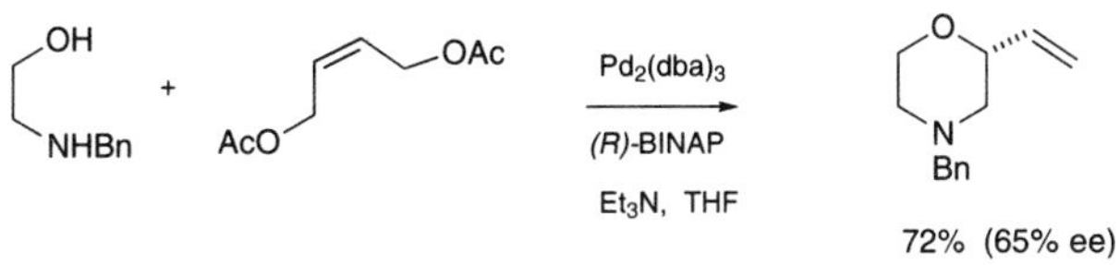

72%   (65% ee)

[1] F. Ozawa, Y. Kobatake, and T. Hayashi, *TL* **34**, 2505 (1993).
[2] F. Ozawa, A. Kubo, Y. Matsumoto, T. Hayashi, E. Nishioka, K. Yanagi, and K. Moriguchi, *OM* **12**, 4188 (1993).
[3] Y. Uozumi, A. Tanahashi, and T. Hayashi, *JOC* **58**, 6826 (1993).

Platinum complexes.

Baeyer–Villiger oxidation.[1]   In kinetic resolutions of chiral ketones (as racemic mixtures) up to 58% ee has been observed. The configuration of the migrating carbon is retained.

[1] A. Gusso, C. Baccin, F. Pinna, and G. Strukul, *OM* **13**, 3442 (1994).

Rhodium(I) complexes.

Intramolecular hydrosilylation.[1]   The yields and ee values of this reaction are dependent on substrate structures and solvents. There are subtle mechanistic variations that are not clearly understood.

[1] X. Wang and B. Bosnich, *OM* **13**, 4131 (1994).

Ruthenium(II) complexes.

Aymmetric hydrogenation.[1]   An extension of the general method to various α-amino α,ω-dicarboxylates has been accomplished, furnishing products with 70–98% ee.

A ruthenium(II) complex containing an octahydro-BINAP ligand[2] has been tested for its effectiveness in inducing enantioselective hydrogenation of conjugated acids.

Reduction of carbonyl groups.   Aldehydes and ketones are subjected to enantioselective reduction. Hydrogenation of benzaldehyde-α-d,[3] α-alkoxyketones[4,5] or β-ketoesters[6,7,8] can be accomplished using either the Ru dihalide complexes or some modified forms. α-Ketoesters[9] are also similarly reduced.

[1] T. Pham and W. D. Lubell, *JOC* **59**, 3676 (1994).
[2] X. Zhang, T. Uemura, K. Matsumura, N. Sayo, H. Kumobayashi, and H. Takaya, *SL* 501 (1994).
[3] T. Ohta, T. Tsutsumi, and H. Takaya, *JOMC* **484**, 191 (1994).
[4] E. Cesarotti, P. Antognazza, M. Pallavincini, and L. Villa, *HCA* **76**, 2344 (1993).
[5] E. Cesarotti, P. Antognazza, A. Mauri, M. Pallavincini, and L. Villa, *HCA* **75**, 2563 (1992).
[6] J. P. Genet, C. Pinel, V. Ratovelomanana-Vidal, S. Mallart, X. Pfister, L. Bischoff, M. C. C. de Andrade, S. Darses, C. Galopin, and J. A. Laffitte, *TA* **5**, 665, 675 (1994).

[7] J. B. Hoke, L. S. Hollis, and E. W. Stern, *JOMC* **455**, 193 (1993).
[8] L. Shao, H. Kawano, M. Saburi, and Y. Uchida, *T* **49**, 1997 (1993).
[9] K. Mashima, K.-H. Kusano, N. Sato, Y. Matsumura, K. Nozaki, H. Kumobayashi, N. Sayo, Y. Hori, T. Ishizaki, S. Akutagawa, and H. Takaya, *JOC* **59**, 3064 (1994).

1,1'-Bi-2,2'-naphthol (BINOL) boronates.

Hetero-Diels–Alder reactions.[1]   The cycloaddition of Danishefsky's diene (and analogs) to chiral imines catalyzed by boronates derived from commercially available BINOL ligands proceeds with good diastereoselectivity. The double asymmetric induction operates for matching pairs, which exhibit fast reaction rates. On the other hand, the reaction of achiral substrates in the presence of chiral binol-boronates gives rise to products in good ee.[2]

20 - 63%
(99:1–86:14)

The $(RO)_4BH$ species in which the four oxygen atoms belong to the chiral 3,3'-bis(o-hydroxyphenyl)-BINOL proves to be an extremely selective catalyst for the Diels–Alder cycloaddition.[3] Almost exclusively one product is obtained in a favorable case.

β-Amino esters.[4]   By means of double stereodifferentiation using chiral imines and a chiral BINOL-boronate catalyst, the condensation with ketene silyl acetals is a simple method for the synthesis of β-amino esters in optically active forms. α-Hydroxy-β-amino esters are similarly accessible.[5]

[1] K. Hattori and H. Yamamoto, *SL* 129 (1993).
[2] K. Hattori and H. Yamamoto, *T* **49**, 1749 (1993).
[3] K. Ishihara and H. Yamamoto, *JACS* **116**, 1561 (1994).
[4] K. Hattori, M. Miyata, and H. Yamamoto, *JACS* **115**, 1151 (1993).
[5] K. Hattori and H. Yamamoto, *T* **50**, 2785 (1994).

1,1'-Bi-2,2'-naphthol–lanthanide complexes. 17, 28–30

Nef reaction.[1,2]   Optically active β-hydroxy nitroalkanes are obtained in the condensation of nitromethane with aldehydes using a BINOL-lanthanum complex as catalyst.

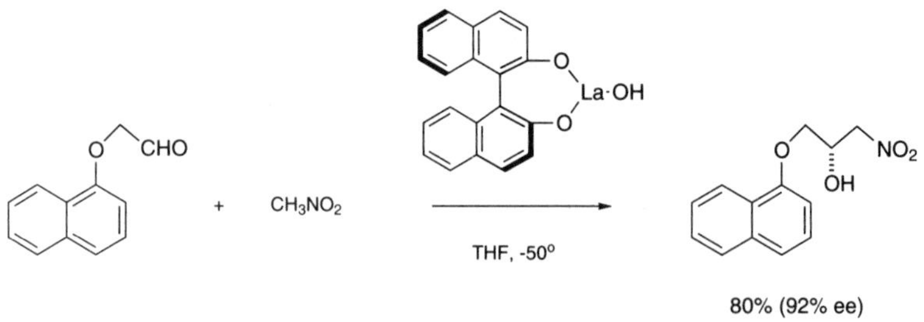

80% (92% ee)

Michael reactions.[2]   A BINOL complex prepared from $(i\text{-PrO})_3\text{La}$ promotes enantioselective Michael reactions with excellent results.

Alkylation.[3]   Organolanthanide reagents derived from lanthanide chlorides, alkyllithiums, and BINOL add to aldehydes with good stereoselectivity.

Diels–Alder reactions.[4]   The version with inverse electron demand involving α-pyrones and vinyl ethers is subjected to asymmetric induction by a BINOL-ytterbium complex.

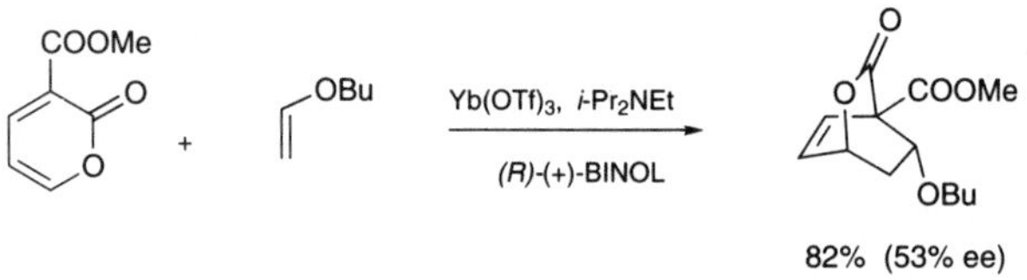

82% (53% ee)

[1] H. Sasai, T. Suzuki, N. Itoh, and M. Shibasaki, *TL* **34**, 851 (1993).
[2] H. Sasai, T. Arai, and M. Shibasaki, *JACS* **116**, 1571 (1994).
[3] K. Chibale, N. Greeves, L. Lyford, and J. E. Pease, *TA* **4**, 2407 (1993).
[4] I. E. Marko, G. R. Evans, J.-P. Declercq, J. Feneau-Dupont, and B. Tinant, *BSCB* **103**, 295 (1994).

1,1'-Bi-2,2'-naphthol–lithium aluminum hydride. 16, 133

Enantioselective reductions.[1]   Lactones and hydroxylactams are obtained in chiral forms when *meso*-cyclic 1,2-carboxylic anhydrides and dicarboximides are treated with the complex at low temperature.

[1] K. Matsuki, H. Inoue, A. Ishida, M. Takeda, M. Nakagawa, and T. Hino, *CPB* **42**, 9 (1994).

1,1'-Bi-2,2'-naphthol–tin(IV) chloride.

Enantioselective protonation.[1]   Cleavage of enol silyl ethers and ketene bis(trialkylsilyl) acetals by the complex leads to chiral ketones and esters.

[1] K. Ishihara, M. Kaneeda, and H. Yamamoto, *JACS* **116**, 11179 (1994).

1,1'-Bi-2,2'-naphthol/titanium complexes. 15, 26–27; 16, 24–25; 17, 28–30

In addition to the chiral BINOL-ligated dichlorotitanium complex, the variants using diisopropoxytitanium and oxotitanium species are shown to be effective to different degrees.

Ene reactions.[1]   The enantioselective synthesis of α-hydroxy esters from ene reaction of glyoxylate is effective using vinyl chalcogenides[2] and allylsilanes.[3] The reaction of fluoral[4] also attains a high level of enantio- and diastereoselectivities. Chiral catalysts that contain substituents in the BINOL moiety have been evaluated.

Allylation.   Both allylstannanes[5] and silanes[6] transfer their allyl groups to carbonyl substrates in enantioselective and diastereoselective manners. The facile preparation[7] of a very effective catalyst from BINOL and $(i\text{-PrO})_4$Ti in a 2:1 ratio without the need for stirring, heating or cooling is most advantageous in future applications of the methodology. Under such conditions homopropargylic alcohols are also obtained by substituting the allyl reagent with allenyltributylstannane.[8]

Aldol condensation.[9]   The formation of β-siloxy ester derivatives from ketene silyl acetals and aldehydes may actually be an ene-type reaction involving silyl group migration.

Michael reaction.[10]   Using enones as acceptors, enol silyl ethers as nucleophiles, and a chiral BINOL-TiO catalyst, the Michael reaction takes place at $-78°C$. The ee values range from moderate to highly respectable (36–90%).

[1] M. Terada, Y. Motoyama, and K. Mikami, *TL* **35**, 6693 (1994).
[2] M. Terada, S. Matsukawa, and K. Mikami, *CC* 327 (1993).
[3] K. Mikami and S. Matsukawa, *TL* **35**, 3133 (1994).
[4] K. Mikami, T. Yajima, M. Terada, E. Kato, and M. Maruta, *TA* **5**, 1087 (1994).

[5] A. L. Costa, M. G. Piazza, E. Tagliavini, C. Trombini, and A. Umani-Ronchi, *JACS* **115**, 7001 (1993).
[6] S. Aoki, K. Mikami, M. Terada, and T. Nakai, *T* **49**, 1783 (1993).
[7] G. E. Keck and L. S. Geraci, *TL* **34**, 7827 (1993).
[8] G. E. Keck, D. Krishnamurthy, and X. Chen, *TL* **35**, 8323 (1994).
[9] K. Mikami and S. Matsukawa, *JACS* **116**, 4077 (1994).
[10] S. Kobayashi, S. Suda, M. Yamada, and T. Mukaiyama, *CL* 97 (1994).

Bis(acetonitrile)dichloropalladium(II). 14, 35–36; 15, 28–29; 16, 25–26; 17, 30–31

Allylic rearrangement. This suprafacial rearrangement finds use in the preparation of 4-acetoxy-2-alkenonitriles[1] and δ-hydroxy allylic phosphine oxides.[2]

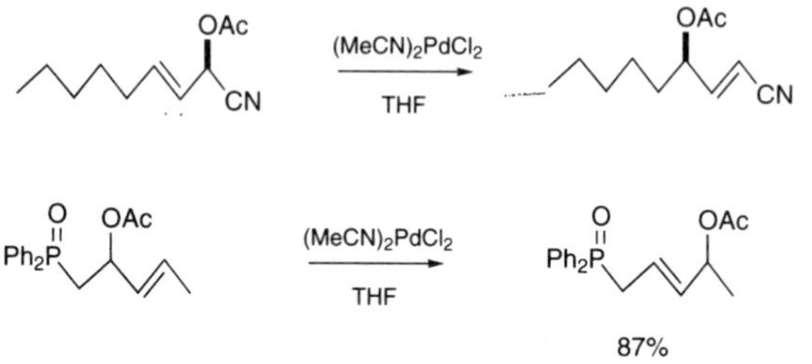

Cyclization. The catalyst induces *endo–trig* cyclization of 2-hydroxy-3-butenylamine derivatives at room temperature to afford 3-pyrrolines.[3] An oxidative cyclization of tertiary *o*-allylbenzylamines[4] occurs when Ph_3P is added after complexation.

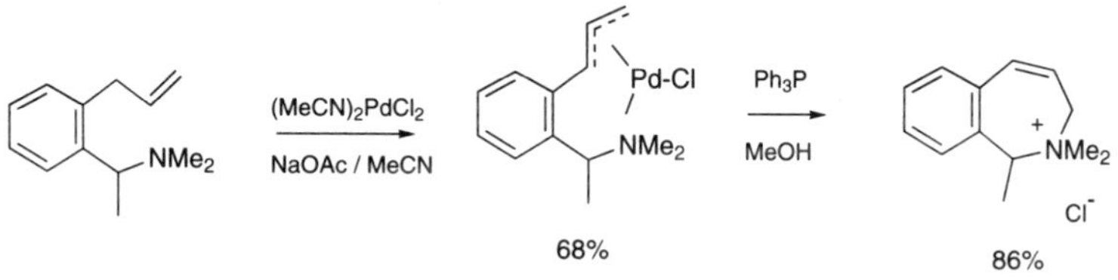

Cross couplings. The catalyzed coupling of haloalkenes with organometallic reagents gives rise to various functionalized alkenes, including enynes,[5] 3-substituted 3-butenoic acids,[6] and 1,4-dienes.[7]

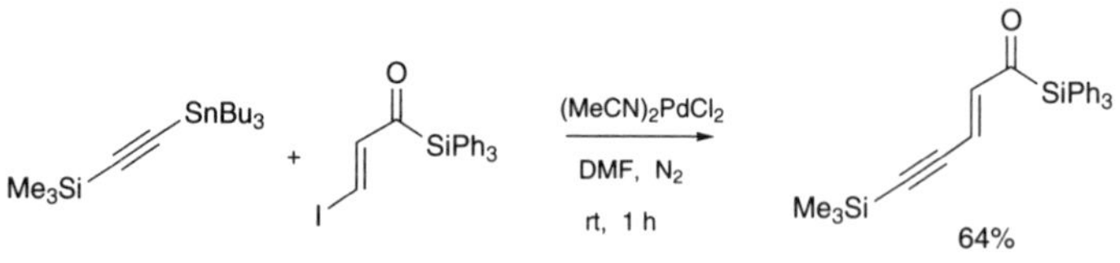

[1] H. Abe, H. Nitta, A. Mori, and S. Inoue, *CL* 2443 (1992).
[2] J. Clayden and S. Warren, *JCS(P1)* 2913 (1993).
[3] M. Kimura, H. Harayama, S. Tanaka, and Y. Tamaru, *CC* 2531 (1994).
[4] P. A. van der Schaaf, J.-P. Sutter, M. Grellier, G. P. M. van Mier, A. L. Spek, G. van Koten, and M. Pfeffer, *JACS* **116**, 5143 (1994).
[5] A. Degl'Innocenti, A. Capperucci, L. Bartoletti, A. Mordini, and G. Reginato, *TL* **35**, 2081 (1994).
[6] A. Duchene, M. Abarbri, J.-L. Parrain, M. Kitamura, and R. Noyori, *SL* 524 (1994).
[7] M. Kosugi, T. Sakaya, S. Ogawa, and T. Migita, *BCSJ* **66**, 3058 (1993).

Bis(acetonitrile)chloronitropalladium(II)–copper(II) chloride–oxygen.

Oxidative cyclization. Homoallylic and homopropargylic alcohols are converted to γ-lactols[1] and γ-lactones,[2] respectively.

[1] T. M. Meulemans, N. H. Kiers, B. L. Feringa, and P. W. N. M. van Leeuwen, *TL* **35**, 455 (1994).
[2] P. Compain, J.-M. Vatele, and J. Gore, *SL* 943 (1994).

Bis(allyl)di-μ-chlorodipalladium.

Cross coupling.[1] Biaryls are obtained from aryl halides and aryl(halo)silanes in the presence of the palladium complex and KF.

vic-Bissilylation.[2] Disilanes are split and add to alkenes and alkynes to give the 1,2-disilyl derivatives.

Allylic substitution.[3] Di-Boc-allylamines are readily obtained from allyl acetates. The products are converted into protected glycine esters on ozonolysis.

Many chiral ligands have been investigated to assist the substitution in an enantioselective sense. These bidentate ligands usually possess an N,P-,[4-6] N,S-,[7] or O,P-donor pair.[8]

Reductive cleavage of β-(N-Boc-aziridin-2-yl) α,β-unsaturated esters.[9] Using formic acid as hydrogen source, the reductive cleavage gives a mixture of conjugated and deconjugated esters. The catalyst is bis(2-methallyl)di-μ-chlorodipalladium.

[1] Y. Hatanaka, K.-i. Goda, Y. Okahara, and T. Hiyama, *T* **50**, 8301 (1994).
[2] F. Ozawa, M. Sugawara, and T. Hayashi, *OM* **13**, 3237 (1994).
[3] R. Jumnah, J. M. J. Williams, and A. C. Williams, *TL* **34**, 6619 (1993).
[4] B. M. Trost and M. G. Organ, *JACS* **116**, 10320 (1994).

[5] P. von Matt and A. Pfaltz, *ACIEE* **32**, 566 (1993).
[6] P. von Matt, O. Loiseliur, G. Koch, A. Pfaltz, C. Lefeber, T. Feucht, and G. Helmchen, *TA* **5**, 573 (1994).
[7] J. V. Allen, S. J. Coote, G. J. Dawson, C. G. Frost, C. J. Martin, and J. M. J. Williams, *JCS(P1)* 2065 (1994).
[8] C. G. Frost and J. M. J. Williams, *SL* 551 (1994).
[9] A. Satake, I. Shimizu, and A. Yamamoto, *SL* 64 (1995).

Bis(arenesulfonyl)methanes.

Aldehyde synthesis.[1]   Aldehydes are obtained through alkylation under solid–liquid biphasic conditions followed by reduction with LiAlH$_4$ and Hg(II)-promoted hydrolysis. Thus the bissulfonyl methanes are a formyl anion synthon.

[1] Y.-P. Wang and X. Huang, *YH* **13**, 253 (1993).

Bis(benzonitrile)dichloropalladium(II). 13, 34; 15, 29

Cleavage of propargyl vinyl ethers.[1]   The selective cleavage (for which other Pd(II) reagents are also effective) in the presence of other ethers (e.g., acetals) is useful for synthetic manipulations. Using this methodology, the preparation of the very sensitive *O*-protected γ-hydroxy β-keto esters is simplified, using the adducts of the alkynoic esters with propargyl alcohol as precursors.

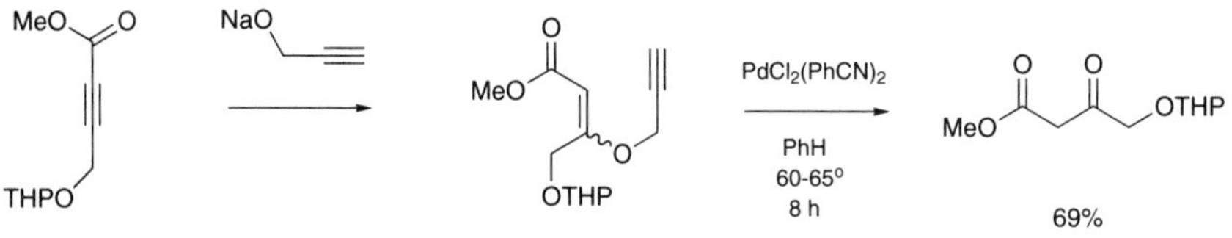

Allylation of aldehydes with allyl alcohols.[2]   Umpolung of allylic alcohols (e.g., isoprenol) and the direct nucleophilic attack on aldehydes is realized in the presence of (PhCN)$_2$PdCl$_2$ and SnCl$_2$.

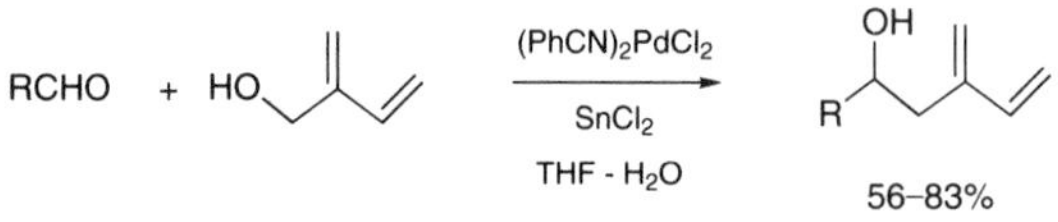

Cyclization of allyl propynoates.[3,4]   α-Chloromethylene-γ-lactones are formed. The nature of the β-group is influenced by additives and the original substitution pattern.

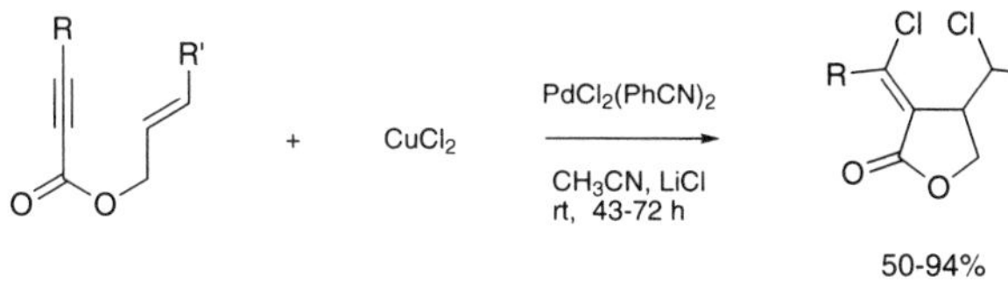

50-94%

ArSO$_2$Cl → ArI.[5]   This transformation requires ZnI$_2$ or KI as iodide source, and also LiCl and (iPrO)$_4$Ti.

Arenesulfonyl isocyanates.[6]   Arenesulfonylimino derivatives of iodobenzene or diphenyl selenide undergo group exchange with CO in the presence of the palladium catalyst.

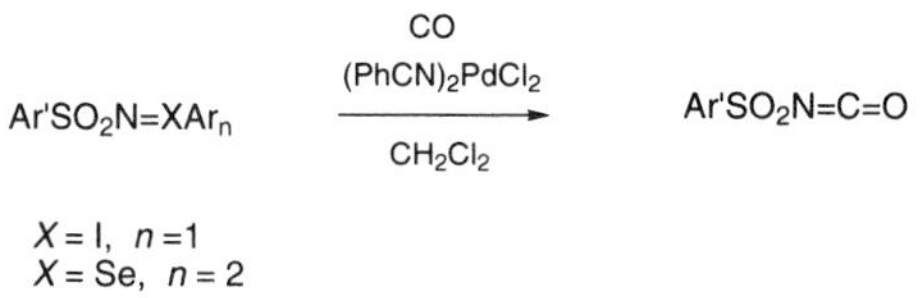

X = I, n = 1
X = Se, n = 2

Imidazolidin-2-thiones.[8]   Aziridines condense with sulfur diimides under the influence of (PhCN)$_2$PdCl$_2$.

[1] G. S. Sarin, *TL* **34**, 6309 (1993).
[2] Y. Masuyama, M. Fuse, and Y. Kurusu, *CL* 1199 (1993).
[3] S. Ma and X. Lu, *JOC* **58**, 1245 (1993).
[4] S. Ma and X. Lu, *JOMC* **447**, 305 (1993).
[5] T. Satoh, K. Itoh, M. Miura, and M. Nomura, *BCSJ* **66**, 2121 (1993).
[6] G. Besenyei and L. I. Simandi, *TL* **34**, 2839 (1993).
[7] G. Besenyei, S. Nemeth, and L. I. Simandi, *TL* **35**, 9609 (1994).
[8] J.-O. Baeg and H. Alper, *JACS* **116**, 1220 (1994).

Bis(benzyloxy)diethylaminophosphine.

Glycosyl phosphites.[1]   The diethylamino group of the reagent is exchangeable; thus on reaction with sugars in the presence of 1,2,4-triazole, glycosyl phosphites are formed. These products are not only useful as glycosylation agents, they can be oxidized to the phosphates, which are precursors of sugar nucleotides.

[1] M. M. Sim, H. Kondo, and C.-H. Wong, *JACS* **115**, 2260 (1993).

Bis(bromomagnesium) sulfide.

Sulfides.   The reagent, prepared by saturating an ethereal solution of EtMgBr with anhydrous H$_2$S, reacts with organic halides to furnish sulfides.

[1] A. N. Nodugov and N. N. Pavlova, *JOCU* **28**, 1103 (1992).

Bis(*t*-butoxycarbonyl) oxide. [Di-*t*-butyl pyrocarbonate]

Derivatization of amines and alcohols.   Attachment of the Boc group to ary-lamines[1] is achieved by reaction with sodium hexamethyldisilazide and Boc₂O. Both nitrogen and oxygen atoms of hydroxylamine are derivatized in a biphasic system in the presence of Et₃N as base.[2]

Cyclic carbodiimides.[3]   Bis(iminophosphoranes) behave as nucleophiles toward Boc₂O. After an isocyanate is formed, a Wittig reaction follows. Either an intra-molecular or intermolecular process predominates according to the influence of ring strain.

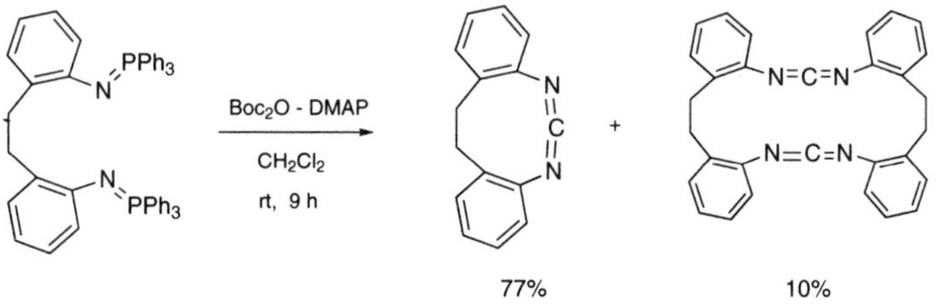

[1] T. A. Kelly and D. W. McNeil, *TL* **35**, 9003 (1994).
[2] M. A. Staszak and C. N. Doecke, *TL* **34**, 7043 (1993).
[3] P. Molina, M. Alajarin, P. Sanchez-Andrada, J. Elguero, and M. L. Jimeno, *JOC* **59**, 7306 (1994).

N,O-Bis(*t*-butoxycarbonyl)hydroxylamine.

Synthesis of N-alkylhydroxylamines and hydroxamic acids.[1]   *N*-Alkylation is conveniently achieved under phase-transfer conditions (96–98% yield), whereas acy-lation with an acyl chloride proceeds in the presence of Et₃N in dichloromethane. The Boc groups are subsequently removed on treatment with trifluoroacetic acid.

[1] M. A. Staszak and C. W. Doecke, *TL* **35**, 6021 (1994).

[*N,N'*-Bis(3,5-di-*t*-butylsalicylidene)-1,2-cyclohexanediaminato(2-)] manganese(III) chloride.

Enantioselective epoxidation.[1]   This effective catalyst (**1**) is readily prepared in a 70–100 kg scale from the resolved diamine (with tartaric acid), the salicylaldehyde (from formylation of the phenol by Duff reaction), and manganese acetate, followed by anion exchange by treatment with aqueous NaCl.

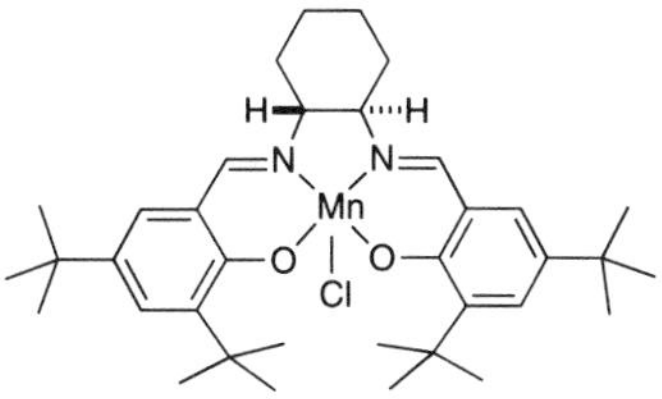

(1)

[1] J. F. Larrow, E. N. Jacobsen, Y. Gao, Y. Hong, X. Nie, and C. M. Zepp, *JOC* **59**, 1939 (1994).

Bis(chlorodibutyltin) oxide.

Diol monoesters.[1]   1,*n*-Diacetates ($n = 2, 3, 4$) are converted to the monoesters by transesterification to methanol in the presence of the tin oxide. The catalytic effect arises from cooperation of two different tin metal atoms held in proximity. For long-chain diacetates ($n > 5$) the selectivity vanishes because the template effect cannot operate.

[1] J. Otera, N. Dan-oh, and H. Nozaki, *T* **49**, 3065 (1993).

Bis(*sym*-collidine)iodine(I) perchlorate. 15, 30; 17, 155

Silylation.[1]   Transsilylation from a 4-pentenyl silyl ether to an alcohol involves iodination of the double bond, rendering the pentenyloxy group nucleofugal (leaving as 2-iodomethyltetrahydrofuran).

Iodolactonization.[2]   Medium-sized (7- to 12-membered) ring formation from alkenoic acids is favored by the presence of an in-chain oxygen atom.

Glycosylation.[3]   Phenyl selenoglycosides are activated by the iodinating agent. Dissacharides are formed at room temperature within minutes.

[1] C. Colombier, T. Skrydstrup, and J.-M. Beau, *TL* **35**, 8167 (1994).
[2] B. Simonot and G. Rousseau, *JOC* **59**, 5912 (1994).
[3] H. M. Zuurmond, P. H. van der Meer, P. A. M. van der Klein, G. A. van der Marel, and J. H. van Boom, *JCC* **12**, 1091 (1993).

Bis(1,5-cyclooctadiene)nickel(0). 13, 35; 14, 36−37; 15, 30−32, 131−132; 16, 19; 17, 32

Bromoalkene + alkene cyclization.[1]   A nitrogen atom assists the intramolecular cyclization, making methyleneazacyclic compounds readily available.

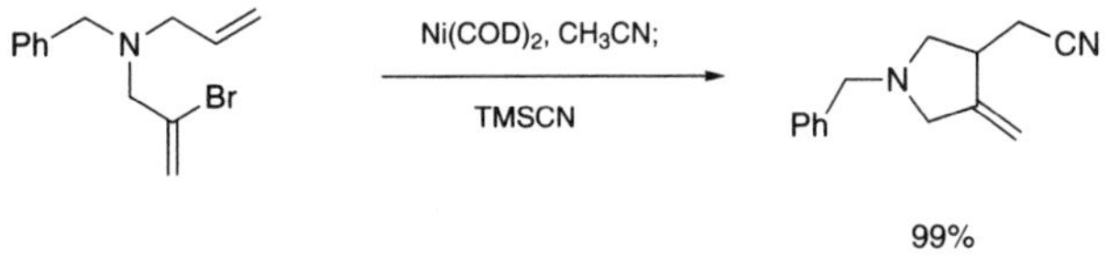

99%

Intramolecular allylation.[2]   Certain dienic aldehydes undergo cyclization to give 2-alkenylcycloalkanols in the presence of Ni(cod)$_2$, Et$_3$SiH, and Ph$_3$P. π-Allylnickel nucleophiles are generated from reaction of the diene units with the silylnickel hydride. Cycloalkenols with 5-, 6-, and 7-membered rings are readily obtained.

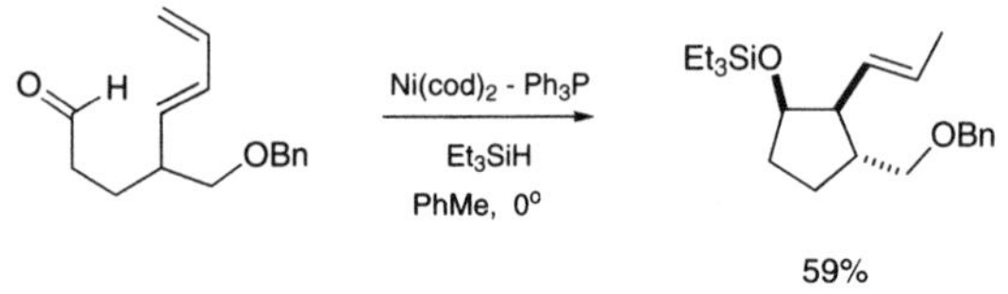

59%

[1] D. Sole, Y. Cancho, A. Llebaria, J. M. Moreto, and A. Delgado, *JACS* **116**, 12133 (1994).
[2] Y. Sato, M. Takimoto, K. Hayashi, T. Katsuhara, K. Takagi, and M. Mori, *JACS* **116**, 9771 (1994).

Bis(cyclopentadienyl)iron hexafluorophosphate.

Acyliminium ions.[1]   Oxidative destannylation of *N*-(α-stannylalkyl) amides leads to reactive electrophiles, which can react with enol silyl ethers, allylsilanes, allylstannanes, and trimethylsilyl cyanide. The stannyl group lowers the oxidation potential of the amidic nitrogen, without which the acyliminium ions do not form under the oxidation conditions.

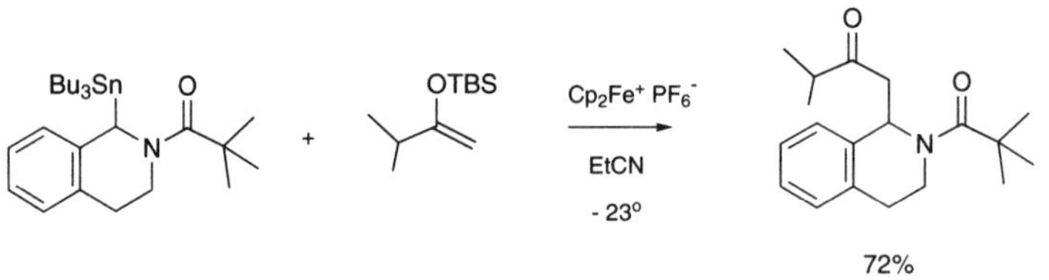

72%

[1] K. Narasaka, Y. Kohno, and S. Shimada, *CL* 125 (1993).

Bis(cyclopropyl)titanocene.

Alkylidene- and vinylcyclopropanes.[1]   The reagent effects cyclopropylidenation of carbonyl compounds including esters and lactones. Reaction with alkynes gives vinylcyclopropanes on acidic workup.

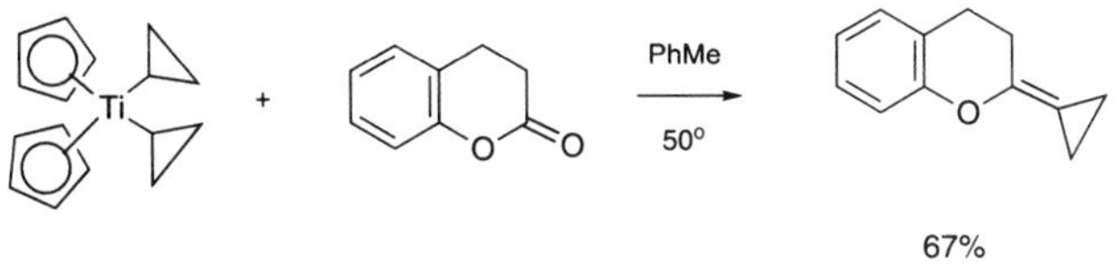

67%

[1] N. A. Petasis and E. I. Bzowej, *TL* **34**, 943 (1993).

Bis(dibenzylideneacetone)palladium–oxygen.

Oxidation of allylic alcohols.[1]   The catalyst system (dba)$_2$Pd-Ph$_3$P in the presence of oxygen is effective to oxidize allylic alcohols. Interestingly, with exclusion of O$_2$ and in the presence of an acid, certain substrates containing a remote double bond undergo cyclodehydration. The oxidation can also be effected with (Ph$_3$P)$_4$Pd, NH$_4$PF$_6$, and O$_2$.

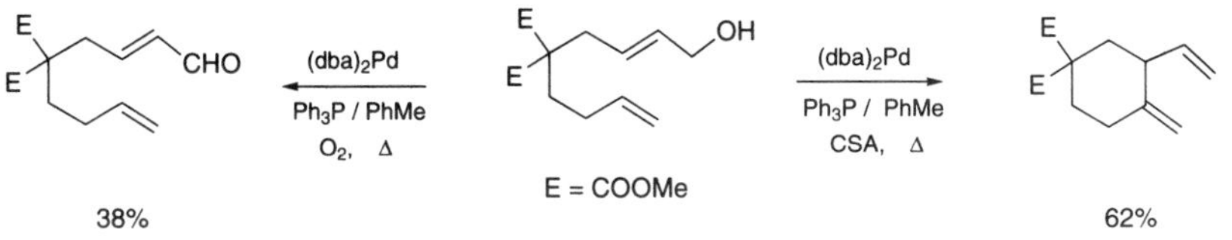

38%   E = COOMe   62%

[1] E. Gomez-Bengoa, P. Noheda, and A. M. Echavarren, *TL* **35**, 7097 (1994).

[Bis(diphenylphosphino)alkane]bis(2-methallyl) ruthenium complexes.

(Z)-Enol esters.[1]   The stereoselective addition of carboxylic acid to a terminal alkyne linkage is catalyzed by ruthenium complexes. Bidentate phosphine ligands separated by 2, 3, or 4 methylene groups are all effective.

[1] H. Doucet, J. Höfer, C. Bruneau, and P. H. Dixneuf, *CC* 850 (1993).

2,4-Bis(4-methoxyphenyl)-1,3,2,4-dithiadiphosphetane 2,4-disulfide (Lawesson's reagent). 13, 38–39; **15**, 37; **16**, 37–38

Thiols from alcohols.[1]   In toluene or DME the Lawesson's reagent directly converts aralkyl and cycloalkyl alcohols to thiols.

Perfluoroalkyl amines.[2]   Thionation of perfluoroacylamines followed by reaction with NBS-Bu$_4$NH$_2$F$_3$ constitutes a convenient method for the preparation.

β-D-Ribofuranosides.[3]   Starting from ribofuranoses and alcohols, the glycosidation is mediated by a mixture of $AgClO_4$ and Lawesson's reagent or $Ph_2Sn=S$.

[1] T. Nishio, *JCS(PI)* 1113 (1993).
[2] M. Kurobashi and T. Hiyama, *TL* **35**, 3983 (1994).
[3] N. Shimomura and T. Mukaiyama, *CL* 1941 (1993).

Bismuth. 13, 39

Allylation.[1]   Carbonyl compounds and imines are attacked to give homoallylic alcohols and amines by a mixture of bismuth and allyl bromide in acetonitrile. The presence of Bu_4NBr is required, whereas Me_3SiCl and NaI are less effective. Generally, the yields range from 85 to 95%. The use of tantalum instead of bismuth works only for imines, and with lower yields.

[1] P. J. Bhuyan, D. Prajapati, and J. S. Sandhu, *TL* **34**, 7975 (1993).

Bismuth(III) chloride. 15, 37

Chlorination of alcohols.[1]   Bismuth(III) chloride is an excellent catalyst for chlorination of alcohols with chlorosilanes. Thus *t*-butyl chloride is obtained quantitatively at room temperature.

Reduction of nitroarenes.[2]   This reagent in combination with zinc is a mild reducing agent, converting nitroarenes to azoxyarenes.

[1] M. Labrouillere, C. Le Roux, H. Gaspard-Iloughmane, and J. Dubac, *SL* 723 (1994).
[2] H. N. Borah, D. Prajapati, J. S. Sandhu, and A. C. Ghosh, *TL* **35**, 3167 (1994).

Bismuth(III) chloride–metal iodide.

Aldol condensation.   Silyl enol ethers react with carbonyl acceptors to give β-siloxy ketones and esters.[1] As BiX_3 is also a catalyst for the replacement of β-siloxy groups, prolonged reaction times can be applied to prepare β-halo carbonyl compounds directly.[2] The reaction is enhanced by ultrasound.

[1] C. Le Roux, H. Gaspard-Iloughmane, J. Dubac, J. Jaud, and P. Vignaux, *JOC* **58**, 1835 (1993).
[2] C. Le Roux, H. Gaspard-Iloughmane, and J. Dubac, *JOC* **59**, 2238 (1994).

Bismuth(III) mandelate–dimethyl sulfoxide.

Epoxide cleavage.   Carboxylic acids are obtained from this reaction.[1] It shows chemoselectivity in that alkenes and alcohols are inert. Bismuth(III) acetate and oxide are not effective. In aryl epoxides, the presence of electron-donating groups favor the oxidation.[2]

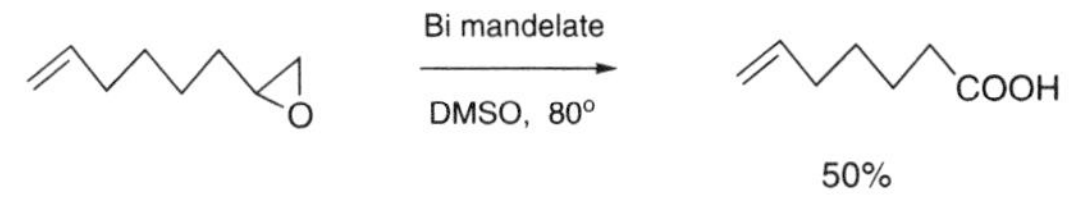

[1] T. Zevaco, E. Dunach, and M. Postel, *TL* **34**, 2601 (1993).
[2] V. Le Boisselier, E. Dunach, and M. Postel, *JOMC* **482**, 119 (1994).

Bis(pyridine)iodine(I) tetrafluoroborate–tetrafluoroboric acid.

Dimerization of 1-iodoalkynes.[1]   Head-to-tail coupling gives 1,1-diiodoenynes.

Iodination.[2]   Arenes undergo iodination in CH_2Cl_2. Trifluoroacetic acid or trifluoromethanesulfonic acid may be used instead of tetrafluoroboric acid as catalyst.

[1] J. Barluenga, L. M. Gonzalez, I. Llorente, and P. J. Campos, *ACIEE* **32**, 893 (1993).
[2] J. Barluenga, J. M. Gonzalez, M. A. Garcia-Martin, P. J. Campos, and G. Asensio, *JOC* **58**, 2058 (1993).

Bis(tetrabutylammonium)cerium(IV) nitrate.

Oxidative cross-coupling.[1] α-Stannylalkanoic esters and amides undergo oxidation, and the resulting free radicals can be trapped in situ by electron-rich alkenes such as silyl enol ethers and allylsilanes. Thus γ-keto esters are accessible by this reaction. Note that α-germanylalkanoic esters are less reactive toward the oxidant and the α-silylalkanoic esters do not undergo oxidation at all.

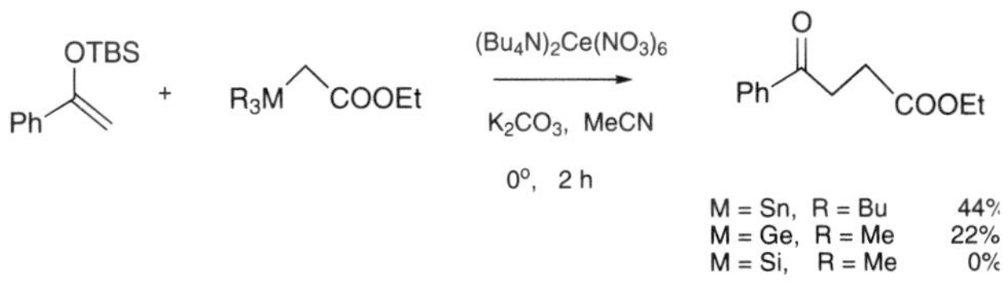

[1] Y. Kohno and K. Narasaka, *CL* 1689 (1993).

Bis(tributylstannyl)acetylene.

Ethynylation of azaaromatics.[1]   After activation of the system by a chloroformate ester, pyridine, quinoline, pyridazine, and analogous heterocycles are attacked by the tin compound. α-Chloroethyl chloroformate is particularly effective as the activator.[2]

[1] T. Itoh, H. Hasegawa, K. Nagata, M. Okada, and A. Ohsawa, *T* **50**, 13089 (1994).
[2] T. Itoh, H. Hasegawa, K. Nagata, M. Okada, and A. Ohsawa, *SL* 557 (1994).

Bis(tributyltin) oxide. 13, 41–42; 15, 39

Transmetallation.[1] Catalyzed by Bu$_4$NF, benzyl-, allyl-, and alkynylsilanes are converted to the tributyltin derivatives by the tin oxide in THF. Yields are excellent (10 examples, 95–99% yield).

3-Alken-2-ones.[2] Diketene undergoes ring opening to afford the stannyl ester of β-stannyloxybutenoic acid. Addition of an aldehyde and HMPA induce a tandem aldolization and decarboxylation.

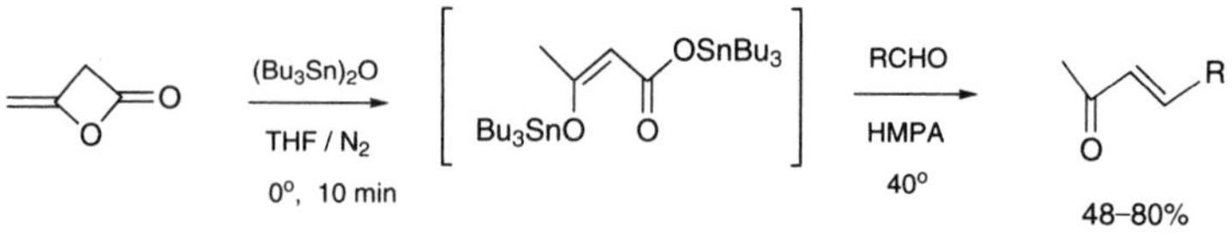

2-Alkyl-4-oxopentanals.[3] In the presence of LiBr, the tin enolate derived from diketene and bis(tributyltin) oxide reacts with α-bromoaldehydes in a chemoselective manner. It involves displacement of the bromine and decarboxylation.

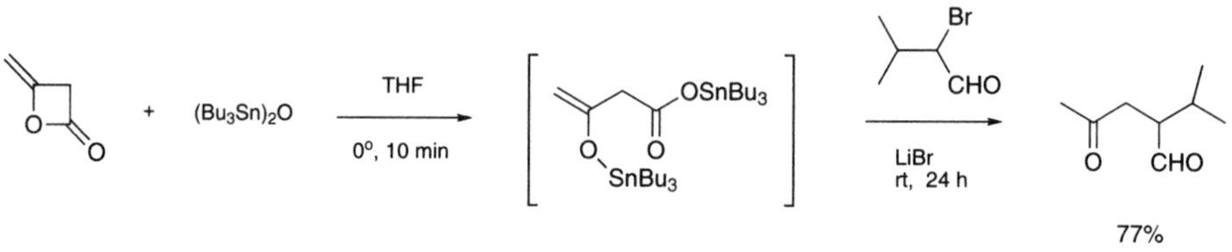

1,5-Diketones.[4] The diketene–tin oxide adduct is a useful Michael donor that reacts with enones to give 1,5-diketones. Note that simple stannyl enol ethers do not undergo the same reaction; thus the coordination of the O-stannyl group by the ester carbonyl must be important.

Macrolactonization.[5] Bis(tributyltin) oxide, as well as many other organotin(IV) compounds, catalyze the formation of 11-, 13-, 14-, 16-, and 17-membered lactones from 2,2,2-trifluoroethyl ω-hydroxyalkanoates in hot octane. The yields range from 21% to 81%.

[1] B. P. Warner and S. L. Buchwald, *JOC* **59**, 5822 (1994).
[2] I. Shibata, M. Nishio, A. Baba, and H. Matsuda, *CL* 1219 (1993).
[3] M. Yasuda, M. Nishio, I. Shibata, A. Baba, and H. Matsuda, *JOC* **59**, 486 (1994).
[4] I. Shibata, M. Nishio, A. Baba, and H. Matsuda, *CC* 1067 (1993).
[5] J. D. White, N. J. Green, and F. R. Fleming, *TL* **34**, 3515 (1993).

Bis(2,2,2-trichloroethyl) azodicarboxylate.

Amination of arenes.[1] With ZnI_2 as the catalyst, the reagent is an adequate electrophilic reagent for many arenes. The products are cleaved in one step to provide the amines by treatment with zinc dust in acetic acid.

[1] H. Mitchell and Y. Leblanc, *JOC* **59**, 682 (1994).

Bis(trichloromethyl) carbonate = [triphosgene].

Dehydration.[1] Aldoximes and carboxamides are dehydrated in the presence of Et_3N to give nitriles in good yields.

Chlorination of phosphines.[2] Both trialkyl- and triarylphosphines afford the dichlorides on reaction with triphosgene, which can then be used to convert epoxides to 1,2-dichloroalkanes, aldehydes to *gem*-dichlorides, amides to nitriles, formamides to isonitriles, ureas to carbodiimides, and amines to triphenylphosphoranylimines.

[1] D. P. Sahu, *IJC(B)* **32B**, 385 (1993).
[2] A. Wells, *SC* **24**, 1715 (1994).

N,N-Bis(trifluoromethanesulfonyl)benzenesulfonodiimidoyl chloride.

Aromatic chlorination.[1] This reagent $PhS(=NSO_2CF_3)_2Cl$ chlorinates electron-deficient arenes at room temperature.

[1] R. Yu. Garlyauskajte, S.V. Sereda, and L. M. Yagupolskii, *T* **50**, 6891 (1994).

Bis(trifluoromethyl) telluride ($(CF_3)_2Te$.

Trifluoromethylation.[1] In a sealed tube aniline undergoes nuclear trifluoromethylation. Three different positional isomers are obtained in about equal amounts. Arenethiols are *S*-trifluoromethylated under ultraviolet irradiation.

[1] D. Naumann, S.V. Pazenok, and V. Turra, *RJOC* **29**, 128 (1993).

1,2-Bis(trimethylsiloxy)cyclobutene.

1,1-Butanediacylation. For the synthesis of 2,2-disubstituted 1,3-cyclopentanediones in one pot BF_3 etherate is used to promote the condensation between the cyclobutene with acetals or ketones.[1] In the reaction with ketones, the addition of a small amount of water after the initial step is necessary to assist the rearrangement and to render the reversible reaction insignificant. Steric hindrance and conjugate double bonds disfavor the process.[2]

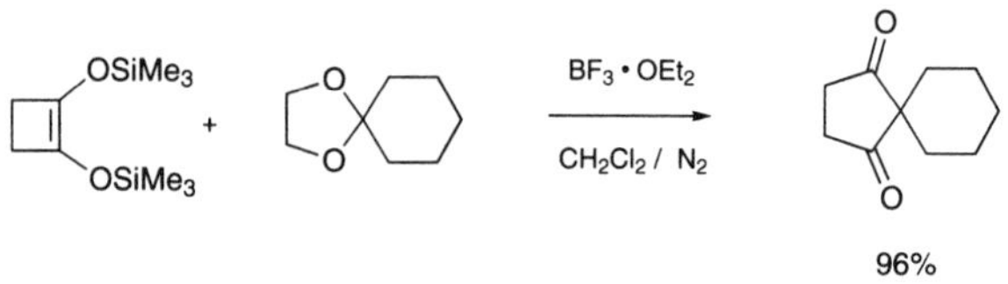

96%

[1] Y.-J. Wu, D.W. Strickland, T. J. Jenkins, P.-Y. Liu, and D. J. Burnell, *CJC* **71**, 1311 (1993).
[2] T. J. Jenkins and D. J. Burnell, *JOC* **59**, 1485 (1994).

N,O-Bis(trimethylsilyl)hydroxylamine Me₃SiNHOSiMe₃.

Primary amines.[1]   The reagent behaves as an electrophile toward higher-order organocuprates, transferring an $NHSiMe_3$ moiety to one of the anionic ligands.

β-Hydroxylamino esters.[2]   Alkylidenemalonic esters undergo Michael addition with the bissilylated hydroxylamine, while the corresponding acrylic esters react very sluggishly. The addition is subject to 1,2-asymmetric induction by an allylic substituent.

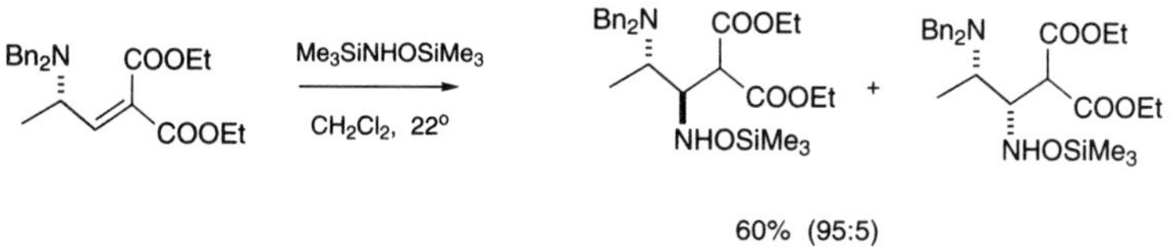

60% (95:5)

[1] A. Casarini, P. Dembech, D. Lazzari, E. Marini, G. Reginato, A. Ricci, and G. Seconi, *JOC* **58**, 5620 (1993).
[2] M. T. Reetz, D. Röhrig, K. Harms, G. Frenking, and F. Kayser, *TL* **35**, 8765, 8769 (1994).

Bistrimethylsilylmethylamine.

Stable imines.   Imines derived from aliphatic amines usually have the tendency to isomerize to the enamine form. However, the bissilylated methylimines are stable[1]; therefore, they are potentially useful synthetic intermediates.

The conjugate base of such imines reacts with electrophiles in accordance with the nature of the reagent.[2]

[1] A. Capperucci, A. Ricci, G. Seconi, J. Dunogues, S. Grelier, J.-P. Picard, C. Palomo, and J.-M. Aizpurua, *JOMC* **458**, C1 (1993).
[2] A. Ricci, A. Guerrini, G. Seconi, A. Mordini, T. Constantieux, J.-P. Picard, J.-M. Aizpurua, and C. Palomo, *SL* 955 (1994).

N,O-Bis(trimethylsilyl)acetamide.

TMSOTf generation.[1] Reactions that require catalysis of TMSOTf can be promoted by the combination of triflic acid and bis(trimethylsilyl)acetamide [or bis(trimethylsilyl)urea].

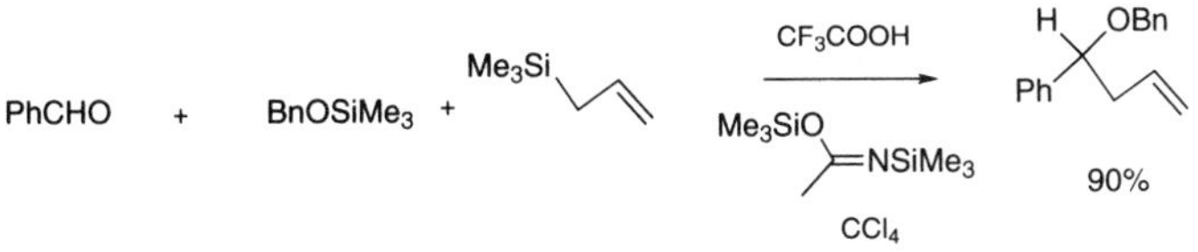

[1] M. El Gihani and H. Heaney, *SL* 433 (1993).

Bis(trimethylsilyl) sulfate–silica.

Thioacetalization.[1] In the reaction with 1,2-ethanedithiol, aldehydes and ketones are converted to dithioacetals very rapidly at room temperature (17 examples, 75–99% yield).

[1] H. K. Patney, *TL* **34**, 7127 (1993).

Bis(trimethylsilyl) selenide. 14, 51

Selenoxo esters and amides.[1] By virtue of silicon's high oxaphilicity and the weak Si–Se bond, the reagent is capable of replacing oxygen atoms in esters and amides by selenium atoms. $BF_3 \cdot OEt_2$ is a suitable catalyst.

[1] Y. Takikawa, H. Watanabe, R. Sasaki, and K. Shimada, *BCSJ* **67**, 876 (1994).

Bis(trimethylsilyl) sulfide.

Thionation of carbonyl compounds.[1] Thiones are formed under very mild conditions. Importantly, side reactions are minimized due to the possible stoichiometric control of the reagent.

Thioamides.[2] By activation of the substrate with phosphoryl chloride before treatment with the silyl sulfide, amides and lactams can be converted to the thio analogs under extremely mild (−78°C) conditions without the need for a catalyst.

Diallyl sulfides.[3] A one-step preparation from allylic alcohols involves BF_3 catalysis.

[1] A. Degl'Innocenti, A. Capperucci, A. Mordini, G. Reginato, A. Ricci, and F. Cerreta, *TL* **34**, (1993).
[2] D. C. Smith, S. W. Lee, and P. L. Fuchs, *JOC* **59**, 348 (1994).
[3] S.-C. Tsay, G. L. Yep, B.-L. Chen, L. C. Lin, and J. R. Hwu, *T* **49**, 8969 (1993).

Bis(triphenylphosphonio)oxide bistriflate $(Ph_3P^+)_2O$ 2 TfO$^-$.

Alkanes.[1]   The phosphonium salt (from Ph_3PO + Tf_2O) activates alcohols for direct deoxygenation with $NaBH_4$.

[1] J. B. Hendrickson, M. Singer, and M. S. Hussoin, *JOC* **58**, 6913 (1993).

Bis(triphenylstannyl) chalcogenides.

Desulfurization.[1]   Diorganotrisulfides undergo monodesulfurization in a non-concerted process. Reaction temperatures can be regulated by considering the relative reactivities. The tellurium reagent is the most reactive one.

[1] C. J. Li and D. N. Harpp, *TL* **34**, 903 (1993).

9-Borabicyclo[3.3.1]nonane, B-functionalized. 13, 249

Enolboration.[1]   With Et_3N the enolboration of ketones proceeds with stereoselectivities following the leaving abilities of the substituents on boron. A better leaving group favors formation of (Z)-enol borinates, whereas a poorer leaving group leads to more (E)-enol borinates.

2-Amino alcohols.[2]   In epoxide opening with lithium amides, a catalytic amount of B-bromo-9-BBN suppresses formation of allylic alcohols.

[1] H. C. Brown, K. Ganesan, and R. K. Dhar, *JOC* **58**, 147 (1993).
[2] C. E. Harris, G. B. Fisher, D. Beardsley, L. Lee, C. T. Goralski, L. W. Nicholson, and B. Singaram, *JOC* **59**, 7746 (1994).

Borane.

Carbonyl group reduction.[1]   With $LiBH_4$ as catalyst the reduction prevails over hydroboration of unsaturated carbonyl compounds. Conjugated alkenones give allylic alcohols.

N-Alkylindoles from N-acylisatins.[2]   The reductive aromatization occurs at room temperature in good yields (72–86%).

[1] A. Arase, M. Hoshi, T. Yamakai, and H. Nakanishi, *CC* 855 (1994).
[2] A. C. Pinto, F. S. Q. da Silva, and R. B. da Silva, *TL* **35**, 8923 (1994).

Borane-amines. 13, 42

Reductive amination with piperidines.[1]   Treatment of a piperidine and an aldehyde with the borane–pyridine complex gives the tertiary amine.

S_R2' reaction catalyst.[2]   Borane–quinuclidine is a catalyst for the substitution. Epoxide formation is observed from allylic *t*-butyl peroxides.

[1] A. E. Moormann, *SC* **23**, 789 (1993).
[2] H.-S. Dang and B. P. Roberts, *JCS(PI)* 891 (1993).

Borane–dimethyl sulfide. **14**, 53; **15**, 44; **17**, 50–51

3-Hydroxypropyl ethers.[1]   With TMSOTf as catalyst, 1,3-dioxanes undergo reductive ring opening very cleanly. Both C–O bonds of 1,3-dioxolanes may be cleaved completely under the same conditions.

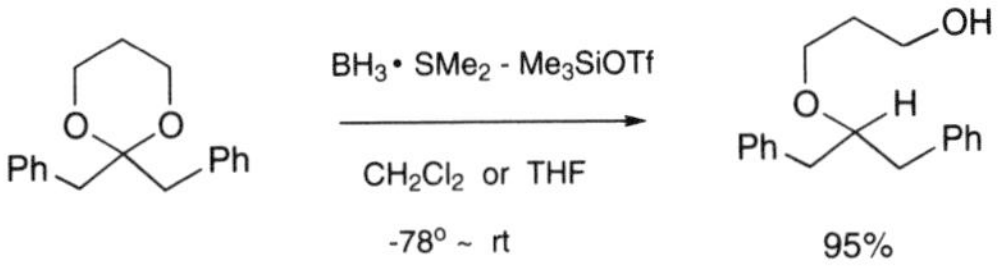

Erythro-1,2-diols.[2]   Hydroboration of 2-substituted 4,5-dihydrofurans followed by regioselective ring cleavage affords the diols in useful yields.

[1] B. Bartels and R. Hunter, *JOC* **58**, 6756 (1993).
[2] R. Amouroux, A. Slassi, and C. Saluzzo, *H* **36**, 1965 (1993).

Boron tribromide. **13**, 43; **14**, 53–54

Bromination of alcohols.[1]   Primary, secondary, and tertiary alcohols are converted to bromides.

(Z)-2-Bromoalkenyl boronates.[2]   These compounds are obtained by treatment of alkynes with BBr$_3$ and subsequent quenching with an alcohol. As both the bromine atom and the boron residue can be selectively replaced, the difunctionalized alkenes are useful synthetic intermediates.

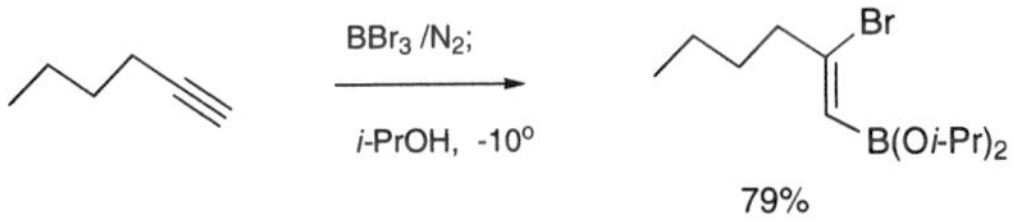

[1] J. D. Pelletier and D. Poirier, *TL* **35**, 1051 (1994).
[2] K. K. Wang and Z. Wang, *TL* **35**, 1829 (1994).

Boron trichloride. **13**, 43; **14**, 54; **15**, 44

Activation of boronic acids.   Alkylboronic acids are converted to dichloroboranes, as reactive intermediates. For example, azidoalkylboronic acids give cyclic azacycles (azetidines, pyrrolidines, piperidines).[1] An alternative method[2] for the heterocycle synthesis involves reaction of a bromoalkylalkene with BCl$_3$-Et$_3$SiH and an organoazide.

Borodesilylation.[3]   The element exchange on the furan ring enables the preparation of aryl derivatives by the Suzuki coupling. The silyl compounds are much more readily available.

45–97%

Selective ether cleavage.[4]   The $BCl_3 \cdot SMe_2$ complex is useful for selective cleavage. Thus primary and secondary benzyl ethers are removed in the presence of a TBDPS group. Note that BF_3 etherate–SMe_2 effects predominantly desilylation (due to F^-). The cleavage of trityl ethers without affecting benzyl ethers is also achievable.

[1] J. M. Jego, B. Carboni, and M. Vaultier, *BSCF* 554 (1992).
[2] J. M. Jego, B. Carboni, A. Youssofi, and M. Vaultier, *SL* 595 (1993).
[3] Z. Z. Song, Z. Y. Zhou, T. C. W. Mak, and H. N. C. Wong, *ACIEE* **32**, 432 (1993).
[4] M. S. Congreve, E. C. Davison, M. A. M. Fuhry, A. B. Holmes, A. N. Payne, R. A. Robinson, and S. E. Ward, *SL* 663 (1993).

Boron trifluoride etherate. 13, 43–47; 14, 54–56; 15, 45–47; 16, 44–47; 17, 52–53

HF generation.[1]   Treatment of $BF_3 \cdot OEt_2$ with 4-methoxysalicylaldehyde generates HF, which can be used to deprotect silylated alcohols.

Tritiation of arenes.[2]   Introduction of tritium labels to activated arenes is conveniently performed at room temperature by treatment with $BF_3 \cdot OEt_2$ and T_2O.

Fluoroarenes.   Reaction of aryllead triacetates with $BF_3 \cdot OEt_2$ at room temperature gives the fluoroarenes.[3] The Schiemann reaction can be effected at lower temperatures[4] than usual when arenediazonium salts are heated with $BF_3 \cdot OEt_2$. Under photochemical conditions a reaction temperature of 20–37°C is adequate.

Ring expansion of epoxides and oxetanes.[5]

m = 1, 2
n = 0, 1

Isomerization of glycidic esters.[6] $BF_3 \cdot OEt_2$-induced formation of α-hydroxy-β,γ-unsaturated esters opens ways to other difunctional compounds such as alkenyl epoxides, which are not directly available from the corresponding dienes.

Primary alcohol protection.[7]   *t*-Pentylation is limited to primary alcohols when they are treated with 2-methylbutene and $BF_3 \cdot OEt_2$.

Cumulenyl aldehydes.[8]   Exposure of epoxypropargyl alcohols to $BF_3 \cdot OEt_2$ promotes rearrangement of the epoxide moiety to the aldehyde and dehydration. A portion of the initial products may only undergo the prototropic shift.

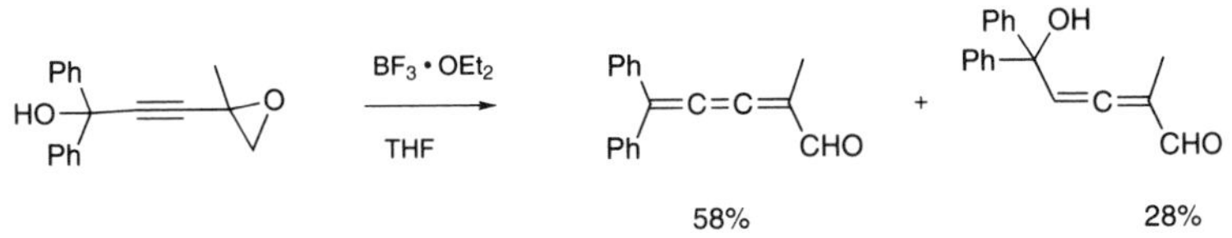

Fused ring C-glycosides.[9]   Ionization of glycosides in the presence of appropriate alkenes leads to C–C bond formation. When C-2 is protected by an acid- cleavable group, formation of a tetrahydrofuran ring ensues. Since the alkene attacks from the axial direction, the C-2 oxygen function must be equatorial in order to form the new ring.

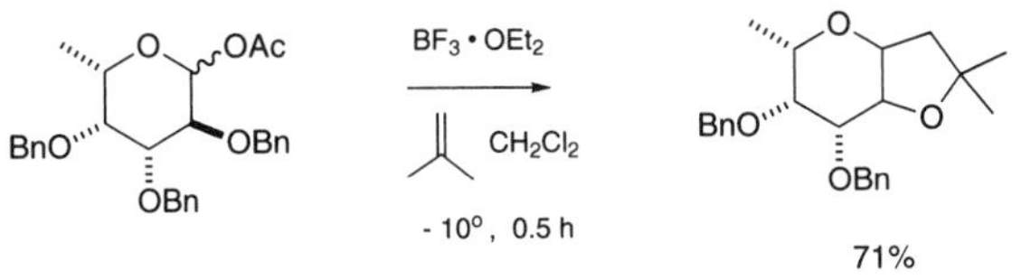

Protecting group switch.[10]   2-(Trimethylsilyl)ethyl glycosides are directly converted to 1-O-acyl sugars by reaction with carboxylic anhydrides. Boron trifluoride etherate has a dual role in activating the anhydride and providing a desilylating agent (possibly F^-).

Allylation.   Reactions of allyltin[11,12] and allylsilane reagents[13] with aldehydes are greatly influenced by the nature of the Lewis acid catalyst, with respect to diastereoselectivity. The stereocontrol is important for application of the reaction to polyol synthesis.

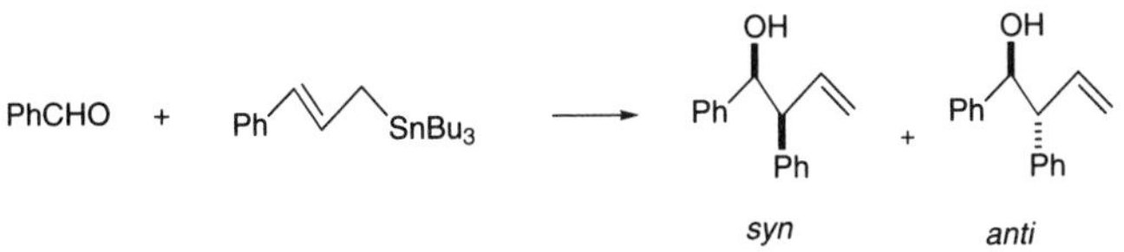

	syn	anti
$ZnCl_2$ / Et_2O, rt	90%	19:8
$BF_3 \cdot OEt_2$ / CH_2Cl_2, -78°	92%	98:2

4-alkynyl ketones.   The Mukaiyama condensation using $BF_3 \cdot OEt_2$ as catalyst is sufficiently mild to allow the presence of a chalcogen substituent at the terminal carbon of propargyl acetals.[14] An intramolecular Nicholas reaction followed by demetallation with $Fe(NO_3)_3$ constitutes a convenient method for accessing ketones with an ethynyl pendant.[15]

73%

Aldolization-fragmentation.[16]   When ethylene glycol is present, the acetal derived from the 2-hydroxyalkylcycloalkanone is liable to acid-catalyzed fragmentation to give an unsaturated ester.

61%

Activation of imines.[17]   The electrophilic reaction of imines with organometallic reagents such as RMgX, RLi, R_2CuLi is promoted by a Lewis acid ($BF_3 \cdot OEt_2$). A flexible synthesis of 1-substituted 1,2,3,4-tetrahydro-β-carbolines is based on the process.

Mannich-type reaction.[18]   By using a carbamate ester in place of a secondary amine, the condensation is no longer limited to the very few aldehydes (especially formaldehyde). If $BF_3 \cdot OEt_2$ is replaced by tosic acid, a higher reaction temperature is needed.

Aza-Cope rearrangement.[19]   The rearrangement of N-alkyl-N-allylanilines is promoted by Lewis acids. In comparison with $ZnCl_2$, the catalytic effect of $BF_3 \cdot OEt_2$ is superior.

Pyridine ring formation.[20]   Thermolysis of oxime O-allyl ethers in the presence of $BF_3 \cdot OEt_2$ leads to pyridine derivatives.

Hetero-Diels–Alder reaction.[21]   3,4-Bis(trimethylsilyl)-5,6-dihydro-2H-pyrans are accessible from a catalyzed cycloaddition under very mild conditions. The diene is prepared in 44–47% yield from the reaction of 1,4-ditosyloxy-2-butyne with trimethylsilylmethylmagnesium chloride.

[3+2]Cycloaddition.[22]   The catalyzed reaction between an electron-deficient alkene and an allylsilane establishes four stereocenters in a cyclopentane ring. It involves migration of the silyl group.

[1] S. Mabic and J.-P. Lepoittevin, SL 851 (1994).
[2] P. McGeady and R. Croteau, CC 774 (1993).

[3] G. De Meio, J. Morgan, and J. T. Pinhey, *T* **49**, 8129 (1993).
[4] K. Shinhama, S. Aki, T. Furuta, and J. Minamikawa, *SC* **23**, 1577 (1993).
[5] A. Itoh, Y. Hirose, H. Kashiwagi, and Y. Masaki, *H* **38**, 2165 (1994).
[6] I. Bhattacharya, K. Shah, P. S. Vankar, and Y. D. Vankar, *SC* **23**, 2405 (1993).
[7] B. Figadere, X. Franck, and A. Cave, *TL* **34**, 5893 (1993).
[8] X. Wang, B. Ramos, and A. Rodriguez, *TL* **35**, 6977 (1994).
[9] D. E. Levy, F. Dasgupta, and P. C. Tang, *TA* **5**, 2265 (1994).
[10] U. Ellervik and G. Magnusson, *ACS* **47**, 826 (1993).
[11] Y. Nishigaichi and A. Takuwa, *CL* 1429 (1994).
[12] J. A. Marshall, J. A. Jablonowski, and G. P. Luke, *JOC* **59**, 7825 (1994).
[13] J. S. Panek and P. F. Cirillo, *JOC* **58**, 999 (1993).
[14] M. Yoshimatsu, H. Shimizu, and T. Kataoka, *CC* 149 (1995).
[15] E. Tyrrell, P. Heshmati, and L. Sarrazin, *SL* 769 (1993).
[16] S. Nagumo, A. Matsukuma, H. Suemune, and K. Sakai, *T* **49**, 10501 (1993).
[17] T. Kawate, M. Nakagawa, H. Yamazaki, M. Hirayama, and T. Hino, *CPB* **41**, 287 (1993).
[18] W. ten Hoeve and H. Wynberg, *SC* **24**, 899 (1994).
[19] L. G. Beholz and J. R. Stille, *JOC* **58**, 5095 (1993).
[20] J. Koyama, T. Ogura, K. Tagahara, M. Miyashita, and H. Irie, *CPB* **41**, 1297 (1993).
[21] C. Brouard, J. Pornet, and L. Miginiac, *SC* **24**, 3047 (1994).
[22] J. S. Panek and N. F. Jain, *JOC* **58**, 2345 (1993).

Boron trifluoride etherate–thiol.

Cleavage of THP-ethers.[1]   The boron trifluoride etherate–ethanethiol mixture cleaves THP ethers at low temperatures without affecting the methylenedioxy group, acetonides, mesitaldehyde acetals, methoxymethyl ethers, and *t*-butyldimethylsilyl ethers.

Allylic sulfides.[2]   The direct synthesis from allylic alcohols such as geraniol is accomplished with BF_3–OEt_2 and a thiol.

[1] K. P. Nambiar and A. Mitra, *TL* **35**, 3033 (1994).
[2] S.-C. Tsay, L.-C. Lin, P. A. Furth, C. C. Shum, D. B. King, S. F. Yu, B.-L. Chen, and J. R. Hwu, *S* 329 (1993).

Boron trifluoride-nitromethane.

Polyene cyclization.[1]   The catalyst enables an efficient synthesis of precursors of taxodione. It is vastly more effective than $BF_3 \cdot OEt_2$ or other Lewis acids.

95%

[1] S. R. Harring and T. Livinghouse, *T* **50**, 9229 (1994).

Bromine. **13**, 47; **14**, 56–57; **15**, 47

Bromodeselenenylation.[1]   γ-Phenylseleno α,β-unsaturated esters undergo bromination at the α-carbon while detaching the selenenyl group and deconjugating the double bond.

Bromination of deactivated arenes.   Good yields are obtained when the bromination is carried out in concentrated nitric acid and sulfuric acid at temperatures between 80 to 85°C.[2] Alternatively, using BrF_3 as catalyst,[3] the reaction temperature can be lowered to 0–10°C.

Stilbene → benzil.[4]   An economical method for the transformation uses bromine and sulfuric acid in refluxing acetic acid.

[1] J.-F. Duclos, F. Outurquin, and C. Paulmier, *TL* **34**, 7417 (1993).
[2] A. M. Andrievskii, M. V. Gorelik, S. V. Avidon, and E. Sh. Al'tman, *RJOC* **29**, 1519 (1993).
[3] S. Rozen and O. Lerman, *JOC* **58**, 239 (1993).
[4] M. S. Yusubov, E. A. Krasnokutskaya, and V. D. Filimonov, *JOCU* **28**, 670 (1992).

Bromine-amine.

Oxidation.[1]   Primary and secondary alcohols are readily oxidized to carbonyl compounds by bromine-hexamethylenetetramine (88–96% yield).

Bromination.[2]   Bromine complexed to polymeric pyridine has been used to brominate aromatic compounds.

[1] I. Yavari and A. Shaabani, *JCR(S)* 274 (1994).
[2] M. Zupan and N. Segatin, *SC* **24**, 2617 (1994).

Bromine trifluoride.

Fluorination.   In a freon solvent, this powerful fluorinating agent is able to convert compounds containing C=N (ketone oxime ethers, hydrazones)[1] and C=S groups to *gem*-difluorides.[2] Trifluoromethyl derivatives are directly obtained from alkyl dithioates.[3] Caution must be paid to avoid the use of oxygenated solvents with which BrF_3 reacts violently.

[1] S. Rozen, E. Mishani, and A. Bar-Haim, *JOC* **59**, 2918 (1994).
[2] S. Rozen and E. Mishani, *CC* 1761 (1993).
[3] S. Rozen and E. Mishani, *CC* 2081 (1994).

Bromoform.

Oxidative bromination of uracils.[1]   Uracils form 5-bromo derivatives in excellent yields (83–95%) on heating with bromoform in the presence of oxygen.

Dibromocyclopropanation.   This well-known reaction has been applied to unsaturated acylphosphoranes[2] with good results. It also shows selectivity for silyldienes.[3]

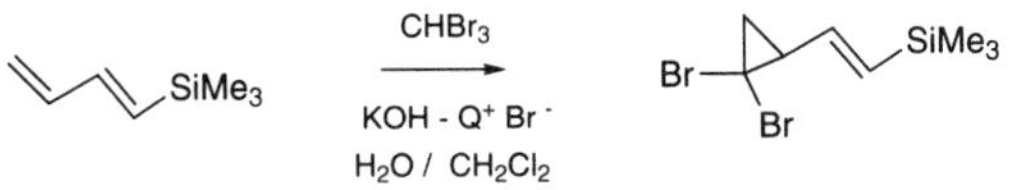

$$\text{CHBr}_3$$
$$\text{KOH - Q}^+ \text{Br}^-$$
$$\text{H}_2\text{O} / \text{CH}_2\text{Cl}_2$$

63%

[1] C. Moltke-Leth and K. A. Jorgensen, *ACS* **47**, 1117 (1993).
[2] M. P. Cooke and J. Y. Jaw, *JOC* **58**, 267 (1993).
[3] W.-W. Weng and T.-Y. Luh, *JCS(P1)* 2687 (1993).

N-Bromosuccinimide. **13**, 49; **14**, 57–58; **15**, 50–51; **16**, 47

O-Deprotection. Free-radical-initiated bromination followed by treatment with water removes a benzyl ester.[1] A similar method is also effective to cleave allyl ethers.[2]

Regioselective oxidations. An alkane-1,2-diol can be selectively oxidized to give the ketol (instead of the hydroxy aldehyde) via NBS reaction of the dibutylstannylene acetal.[3]

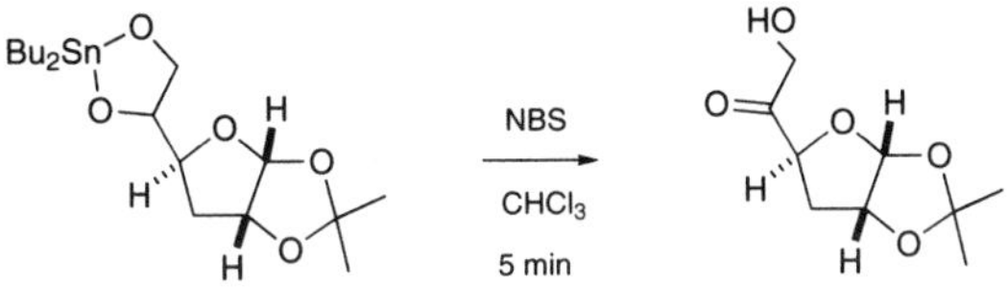

NBS

CHCl₃

5 min

84%

Dehydrogenation of 2-oxazolines.[4] This method involves an allylic bromination–dehydrobromination sequence. Since complications of allylic bromination exist for compounds bearing an alkyl group at C-2, the use of PhCO₃Bu' and CuBr is advisable in such cases.

Transpositional bromination of allylic alcohols.[5] A synthetically significant use of this reaction is the transformation of sulfinylated substrates and subsequent dehydrobromination to acquire 2-sulfinyl-1,3-butadiene.

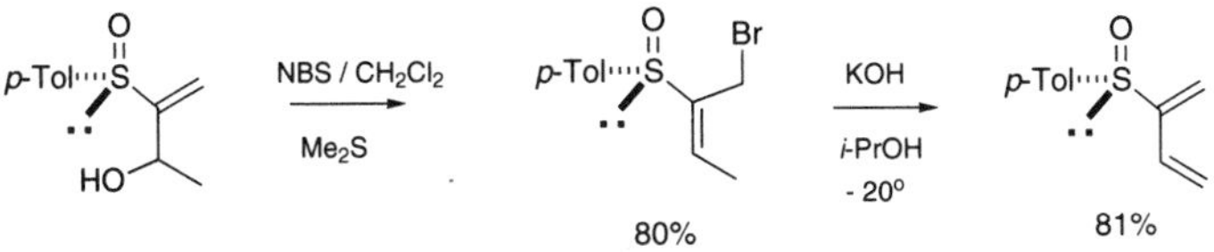

NBS / CH₂Cl₂

Me₂S

80%

(*E:Z* 91:9)

KOH

i-PrOH

- 20°

81%

Aromatic bromination.[6] Activated aromatic and heteroaromatic compounds are brominated with NBS in a biphasic system, with 70% HClO₄ as the catalyst. On the other hand, the bromination of activated benzoic acids is readily achieved in an aqueous base.[7]

Nitrene formation.[8]   2,4-Dinitrobenzenesulfenylnitrene is generated from the sulfenamine by treatment with NBS. Trapping of the nitrene by alkenes gives rise to aziridine derivatives.

2,3-Dihydro-1,4-benzodithiines and -oxathiines.[9]   A remarkable one-pot synthesis of these heterocycles from cyclohexanones and ethanedithiol or 2-mercaptoethanol is mediated by NBS at 0°C.

Z = S   68%
Z = O   82%

Substitution of sulfenyl derivatives.   An α-alkylthio *N*-Boc α-amino acid is easily synthesized from a thiol, *t*-butyl carbamate, and glyoxylic acid. Replacement of the alkylthio group by fluorenylmethylcarbamoyl unit gives a useful derivative for solid-phase synthesis of α-aminoglycine-containing peptides. The replacement is conveniently mediated by NBS.[10]

Thioglycosides are activated for disaccharide formation[11] by means of an NBS–triflate salt combination.

Bromoetherification.[12]   The combination of NBS with an alcohol to functionalize an alkene is analogous to bromohydrination. Interestingly tricyclic skeletons can be constructed from such adducts in one operation based on dehydrobromination, isomerization of an alkyne to an allene, and intramolecular Diels–Alder reaction.

α-Bromomethyl ketones.[13]   Vinyl halides are converted to the more versatile compounds with many other functional groups (e.g., ester) tolerating the reaction conditions. A catalytic amount of HBr is added to the reaction medium. The method is applicable to the preparation of other halomethyl ketones by using the corresponding *N*-halosuccinimides.

[1] M. S. Anson and J. G. Montana, *SL* 219 (1994).
[2] R. R. Diaz, C. R. Melgarejo, M. T. P. Lopez-Espinoza, and I. I. Cubero, *JOC* **59**, 7928 (1994).
[3] X. Kong and T. B. Grindley, *JCC* **12**, 557 (1993).

[4] A. I. Meyers and F. Tavares, *TL* **35**, 2481 (1994).
[5] E. Bonfand, P. Gosselin, and C. Maignan, *TA* **4**, 1667 (1993).
[6] Y. Goldberg and H. Alper, *JOC* **58**, 3072 (1993).
[7] J. Auerbach, S. A. Weissman, J. J. Blacklock, M. R. Angeles, and K. Hoogsteen, *TL* **34**, 931 (1993).
[8] T. Michida and H. Sayo, *CPB* **42**, 27 (1994).
[9] H. Tani, S. Irie, K. Matsumoto, and N. Ono, *H* **36**, 1783 (1993).
[10] D. Qasmi, L. Rene, and B. Badet, *TL* **34**, 3861 (1993).
[11] K. Fukase, A. Hasuoka, and S. Kusumoto, *TL* **34**, 2187 (1993).
[12] J. P. Dulcere, V. Agati, and R. Faure, *CC* 270 (1993).
[13] H. E. Morton and M. R. Leanna, *TL* **34**, 4481 (1993).

N-Bromosuccinimide–iodosylbenzene.

Vinyl bromides.[1]   Unsaturated carboxylic acids undergo decarboxylative bromination on heating with the two reagents. The transformation is reminiscent of the Hunsdiecker reaction.

[1] A. Graven, K. A. Jorgensen, S. Dahl, and A. Stanczak, *JOC* **59**, 3543 (1994).

N-Bromosuccinimide–silver nitrate.

Bromoacetylenes from trimethylsilylacetylenes.[1]   The reagent combination effects the transformation in one step.

[1] T. Nishikawa, S. Shibuya, S. Hosokawa, and M. Isobe, *SL* 485 (1994).

N-Bromosuccinimide–2,2,6,6-tetramethylpiperidine.

Stilbenes.[1]   Dibenzylphosphonium salts are converted to stilbenes in a manner similar to the Ramberg–Bäcklund reaction.

70% (*Z*:*E* 80 : 20)

[1] N. J. Lawrence and F. Muhammad, *TL* **35**, 5903 (1994).

N-Bromosuccinimide–water.

Fragmentation of tosylhydrazones.[1]   Oxabicycloalkenone tosylhydrazones are brominated at nitrogen. The presence of two potential leaving groups (Br, Ts) provokes nucleophilic attack by water at the far end of the conjugated system to complete a fragmentation-prone unit. Accordingly, macrocyclic acetylenic lactones are accessible.

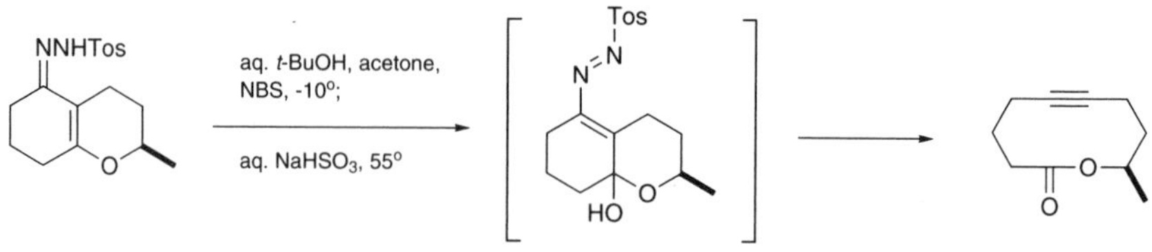

[1] J. R. Mahajan and I. S. Resck, *CC* 1748 (1993).

2,3-Butanedione.

4,5-Dimethylene-2-oxazolidinones.[1]   Condensation of the diketone with *N*-alkyl- or *N*-aryl isocyanates provides an access to the heterocycles in one step.

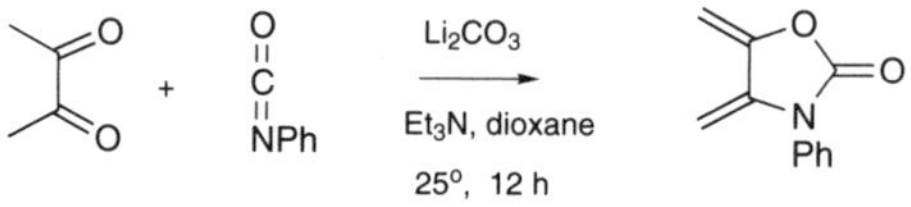

[1] R. Hernandez, J. M. Sanchez, A. Gomez, G. Trujillo, R. Aboytes, G. Zepeda, R. W. Bates, and J. Tamriz, *H* **36**, 1951 (1993).

N-(*t*-Butoxycarbonyloxy)-5-norbornene-*endo*-2,3-dicarboximide (1).

Protection of unhindered amines.[1]   The chemoselectivity of the reagent is shown by the exclusive Boc-derivatization of the CH_2NH_2 group in the presence of several other $CHNH_2$ pendants in an amino sugar.

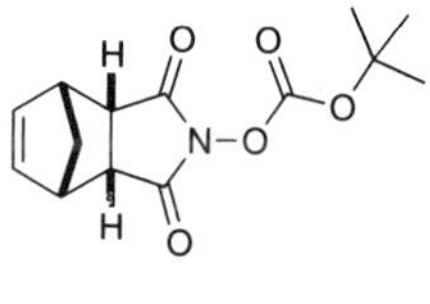

(1)

[1] I. Grapsas, Y. J. Cho, and S. Mobashery, *JOC* **59**, 1918 (1994).

N-(*t*-Butoxycarbonyl)-2-(*t*-butyldimethylsiloxy)pyrrole.

Glycine anion equivalent.[1]   After a Lewis acid–catalyzed condensation with an aldehyde at C-5 and functional group protection, the degradation of the pyrrolinone ring unveils a β-hydroxy-α-amino acid chain. More elaborate systems can be obtained from chiral aldehydes such as glyceraldehyde acetonide.

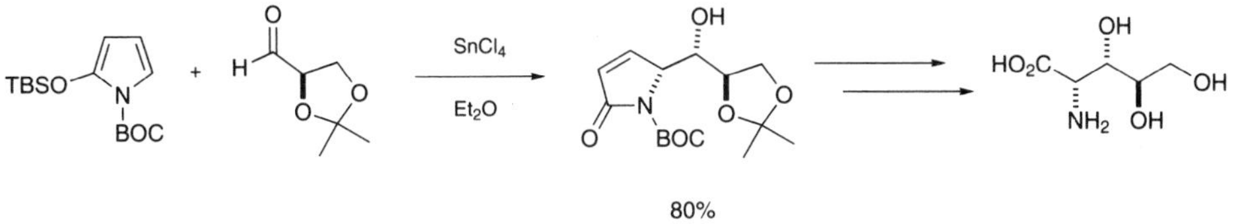

80%

[1] G. Casiraghi, G. Rassu, P. Spanu, and L. Pinna, *TL* **35**, 2423 (1994).

N-(*t*-Butoxycarbonyl)methanesulfonamide, MsNHCOOBu^t.

N-Sulfonylcarbamates.[1] Prepared by reaction of methanesulfonamide with *t*-butylchloroformate (Et₃N-DMAP as base). As a nucleophile in the Mitsunobu reaction, various alkyl groups can be attached to the nitrogen atom.

[1] B. R. Neustadt, *TL* **35**, 379 (1994).

N-Butylanilinomanganese chloride, PhN(Bu)MnCl.

Regioselective enolization.[1] The kinetic ketone enolates are formed with manganese amides. Polyalkylation with such enolater is not observed.

[1] G. Cahiez, B. Figadere, P. Clery, and K. Chau, *TL* **35**, 3065, 3069 (1994).

(*t*-Butyldimethylsilyl)dihalomethyllithium.

Dichloromethylene dianion synthon.[1] Generated by deprotonation (LDA as base), the reagent (TBS)C(Cl)₂Li attacks a carbonyl compound to give an (α-siloxyalkyl)dichloromethyllithium species, as a result of a C-to-O silyl transfer. The new organolithium compound can then react with another electrophile.

Enones from two different aldehydes.[2] The silyldibromolithium reagent stitches two aldehydes on itself, becoming the α-carbon of the resulting enone system. After union with the first aldehyde molecule the nucleophilic site is regenerated by treatment with *sec*-BuLi in a one-pot reaction.

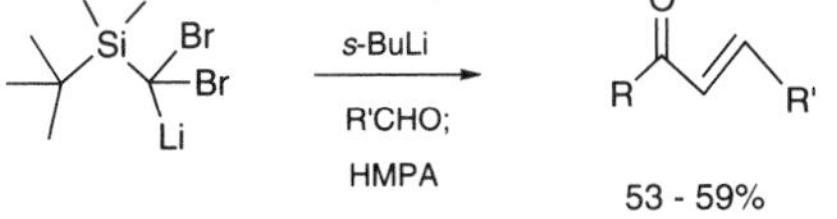

[1] H. Shinokubo, K. Miura, K. Oshima, and K. Utimoto, *TL* **34**, 1951 (1993).
[2] H. Shinokubo, K. Oshima, and K. Utimoto, *TL* **35**, 3741 (1994).

t-Butyldimethylsilyl dimethyl phosphite.

α-Hydroxyphosphonates.[1]   Using a Lewis acid as catalyst, the reaction of the phosphite with aldehydes is rapid, even at $-78°C$. The condensation with α-benzyloxyaldehydes is *syn*-selective.

[1] T. Tokomatsu, Y. Yoshida, and S. Shibuya, *JOC* **59**, 7930 (1994).

t-Butyldimethylsilyl triflate. 13, 50–51; 15, 54–55

β-Siloxyaldehydes.[1]   Chiral epoxy alcohols are readily available (e.g., by Sharpless epoxidation of allylic alcohols). Their transformation via a stereoselective rearrangement–silylation induced by the silyl triflate and *i*-Pr$_2$NEt opens a new way to protected aldols.

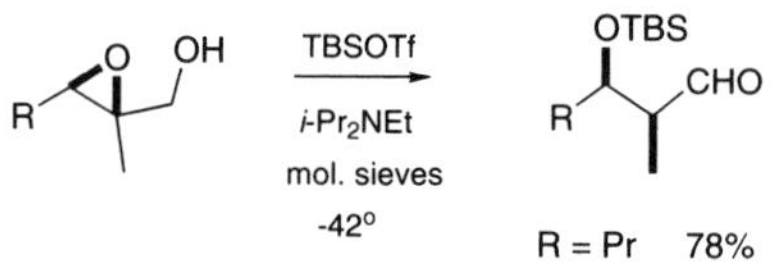

Ether cleavage.[2]   Alkyl *p*-methoxybenzyl ethers are directly converted to TBS-protected alcohols by treatment with TBSOTf and then with Et$_3$N at room temperature. Yields range from fair to excellent (63–91%).

[1] M. E. Jung and D. C. D'Amico, *JACS* **115**, 12208 (1993).
[2] T. Oriyama, K. Yatabe, Y. Kawada, and G. Koga, *SL* 45 (1995).

(*t*-Butyldimethylsilyl)trimethylstannane.

Silastannylation of alkynes.[1]   The partitional difunctionalization is achieved using a catalytic system of Pd(OAc)$_2$-RNC at room temperature (75–99% yield). The many chemoselective coupling procedures available to vinyltin and silicon compounds make these product extremely valuable for synthetic quests.

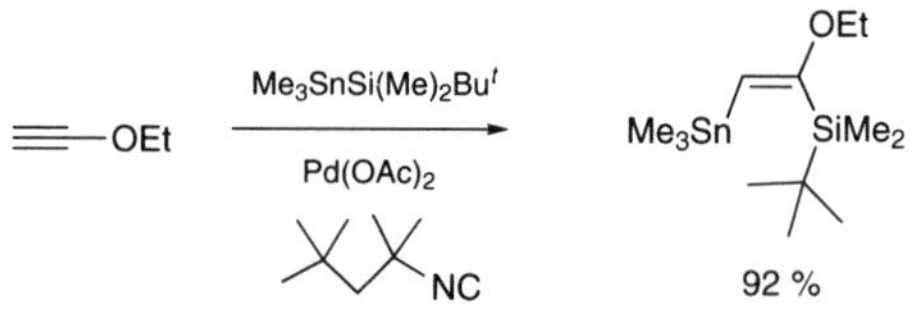

[1] M. Murakami, H. Amii, N. Takizawa, and Y. Ito, *OM* **12**, 4223 (1993).

α-(*t*-Butyldimethylsilyl)vinyl phenyl sulfoxide.

Indole synthesis.[1]   This Michael acceptor reacts with *o*-lithio-*N*-Boc-aniline. The product undergoes a sila–Pummerer rearrangement to induce ring closure. Indoles are obtained on elimination of the [PhSH] element.

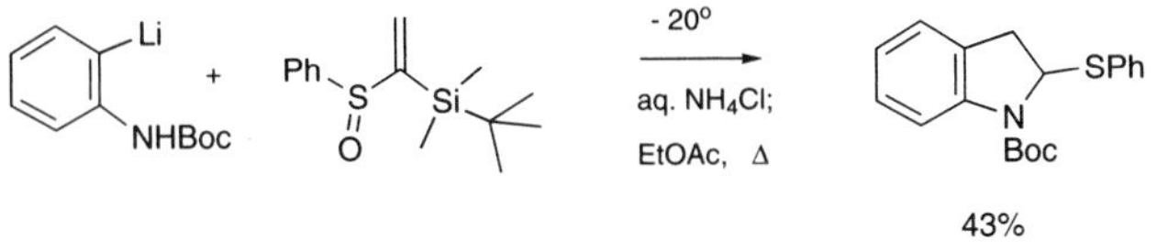

43%

[1] M. Iwao, *H* **38**, 45 (1994).

t-Butyldiphenylsilyl chloride.

Silylation.[1]   Primary and secondary alcohols are rapidly silylated with this reagent in DMF using NH_4NO_3 as catalyst at room temperature.

[1] S. A. Hardinger and N. Wijaya, *TL* **34**, 3821 (1993).

t-Butyl 3-(ethoxycarbonyl)-3-(diethylphosphono)propionate.

Stobbe condensation.[1]   The Emmons–Wadsworth protocol of condensation with an aldehyde is complementary to the classical Stobbe condensation. But here, the product is the diester, which can be hydrolyzed selectively. Thus treatment with CF_3COOH leads to a conjugated ester/saturated acid.

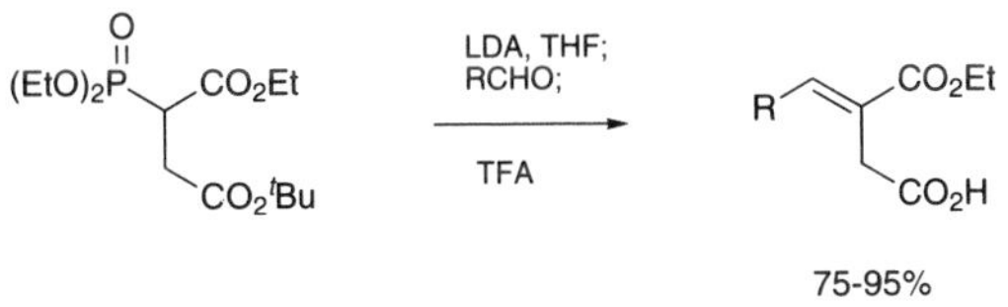

75–95%

[1] W. M. Owton, P. T. Gallagher, and A. Juan-Montesinos, *SC* **23**, 2119 (1993).

t-Butyl hydroperoxide. 16, 53–54; 17, 56–57

Epoxidation.[1]   For epoxidation of electron-deficient alkenes KF supported on alumina is a useful base. Moderate ee values are observed when unfunctionalized alkenes are epoxidized in the presence of a chiral borate derived from dimethyl tartrate.

Oxygenation of carbanions.[2]   The alkoxide form (i.e., *t*-BuOOLi) is capable of transforming carbanions into alcohols (after subsequent protonation). Of particular interest is the formation of acetylene oxides.

$$R\!-\!\!\equiv\!\!-Li \xrightarrow[\text{THF}]{t\text{-BuOOLi}} R\!-\!\!\equiv\!\!-OLi$$

Oxidation of sulfur-containing compounds. A new procedure for oxidation of sulfides to sulfoxides uses *t*-BuOOH in aqueous H_2SO_4 at ambient temperature (95– 100% yield).[4] Interestingly, 4-(methylthio)benzaldehyde is oxidized to the sulfone carboxylic acid in 0.5M NaOH at 70°C, but in the presence of cetyltrimethylammonium sulfate (and 1M NaOH) at room temperature the sulfide remains untouched.[5]

Unsymmetrical peroxides.[6]

62%

Oxidative ring expansion of furans.[7] The transformation into pyranones effected in the presence of camphorsulfonic acid is similar to that involving *t*-BuOOH-VO(acac)$_2$,[8] but there must be also an oxidation at the benzylic position of the original skeleton.

50%

Hydrocarbon oxidation. A metal catalyst is usually required. By adding bis(*t*-riphenylsilyl) chromate,[9] diphenylmethane is oxidized to benzophenone, and vanadium-pillared montmorillonite[10] catalyzes the conversion of arylacetic esters to arylglyoxylic esters by *t*-BuOOH. The presence of calcined $ZnCrO_3$-hydrotalcite enables the selective generation of benzylic hydroperoxides from aralkanes.[11]

The Gif oxidation of aralkanes involving *t*-BuOOH, Fe(NO$_3$)$_3$ · 9H$_2$O, pyridine, HOAc (the so-called GoAgg^{IV} system) gives mixed benzylic peroxides.[12] The addition of MX (M=Li, Na, Et$_3$N; X=Cl, N$_3$, NCS, CN, NO$_2$, ...) enables functionalization of hydrocarbons to afford alkyl chlorides, azides, thiocyanates, cyanides, nitroalkanes, etc.[13]

Using soluble Fe(III) and Cu(II) chelates in the oxidation of hydrocarbons with *t*-BuOOH and air without solvent, it is possible to obtain ketone products.[14] Similar catalytic effects of RuCl$_2$(PPh$_3$)$_3$ have been revealed.[15]

1,3-Dienes give 4-*t*-butylperoxy-2-alkenones on reaction with *t*-BuOOH-pyridinium dichromate.[16]

Oxidation of alcohols. Oxidation systems for alcohols[17] (*t*-BuOOH-RuCl₃), for secondary benzylic alcohols[18] (*t*-BuOOH-CrO₃) and for allylic alcohols[19] (*t*-BuOOH-OsO₄-Et₄NOH) have been developed.

ArNH₂ → ArNO₂.[20] The Fe(III) and Mn(III) complexes of tetraarylporphyrins are very effective catalysts for the oxidation with *t*-BuOOH. 1-Methylimidazole is usually present as an additional ligand for the metal ions.

[1] V. K. Yadav and K. K. Kapoor, *TL* **35**, 9481 (1994).
[2] E. Maniury, H. A. H. Mouloud, and G. G. A. Balavoine, *TA* **4**, 2340 (1993).
[3] M. Julia, V. P. Saint-Jalmes, and J.-N. Verpeaux, *SL* 233 (1993).
[4] F. Fringuelli, R. Pellegrino, and F. Pizzo, *SC* **23**, 3157 (1993).
[5] F. Fringuelli, R. Pellegrino, O. Piermatti, and F. Pizzo, *SC* **24**, 2665 (1994).
[6] M.-J. Bourgeois, E. Montaudon, and B. Maillard, *T* **49**, 2477 (1993).
[7] R. Antonioletti, L. Arista, F. Bonadies, L. Locati, and A. Scettri, *TL* **34**, 7089 (1993).
[8] T.-L. Ho and S. G. Sapp, *SC* **13**, 207 (1983).
[9] J. Muzart and A. N.'Ait Aijou, *SC* **23**, 2113 (1993).
[10] B. M. Choudary, G. V. S. Reddy, and K. K. Rao, *CC* 323 (1993).
[11] B. M. Choudary, N. Narender, and V. Bhuma, *SL* 641 (1994).
[12] F. Minisci, F. Fontana, S. Araneo, and F. Recupero, *CC* 1823 (1994).
[13] D. H. R. Barton and W. Chavasiri, *T* **50**, 19, 31, 47 (1994).
[14] D. H. R. Barton, S. D. Beviere, and D. R. Hill, *T* **50**, 2665 (1994).
[15] S.-I. Murahashi, Y. Oda, T. Naota, and T. Kuwabara, *TL* **34**, 1299 (1993).
[16] S. Bhat, N. Chidambaram, and S. Chandrasekaran, *CC* 651 (1993).
[17] D. D. Agarwal, R. Jain, P. Sangha, and R. Rastogi, *IJC(B)* **32B**, 381 (1993).
[18] J. Muzart and A. N.'Ait Aijou, *S* 786 (1993).
[19] C. Beck and K. Seifert, *TL* **35**, 7221 (1994).
[20] S. Tollari, D. Vergani, S. Banfi, and F. Porta, *CC* 442 (1993).

t-Butyl hydroperoxide–dialkyl tartrate–titanium(IV) isopropoxide. 13, 51–53; 14, 61–62; 15, 55–56; 16, 54–55; 17, 57–58

Epoxides from alkenylsilanols.[1] Enantioselective epoxidation is achieved with the help of Si-OH direction, and the products are desilylated with fluoride ion.

Chiral α-furfuryl amides.[2] Kinetic resolution is achieved by selective conversion of one enantiomer to the dehydropiperid-3-one derivative.

Chiral allenyl sulfones.[3] The oxidation of β-arylselenenyl β,γ-unsaturated sulfones is accompanied by in situ selenoxide elimination.

[1] T.-H. Chan, L. M. Chen, D. Wang, and L. H. Li, *CJC* **71**, 60 (1993).
[2] W.-S. Zhou, Z.-H. Lu, and Z.-M. Wang, *T* **49**, 2641 (1993).
[3] N. Komatsu, T. Murakami, Y. Nishibayashi, T. Sugita, and S. Uemura, *JOC* **58**, 3697 (1993).

t-Butyl hypochlorite. 13, 55

Nitrile oxide formation.[1]   The reaction of aldoxime *O*-tributylstannyl ethers with *t*-BuOCl results in chlorination and then 1,3-elimination. In the presence of a dipolarophile, the isoxazoline or isoxazole cycloadduct is formed.

[1] O. Moriya, H. Takenaka, M. Iyoda, Y. Urata, and T. Endo, *JCS(P1)* 413 (1994).

t-Butyliminoosmium trioxide.

vic-Aminohydroxylation.[1]   The reaction is completely analogous to the dihydroxylation with OsO_4. It is also amenable to asymmetric induction.

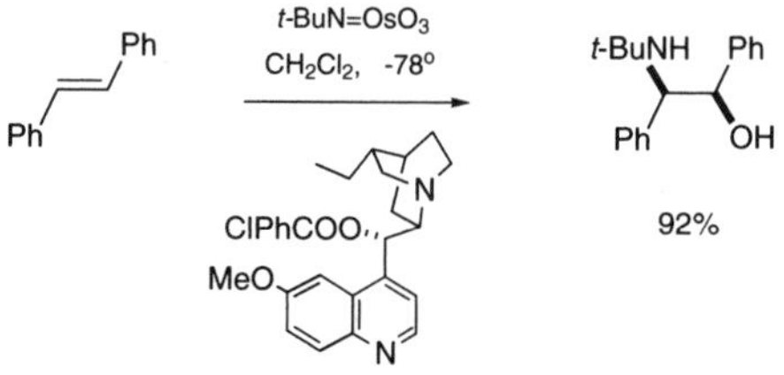

[1] H. Rubinstein and J. S. Svendsen, *ACS* **48**, 439 (1994).

Butyllithium. 13, 56; 14, 63–68; 15, 59–61; 17, 59–60

Monotritylation of symmetrical 1,n-diols.[1]   Deprotonation of the diol in DME using BuLi and subsequent treatment with trityl bromide accomplishes the task.

Propargylic alcohols.[2]   Butyllithium serves as a base and a nucleophile in the three-component reaction, also involving a terminal acetylene and an *N,N*-disubstituted amide.

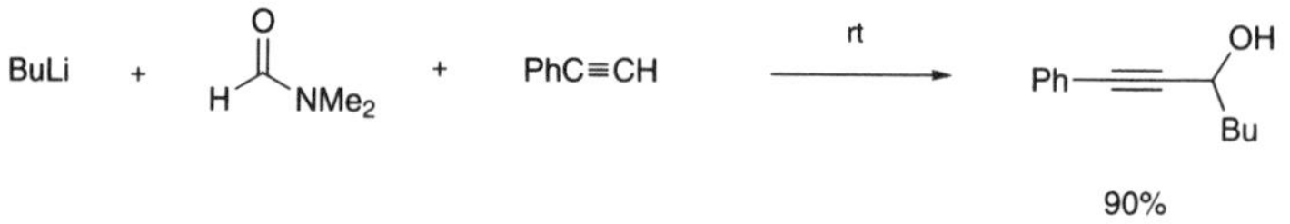

Acetylenes from 1-chlorovinyl sulfoxides.[3]   This high-yield conversion (Fritsch–Wiechell rearrangement) terminates a one-carbon homologation of aldehydes. The method consists of condensation of an aldehyde with chloromethyl phenyl sulfoxide and dehydration prior to the treatment with BuLi.

α-Heterosubstituted alkenyllithiums.   The generation of such sp^2-carbanions and the reaction with carbonyl compounds produce intermediates useful for many synthetic applications. From deprotonated allenyl ethers the reaction gives precursors of 3(2H)-furanones.[4] Asymmetric induction in the reaction by a chiral element present in the allenyl ether has been observed.[5]

Dihydro-1,4-dithiins are liable to deprotonation at the double bond.[6] However, in a subsequent reaction with aldehydes the addition of 20 mol % $(i\text{-PrO})_4$Ti is essential, since $TiCl_4$ is too acidic for the purpose.

A route to 4-substituted imidazoles[7] involving C-2 silylation of a sulfamoylimidazole and subsequent lithiation at C-4 (sec-BuLi can also be used) has been developed.

Lithium–halogen exchange.   Alkenyl iodides (but not the bromides) and BuLi react at room temperature in hydrocarbon solvents such as hexane and benzene (50–100% yield).[8] This method is favorable in comparison with that using t-BuLi because only 1 equivalent of the reagent is needed.

After such an exchange 3-lithio-2H-1-benzopyrans undergo fragmentation to give o-allenylphenols (isolated as mixed carbonates).[9]

The selective exchange of one of two vinylic bromine atoms, as directed by chelation of an allylic methoxyethoxymethyl ether in a chiral carboxyl anion equivalent, enables effective induction of chirality by virtue of the juxtaposed functionalities in the reaction with imines.[10]

Ar = mesityl

62% (92% de)

Lithium–tin exchange.   Organostannanes are valuable precursors of lithium compounds. Regio- and stereoselective functionalization at the site originally occupied by the tin atom of alkenylstannanes, for example, deuteration[11] and sulfenylation[12] is possible. N-(α-Tributylstannylalkyl)imines[13] and 2-tributylstannyl-N-trityl-aziridines[14] afford the corresponding lithium derivatives, respectively, useful in the construction of five-membered heterocycles and functionalized aziridines. Homoallylamines have been obtained from aza-Wittig rearrangement of allyl(trialkylstannylmethyl)amines.[15]

Isoxazoles from oximes and esters.[16]   Dianions from oximes condense with certain esters, resulting in isoxazoles directly.

Deprotonation of unusual carbon acids.   Lithiation of epoxides[17] as well as of chromium–carbene complexes[18] is a prelude to C–C bond formation. α-Keto acids are converted to α-alkoxyacrylic acids on quenching the dianions with hard carbon electrophiles.[19]

α-Sulfonyl carbanions.   The consecutive alkylation of methyl phenyl sulfone anion with an *N*-tosylaziridine and an aldehyde results in a trifunctional compound.[20] For access to a homoallylic amine (derivative) it only requires elimination of the PhSO$_3$H element.

An α-sulfonyl carbanion (and other stabilized carbanions, e.g., phosphorus ylides) can add to a thio-substituted triple bond to furnish 5-membered ring products.[21] The corresponding alkoxy compound behaves similarly.

Wittig and related reactions.   2-Alkenyl-2-methoxycyclopropyl ylides afford methylenecyclopentenes on treatment with BuLi, due to skeletal reorganization of the initial products.[22]

51% (R=*n*-C$_6$H$_{13}$, overall yield)

Cyclic phosphonamide α-carbanions are excellent olefination reagents particularly useful for enolizable carbonyl compounds.[23] *P*-stabilized allyl anions have found application as Michael donors.[24]

[1] D. Komiotis, B. L. Currie, G. C. LeBreton, and D. L. Venton, *SC* **23**, 531 (1993).
[2] J. R. Hwu, G. H. Hakimelahi, F. F. Wong, and C. C. Lin, *ACIEE* **32**, 608 (1993).
[3] T. Satoh, Y. Hayashi, and K. Yamakawa, *BCSJ* **66**, 1866 (1993).
[4] S. Hormuth and H.-U. Reissig, *JOC* **59**, 67 (1994).
[5] P. Rochet, J.-M. Vatele, and J. Gore, *SL* 105 (1993).
[6] R. Caputo, C. Ferreri, L. Longobardo, D. Mastroianni, G. Palumbo, and S. Pedatella, *SC* **24**, 1223 (1994).
[7] R. C. Vollinga, W. M. P. B. Menge, and H. Timmerman, *RTC* **112**, 123 (1993).
[8] T. Yokoo, H. Shinokubo, K. Oshima, and K. Utimoto, *SL* 645 (1994).

[9] C. D. Gabbutt, J. D. Hepworth, B. M. Heron, M. M. Rahman, *JCS(P1)* 1733 (1994).

[10] M. Braun and K. Opdenbusch, *ACIEE* **32**, 578 (1993).

[11] S. Casson and P. Kocienski, *S* 1133 (1993).

[12] Y. Zhao, P. Quayle, E. A. Kuo, S. Rahman, and E. L. M. Ward, *TL* **35**, 3797 (1994).

[13] W. H. Pearson and E. P. Stevens, *TL* **35**, 2641 (1994).

[14] E. Vedejs and W. O. Moss, *JACS* **115**, 1607 (1993).

[15] I. Coldham, *JCS(P1)* 1275 (1993).

[16] Y. He and N.-H. Lin, *S* 989 (1994).

[17] D. Grandjean, P. Pale, and J. Chuche, *TA* **4**, 1991 (1993).

[18] L. Lattuada, E. Licandro, S. Maiorana, and A. Papagni, *G* **123**, 31 (1993).

[19] R. B. Bates and S. Caldera, *JOC* **58**, 6920 (1993).

[20] M. B. Berry, D. Craig, and P. S. Jones, *SL* 513 (1993).

[21] R. L. Funk, G. L. Bolton, K. M. Brummond, K. E. Ellestad, and J. B. Stallman, *JACS* **115**, 7023 (1993).

[22] P. H. Lee, J. S. Kim, Y. C. Kim, and S. Kim, *TL* **34**, 7583 (1993).

[23] S. Hanessian, Y. L. Bennani, and Y. Leblanc, *H* **35**, 1411 (1993).

[24] S. Hanessian, A. Gomtsyan, A. Payne, Y. Herve, and S. Beaudoin, *JOC* **58**, 5032 (1993).

s-Butyllithium. 14, 69; 16, 56

Amide polyanions. The chelating power of an amidic oxygen atom often makes it possible to generate a carbanion at the α-position to the nitrogen.[1] *s*-Butyllithium seems to be sufficiently strong as a base yet low enough in nucleophilicity not to attack on the carbonyl. With an electron-withdrawing group present in the acyl moiety a remote (stabilized) carbanion can be generated. Furthermore, chiral species are created in the presence of (-)-sparteine.[2]

Regioselective alkylation. Through carbamation and assisted deprotonation it is possible to alkylate an alcohol at the α-position. A proximal amino group can exert regio- and stereocontrol in the alkylation of a diol.[3]

Propargylic alcohols can be synthesized by alkylation of silylmethylacetylenes and oxidative desilylation. This reaction is rendered enantioselective by anchoring the silyl group to a prolinol methyl ether.[4]

Alkylative deoxygenation of epoxides.[5]   This transformation to alkenes is actually quite common and not limited to s-BuLi.

Peterson olefination reaction.[6]   The application to reaction of an N-(trimethylsilylmethyl) formamidine with carbonyl compounds provides enamines that are useful for transformation into homologous nitriles via the N,N-dimethylhydrazones.

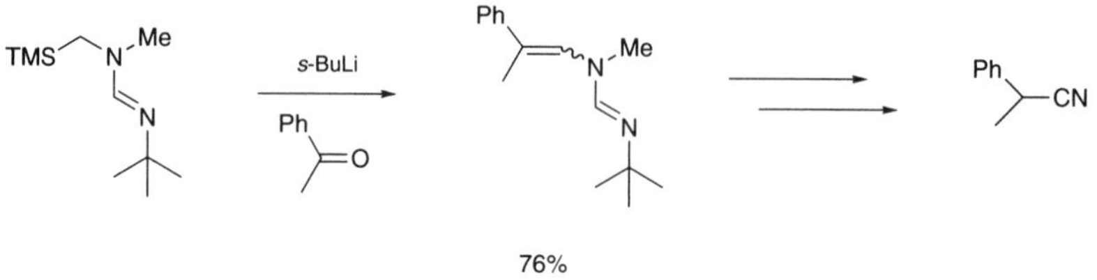

76%

[1] J. E. Resek and P. Beak, *TL* **34**, 3043 (1993).
[2] P. Beak and H. Du, *JACS* **115**, 2516 (1993).
[3] W. Guarnieri, M. Grehl, and D. Hoppe, *ACIEE* **33**, 1734 (1994).
[4] R. C. Hartley, S. Lamothe, and T.-H. Chan, *TL* **34**, 1449 (1993)
[5] E. Doris, L. Dechoux, and C. Mioskowski, *TL* **35**, 7943 (1994).
[6] B. Santiago and A. I. Meyers, *TL* **34**, 5839 (1993).

s-Butyllithium–potassium t-butoxide.

1-Alkoxy-1,3-dienes.[1]   α,β-Unsaturated acetals undergo 1,4-elimination on treatment with this combination of strong bases. Since the alkoxydienes can be deprotonated again by the same base, a one-pot umpolung alkylation is also possible (e.g., starting with 2-alkenyl-1,3-dioxanes).[2]

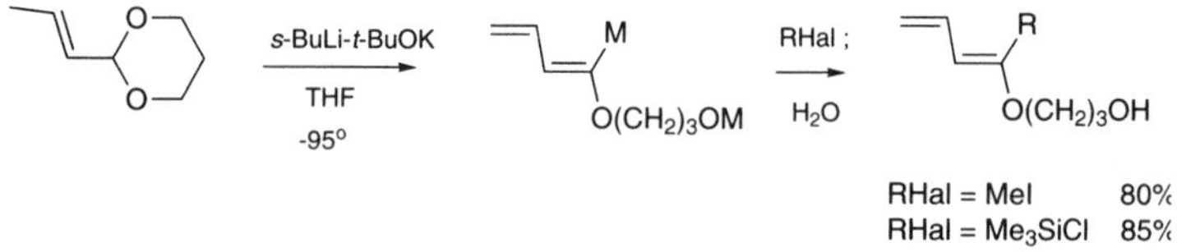

RHal = MeI   80%
RHal = Me₃SiCl   85%

[1] C. Prandi and P. Venturello, *T* **50**, 12463 (1994).
[2] C. Canepa, C. Prandi, L. Sacchi, and P. Venturello, *JCS(P1)* 1875 (1993).

s-Butyllithium–N,N,N',N'-tetramethylethylenediamine.

Amide α-anions.   The Boc-protected secondary amines are readily converted to α-lithio derivatives,[1] which behave in the same way as other organolithium reagents. N-Boc-2-methyltetrahydro-1,3-oxazine undergoes regioselective deprotonation at C-4.[2]

Lithiation at the methyl group[3] is favored over other *N*-alkyl substituents, except for those bearing carbanion-stabilizing groups. Azacycles (3-, 5-, 6-membered) are formed[4] when *N*-Boc-benzyl(chloroalkyl)amines are treated with *s*-BuLi–TMEDA.

Directed metallation vs. rearrangement of O-arylmethyl carbamates. The selection of the two pathways (site of deprotonation) is dependent on the substitution pattern at the benzylic position.[5]

R = H   80%   R = Bu   91%

o-Alkylation of benzoic acid.[6] Alkoxy and amide groups are the most popular *o*-directing substituents on an aromatic ring to achieve nuclear metallation. However, the carboxyl group is also effective.

[1] P. Beak and W. K. Lee, *JOC* **58**, 1109 (1993).
[2] P. Beak and E. K. Yum, *JOC* **58**, 823 (1993).
[3] V. Snieckus, M. Rogers-Evans, P. Beak, W. K. Lee, E. K. Yum, and J. Freskos, *TL* **35**, 4067 (1994).
[4] P. Beak, S. Wu, E. K. Yum, and Y. M. Jun, *JOC* **59**, 276 (1994).
[5] P. Zhang, and R. E. Gawley, *JOC* **58**, 3223 (1993).
[6] J. Mortier, J. Moyroud, B. Bennetau, and P. A. Cain, *JOC* **59**, 4042 (1994).

t-Butyllithium. 13, 58; 15, 64–65; 16, 56–57

Deprotonation. The presence of a chelator unit in the molecule facilitates and directs the deprotonation. For example, The H-6 of 2-*t*-butyl-1-methoxymethyl-3-methyl-1,2,3,4-tetrahydropyrimidin-4-one,[1] H-2 of *N*-methylindole[2] and indole-1-carboxylic acid,[3] H-4 of 1-triisopropylsilylgramine[4] are selectively removed.

Even a remote Boc-NH unit exerts a directing effect on the lithiation at a benzylic position.[5] Cyclic phosphonamidates form stabilized carbanions that react with electrophiles stereoselectively. A bulky (e.g., *i*-Pr) group on the nitrogen atom impedes the approach of the reagents from its side.[6]

80%

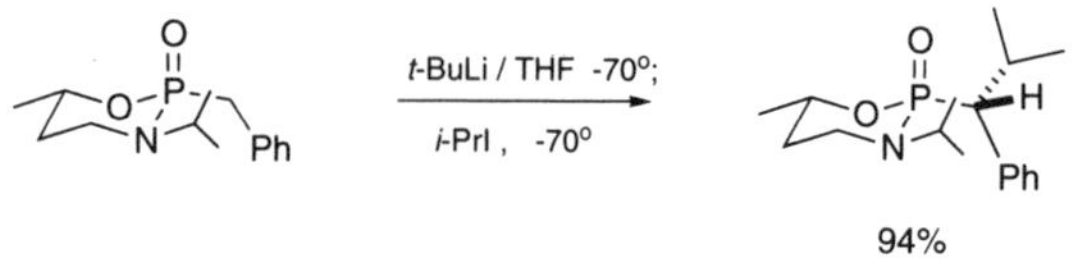

94%

Halogen–lithium exchange.   Special uses of this method include functionalization of bromoallylamine,[7] preparation of α-silylaldehydes,[8] synthesis of alkenyl triflones,[9] trifluoromethyl ketones,[10] alkylidenecyclopentanes[11,12] and diaryl tellurides.[13]

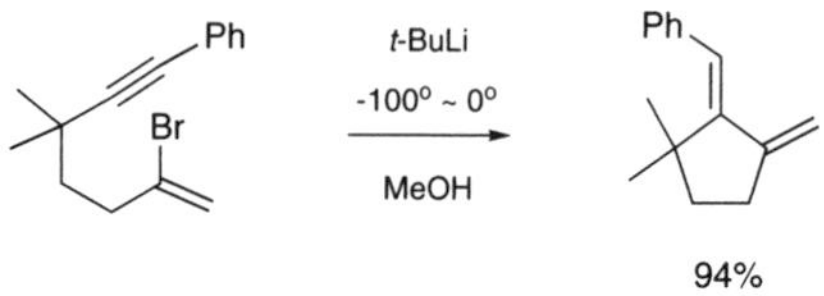

94%

1-Ethynyl ethers.[14]   A synthesis from an acetic ester involves phosphorylation of
the enolate [LDA; (EtO)$_2$POCl] and using *t*-BuLi for the elimination.

[1] K. S. Chu and J. P. Konopelski, *T* **49**, 9183 (1993).
[2] M. Ishikura and M. Terashima, *JOC* **59**, 2634 (1994).
[3] T. Sakamoto, Y. Kondo, N. Takazawa, and H. Yamanaka, *H* **36**, 941 (1993).
[4] M. Iwao, *H* **36**, 29 (1993).
[5] R. D. Clark and Jahangir, *T* **49**, 1351 (1993).
[6] S. E. Denmark and C.-T. Chen, *JOC* **59**, 2922 (1994).
[7] J. Barluenga, R.-M. Canteli, and J. Florez, *JOC* **59**, 602, 1586 (1994).
[8] L. Duhamel, J. Gralak, and A. Bouyanzer, *CC* 1763 (1993).
[9] A. Mahadevan and P. L. Fuchs, *TL* **35**, 6025 (1994).
[10] I. Villuendas, A. Parrilla, and A. Guerrero, *T* **50**, 12673 (1994).
[11] T. V. Ovaska, R. R. Warren, C. E. Lewis, N. Wachter-Jurcsak, and W. Bailey, *JOC* **59**, 5868 (1994).
[12] M. P. Cooke, *JOC* **59**, 2930 (1994).
[13] L. Engman and D. Stern, *OM* **12**, 1445 (1993).
[14] J. A. Cabezas and A. C. Oehlschlager, *JOC* **59**, 7523 (1994).

t-Butyllithium–hexamethylphosphoric triamide.

Disilylalkyllithiums.[1]   A reaction scheme for C–C bond formation that involves
umpolung of an oxygenated carbon atom may exploit a *gem*-disilyl compound as
starting material, although its availability is the primary consideration. Alkylation
and oxidation protocols are routine.

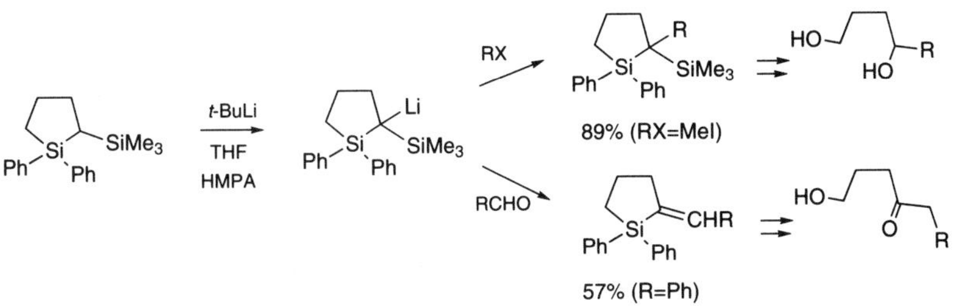

89% (RX=MeI)

57% (R=Ph)

[1] K. Matsumoto, T. Yokoo, K. Oshima, K. Utimoto, and N. A. Rahman, *BCSJ* **67**, 1694 (1994).

t-Butyllithium–potassium hydride.

Homologation of acids (esters).[1]   α-Chloro-α-phenylsulfinyl ketones are the condensation products of esters with chloromethyl phenyl sulfoxide. On treatment with *t*-BuLi–KH these ketones are converted to potassium alkynoxides, which can be transformed into acids through hydration.

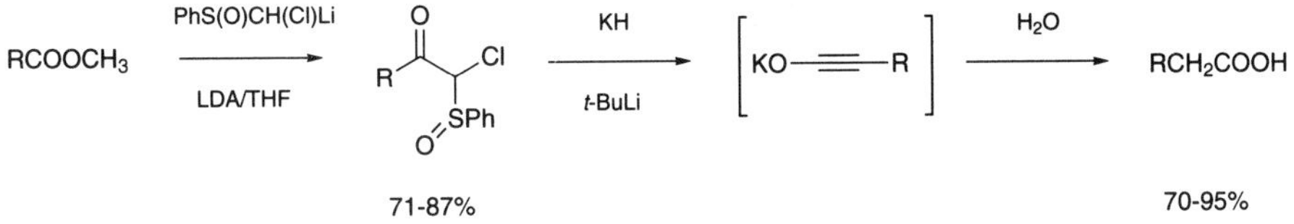

71-87%   70-95%

[1] T. Satoh, Y. Mizu, Y. Hayashi, and K. Yamakawa, *TL* **35**, 133 (1994).

t-Butyllithium–*N*,*N*,*N*′,*N*′-tetramethylethylenediamine.

Directed lithiation.[1]   The *o*-direction of (*Z*)-β-lithiostyrene to give a dilithio compound with two proximal metal atoms is remarkable.

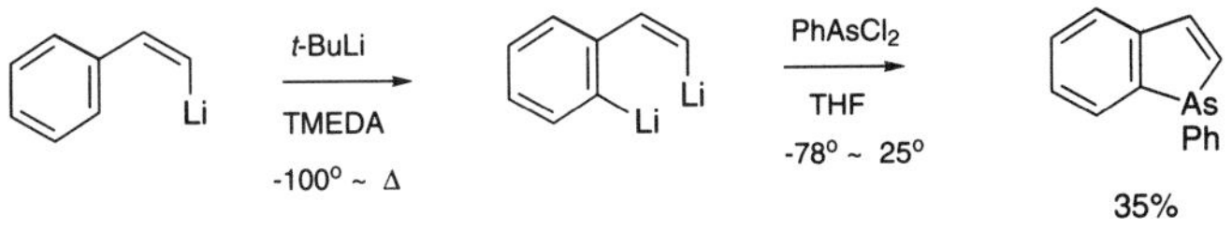

35%

A tandem elimination and deprotonation occurs with 2,2-difluoroethyl *N*,*N*-diisopropylcarbamate.[2] By virtue of this behavior the carbamate is a synthetic equivalent of α-fluoroacetyl anion.

[1] A. J. Ashe and P. M. Savla, *JOMC* **461**, 1 (1993).
[2] J. A. Howarth, W. M. Owton, and J. M. Percy, *SL* 503 (1994).

t-Butyl 2,3-dioxo-4-pentynoate.

Dienophile.[1]   This polycarbonyl compound forms Diels–Alder adducts, which are versatile intermediates for various syntheses. For example, an indoxyl derivative is accessible on reaction of its butadiene adduct with a primary amine and dehydrogenation.

[1] H. H. Wasserman and C. A. Blum, *TL* **35**, 9787 (1994).

2-*t*-Butylthio-1-azo-(4′-methylbenzene).

α-p-tolylhydrazonation.[1]   Azo group transfer to ketone enolates with in situ prototropic shift is accomplished at room temperature. In many cases the yields are excellent.

[1] C. Dell'Erba, M. Novi, G. Petrillo, and C. Tavani, *T* **50**, 11239 (1994).

n-Butyltin trichloride. 15, 65–66

(Z)-Homoallylic alcohols.   The (Z)-selective reaction of allyltin reagents with carbonyl compounds is catalyzed by BuSnCl$_3$.[1,2] The nucleophiles are actually the 1,3-transposed allyl(butyl)tin dichlorides. Operationally, it is important that the allylstannanes (*E,Z* mixtures) must be introduced last in order to achieve good stereoselectivities.

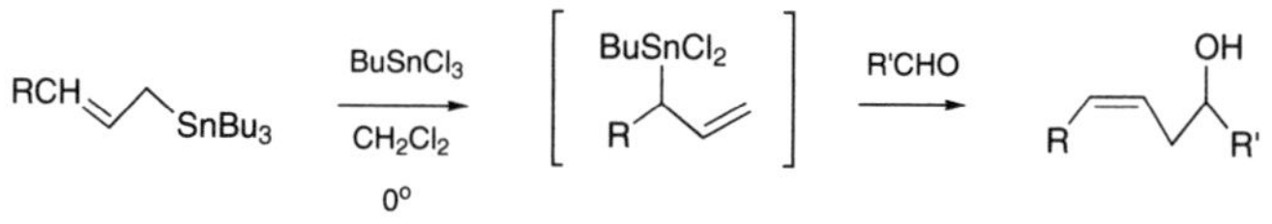

[1] H. Miyake and K. Yamamura, *CL* 1473 (1993).
[2] H. Miyake and K. Yamamura, *CL* 897 (1994).

t-Butyl trimethylsilyl peroxide, *t*-BuOOSiMe$_3$.

Aldehydes from sulfones.[1]   Sulfone α-carbanions undergo trimethylsiloxylation in good yields. On silica gel chromatography the products are decomposed to furnish aldehydes.

[1] F. Chemla, M. Julia, and D. Uguen, *BSCF* 547 (1993).

C

Cadmium. 13, 60

Deprotection. For removal of the 2,2,2-trichloroethoxycarbonyl group from a protected amino acid, Cd in 1:1 HOAc-DMF is a reagent superior to Zn.[1] A didehydroindolizidine has been synthesized[2] directly from a carbamate containing a diketo chain involved liberating the amine with sonication.

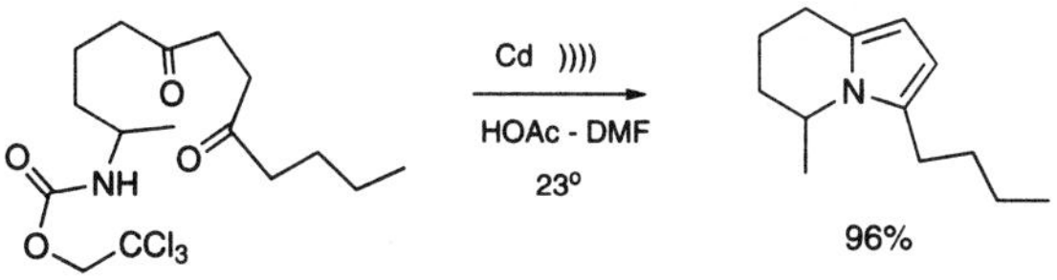

Cinnamic esters.[3] The synthesis by heating a mixture of ArCHO, BrCH$_2$-COOMe, and Bu$_3$As is catalyzed by Cd (10 mol %).

[1] G. Hancock, I. J. Galpin, and B. A. Morgan, *TL* **23**, 249 (1982).
[2] A. M. Castano, J. M. Cuerva, and A. M. Echavarren, *TL* **35**, 7435 (1994).
[3] J. Zheng and Y. Shen, *SC* **24**, 2069 (1994).

Cadmium iodide.

Knoevenagel condensation.[1] CdI$_2$ is a useful catalyst.

[1] D. Prajapati and J. S. Sandhu, *JCC* **12**, 739 (1993).

Calcium borohydride.

Reduction of esters.[1] With an alkene as catalyst, esters are reduced to the alcohols with this borohydride. Intermediate borates formed during reduction of aryl esters are converted to aldehydes on treatment with NaOCl in good yields.

[1] S. Narasimhan, K. G. Prasad, and S. Madhavan, *SC* **25**, 1689 (1995).

Calcium hydride.

Aldehyde enamines.[1] Good yields and purity of enamines are obtained with CaH$_2$ as the condensating agent. Cyclohexane is used as a diluent to moderate the exothermic reaction. The optimal ratio of CaH$_2$ and R$_2$NH to RCHO is 1.2 equivalents each.

[1] G. B. Fisher, L. Lee, and F. W. Klettke, *SC* **24**, 1541 (1994).

Calcium hypochlorite.

RCH₂OH → RCOOMe.[1]   Methyl esters are produced (82–97% yield) on treatment of primary alcohols with $Ca(OCl)_2$ and HOAc in MeCN/MeOH and molecular sieves in the dark.

Stabilized α-chloromethylenetriphenylphosphoranes.[2]   A convenient preparation of $Ph_3P=C(Cl)R$, where R=CN, Ac, COOEt, is by treatment of the phosphonium salts with bleaching powder in EtOH containing HCl.

[1] C. E. McDonald, L. E. Nice, A. W. Shaw, and N. B. Nestor, *TL* **34**, 2741 (1993).
[2] B. Schäfer, *T* **49**, 1053 (1993).

Calcium nitrate.

Nitrophenols.[1]   In aqueous sulfuric acid at low temperature $Ca(NO_3)_2$ is comparable to $NaNO_3$ as nitrating agent for substituted phenols.

[1] S. C. Bisarya, S. K. Joshi, and A. G. Holkar, *SC* **23**, 1125 (1993).

Camphorsulfonic acid. **13**, 62–64; **14**, 71–72; **15**, 68; **16**, 58

Acetalization.[1]   This sulfonic acid (CSA) is a very useful catalyst for many reactions. It has been employed in the selective protection of diequatorial diols.[2]

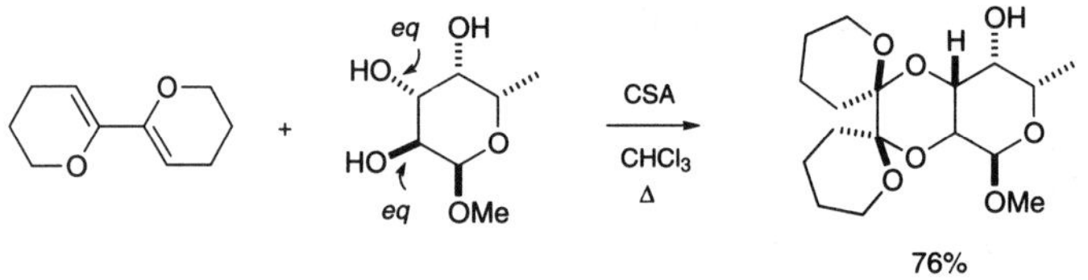

76%

Diels–Alder reactions.[3]   In an intramolecular cycloaddition approach to construct the pentacyclic skeleton of eburnamonine the catalytic system of $LiClO_4$–Et_2O is insufficient. The addition of 10 mol % of CSA is required.

Benzannulation.[4]

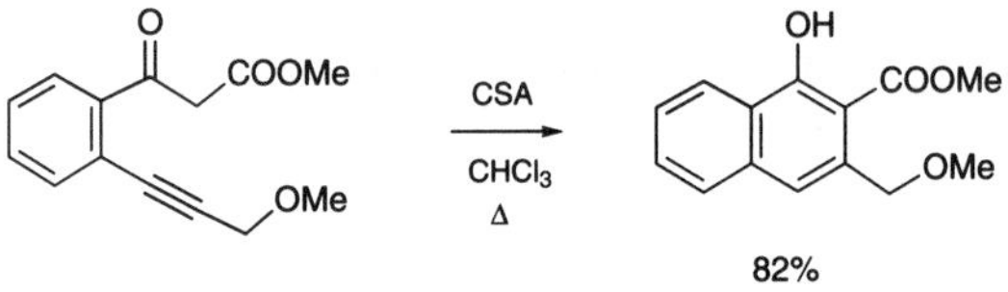

82%

[1] S. V. Ley, G.-J. Boons, R. Leslie, M. Woods, and D. M. Hollinshead, *S* 689 (1993).
[2] S. V. Ley, H. W. M. Priepke, and S. L. Warriner, *ACIEE* **33**, 2290 (1994).
[3] M. D. Kaufman and P. A. Grieco, *JOC* **59**, 7197 (1994).
[4] M. A. Ciufolini and T. J. Weiss, *TL* **35**, 1127 (1994).

Carbon disulfide.

Activation of N-CH₂.[1]   Derivatization of indolines with CS_2/BuLi accomplishes two purposes. The N–H bond is protected against electrophiles in the alkylation at C-2. More importantly, the lithiodithiocarboxyl function directs deprotonation at the desired position.

[1] H. Ahlbrecht and C. Schmitt, *S* 983 (1994).

Carbonyldiimidazole. 13, 66; 16, 64

α-Diketones.[1]   This synthesis starting from α,β-dihydroxy ketones proceeds via the cyclic carbonates. The liberated imidazole promotes β-elimination and thence decarboxylation. *syn*-Diols react faster than *anti*-diols.

[1] S.-K. Kang, D.-C. Park, H.-S. Rho, and S.-M. Han, *SC* **23**, 2219 (1993).

Catecholborane. 16, 65–66; 17, 67–68

Generated by passing diborane (ex $NaBH_4 + I_2$) to a benzene suspension of catechol at room temperature.[1]

Catalyzed hydroborations.   Starting from 1-alkynes, catecholborylalkenes are prepared in the presence of 10 mol % $H_3B \cdot NEt_2Ph$ or $H_3B \cdot THF$.[1]

Many catalysts have been discovered for the hydroboration with catecholborane. Sonication with activated nickel powder[2] (obtained from Li reduction of NiI_2 in ether) significantly increases the reactivity. In the presence of $(Ph_3P)_3RhCl$ the regioselectivity for the hydroboration of allylic sulfones[3] is enhanced. The catalyzed reactions using bis(mesityl)niobium[4] are different from those using Cp_2TiCl_2[5] because the major hydroborating agent for the former reactions actually is borane.

α-Halo boronic acids.[6]   These substances are made from 1-haloalkenes and catecholborane without solvent.

[1] Y. Suseela and M. Periasamy, *JOMC* **450**, 47 (1993).
[2] G.W. Kabalka, C. Narayana, and N. K. Reddy, *SC* **24**, 1019 (1994).
[3] X.-L. Hou, D.-G. Hong, Y.-L. Guo, and L.-X. Dai, *TL* **34**, 8513 (1993).
[4] K. Burgess, M. Jaspars, and W. A. van der Donk, *TL* **34**, 6813 (1993).
[5] K. Burgess, M. Jaspars, and W. A. van der Donk, *TL* **34**, 6817 (1993).
[6] S. Elgendy, G. Claeson, V.V. Kakkar, D. Green, G. Patel, C. A. Goodwin, J. A. Baban, M. F. Scully, and J. Deadman, *T* **50**, 3803 (1994).

Cerium(IV) ammonium nitrate. 13, 67–68; 14, 74–75; 15, 70–72; 16, 66; 17, 68

Tetrahydropyranyl ethers.[1]   The condensation of alcohols with dihydropyran is catalyzed by CAN.

Methoxybenzaldehydes.[2]   Benzyldimethylamines methoxylated at the *o*- and/or *p*-position are oxidized at room temperature.

Allyl ethers.   Conversion of allylic alcohols to ethers (and ether exchange also) and transpositional allylic methoxylation have been effected by the CAN-ROH system.[3,4] Cephem sulfoxides undergo alkoxylation at the α-position of the ester.[5] The β-methoxyalkyl phenyl selenides obtained from CAN oxidation of diphenyl diselenide in the presence of alkenes in methanol are potential intermediates of allyl ethers.[6]

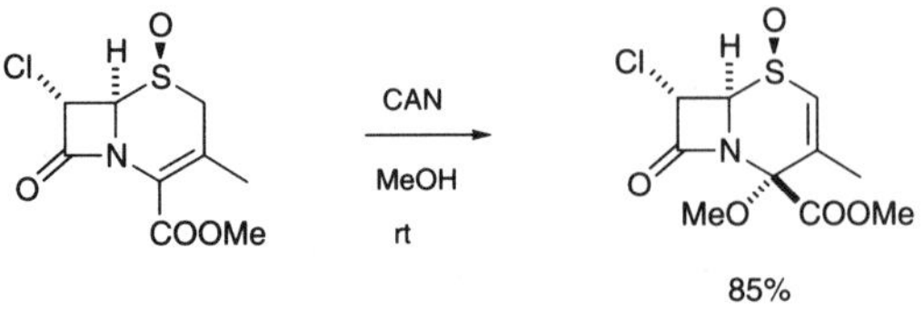

Nitration.   Carbazoles are nitrated[7] using CAN as the reagent. Alkenes are converted to nitroalkenes[8] by CAN in the presence of $NaNO_2$ and HOAc, whereas epoxides give β-nitrato alcohols[9] on reaction with $CAN-NH_4NO_3$.

Functionalization of alkenes.   CAN oxidation is often employed in the generation of reactive species for addition to alkenes. Thus an azido radical created in the presence of α-methoxy α,β-unsaturated nitriles is trapped.[10] The adducts are useful precursors of α-azido carboxylic acids, and thence α-amino acids.

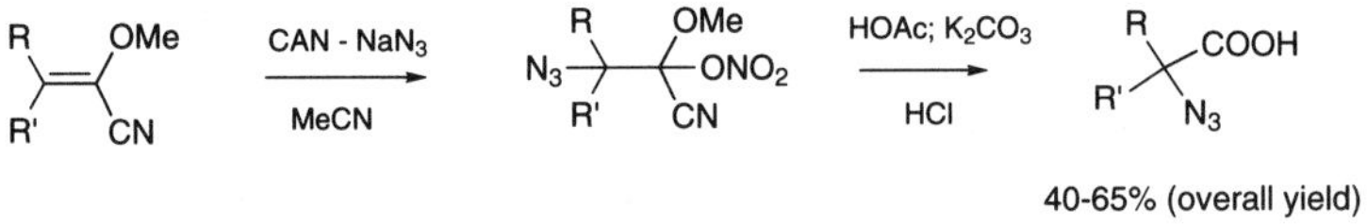

Bicyclic lactones are produced[11] when dihydropyran and malonic monomethyl ester are treated with CAN and $Cu(OAc)_2$, preferably with ultrasound irradiation.

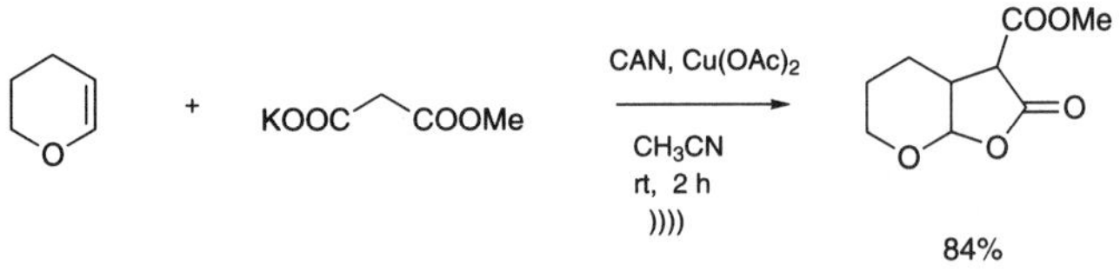

Cleavage of cyclopropyl sulfides.   Oxidative desulfurization of these compounds gives ketones.[12]

Oxidative coupling.   Methods for ketone synthesis by trapping electrophilic species generated by CAN oxidation with silyl enol ethers are quite useful. 2-Tributylstannyl-1,3-dithiane[13] and 1-trimethylsiloxy-1,3-butadiene[14] are exemplary substrates.

69%

[1] G. Maity and S. C. Roy, *SC* **23**, 1667 (1993).
[2] I. Badea, P. Cotelle, and J.-P. Catteau, *SC* **24**, 2011 (1994).
[3] N. Iranpoor and E. Mothaghineghad, *T* **50**, 1859 (1994).
[4] N. Iranpoor and E. Mothaghineghad, *T* **50**, 7299 (1994).
[5] M. Alpegiani, P. Bissolino, D. Borghi, and E. Perrone, *SL* 233 (1994).
[6] C. Bosman, A. D'Annibale, S. Resta, and C. Trogolo, *TL* **35**, 6525 (1994).
[7] M. Chakrabarty and A. Batabyal, *SC* **24**, 1 (1994).
[8] J. R. Hwu, K.-L. Chen, and S. Ananthan, *CC* 1425 (1994).
[9] N. Iranpoor and P. Salehi, *T* **50**, 909 (1994).
[10] D. L. J. Clive and N. Etkin, *TL* **35**, 2459 (1994).
[11] D. D'Annibale and C. Trogolo, *TL* **35**, 2083 (1994).
[12] Y. Takemoto, T. Ohra, S. Ruruse, H. Koike, and C. Iwata, *CC* 1529 (1994).
[13] K. Narasaka, N. Arai, and T. Okauchi, *BCSJ* **66**, 2995 (1993).
[14] A. B. Paolobelli, D. Latini, and R. Ruzziconi, *TL* **34**, 721 (1993).

Cerium(III) chloride. **14**, 75–77; **15**, 72–73; **16**, 67–68

β-Hydroxy amides.[1] Cerium(III) enolates of amides show better reactivities than the Li enolates toward ketones and aldehydes. Reaction with camphor gives >94% yield of the aldol.

Modified Reformatsky reaction.[2] α-Bromoesters are probably converted to Ce(III) enolates by the $PhTeLi–CeCl_3$ combination, which makes them reactive towards carbonyl compounds.

[1] X. Shang and H.-J. Liu, *TL* **35**, 2485 (1994).
[2] S. Fukazawa and K. Hirai, *JCS(PI)* 1963 (1993).

Cesium carbonate. **14**, 77–78; **15**, 73–75

Alkylations. Cs_2CO_3 has shown some unique properties as a base. It promotes O-alkylation of 2-pyrimidinone,[1] diastereoselective Michael addition,[2] and tandem alkylation/Wittig reaction[3] to form ethoxycyclopentadienes with good results.

Carbamates: Formation and hydrolysis. Amines form carbamic acid salts on reaction with CO_2 in the presence of Cs_2CO_3.[4] Carbamate esters are isolated after alkylation. When the alkylating agent is an α-bromoalkyl epoxide, 5-(α-hydroxy-alkyl)oxazolidin-2-one is the product.[5] Apparently, epoxide opening by the amine precedes N-carboxylation.

R = Bn, Pr, i-Pr, CH_2Br

Bis-urethanes derived from propylenediamines are converted to cyclic ureas[6] on treatment with Cs_2CO_3.

Nitrene formation.[7] Ethyl *N*-tosylcarbamate undergoes α-elimination (-TsOH) on treatment with Cs_2CO_3 at room temperature. The nitrene cycloadduct with cyclohexene is produced in 79% yield under these conditions.

[1] B. S. Moller, M. L. Falck-Pedersen, T. Benneche, and K. Undheim, *ACS* **46**, 1219 (1992).
[2] N. Ouvrard, J. Rodriguez, and M. Santelli, *ACIEE* **31**, 1651 (1992).
[3] M. Hatanaka, Y. Himeda, R. Imashiro, Y. Tanaka, and I. Ueda, *JOC* **59**, 111 (1994).
[4] K. J. Butcher, *SL* 825 (1994).
[5] M. Yoshida, M. Ohshima, and T. Toda, *H* **35**, 623 (1993).
[6] K. J. Fordon, C. G. Crane, and C. J. Burrows, *TL* **35**, 6215 (1994).
[7] M. Barani, S. Fioravanti, L. Pellacani, and P. A. Tardella, *T* **50**, 11235 (1994).

Cesium fluoride. 13, 68; **14,** 79; **15,** 75–76; **16,** 69–70; **17,** 68

Desilylations. Silyl groups attached to various kinds of atoms can be removed by CsF, leaving anionic species ready for reactions. Unusual routes to dienals,[1] allylic alcohols,[2] 2,5-dihydrofurans,[3] and tris(perfluoroalkyl) carbinols[4] have been developed based on this reaction.

The CsF–Si(OEt)$_4$ system, which operates as a base for the conjugate addition[5] of amides to unsaturated esters, probably releases ethoxide ions.

α-Fluoroalkanoic acids.[6] When trichloromethyl carbinols, the adducts of aldehydes and chloroform, are treated with CsF, Bu$_4$NF, and Et$_3$N, a fluorine atom is introduced to the position originally occupied by the OH group. Saponification then delivers the fluorinated carboxylic acids in 76–100% yield.

Trimerization of aryl isocyanates.[7]

[1] M. Bellassoued and M. Salemkour, *TL* **34**, 5281 (1993).
[2] A. B. Muccioli and N. S. Simpkins, *JOC* **59**, 5141 (1994).
[3] M. Hojo, M. Ohkuma, N. Ishibashi, and A. Hosomi, *TL* **34**, 5943 (1993).
[4] G. J. Chen, L. S. Chen, K. C. Eapen, and W. E. Ward, *JFC* **69**, 61 (1994).

[5] K. H. Ahn and S. J. Lee, *TL* **35**, 1875 (1994).
[6] J. E. Oliver, R. M. Waters, and W. R. Lusby, *S* 273 (1994).
[7] Y. Nambu and T. Endo, *JOC* **58**, 1932 (1993).

Chiral auxiliaries and catalysts.

Small ring synthesis. Simmons–Smith reactions using CH_2I_2-Et_2Zn as reagent are enantioselective in the presence of chiral *trans*-1,2-cyclohexanedisulfonamides (**1**).[1,2]

Formation of cyclopropanecarboxylic esters by capture of the metal-carbenoid derived from diazoacetates with alkenes is rendered enantioselective by a chiral 2,2'-bipyridine (**2**) [complexed to Cu][3] and a bis(oxazolinyl)pyridine (**2a**) [complexed to Ru].[3a]

The [2+2]cycloaddition of ketene with aldehydes is subject to asymmetric induction by chiral organoaluminum complexes, for example, of C_2-symmetrical 1,2-bis-sulfonamides (**3**).[4]

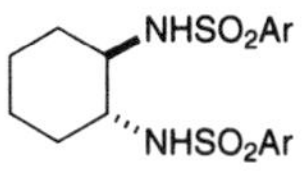

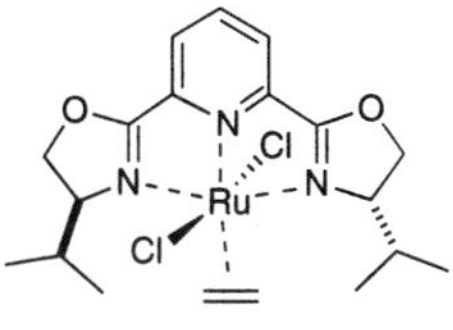

1,3-Dipolar cycloadditions. In the reaction of allylic alcohols with nitrile oxides cocomplexation of the alcohol and diisopropyl tartrate to Zn directs the steric course in the formation of 2-isoxazolines.[5] Bonding of the nitrones that participate in cycloadditions to the boron atom of a chiral oxazaborolidine (**4**)[6] through their oxygen atoms is important to determine the transition states leading to isoxazolidine products.

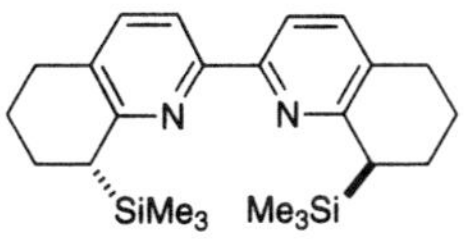

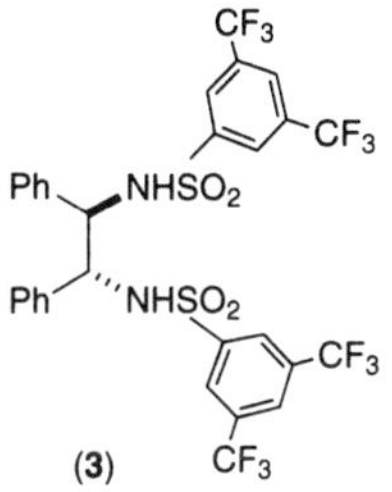

The monoacetonide (**5**)[7] of the tetraphenyltetraol (Seebach's TADDOL) derived from tartaric acid has been used to modify many Lewis acids. For example, the dichlorotitanium dialkoxide is able to catalyze asymmetric cycloadditions between nitrones and alkenes.

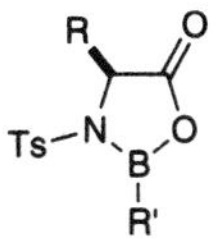

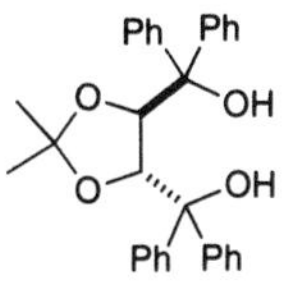

Diels–Alder reactions. Many excellent Lewis acid catalysts for the Diels–Alder reaction are formed from enantiomeric binaphthols and substituted binaphthols. Ytterbium-,[8] aluminum-,[9] and titanium-based[10] species have been developed. A chiral iron catalyst[11] is also quite effective, but a zirconocene-based catalyst[12] is disappointing in terms of asymmetric induction.

A tricyclic, C_2-symmetrical polyoxa diol auxiliary can be used to attach acryloyl groups to form chiral dienophiles (**6**).[13]

An acyloxyborane catalyst (**7**) made from a tartaric acid derivative promotes cyclocondensation of Danishefsky's diene and aldehydes to construct substituted dihydropyran-4-ones.[14]

The dibromotitanium derivative of TADDOL is a catalyst for the hetero-Diels–Alder reaction between enones and vinyl ethers.[15]

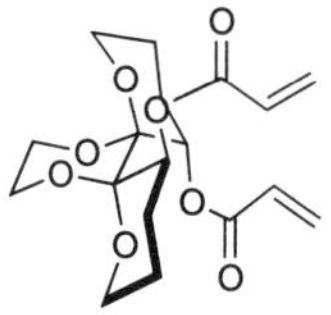

(6) 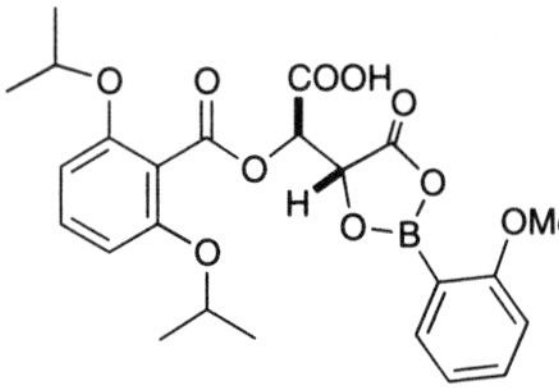   (7)

Oxidations. Much effort has been spent in finding conditions for enantioselective epoxidation of different kinds of alkenes. The most successful catalysts seem to be C_2-symmetrical (salen)Mn complexes (**8**) (**9**) (**10**). Oxidants such as 4-phenylpyridine oxide in combination with NaOCl[16,17] or PhIO,[18] O_2/pivalaldehyde with[19] or without[20] N-methylimidazole, as well as periodate ion,[21] are effective. A titanocene catalyzes epoxidation of several alkenes by t-BuOOH with moderate ee.[22]

Optically active lactones are formed by an air oxidation using a Cu(II) complex of a 2-(o-hydroxyaryl)-4-t-butyloxazoline (**11**).[23]

(Salen)Mn complexes[24] and a camphorsultim (**12**)[25] have been employed in asymmetric oxidation of sulfides to chiral sulfoxides.

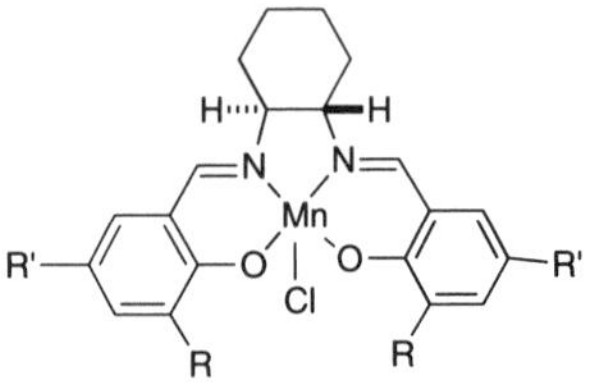

(8) 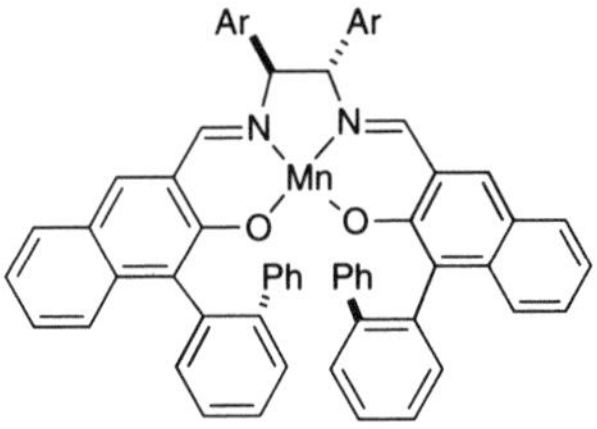   (9) 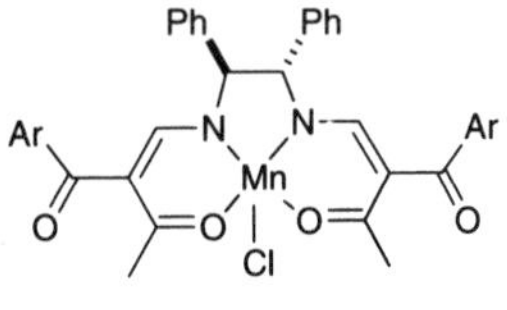   (10)

R = R' = t-Bu
R = t-Bu, R' = Me

(11)

(12)

Dihydroxylation.   Besides the enormously popular and effective cinchona alkaloid-based chiral auxiliaries several C_2-symmetrical diamines (13),[26] (14),[27] and (15)[28] have been developed to direct alkene dihydroxylation with OsO_4. These efforts are probably overwhelmed by the Sharpless protocols because the approaches are not catalytic with respect to the most expensive and toxic reagent.

(13)   (14)   (15)

Homogeneous hydrogenation.   Using Rh(cod)Cl dimer as the catalyst and chiral diphosphine ligands, the asymmetric hydrogenation of dehydroamino acid derivatives[29] and itaconic acid[30] is accomplished. Related catalysts are successfully applied in the reduction of peptides on polymer support[30a] and of bisdehydrodipeptides.[30b] Other cationic Rh complexes containing modified DIOPS,[30c] propraphos,[30d] and ferrocenyldiphosphine[30e] enjoy a different degree of success.

Reductions with borane.   Many chiral 1,2-amino alcohols,[31-34] which form 1,3,2-oxazaborolidines in situ, are worthy of evaluation for their effect on ketone reduction by borane. The most convenient source of such amino alcohols is provided by the natural amino acids; therefore, it is not surprising that various derivatives of phenylglycine, phenylalanine, methionine, cysteine, and proline have been made for that purpose. Resolved synthetic amino alcohols that have been used include *cis*-1-amino-2-indanol[35] and *cis*-1,2-diphenyl-2-aminoethanol.[36] Analogous chiral auxiliaries are obtained from 2-arylaminomethylpiperidines,[37] as β-hydroxysulfoximines (16) (which can be used to reduce ketones[38] and imines[39] by nascent borane generated in situ from $NaBH_4$–Me_3SiCl).[40]

Bicyclic 1,3,2-oxazaborolidines synthesized from proline are the original "chemzymes." The crystalline BH_3 adduct of the 1-methyl derivative can be prepared from

Me$_2$S · BH$_3$ in toluene or xylene at room temperature.[41] Results from reduction of diethyl α-ketophosphates[42] in the presence of the bicyclic 1-butyl oxazaborolidine, and of numerous ketones using a monocyclic catalyst (17),[43] are quite respectable.

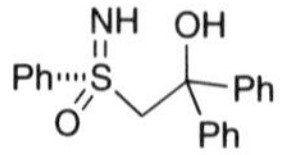

(16) 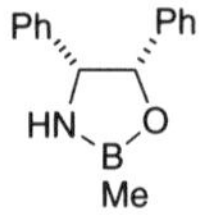   (17)

A tetracyclic oxazaborolidine derived from tryptophol[44] can be used to reduce ketones enantioselectively. It is not effective for imines.

Long-range mediation by a chiral boronate (1,7-asymmetric induction)[45] in ketone reduction by borane is feasible. Enantioselective reduction of imines[46] with stoichiometric quantities of a cyclic dialkoxyborane prepared from tartaric acid has been reported.

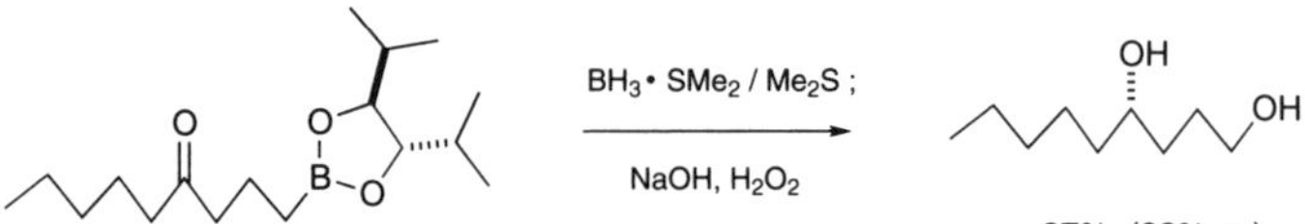

87% (92% ee)

Amino diol auxiliaries can be highly effective or modest in their effects on modifying the steric course of ketone reduction by LiAlH$_4$. The C_2-symmetrical diethanolamine (18)[47] gives good results, whereas a ligand (19) prepared from (-)-β-pinene performs poorly.[48]

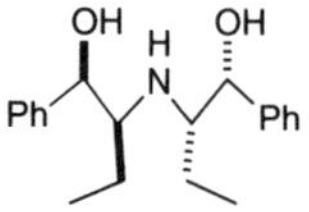

(18) 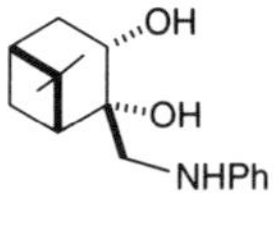   (19)

The rhodium(I)-catalyzed reduction of ketones with silanes is also subject to enantioselection by chiral ligands, which can be a bipyridine derivative,[49] cyclic phosphonites and phosphites,[50] or diferrocenyl dichalcogenides.[51]

α-Cyanohydrination.   Titanium(IV) complexes with chiral salen ligands such as (20)[52] are useful to induce enantioselective formation of *O*-silyl α-cyanohydrins from carbonyl compounds and Me$_3$SiCN. Another catalyst is a chloromagnesium complex of a C_2-symmetrical bis(oxazolin-2-yl)methylene derivative (**21**).[53] This system has separate binding sites for the aldehyde and cyanide components.

(**20**)   (**21**)

Transpositional reduction of allylic carbonates. Pd-catalyzed S_N2' reduction (HCOOH as hydride donor) of allylic carbonates proceeds enantioselectively in the presence of chiral ligands based on binaphthyl.[54] Chiral allylsilanes are readily available in this manner using an MOP-phen ligand.[55]

Addition of R$_2$Zn to aldehydes. This seems to be the standard reaction for testing the asymmetric induction efficiency of chiral ligands. Thus, various β-hydroxy tertiary amines have been derived from proline,[56] valine,[57] norephedrine,[58,59] a bicyclic proline analog,[60] norproline analog,[61] (OC)$_3$Cr-complexed norephedrine,[62] polymer-bound norephedrine,[63] and m-xylylenediamine-linked units.[64] A hydroxymethyloxazoline ligand[65] belonging to this category shows a lower efficiency.

The catalytic efficiencies of some 1,3- and 1,4-amino alcohols remain high. Except for one ligand derived from hydroxyproline,[66] most others[67–69] are based on ferrocene. A similar ligand is a Cr(CO)$_3$ complex.[70] β-Hydroxy ethers[71] and β-amino thiols[72] are also attracting a great deal of attention.

The reaction in the presence of a 1,2-diamine ligand requires a Lewis acid. Perhaps the most popular combination comprises *trans*-1,2-bistriflamidocyclohexane and (i-PrO)$_4$Ti. Many polyfunctional compounds can be prepared in chiral form by this method.[73,74] The organozinc reagents attack conjugated aldehydes in the 1,2-fashion.[75–77] However, the chemoselectivity can be changed by using NiCl$_2$ instead of (i-PrO)$_4$Ti in the catalytic system.[78]

Dialkyl thiophosphoramidates[79] bearing a β-hydroxy group are also highly enantioselective catalysts, and alkylbenzenes exhibit special effects in the reaction of N-diphenylphosphinoylimines in the presence of polystyrene-supported ephedrine.[80] N,N-Diallylnorephedrine has been chosen as catalyst in the Reformatsky reaction.[81]

Other organometallic additions. Alkyl- and aryltitanium alkoxides add to aldehydes in the presence of Ti-TADDOLate[82] in a highly enantioselective manner. The cleavage of allyl ethers[83] by Grignard reagents generates a new allylic stereocenter the absolute configuration of which can be controlled by a chiral Zr catalyst.

Other effective catalysts for similar reactions include R_3P-ligated nickel[84] and palladium derivatives.[85,86] Chiral diamines[87] mediate Grignard reactions with aldehydes.

Two different tactics for the allylation of aldehydes with allylsilanes[88] and allylstannanes[89] involve the use of a chiral Lewis acid catalyst and the attachment of the reagent to a chiral template, respectively.

The combination of a bis(amide/phosphine) and π-allylpalladium chloride can be used to achieve asymmetric allylic alkylation[90] and deracemization of cyclic allyl esters.[91]

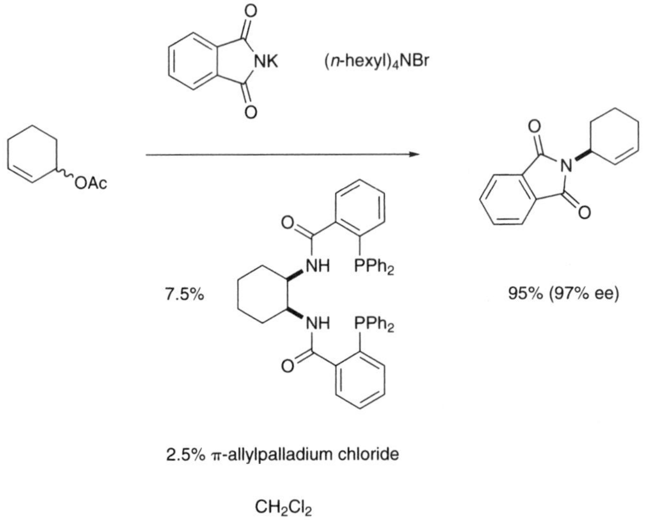

Aldolization and related reactions. Tartaric acid–derived acyloxyborane complexes are shown to be useful catalysts for asymmetric aldol reactions. (*S*)-4-Isopropyl-3-tosyl-1,3,2-oxazaborolidin-5-one is an excellent catalyst, not only for the aldolization[92,93] between a silyl enol ether and an aldehyde; it also reduces the products to afford *syn*-1,3-diols.[94]

Asymmetric aldolization of α-isocyanoacetamide and fluorinated benzaldehydes has been realized with a gold(I) salt and a ferrocenyl amine–phosphine ligand.[95] (Salen)-Ti complexes serve well in catalyzing the condensation of diketene with aldehydes.[96] A camphor lactam[97] is an adequate chiral auxiliary as its derived imide undergoes asymmetric aldol reactions.

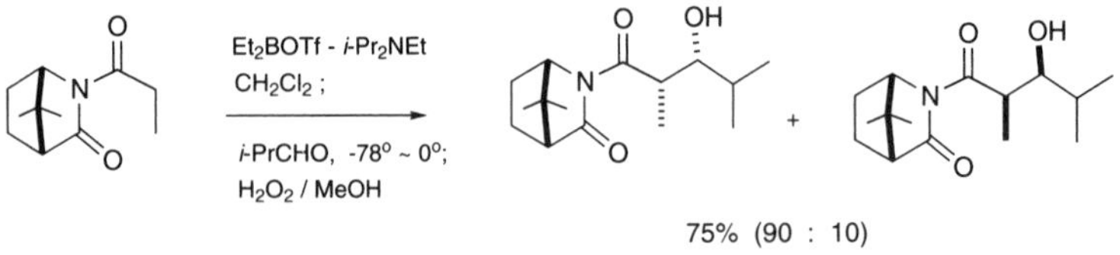

An extensive study on Sn(OTf)$_2$-catalyzed asymmetric aldol reactions[98] using chiral diamine ligands derived from proline has been carried out. Of particular significance is that the enantioselectivity can be switched completely by slight changes of the ligand structures.

The DIPT-Ti(OPr^i)$_2$ has been used in hydrophosphonylation of aromatic aldehydes.[99]

Michael reaction. Different ligands for copper catalysts are available. For example, an oxazoline prepared from valine,[100] a phosphine from proline,[101] and a carbohydrate-based thiol[102] have been tested.

[1] N. Imai, K. Sakamoto, H. Takahashi, and S. Kobayashi, *TL* **35**, 7045 (1994).

[2] N. Imai, H. Takahashi, and S. Kobayashi, *CL* 177 (1994).

[3] K. Ito and T. Katsuki, *TL* **34**, 2661 (1993); *SL* 638 (1993).

[3a] H. Nishiyama, Y. Itoh, H. Matsumoto, S.-B. Park, and K. Itoh, *JACS* **116**, 2223 (1994).

[4] Y. Tamai, H. Yoshiwara, M. Someya, J. Fukumoto, and S. Miyano, *CC* 2281 (1994).

[5] Y. Ukaji, K. Sada, and K. Inomata, *CL* 1847 (1993).

[6] J.-P. G. Seerden, A.W. A. Scholte op Reimer, and H.W. Scheeren, *TL* **35**, 4419 (1994).

[7] K.V. Gothelf and K. A. Jorgensen, *JOC* **59**, 5687 (1994).

[8] S. Kobayashi, I. Hachiya, H. Ishitani, and M. Araki, *TL* **34**, 4535 (1993).

[9] K. Maruoka, A. B. Concepcion, and H. Yamamoto, *BCSJ* **65**, 3501 (1992).

[10] K. Maruoka, N. Murase, and H. Yamamoto, *JOC* **58**, 2938 (1993).

[11] E. P. Kündig, B. Bourdin, and G. Bernardinchi, *ACIEE* **33**, 1856 (1994).

[12] Y. Hong, B. A. Kuntz, and S. Collins, *OM* **12**, 964 (1993).

[13] B. C. B. Bezuidenhoudt, G. H. Castle, S.V. Ley, and J.V. Geden, *TL* **35**, 7447, 7451 (1994).

[14] Q. Gao, K. Ishihara, T. Maruyama, M. Mouri, and H. Yamamoto, *T* **50**, 979 (1994).

[15] E. Wada, H. Yasuoka, and S. Kanemasa, *CL* 1637 (1994).

[16] B. D. Brandes and E. N. Jacobsen, *JOC* **59**, 4378 (1994).

[17] S. Chang, R. M. Held, and E. N. Jacobsen, *TL* **35**, 669 (1994).

[18] H. Sasaki, R. Irie, T. Hamada, K. Suzuki, and T. Katsuki, *T* **50**, 11827 (1994).

[19] T. Yamada, K. Imagawa, T. Nagata, and T. Mukaiyama, *BCSJ* **67**, 2248 (1994).

[20] T. Nagata, K. Imagawa, T. Yamada, and T. Mukaiyama, *CL* 1259 (1994).

[21] P. Pietikainen, *TL* **36**, 319 (1995).

[22] S. L. Colletti and R. L. Halterman, *JOMC* **455**, 99 (1993).

[23] C. Bolm, G. Schlingloff, and K. Weickhardt, *ACIEE* **33**, 1848 (1994).

[24] K. Noda, N. Hosoya, R. Irie, Y. Yamashita, and T. Katsuki, *T* **50**, 9609 (1994).

[25] P. C. B. Page, J. P. Heer, D. Bethell, E. W. Collington, and D. M. Andrews, *TL* **35**, 9629 (1994).

[26] S. Hanessian, P. Meffre, M. Girard, S. Beaudoin, J.-Y. Sanceau, and Y. Bennani, *JOC* **58**, 1991 (1993).

[27] T. Oishi, K. Iida, and M. Hirama, *TL* **34**, 3573 (1993).

[28] M. Nakajima, K. Tomioka, Y. Iitaka, and K. Koga, *T* **49**, 10793 (1993).

[29] K. Inoguchi, N. Fujie, K. Yoshikawa, and K. Achiwa, *CPB* **40**, 2921 (1992).

[30] T. Morimoto, M. Chiba, and K. Achiwa, *CPB* **41**, 1149 (1993).

[30a] I. Ojima, C.-Y. Tsai, and Z. Zhong, *TL* **35**, 5785 (1994).

[30b] S. El Baba, K. Sartor, J.-C. Poulin, and H. B. Kagan, *BSCF* **131**, 525 (1994).

[30c] T. Morimoto, M. Chiba, and K. Achiwa, *CPB* **40**, 2894 (1992).

[30d] S. Taudien, K. Schinkowski, and H.-W. Krause, *TA* **4**, 1983, 73 (1993).

[30e] A. Togni, C. Breutel, A. Schnyder, F. Spindler, H. Landert, and A. Tijani, *JACS* **116**, 4062 (1994).

[31] C. Dauelsberg and J. Martens, *SC* **23**, 2091 (1993).

[32] T. Mehler and J. Martens, *TA* **4**, 1983, 2299 (1993).

[33] Y. H. Kim, D. H. Park, and I. S. Byun, *JOC* **58**, 4511 (1993).

[34] M. Periasamy, J. V. B. Kanth, and A. S. B. Prasad, *T* **50**, 6411 (1994).

[35] Y. Hong, Y. Gao, X. Nie, and C. M. Zepp, *TL* **35**, 6631 (1994).

[36] G. J. Quallich and T. M. Woodall, *SL* 929 (1993).

[37] O. Fröhlich, M. Bonin, J.-C. Quiron, and H.-P. Husson, *TA* **4**, 2333 (1993).

[38] C. Bolm and M. Felder, *TL* **34**, 6041 (1993).

[39] C. Bolm and M. Felder, *SL* 655 (1994).

[40] C. Bolm, A. Seeger, and M. Felder, *TL* **34**, 8079 (1993).

[41] D. J. Mathre, A. S. Thompson, A. W. Douglas, K. Hoogsteen, J. D. Carroll, E. G. Corley, and E. J. J. Grabowski, *JOC* **58**, 2880 (1993).

[42] T. Gajda, *TA* **5**, 1965 (1994).

[43] G. J. Quallich and T. M. Woodall, *TL* **34**, 4145 (1993).

[44] M. Nakagawa, T. Kawate, T. Kakikawa, H. Yamada, T. Matsui, and T. Hino, *T* **49**, 1739 (1993).

[45] G. A. Molander and K. L. Bobbitt, *JACS* **115**, 7517 (1993).

[46] M. Nakagawa, T. Kawate, T. Kakikawa, H. Yamada, T. Matsui, and T. Hino, *T* **49**, 1739 (1993).

[47] E. F. J. de Vries, J. Brusse, C. G. Kruse, and A. van der Gen, *TA* **5**, 377 (1994).

[48] T.-J. Lu and S.-W. Liu, *JCCS(T)* **41**, 467 (1994).

[49] H. Nishiyama, S. Yamaguchi, S.-B. Park, and K. Itoh, *TA* **4**, 143 (1993).

[50] J.-i. Sakaki, W. B. Schweizer, and D. Seebach, *HCA* **76**, 2654 (1993).

[51] H. Nishibayashi, J. D. Singh, K. Segawa, S.-i. Fukuzawa, and S. Uemura, *CC* 1375 (1994).

[52] M. Hayashi, Y. Miyamoto, T. Inoue, and N. Oguni, *JOC* **58**, 1515 (1993).

[53] E. J. Corey and Z. Wang, *TL* **34**, 4001 (1993).

[54] T. Hayashi, H. Iwamura, M. Naito, Y. Matsumoto, Y. Uozumi, M. Miki, and K. Yanagi, *JACS* **116**, 775 (1994).

[55] T. Hayashi, H. Iwamura, and Y. Uozumi, *TL* **35**, 4813 (1994).

[56] M. Watanabe and K. Soai, *JCS(P1)* 3125 (1994).

[57] P. Delair, C. Einhorn, J. Einhorn, and J. L. Luche, *T* **51**, 165 (1995).

[58] K. Soai, C. Shimada, M. Takeuchi, and M. Itabashi, *CC* 567 (1994).

[59] K. Soai, T. Hayase, K. Takai, and T. Sugiyama, *JOC* **59**, 7908 (1994).

[60] S. Walbaum and J. Martens, *TA* **4**, 637 (1993).

[61] W. Behnen, T. Mehler, and J. Martens, *TA* **4**, 1413 (1993).

[62] G. B. Jones and S. B. Heaton, *TA* **4**, 261 (1993).

[63] M. Watanabe and K. Soai, *JCS(P1)* 837 (1994).

[64] J. M. Andres, M. A. Martinez, R. Pedrosa, and A. Perez-Encabo, *TA* **5**, 67 (1994).

[65] J. V. Allen, C. G. Frost, and J. M. J. Williams, *TA* **4**, 649 (1993).

[66] T. Mehler, J. Martens, and S. Wallbaum, *SC* **23**, 2691 (1993).

[67] G. Nicolosi, A. Patti, R. Morrone, and M. Piattelli, *TA* **5**, 1639 (1994).

[68] M. Watanabe, M. Komota, M. Nishimura, S. Araki, and Y. Butsugan, *JCS(P1)* 2193 (1993).

[69] H. Wally, M. Widhalm, W. Weissensteiner, and K. Schlogl, *TA* **4**, 285 (1993).

[70] M. Uemura, R. Miyake, K. Nakayama, M. Shiro, and Y. Hayashi, *JOC* **58**, 1238 (1993).

[71] M. Ishizaki, K. Fujita, M. Shimamoto, and O. Hoshino, *TA* **5**, 411 (1994).

[72] J. Kang, J. W. Lee, and J. I. Kim, *CC* 2009 (1994).

[73] C. Eisenberg and P. Knochel, *JOC* **59**, 3760 (1994).

[74] L. Schwink and P. Knochel, *TL* **35**, 9007 (1994).

[75] R. Ostwald, P.-Y. Chavant, H. Stadtmuller, and P. Knochel, *JOC* **59**, 4143 (1994).

[76] S. Vettel and P. Knochel, *TL* **35**, 5849 (1994).

[77] K. Soai and K. Takahashi, *JCS(P1)* 1257 (1994).

[78] M. Asami, K. Usui, S. Higuchi, and S. Inoue, *CL* 297 (1994).

[79] K. Soai, Y. Hirose, and Y. Ohno, *TA* **4**, 1473 (1993).

[80] K. Soai, T. Suzuki, and T. Shono, *CC* 317 (1994).

[81] K. Soai, A. Oshio, and T. Saito, *CC* 811 (1993).

[82] B. Weber and D. Seebach, *T* **50**, 7473 (1994).

[83] J. P. Morken, M. T. Didiuk, and A. H. Hoveyda, *JACS* **115**, 6997 (1993).

[84] A. F. Indolese and G. Consiglio, *OM* **13**, 2230 (1994).

[85] J. Sprinz and G. Helmchen, *TL* **34**, 1769 (1993).

[86] P. V. Matt, O. Loiseleur, G. Koch, A. Pfaltz, C. Lefeber, T. Feucht, and G. Helmchen, *TA* **5**, 573 (1994).

[87] M. Nakajima, K. Tomioka, and K. Koga, *T* **49**, 9751 (1993).

[88] K. Ishihara, M. Mouri, Q. Gao, T. Maruyama, K. Furuta, and H. Yamamoto, *JACS* **115**, 11490 (1993).

[89] W. R. Roush and N. S. Van Nieuwenhze, *JACS* **116**, 8536 (1994).

[90] B. M. Trost and R. C. Bunt, *JACS* **116**, 4089 (1994).

[91] B. M. Trost and M. G. Organ, *JACS* **116**, 10320 (1994).

[92] K. Ishihara, T. Maruyama, M. Mouri, Q. Gao, K. Furuta, and H. Yamamoto, *BCSJ* **66**, 3483 (1993).

[93] M. Sato, S. Sunami, Y. Sugita, and C. Kaneko, *CPB* **42**, 839 (1994).

[94] Y. Kaneko, T. Matsuo, and S. Kiyooka, *TL* **35**, 4107 (1994).

[95] V. A. Soloshonok and T. Hayashi, *TA* **5**, 1091 (1994).

[96] M. Hayashi, T. Inoue, and N. Oguni, *CC* 341 (1994).

[97] R. K. Boeckman, Jr., A. T. Johnson, and R. A. Musselman, *TL* **35**, 8521 (1994).

[98] S. Kobayashi and M. Horibe, *ACS* **48**, 9805 (1994).

[99] T. Yokomatsu, T. Yamagishi, and S. Shibuya, *TA* **4**, 1779 (1993).

[100] Q.-L. Zhou and A. Pfaltz, *TL* **34**, 7725 (1993).

[101] M. Kanai and K. Tomioka, *TL* **35**, 895 (1994).

[102] M. Spescha and G. Rihs, *HCA* **76**, 1219 (1993).

Chlorine. 17, 198–199

gem-Dichloroalkanes.[1]   Oximes are transformed into *gem*-dichlorides by chlorine with $AlCl_3$ as catalyst.

Oxidative displacement of iodoalkanes.[2]   The combination of chlorine and an alcohol converts an iodo compound into an ether.

[1] M. Tordeux, K. Boumizane, and C. Wakselman, *JOC* **58**, 1939 (1993).
[2] C. Bayle and A. Gadelle, *TL* **35**, 2335 (1994).

2-(Chloroacetoxymethyl)benzoyl chloride.

Protection of alcohols.[1]   This reagent is suitable for acylation of the hydroxyl group at C-2 of glycosyl donors. Esters are formed by conventional manipulation and their clearage is triggered by phthalide formation

[1] T. Ziegler and G. Pantkowski, *LA* 659 (1994).

1-Chloro-1-alkylsilacyclobutanes.

Enoxysilacyclobutanes.   These compounds can be prepared[1] by Wurtz coupling of 3-chloropropyltrichlorosilane with Mg in ether. Introduction of one alkyl group is accomplished by reaction with an organolithium reagent, and the silyl chloride can then be used for the formation of silyl enol ethers.[2] Such *O*-silyl ketene acetals are extremely reactive in aldol condensations with aldehydes *without* catalysts. The reaction is *syn*-selective. An asymmetric version[3] uses silyl ketene acetals bearing a chiral *Si*-alkoxy (e.g., 8-phenylmenthoxy) group instead of an alkyl substituent.

45 - 68% (90 - 94% ee)

The effects of the silacyclobutane moiety on the derived silyl enol ethers in aldol reactions were discovered earlier.[4] It was also suggggested that these reactions proceed via pentacoordinate silicon species and boatlike transition states.

[1] S. E. Denmark, B. D. Griedel, D. M. Coe, and M. E. Schnute, *JACS* **116**, 7026 (1994).
[2] S. E. Denmark, B. D. Griedel, and D. M. Coe, *JOC* **58**, 988 (1993).
[3] S. E. Denmark and B. D. Griedel, *JOC* **59**, 5136 (1994).
[4] A. G. Myers, S. E. Kephart, and H. Chen, *JACS* **114**, 7922 (1992).

Chloroborane.

Hydroboration.[1]   The advantage of this borane (as dimethyl sulfide complex) in hydroboration is that the adducts can be oxidized directly to carboxylic acids in excellent yields. Thus a synthesis of γ-valerolactone from 4-(*p*-nitrobenzoyloxy)-1-pentene can be easily achieved. This particular ester is used in this case because it is reduced very slowly (in comparison with the acetate).

$$\text{OCOR} \xrightarrow[\text{CrO}_3, \text{AcOH/H}_2\text{O, rt, 1 h}]{\text{BH}_2\text{Cl} \cdot \text{SMe}_2, \text{CH}_2\text{Cl}_2, \text{rt, 2 h;}} \text{OCOR} \quad \text{COOH}$$

84% (R=p-O$_2$NC$_6$H$_4$)

[1] H. C. Brown, S. V. Kulkarni, and U. S. Racherla, *JOC* **59**, 365 (1994).

2-Chloro-2,3-dihydro-1,3,4,2-oxadiazaphosphole.

Amides.[1] This is a new condensating agent for the union of carboxylic acids and amines at 0°C. Pyridine can be used as HCl scavenger.

[1] H. Kimura, H. Konno, and N. Takahashi, *BCSJ* **66**, 327 (1993).

Chlorodiphenylphosphine.

Aziridine derivatives.[1] The selective reaction of 2-amino alcohols with Ph$_2$PCl, followed by a subsequent O-tosylation and a NaH induced cyclization allows for the preparation of N-diphenylphosphinyl aziridines.

α-Alkoxy allylphosphine oxides.[2] Allylic acetals react with Ph$_2$PCl by replacing one of the alkoxy groups with the diphenylphosphinoyl residue. The phosphine oxide derived from acrolein diethyl acetal has been transformed into (E)-enol ethers of protected 4-amino aldehydes.

[1] H. M. I. Osborn, J. B. Sweeney, and B. Howson, *SL* 145 (1994).
[2] S. K. Armstrong, E. W. Collington, and S. Warren, *JCS(PI)* 515 (1994).

α-Chloroethyl chloroformate.

De-N-benzylation.[1] Tertiary benzylamines are transformed into the O-(α-chloroethyl)carbamates in refluxing 1,2-dichloroethane. These carbamates are readily hydrolyzed by heating with methanol, giving the secondary amine hydrochlorides.

[1] B. V. Yang, D. O'Rourke, and J. Li, *SL* 195 (1993).

Chloroiodomethane.

Simmons–Smith reaction.[1] ClCH$_2$I delivers its methylene group to allylic ethers in the presence of Et$_2$Zn.

Homologation reactions. ClCH$_2$I undergoes I/Li exchange with an alkyllithium reagent and the chloromethyllithium thus obtained can be used to convert esters into chloromethyl ketones,[2] Fischer carbene complexes into methyl ketones,[3] and alkenylboronates into allylboronates.[4]

70% (E : Z   98 : 2)

[1] A. B. Charette and J.-F. Marcoux, *TL* **34**, 7157 (1993).
[2] J. Barluenga, B. Baragana, A. Alonso, and J. M. Concellon, *CC* 969 (1994).
[3] J. Barluenga, P. L. Bernard, and J. M. Concellon, *TL* **35**, 9471 (1994).
[4] H. C. Brown, A. S. Phadke, and N. G. Bhat, *TL* **34**, 7845 (1993).

Chloromethyl chlorosulfate.

Chloromethyl esters.[1]   The reagent $ClCH_2OSO_2Cl$ is useful to derivatize N-protected amino acids under phase-transfer conditions.

[1] N. Harada, M. Hongu, T. Tanaka, T. Kawaguchi, T. Hashiyama, and K. Tsujihara, *SC* **24**, 767 (1994).

Chloromethyl methyl ether. 13, 76; 14, 5–6

A convenient method for its synthesis[1] is by heating an acid chloride with methylal. The reagent is commonly employed for alcohol protection.

[1] R. J. Linderman, M. Jaber, and B. D. Griedel, *JOC* **59**, 6499 (1994).

1-Chloromethyl-4-fluoro-1,4-diazabicyclo[2.2.2]octane bis(tetrafluoroborate).

Fluorination.[1]   This salt is an effective fluorinating agent for enol acetates, silyl enol ethers, phenyl-substituted alkenes, mildly activated arenes, sulfides bearing α-H atoms, and certain carbanions.

The introduction of fluorine into β-dicarbonyl compounds,[2] as well as the replacement of vinylic tributylstannyl groups,[3] have also been reported.

Oxidation.[4]   Benzylic alcohols and benzaldehydes are oxidized. To transform the alcohols into acids, two equivalents of the reagent and longer reaction times are required.

[1] G. S. Lal, *JOC* **58**, 2791 (1993).
[2] R. E. Banks, N. J. Lawrence, and A. L. Popplewell, *CC* 343 (1994).
[3] D. P. Matthews, S. C. Miller, E. T. Jarvi, J. S. Sabol, and J. R. McCarthy, *TL* **34**, 3057 (1993).
[4] R. E. Banks, N. J. Lawrence, and A. L. Popplewell, *SL* 831 (1994).

N-Chloromethylphthalimide.

Thiol protection.[1]   The phthalimidomethyl group, introduced to thiols under mild reaction conditions using this reagent, is removed by treatment with hydrazine hydrate followed by mercuric or cupric acetate.

[1] Y.-D. Gong and N. Iwasawa, *CL* 2139 (1994).

m-Chloroperoxybenzoic acid. **13**, 76–79; **14**, 84–87; **15**, 86; **16**, 80–83; **17**, 76

Oxidation of organochalcogenides.[1]   5-Ethylthiofuran-2(5*H*)-ones are oxidized at −50°C, to the sulfoxides or sulfones according to the stoichiometry of the peracid used. Tellurides are converted directly to alkenes[2] at room temperature. Triethylamine has a remarkable effect on the oxidation.

α-Chloro enones.[3]   A convenient method for the introduction of a chlorine atom to the α-position of enones is by treatment with HCl-MCPBA in DMF.

[1] F. Farina, M.V. Martin, R.M. Martin-Aranda, and A. Martinez de Guerenu, *SC* **23**, 459 (1993).
[2] Y. Nishibayashi, N. Komatsu, K. Ohe, and S. Uemura, *JCS(PI)* 1133 (1993).
[3] Y.M. Kim, K.H. Chung, J.N. Kim, and E.K. Ruy, *S* 283 (1993).

Chloro(pyridine)cobaloxime(III).

Free radical cyclizations.[1]   Cobaloxime in combination with a sacrificial Zn foil anode has been used to effect the reductive generation of carbon-centered radicals from bromoacetals in electrochemical processes. When an unsaturated side chain is present in such acetals, cyclization may occur.

[1] T. Inokuchi, H. Kawafuchi, K. Aoki, A. Yoshida, and S. Torii, *BCSJ* **67**, 595 (1994).

N-Chlorosuccinimide. **13**, 79–80; **15**, 86–88

Dehydrogenation of sulfenamides.[1]   The expedient transformation to sulfenimines is applicable to the synthesis of an electrophilic glycine synthon appropriate for the elaboration of other amino acids.

Chlorodehydrogenation of α-phenylthio amides.[2]   (Z)-3-Chloro-2-phenylthioacrylamides are formed through a reaction sequence involving *S*-chlorination, dehydrochlorination, and chlorination.

57–80%

Degradative chlorination.   α-Sulfonyl esters undergo α,α-dichlorination, but in the presence of NaOMe (in methanol) loss of the ester group seems inevitable.[3] Chlorination of *N*-tosyl-*N*-alkylhydrazines under photochemical conditions results in the formation of chloroalkanes.[4]

86%

Activation of Se-arsanyl selenoesters.[5] Chlorination renders arsanyl selenoesters electrophilic, enabling them to react with silyl enol ethers to form *Se*-(β-oxoalkyl) selenoesters.

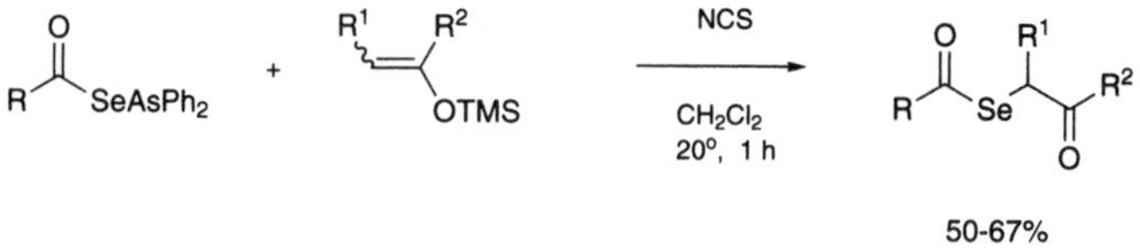

50-67%

Cyclic ethers from lactols.[6] Aldol condensation of lactols with enol acetates is promoted by a combination of NCS and $SnCl_2$.

[1] R. G. Lovey and A. B. Cooper, *SL* 167 (1994).
[2] A. R. Maguire, M. E. Murphy, M. Schaeffer, and G. Ferguson, *TL* **36**, 467 (1995).
[3] R. F. Langler and J. L. Steeves, *AJC* **47**, 1641 (1994).
[4] L. R. Collazo, F. S. Guziec, W.-X. Hu, and R. Pankayatselvan, *TL* **35**, 7911 (1994).
[5] T. Kanda, K. Mizoguchi, T. Koike, T. Murai, and S. Kato, *CC* 1631 (1993).
[6] Y. Masuyama, Y. Kobayashi, and Y. Kurusu, *CC* 1123 (1994).

Chlorosulfonyl isocyanate. 13, 80−81

Nitriles.[1] A preparation of nitriles from carboxylic acids involves reaction with $ClSO_2N=C=O$ and then with Et_3N. A similar procedure is also useful for the introduction of cyano groups into certain heterocycles.

2-Iminopyrrolidines.[2] *N*-Substituted pyrrolidinones undergo cycloaddition with $ClSO_2N=C=O$, and the products decompose on heating with water to afford the amidines.

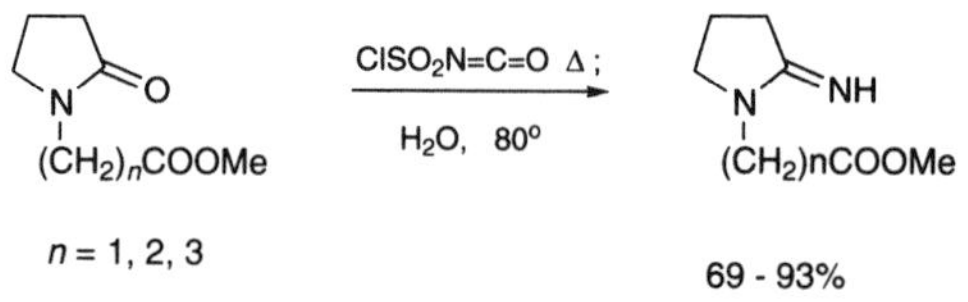

$n = 1, 2, 3$

69 - 93%

[1] H. Vorbrüggen and K. Krolikiewicz, *T* **50**, 6549 (1994).
[2] P. M. Burden and R. D. Allan, *SC* **23**, 1195 (1993).

Chromium(II) acetate.

Cyclization of ω-bromoalkenes.[1] The Cr(II)-promoted radical formation is definitely better than that initiated by Bu_3SnH. The tin compounds are quite toxic and much more difficult to separate from the products.

Denitroacetoxylation.[2] Vicinal elimination of the two functional groups with $Cr(OAc)_2$ in the presence of 2,2'-dipyridyl has the implication that a ketone can be extended by a functionalized isopropene unit. The substrates are prepared by a Nef reaction and dehydration, followed by reaction with formaldehyde and acetylation.

[1] C. Hackmann and H.-J. Schäfer, *T* **49**, 4559 (1993).
[2] B. Barlaam, J. Boivin, and S. Z. Zard, *TL* **34**, 1023 (1993).

Chromium–carbene complexes. 13, 82–83; 14, 91–93; 15, 93–95; 16, 88–92; 17, 80–84

1,4-Dicarbonyl compounds. 1-Oxidoalkylidenechromium(0) complexes react as acylanions toward Michael acceptors. The reaction is induced by light[1] or by a copper complex.[2]

Cyclization and cycloaddition. Aminocarbene complexes behave like aminoketenes under photochemical conditions. With the participation of a benzene ring, cyclization leads to aminonaphthols,[3] whereas the amino ketenes can also be trapped with imines to form β-lactams.[4] Intramolecular capture of analogous ketenes by an aldehyde leads to β-lactones.[5] The intermolecular process gives much inferior results.

α-Amino acids.[6] Alkylation of the carbanions derived from aminocarbenes followed by photocarbonylation generates protected amino acids. When the amino group is chiral, its stereocenter(s) can exert great steric influences on the formation of two new C–C bonds.

Michael acceptors. Some unsaturated carbene complexes of Cr(CO)₄ exhibit acceptor characteristics. Interestingly, they undergo Michael additions[7] with very high diastereoselectivities.

81%

(*syn : anti* > 99.5 : 0.5)

[1] B. C. Soderberg, D. C. York, T. R. Hoyle, G. M. Rehberg, and J. A. Suriano, *OM* **13**, 4501 (1994).
[2] H. Sakurai and K. Nakasaka, *CL* 2017 (1994).
[3] C. A. Merlic, D. Xu, and B. G. Gladstone, *JOC* **58**, 538 (1993).
[4] P.-J. Colson and L. S. Hegedus, *JOC* **58**, 5918 (1993).
[5] P.-J. Colson and L. S. Hegedus, *JOC* **59**, 4972 (1994).

[6] C. Schmeck and L. S. Hegedus, *JACS* **116**, 9927 (1994).
[7] Y. Shi and W. D. Wulff, *JOC* **59**, 5122 (1994).

Chromium(II) chloride. 13, 84; 14, 94–97; 15, 95–96; 16, 93–94; 17, 84–85

Cyclopropanols.[1] CrCl$_2$ effects a unique transformation of α,β-unsaturated aldehydes. In the product the hydroxyl group is *trans* to the original α-substituent.

63%

Homoallylic alcohols and amines. Carbonyl compounds and allylic bromides react in the presence of Ph$_2$Cr, which is prepared by adding PhMgBr to a CrCl$_2$ suspension in THF containing TMEDA. For enones 1,2-addition is observed.[2] The synthesis of homoallylic amines[3] from imines and allylic bromides requires only CrCl$_2$ and BF$_3$ · OEt$_2$.

Silyl- and stannylalkenes. The addition of α-silyl-[4] and α-stannylalkyl-chromium[5] reagents to carbonyl compounds leads to the alkenes directly. The reagents are formed in situ from the halides and CrCl$_2$.

84%

β-Arylation of electron-deficient alkenes. Aryl radicals are generated from aryl halides by reaction with the CrCl$_2$–ethylenediamine complex,[6] and the mode of addition is similar to that of arylcopper reagents.

(E)-Alkenes.[7] Condensation of α-acetoxy bromides with aldehydes by CrCl$_3$-Zn leads to (E)-alkenes. In the synthetic sense this method is complementary to the Wittig reaction.

[1] D. Montgomery, K. Reynolds, and P. Stevenson, *CC* 363 (1993).
[2] P. Wipf and S. Lim, *CC* 1654 (1993).
[3] M. Giammaruco, M. Taddei, and P. Ulivi, *TL* **34**, 3635 (1993).
[4] D. M. Hodgson and P. J. Comina, *TL* **35**, 9469 (1994).
[5] D. M. Hodgson, L. T. Boulton, and G. N. Maw, *TL* **35**, 2231 (1994).
[6] H. I. Tashtoush and R. Sustmann, *CB* **126**, 1759 (1993).
[7] M. Knecht and W. Boland, *SL* 837 (1993).

Chromium(II) chloride–nickel(II) chloride. **14**, 97–98; **15**, 96–97; **17**, 86

Cyclization of o-iodoarylalkynes and -alkenes.[1]   Ring formation involving arylchromium species is catalyzed by Ni(II).

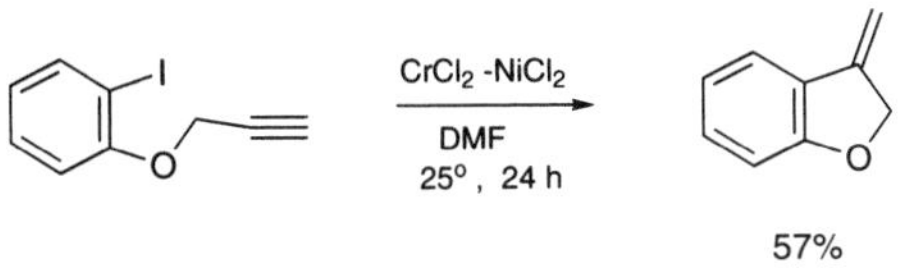

57%

[1] D. M. Hodgson and C. Wells, *TL* **35**, 1601 (1994).

Chromium hexacarbonyl.

(Z)-3-Alkenoic esters.[1]   Selective 1,4-hydrogenation of 2,4-alkadienoic esters to give the deconjugated (Z)-alkenoic esters is readily accomplished using $Cr(CO)_6$ as catalyst. However, high temperatures and pressures are required. For purification [to obtain 99% (Z)-isomers] the products are hydrolyzed in the presence of porcine pancreatic lipase.

2-Vinyl-3-alkenols.[2] η^5-Pentadienylchromium complexes can be generated from pentadienyl acetates with $(OC)_3Cr(PhCOOMe)$, which is obtained from $Cr(CO)_6$ and methyl benzoate. The complexes condense with aldehydes in the presence of a mild base to give the dienols.

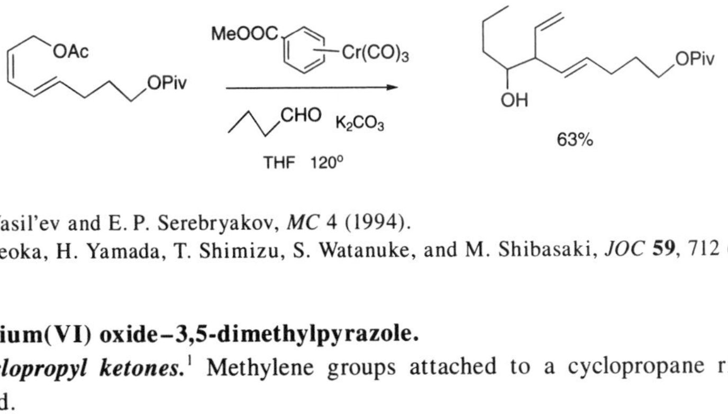

63%

[1] A. A. Vasil'ev and E. P. Serebryakov, *MC* 4 (1994).
[2] M. Sodeoka, H. Yamada, T. Shimizu, S. Watanuke, and M. Shibasaki, *JOC* **59**, 712 (1994).

Chromium(VI) oxide–3,5-dimethylpyrazole.

Cyclopropyl ketones.[1]   Methylene groups attached to a cyclopropane ring are oxidized.

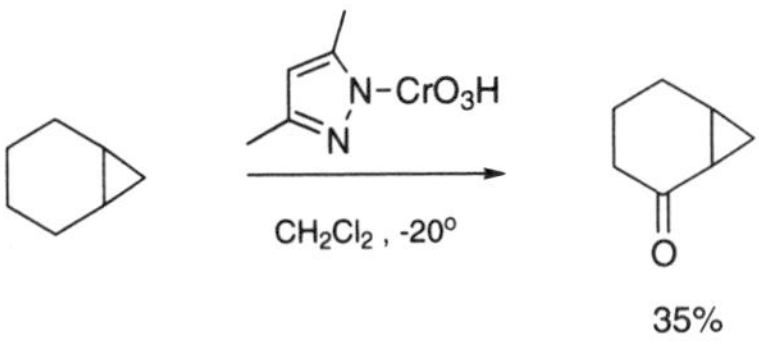

35%

[1] M. G. Banwell, N. Haddad, J. A. Huglin, M. F. MacKay, M. E. Reum, J. H. Ryan, and K. A. Turner, *CC* 954 (1993).

Chromyl diacetate.

Epoxidations.[1] Steroidal epoxides are readily prepared at low temperatures with $CrO_2(OAc)_2$.

[1] L. R. Galagovsky and E. G. Gros, *JCR(S)* 137 (1993).

Chromyl chloride.

Obtained in 70% yield from CrO_3 and $TiCl_4$ in CH_2Cl_2 with ultrasound irradiation[1] for 2.5 h. It can be used in the Etard oxidation or converted into $CrO_2(OBu')_2$ and the cyclic chromate of 2,5-dimethylpentane-2,5-diol for other oxidations.

[1] F. A. Luzzio and W. J. Moore, *JOC* **58**, 512 (1993).

Cob(I)alamin.

Aziridine cleavage.[1] Cob(I)alamin is prepared in situ by Zn reduction of hydroxocob(III)alamin · HCl in the presence of NH_4Cl. Fused *meso*-aziridines are cleaved to chiral allylic amines.

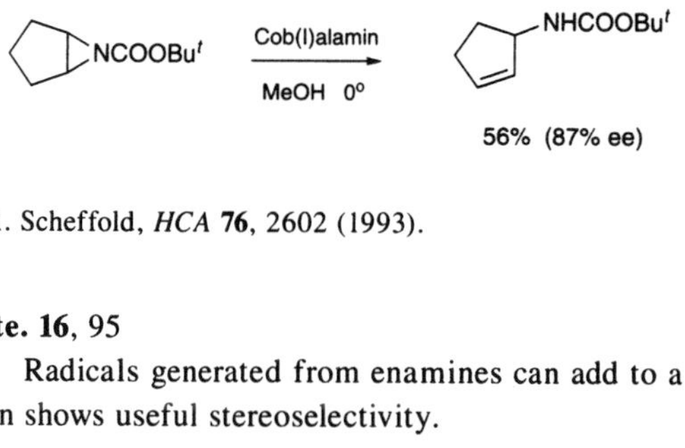

56% (87% ee)

[1] Z.-d. Zhang and R. Scheffold, *HCA* **76**, 2602 (1993).

Cobalt(II) acetate. 16, 95

Cyclization.[1] Radicals generated from enamines can add to a proximal double bond. The reaction shows useful stereoselectivity.

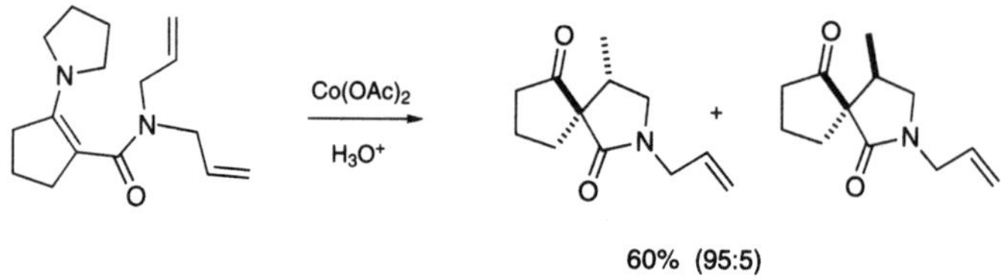

60% (95:5)

Oxidation of alcohols.[2] Aerobic oxidation of alcohols requires $Co(OAc)_2$ as well as $RuCl_3$ and an aldehyde for cooxidation. Yields of acids or ketones are excellent (9 examples, 78–98% yield).

[1] J. Cossy and A. Bouzide, *CC* 1218 (1993).
[2] S.-I. Murahashi, T. Naota, and N. Hirai, *JOC* **58**, 7318 (1993).

Cobalt(II) acetylacetonate. 17, 87

[4π+2π+2π]Cycloadditions.[1] Such cycloadditions can be rendered enantiose-lective by adding a chiral phosphine ligand.

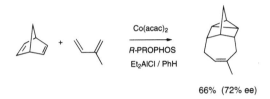

66% (72% ee)

γ-Lactones from tetrahydrofurans.[2] Oxidation with molecular oxygen is ac-complished in the presence of an α-diketone.

Epoxidation.[3] Oxygen is used as oxidant, and propanal diethyl acetal acts as a cooxidant in the presence of 4A-molecular sieves. Silyl ketene acetals give α-siloxy esters as a result of rearrangement.

[1] M. Lautens, W. Tam, and C. Sood, *JOC* **58**, 4513 (1993).
[2] E. Hata, T. Takai, and T. Mukaiyama, *CL* 1513 (1993).
[3] K. Yorozu, T. Takai, T. Yamada, and T. Mukaiyama, *CL* 1579 (1993).

Cobalt(II) bromide–silica.

Dithioacetization.[1] The reaction is usually complete at room temperature within a few minutes.

In situ generation of $Co_2(CO)_8$.[2] Most uses of dicobalt octacarbonyl in organic synthesis are related to its complexation to triple bonds. The complex may be formed by reduction of $CoBr_2$ with Zn dust under a CO atmosphere in the presence of an acetylene. Pauson–Khand reactions can be carried out subsequently.

[1] H. K. Patney, *TL* **35**, 5717 (1994).
[2] M. Periasamy, M. R. Reddy, and A. Devasagayaraj, *T* **50**, 6955 (1994).

Cobalt(II) chloride. 14, 99; 15, 97–98

Allylation of 1,3-dicarbonyl compounds.[1] A catalytic amount of $CoCl_2$ pro-motes the allylation of 1,3-dicarbonyl compounds with allyl acetates in high yields.

Unsymmetrical sulfides.[2] RSH + R′X → RSR′.

β-Acetamido ketones.[3] A three-component condensation involving a ketone, an aldehyde, and acetonitrile appears to involve aldol and Michael reactions and hydration of nitrilium species, which is quite unusual.

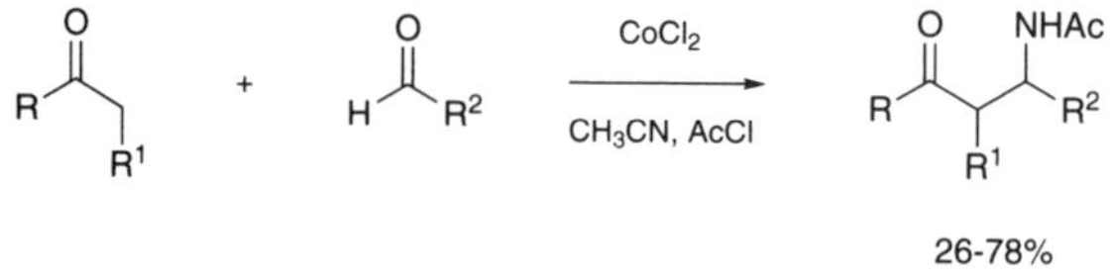

26–78%

[1] G. C. Maikap, M. M. Reddy, M. Mukhopadhyay, B. Bhatia, and J. Iqbal, *T* **50**, 9145 (1994).
[2] S. Chowdhury, P. M. Samuel, I. Das, and S. Roy, *CC* 1993 (1994).
[3] B. Bhatia, M. M. Reddy, and J. Iqbal, *CC* 713 (1994).

Cobalt(II) chloride–oxygen.

Oxidation of aldehydes.　In the presence of butanal, aromatic aldehydes undergo oxidative coupling at room temperature to give benzils (8 examples, 62–75%).[1] The same oxidation can be diverted to the carboxylic acid on addition of Ac_2O to the reaction medium.

A similar process for the oxidation of secondary alcohols is catalyzed by a cobalt–disalen complex.[2]

Oxidative condensation of aldehydes and alkenes.[3]

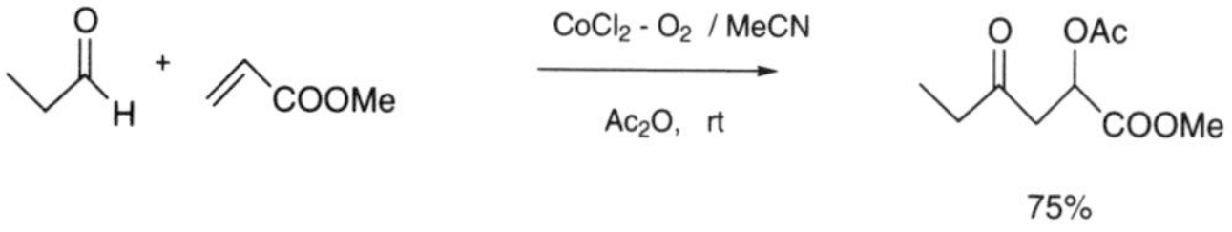

75%

Epoxidation and benzylic oxidation.[4]

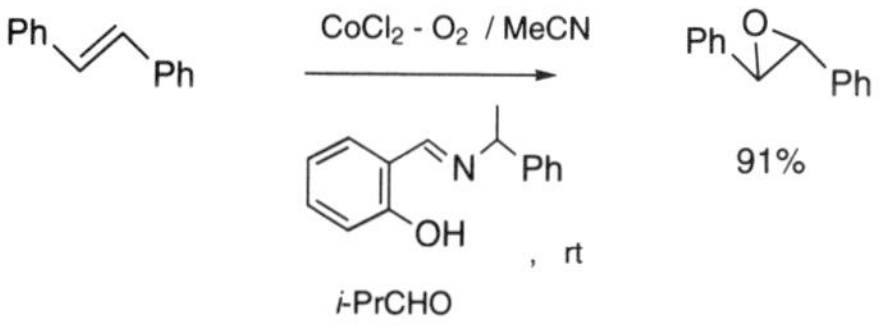

91%

Cleavage of 1,3-dioxolanes.[5]

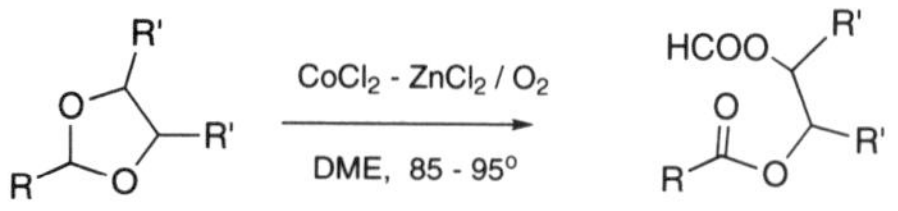

[1] T. Punniyamurthy, S. J. S. Kalra, and J. Iqbal, *TL* **35**, 2959 (1994).
[2] S. J. S. Kalra, T. Punniyamurthy, and J. Iqbal, *TL* **35**, 4847 (1994).
[3] S. Bhatia, T. Punniyamurthy, B. Bhatia, and J. Iqbal, *T* **49**, 6101 (1993).
[4] T. Punniyamurthy, B. Bhatia, and J. Iqbal, *TL* **34**, 4657, 4003 (1993).
[5] P. Li and H. Alper, *CJC* **71**, 84 (1993).

Copper. 15, 99; **16**, 95

o-Halophenylcopper reagents.[1] *o*-Haloiodobenzenes form remarkably stable copper reagents, which can be used at room temperature to introduce alkyl or acyl groups.

N-Phenylanthranilic acids.[2] Copper-catalyzed displacement of halogen in *o*-halobenzoic acids by aniline may be performed in refluxing water.

Ullmann coupling.[3] This classic reaction is still valuable for biphenyl synthesis. Unsymmetrical biphenyls are obtained by tethering the aryl iodides (e.g., by a salicyl alcohol unit), thus rendering the reaction intramolecular.

α-Amino esters and ketones.[4] α-Diazocarbonyl compounds react with tertiary amines in refluxing toluene under the influence of Cu powder to provide the amino ketones or esters in one step. A Stevens rearrangement intervenes in the formation of such products.

R = Ph   70%
R = OEt 63%

Allylation of active methylene compounds.[5] A mixture of Cu and CuClO$_4$ is used to promote *C*-allylation.

[1] G. W. Ebert, D. R. Pfennig, S. D. Suchan, and T. A. Donovan, *TL* **34**, 2279 (1993).
[2] R. F. Pellon, R. Carrasco, and L. Rodes, *SC* **23**, 1447 (1993).
[3] M. Takahashi, T. Oguku, K. Okamura, T. Da-te, H. Ohmizu, K. Kondo, and T. Iwasaki, *JCS(P1)* 1473 (1993).
[4] F. G. West, K. W. Glaeske, and B. N. Naidu, *S* 977 (1993).
[5] J. B. Baruah and A. G. Samuelson, *NJC* **18**, 961 (1994).

Copper(II) acetate.

Dehydrogenation of carboxamido enamines.[1] Conjugated imines are produced in the dehydrogenation, which can also be effected with manganese acetate.

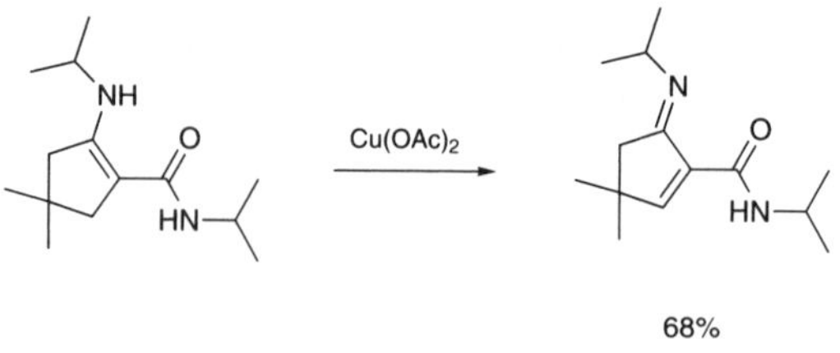

68%

N-Arylpiperidines.[2]   Substitution of arylbismuths by piperidines is catalyzed by Cu(OAc)$_2$.

[1] J. Cossy and A. Bouzide, *TL* **34**, 5583 (1993).
[2] A. Banfi, M. Bartoletti, E. Bellora, M. Bignotti, and M. Turconi, *S* 775 (1994).

Copper(II) acetylacetonate.

Heterocyclizations.   An *n*-donor atom at a proper distance to an α-diazomethyl ketone participates in metal-catalyzed decomposition of the functional group, giving rise to a heterocyclic ketone. For medium azacycle formation[1,2] the soluble Cu(acac)$_2$ complex is much superior to Rh$_2$(OAc)$_4$. Oxacycles are similarly accessible using the hexafluoroacetylacetonate.[3,4]

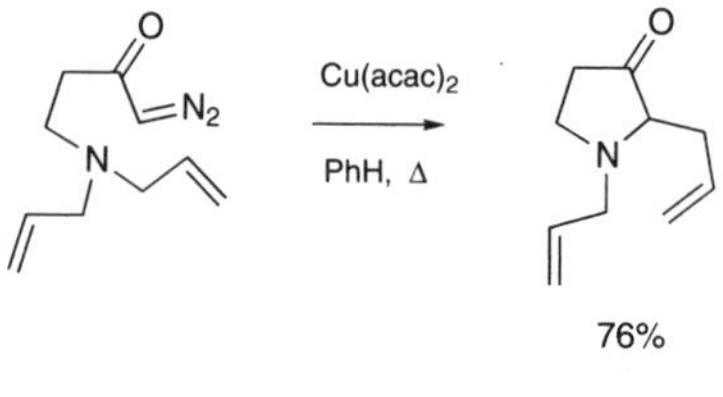

76%

[1] F. G. West, B. N. Naidu, and R. W. Tester, *JOC* **59**, 6892 (1994).
[2] J. S. Clark and P. B. Hodgson, *CC* 2701 (1994).
[3] J. S. Clark, S. A. Krowiak, and L. J. Street, *TL* **34**, 4385 (1993).
[4] J. S. Clark and G. A. Whitlock, *TL* **35**, 6381 (1994).

Copper(I) bora(trisdimethylpyrazole) complex.

Carbenoid and nitrenoid reactions.[1]   Decomposition of diazoacetic esters by the novel Cu complex (**1**) in the presence of alkenes and alkynes gives cyclopropanes and cyclopropenes, respectively. Similarly, a nitrenoid is generated from TsN=IPh, forming aziridines in similar reactions.

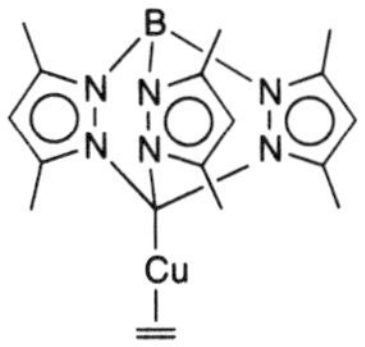

(1)

[1] P. J. Perez, M. Brookhart, and J. L. Templeton, *S* **12**, 261 (1993).

Copper(I) bromide. 14, 99–100; 16, 96

Photoaddition of organohalides to alkenes.[1] $CuBr \cdot PBu_3$ is an effective catalyst for the regioselective photoinduced addition to electron-deficient alkenes. On reduction with zinc dust the overall result is the same as the conjugate addition using an organocuprate. However, in many cases this method is more convenient.

Removal of thiophenoxide.[2]   The $CuBr \cdot SMe_2$ complex can be used as an alternative to CuOTf for this purpose.

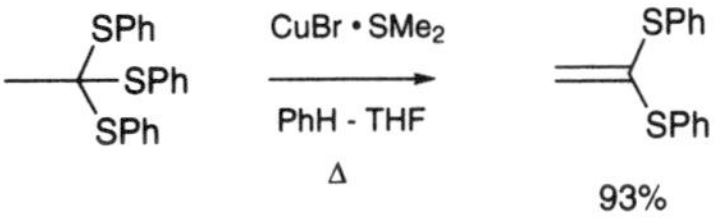

[1] M. Mitani and H. Hirayama, *JCR(S)* 249 (1993).
[2] T. Cohen, C. Shook, and M. Thiruvazhi, *TL* **35**, 6041 (1994).

Copper(II) bromide. 14, 100; 15, 100

Allylic thiocyanates and selenocyanates from silanes.[1]

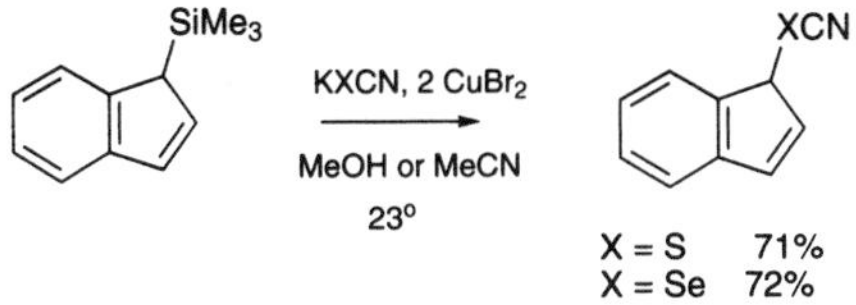

[1] A. S. Guram, *SL* 259 (1993).

Copper(II) carbonate–copper(II) hydroxide.

ArBr → ArOMe.[1] The copper catalyst system is effective for displacement of nonactivated bromide using NaOMe in methanol under CO_2 at 125°C.

[1] D. Nobel, *CC* 419 (1993).

Copper(I) chloride. 13, 85; 15, 101

Propargylamines. The S_N2 displacement of propargyl phosphates or acetates with amines is general for the preparation. (60–95% yield).[1] Another pathway involves oxidative coupling of *N,N*-dimethylanilines with 1-alkynes under oxygen.[2] Other products of this reaction are the *N*-methylanilines and *N*-methylformanilides.

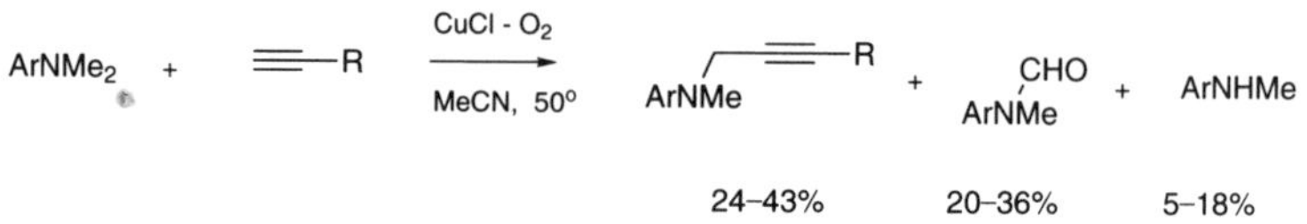

$$24\text{--}43\% \qquad 20\text{--}36\% \qquad 5\text{--}18\%$$

Amidines from amines and nitriles.[3] CuCl activates ordinary nitriles towards attack by amines except those with low nucleophilicity (e.g., $ArNH_2$).

Conjugate addition.[4] Under photoirradiation and CuCl catalysis enones are attacked by silylated nucleophiles (siloxyalkenes or alkylsilanes).

1,2-Dialkylidenecycloalkanes.[5] CuCl brings about rapid intramolecular coupling of vinylic stannanes and iodides in DMF.

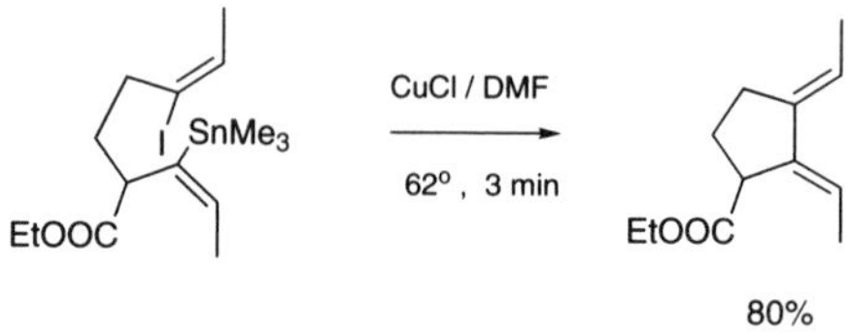

Functionalization of amines via the o-aminobenzamides.[6] Diazotization of the amide and CuCl-promoted decomposition of the diazonium salt in the presence of MeOH result in α-methoxylation of the amide.

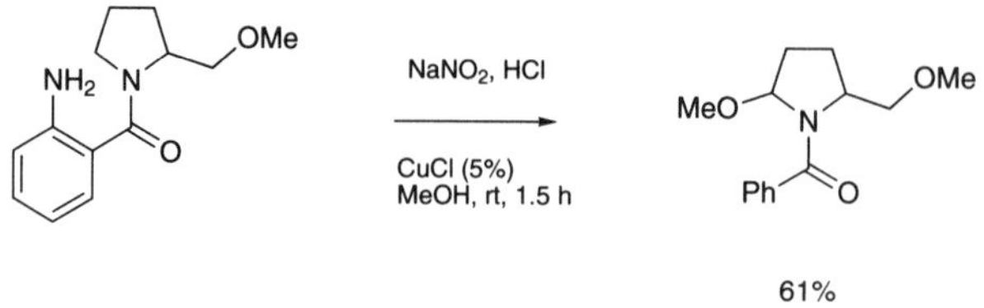

[1] Y. Imada, M. Yuasa, I. Nakamura, and S.-Z. Murahashi, *JOC* **59**, 2282 (1994).
[2] S. Murata, K. Teramoto, M. Miura, and M. Nomura, *JCR(S)* 434 (1993).
[3] G. Rousselet, P. Capdevielle, and M. Maumy, *TL* **34**, 6395 (1993).
[4] M. Mitani and Y. Osakabe, *CC* 1759 (1994).
[5] E. Piers and T. Wong, *JOC* **58**, 3609 (1993).
[6] G. Han, M.C. McIntosh, and S.M. Weinreb, *TL* **35**, 5813 (1994).

Copper(I) chloride–bipyridine.

Heterocyclization. Captodative radicals are generated from α-heterosubstituted α-chloroacetic esters by treatment with CuCl · bipy. Such a radical can be trapped intramolecularly by an alkene (e.g., a 3-butenyl group) attached to the α-heteroatom. Formation of five-membered ring heterocycles is a favorable process. Thus methyl 3-chloromethyltetrahydrofuran-2-carboxylate has been obtained in 75% yield as a mixture of *cis* and *trans* isomers (ratio 64:36).[1] The glycine analog gives both pyrrolidine and piperidine derivatives.[2]

ω-Alkenyl dichloroacetates and trichloroacetates also undergo cyclization to give chlorolactones.[3]

[1] J.H. Udding, K.J.M. Tujip, M.N.A. van Zanden, H. Hiemstra, and W.N. Speckamp, *JOC* **59**, 1993 (1994).
[2] J.H. Udding, K.J.M. Tujip, H. Hiemstra, and W.N. Speckamp, *T* **50**, 1907 (1994).
[3] F.O.H. Pirrung, H. Hiemstra, W.N. Speckamp, B. Kaptein, and H.E. Schoemaker, *T* **50**, 12415 (1994).

Copper(II) chloride. 14, 100

Hydrolysis of oximes and amides. While deoximation[1] is accomplished by heating an oxime or *O*-benzyloxime with $CuCl_2 \cdot 2H_2O$ in MeCN, hydrolysis of amides[2] is carried out in the presence of aqueous glyoxal.

A related reaction is the transformation of 5-aminopyrimidinones into imidazoles[3] in an alcoholic solvent.

59%

Aryl 1,1-dimethylpropargyl ethers.[4] An improved procedure for the etherification of phenols uses $CuCl_2 \cdot 2H_2O$ and DBU in MeCN at 0°C.

Friedel–Crafts benzylation. The catalyst is $CuCl_2$ supported on Al_2O_3.[5]

Chlorination. $CuCl_2$ is a mild chlorinating agent for sodiomalonate esters.[6] The Al_2O_3 supported reagent has also been used in α-chlorination of conjugated carbonyl compounds such as 1,4-naphthoquinones[7] and coumarins[8] in refluxing chlorobenzene.

[1] L. Singh and R. N. Ram, *SC* **23**, 3139 (1993).
[2] L. Singh and R. N. Ram, *JOC* **59**, 710 (1994).
[3] I. Matsuura, T. Ueda, N. Murakami, S.-i. Nagai, A. Nagatsu, and J. Sakakibara, *JCS(P1)* 965 (1993).
[4] J. D. Godfrey, R. H. Mueller, T. L. Sedergran, N. Soundarajan, and V. J. Colandrea, *TL* **35**, 6405 (1994).
[5] M. Kodomari, G. Shimada, and K. Mogi, *NKK* 1137 (1994).
[6] X.-X. Shi and L.-X. Dai, *JOC* **58**, 4596 (1993).
[7] P. K. Singh and R. N. Khanna, *SC* **23**, 2083 (1993).
[8] P. C. Thapliyal, P. K. Singh, and R. N. Khanna, *SC* **23**, 2821 (1993).

Copper(I) cyanide.

Cyanation of 1-alkynes.[1]

$$Ph-\!\!\!\equiv \quad \xrightarrow[\substack{Me_3SiCl \\ DMSO - MeCN - H_2O \\ 50°, \ 60 \ h}]{CuCN - NaI} \quad Ph-\!\!\!\equiv\!\!\!-CN \quad 72\%$$

S_N2' **reaction of allylic phosphates.**[2] The regioselective reaction with Grignard reagents provides a new access to alkenes.

o-Allylation of benzamides.[3] Directed lithiation of benzamides creates a nucleophilic site at an *o*-position, but the lithium species is unsuitable for the allylation. Thus transmetallation with CuCN · LiCl (at −78°C) prior to the reaction with allyl bromide is a necessary step.

[1] F.-T. Luo and R.-T. Wang, *TL* **34**, 5911 (1993).
[2] A. Yanagisawa, N. Nomura, and H. Yamamoto, *SL* 689 (1993).
[3] D. Pini, S. Superchi, and P. Salvadori, *JOMC* **452**, C4 (1993).

Copper(I) iodide. 16, 98

Alkylation of 1-alkynes. Chain extension of 1-alkynes by reaction with propargyl chlorides or tosylates,[1] as well as with aryl and vinyl iodides,[2] is catalyzed by CuI. In addition to the base K_2CO_3, sodium iodide and triphenylphosphine are added in the respective reactions. Carboxylation precedes alkylation with an alkyl halide when the reaction is conducted under CO_2, thus resulting in alkyl 2-alkynoates.[3]

$$Ph-\!\!\!\equiv \ + \ RBr \quad \xrightarrow[\substack{CO_2 \ / \ MeCONMe_2 \\ 100°}]{CuI - K_2CO_3} \quad Ph-\!\!\!\equiv\!\!\!-COOR$$

An addition–dehydration reaction sequence occurs[4] when an arylacetylene and a C,N-diaryl nitrone are mixed with CuI-dppe.

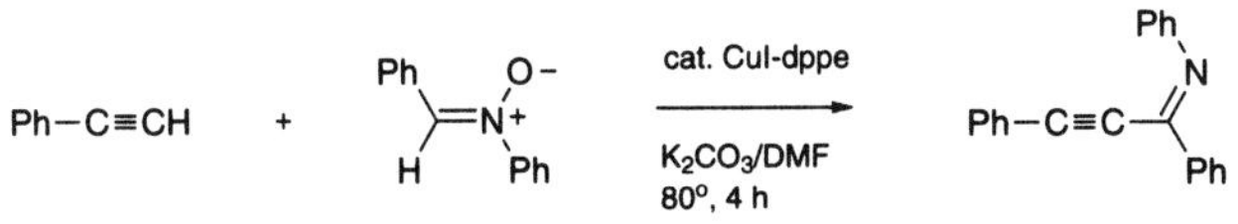

$$Ph-C{\equiv}CH \quad + \quad \underset{H}{\overset{Ph}{=}}N^{+}{\overset{O^{-}}{\underset{Ph}{}}} \quad \xrightarrow[\substack{K_2CO_3/DMF \\ 80^{\circ}, 4\ h}]{\text{cat. CuI-dppe}} \quad Ph-C{\equiv}C{\overset{Ph}{\underset{Ph}{N}}}$$

74%

Cyclopropanation of allyl α-diazoalkanoates.[5]

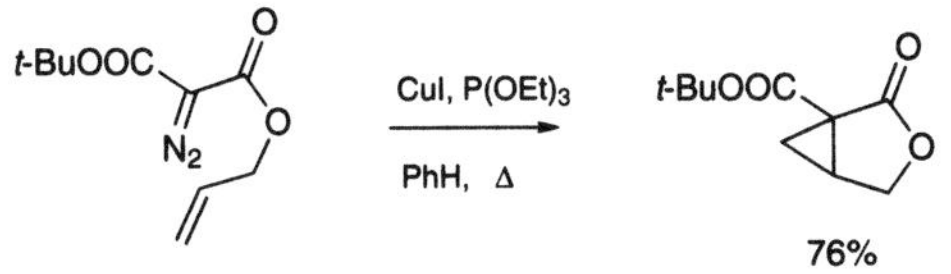

$$\text{t-BuOOC} \xrightarrow[\text{PhH, } \Delta]{\text{CuI, P(OEt)}_3} \text{t-BuOOC}$$

76%

RX → RCF₃. Organic halides are converted into trifluoromethyl compounds by catalyzed reactions with $ClCF_2COOMe$[6] or $FSO_2(CF_2)_2OCF_2COOMe$.[7]

Arylation of active methylene compounds.[8]

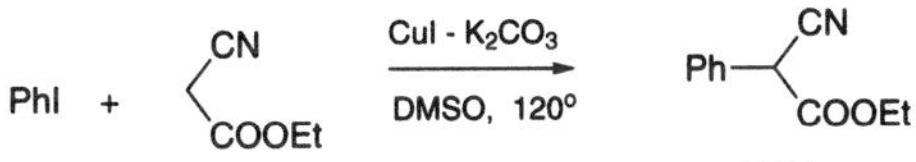

$$PhI \quad + \quad \underset{COOEt}{\overset{CN}{<}} \quad \xrightarrow[\text{DMSO, } 120^{\circ}]{CuI - K_2CO_3} \quad Ph{\overset{CN}{\underset{COOEt}{}}}$$

78%

[1] M. A. Lapitskaya, L. L. Vasiljeva, and K. K. Pivnitsky, *S* 65 (1993).
[2] K. Okuro, M. Furuune, M. Enna, M. Miura, and M. Nomura, *JOC* **58**, 4617 (1993).
[3] Y. Fakue, S. Oi, and Y. Inoue, *CC* 2091 (1994).
[4] K. Okuro, M. Enna, M. Miura, and M. Nomura, *CC* 1107 (1993).
[5] M. P. Koskinen and L. Munoz, *JOC* **58**, 879 (1993).
[6] Q.-Y. Chen and J.-X. Duan, *TL* **34**, 4241 (1993).
[7] Q.-Y. Chen and J.-X. Duan, *CC* 1389 (1993).
[8] K. Okuro, M. Furuune, M. Enna, M. Miura, and M. Nomura, *JOC* **58**, 7606 (1993).

Copper(II) nitrate.

Coupling of alkenylstannanes.[1]

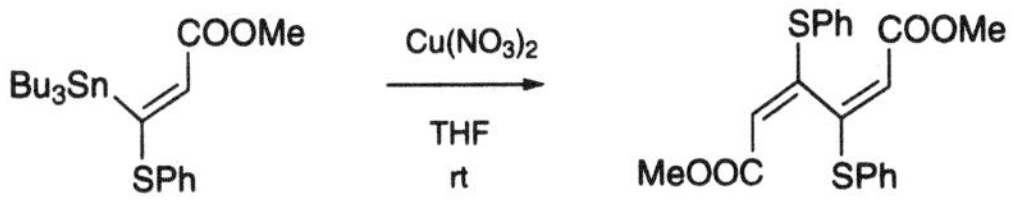

$$\underset{SPh}{\overset{COOMe}{Bu_3Sn}} \xrightarrow[\text{rt}]{Cu(NO_3)_2 \ \ \text{THF}} \underset{MeOOC \quad SPh}{\overset{SPh \quad COOMe}{}}$$

65%

Oxidative cleavage of cycloalkanones.[2]   The formation of acyclic keto acids from α-substituted cyclopentanones and cyclohexanones by $Cu(NO_3)_2$ -mediated autooxidation in aqueous acetic acid is quite efficient (60–96% yield).

[1] R. L. Beddoes, T. Cheeseright, J. Wang, and P. Quayle, *TL* **36**, 283 (1995).
[2] A. Atlamsani and J.-M. Bregeault, *S* 79 (1993).

Copper(II) tetrafluoroborate.

γ-Oxoalkylcoppers.[1]   Trialkylsiloxycyclopropanes undergo ring opening on treatment with $Cu(BF_4)_2$. The alkylcopper species thus produced can be captured with alkynyl sulfones or dimethyl acetylenedicarboxylate. In the absence of such electrophiles, demetallative dimerization results.

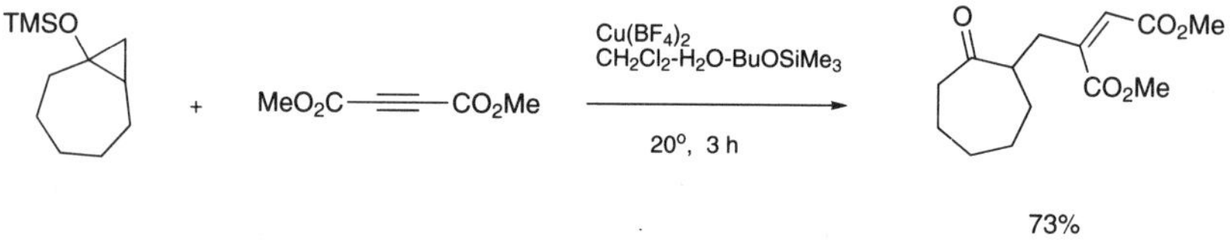

[1] I. Ryu, K. Matsumoto, Y. Kameyama, M. Ando, N. Kusumoto, A. Ogawa, N. Kambe, S. Murai, and N. Sonoda, *JACS* **115**, 12330 (1993).

Cyanoborane–amine.

Carbonyl reduction.[1]   The complexes of BH_2CN with propylamine or butylamine are very efficient and chemoselective reducing agents.

[1] M. K. Das, P. K. Maiti, and A. Bhaumik, *BCSJ* **66**, 810 (1993).

Cyanomethylenetributylphosphorane, $Bu_3P{=}CHCN$.

Esterification.[1]   This stabilized phosphorane effects the condensation of a carboxylic acid with an alcohol similarly to the Mitsunoba reagent.

[1] T. Tsunoda, F. Ozaki, and S. Ito, *TL* **35**, 5081 (1994).

Cyanomethyl formate, $HCOOCH_2CN$.

Formylation.[1]   Alcohols and amines are formylated in good yields at or slightly above room temperature. Imidazole is added during the formylation of alcohols.

[1] J. Deutsch and H.-J. Niclas, *SC* **23**, 1561 (1993).

Cyanotriethylsilane. 17, 89

α-Cyanohydrin triethylsilyl ethers.[1] A convenient preparation method uses
Bu₃SnCN as catalyst.

[1] M. Scholl and G. C. Fu, *JOC* **59**, 7178 (1994).

Cyanotrimethylsilane–titanium(IV) chloride.

α-Amino nitriles.[1]

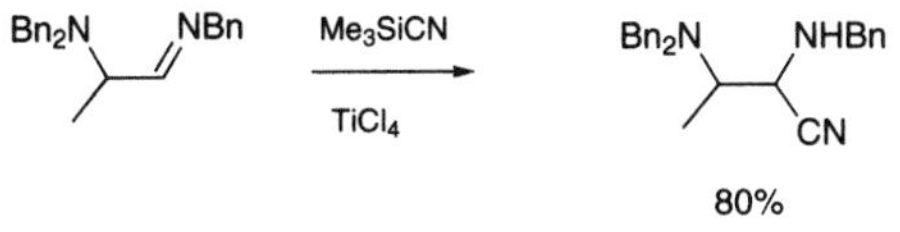

80%

Benzyl cyanides.[2]

93%

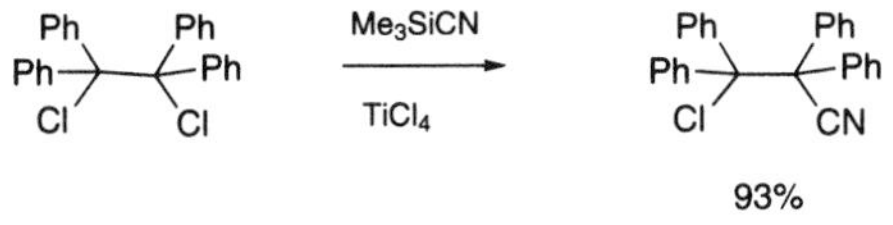

[1] M. T. Reetz, M. Hubel, R. Jaeger, R. Schwickhardt, and R. Goddard, *S* 733 (1994).
[2] H. E. Zieger and S. Wo, *JOC* **59**, 3838 (1994).

β-Cyclodextrin–benzyltriethylammonium chloride.

α-Hydroxy acids.[1] Homologation of aldehydes via trichloromethyl carbinols
requires a strong base to deprotonate chloroform. A defect of this method is related to
the fact that the trichloromethide anion tends to decompose into dichlorocarbene.
Using the β-cyclodextrin/benzyltriethylammonium chloride catalytic system, the
formation of the adducts is facilitated. Saponification of the trichloromethyl
carbinols gives α-hydroxy acids in excellent overall yields.

[1] C.-H. Zhou, D.-Q. Yuan, and R.-G. Xie, *SC* **24**, 43 (1994).

2-Cyclohexenone.

Decarboxylation of amino acids.[1] Amines are obtained on heating α-amino
acids with 2-cyclohexenone (3 examples, 72–82% yield).

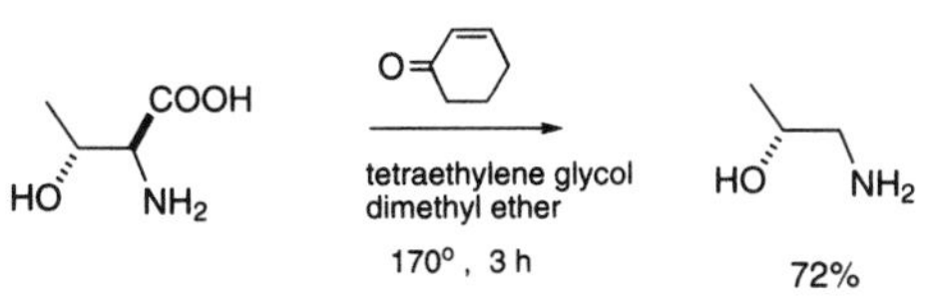

[1] S. Wallbaum, T. Mehler, and J. Martens, *SC* **24**, 1381 (1994).

1,5-Cyclooctadiene(cyclopentadienyl)ruthenium(I) chloride.

1,4-Dienes.[1] Addition of alkenes to alkynes (ene-type reaction) is readily promoted by the Ru complex. Branched products are favored, and functional groups in the substrates are tolerated. Internal alkynes also undergo this addition. The 2:1 adduct from 1,11-dodecadiene and ethyl 4-hydroxypent-2-ynoate is a precursor of ancepsenolide.[2]

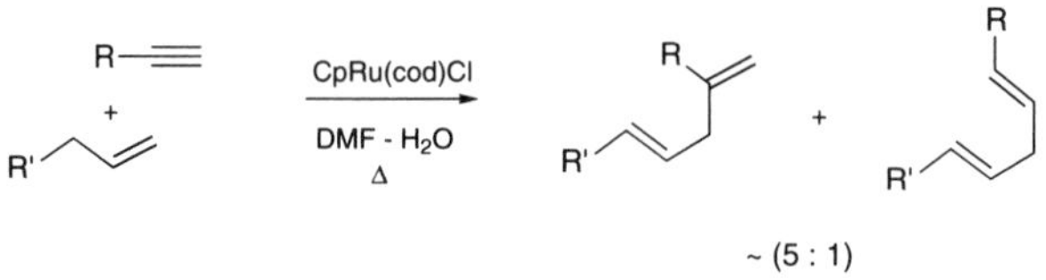

~ (5 : 1)

[4+2]Cycloadditions.[3] 1,5-Cyclooctadiene participates in the cycloaddition with alkynes. The yields of the tricyclic adducts are generally excellent.

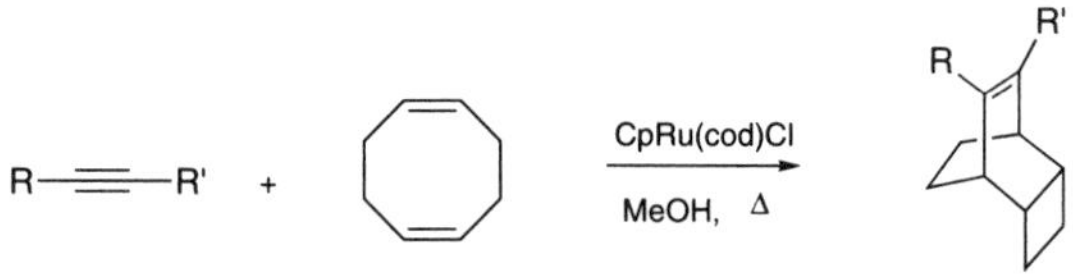

[1] B. M. Trost and A. Indolese, *JACS* **115**, 4361 (1993).
[2] B. M. Trost and T. J. J. Müller, *JACS* **116**, 4985 (1994).
[3] B. M. Trost, K. Imi, and A. F. Indolese, *JACS* **115**, 8831 (1993).

Cyclopentadienylbis(triphenylphosphine)ruthenium(I) chloride.

β,γ-Unsaturated ketones.[1] The reconstitutive addition of 1-alkynes to allylic alcohols is atom-economical. This Ru-catalyzed reaction is applicable to a short rosefuran synthesis.

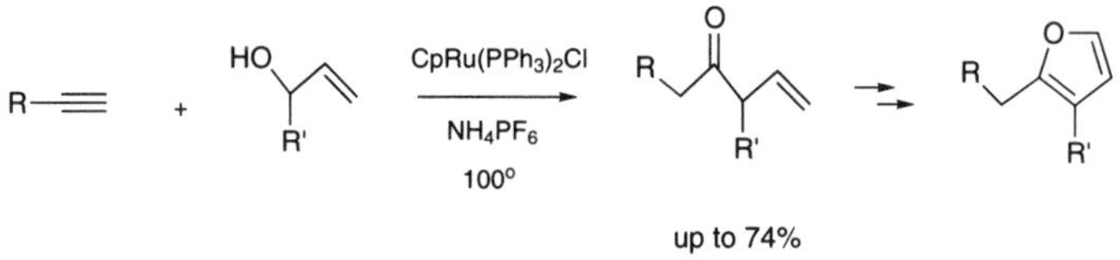

up to 74%

[1] B. M. Trost and J. A. Flygare, *JOC* **59**, 1078 (1994).

D

Dialkyl azodicarboxylates. **15**, 111–112

Allyl carbamates.[1] By catalysis with $SnCl_4$ the reaction temperature of the ene reaction of alkenes and diethyl azodicarboxylate may be lowered to $-60°C$. The adducts are converted to allylic carbamates on reduction with Li in liquid ammonia.

α-Amino β-hydroxy esters.[2] Enolates of the esters react with di-*t*-butyl azodicarboxylate stereoselectively, giving the *anti* products. On reductive cleavage of the N–N bond the reaction sequence accomplishes an electrophilic amination.

[1] M. A. Brimble and C. H. Heathcock, *JOC* **58**, 5261 (1993).
[2] C. Greck, L. Bischoff, F. Ferreira, C. Pinel, E. Piveteau, and J. P. Genet, *SL* 475 (1993).

Dialkyltin dicarboxylates.

Urethanes. The ability of the tin dicarboxylates to promote addition of alcohols to isocyanates suits well for the synthesis of urethanes and in polyurethane technology.[1,2]

[1] S. Roy and K. K. Majumdar, *SC* **24**, 333 (1994).
[2] M. Gormanns and H. Ritter, *T* **49**, 6965 (1993).

Dialkylzincs. **15**, 148; **17**, 96

One new preparation method[1] exploits a boron–zinc exchange.

β-Alkylstyrenes.[2] The nitro substituent in β-nitrostyrenes is replaceable by an alkyl group at room temperature using dialkylzinc reagents.

Organocuprates.[3] The preparation of cuprates from organozincs is viable due to the tolerance of many functional groups (e.g., esters) in the reagents. There are also more synthetic pathways to the organozincs; for example, from an alkene it is possible to achieve hydroboration and boron–zinc exchange.

Sulfinate esters from chlorosulfites.[4]

90% (*R*:*S* 95:5)

Chemoselective reactions with organozincs.[5] The aldehyde group of a keto aldehyde is selectively activated by $TiCl_4$ to engage in reaction with Me_2Zn (in controlled amounts). The same group is completely blocked by the presence of Ph_3P (α-phosphonioalkoxide formation), which permits selective attack of Me_2Zn on the ketone.

[1] F. Langer, J. Waas, and P. Knochel, *TL* **34**, 5261 (1993).
[2] D. Seebach, H. Schäfer, B. Schmidt, and M. Schreiber, *ACIEE* **31**, 1587 (1992).
[3] F. Langer, A. Devasagayaraj, P.-Y. Chavant, and P. Knochel, *SL* 410 (1994).
[4] J. K. Whitesell and M.-S. Wong, *JOC* **59**, 597 (1994).
[5] T. Kauffmann, T. Abel, W. Li, G. Neiteler, M. Schreer, and D. Schwarze, *CB* **126**, 459 (1993).

1,4-Diazabicyclo[2.2.2]octane. 13, 92; 15, 109

Baylis–Hillman reaction.[1,2]   DABCO is an excellent catalyst for the condensation of aldehydes with acrylic esters. The aldehydes include cinnamaldehyde and formylpropionates.

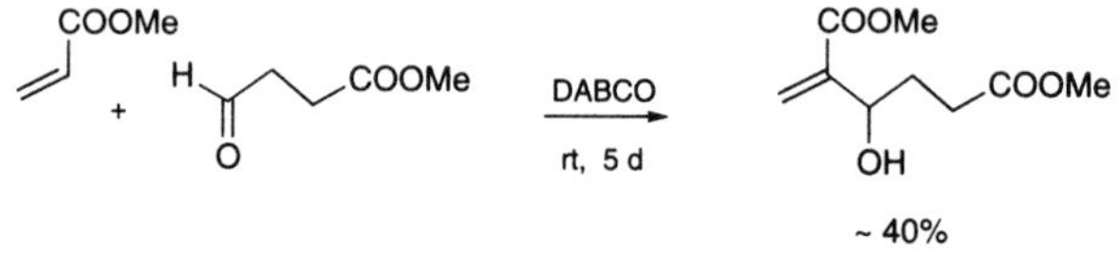

[1] F. R. van Heerden, J. J. Huyser, and C. W. Holzapfel, *SC* **24**, 2863 (1994).
[2] P. Perlmutter and T. D. McCarthy, *AJC* **46**, 253 (1993).

1,8-Diazabicyclo[5.4.0]undec-7-ene. 13, 92; 14, 109; 15, 109–110; 16, 105–106; 17, 99–100

Elimination.[1]   In general, 1-aryl-2-bromoalkynes are formed on treatment of 2,2-dibromostyrenes with DBU-DMSO.

DBU is capable of inducing the elimination of β-acetoxy nitroalkanes and deprotonation of benzyl isocyanoacetate.[2] A tandem Michael–Henry reaction of the resulting species ensues, leading to 3,4-disubstituted pyrrole-2-carboxylic esters.

Certain allylic and benzylic trichloromethyl sulfoxides suffer elimination of chloroform on exposure to tertiary amine bases.[3]

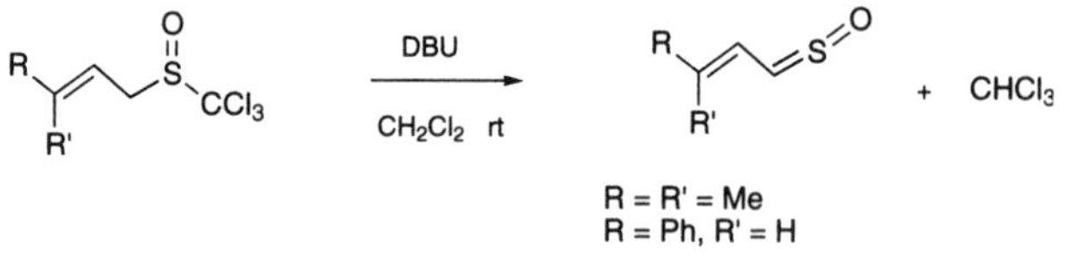

Addition and condensation reactions.   Nitroalkanes and dimethyl maleate afford homologs of dimethyl itaconate.[4]

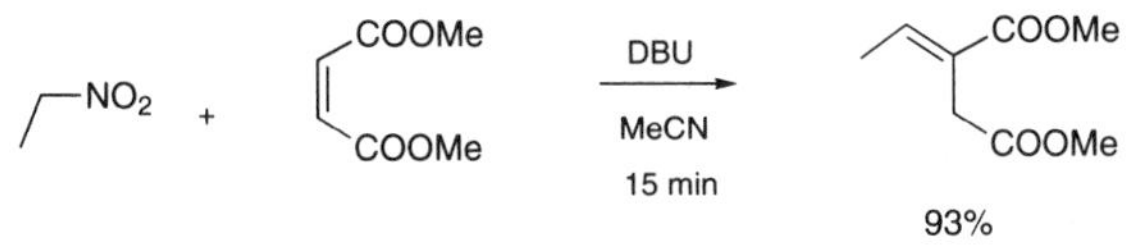

93%

4-Hydroxy-2-cyclopentenones react with diketene under the influence of DBU to give 4-acetonyl-2-cyclopentenones,[5] which is the result of O-acylation, intramolecular Michael reaction and elimination. An unusual condensation reaction of cyclic β-keto esters and araldehydes gives the γ-arylidene derivatives.[6]

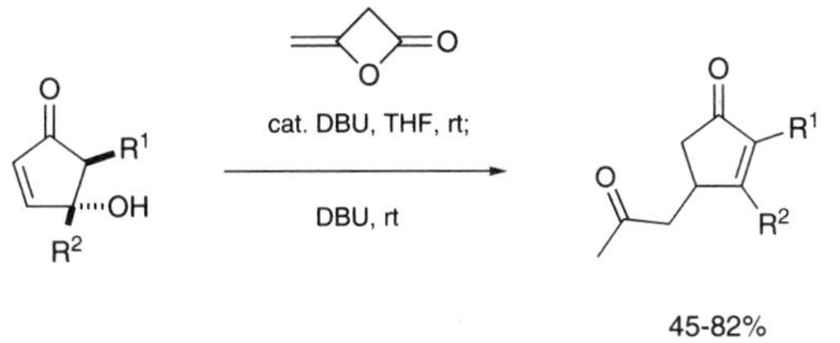

45-82%

[1] V. Ratovelomannana, Y. Rollin, C. Gebehenne, C. Gosimini, and T. Perichon, *TL* **35**, 4777 (1994).
[2] N. Ono, H. Katayama, S. Nishiyama, and T. Ogawa, *JHC* **31**, 707 (1994).
[3] S. Braverman, D. Grinstein, and H. E. Gottlieb, *TL* **35**, 953 (1994).
[4] R. Ballini and A. Rinaldi, *TL* **35**, 9247 (1994).
[5] F. G. West and G. U. Gunawardena, *JOC* **58**, 5043 (1993).
[6] M.-H. Filippini and J. Rodriguez, *CC* 33 (1995).

Diazomethane. **14**, 109–110; **17**, 100

Cyclopropanation. Unsaturated amides derived from Oppolzer's camphorsultam are found to undergo Pd(II)-catalyzed cyclopropanation[1] from the side of the sulfonyl group. For the access of 1-aminocyclopropanecarboxylic acids[2] the reaction of α,β-didehydroamino acid derivatives with diazomethane is most expedient.

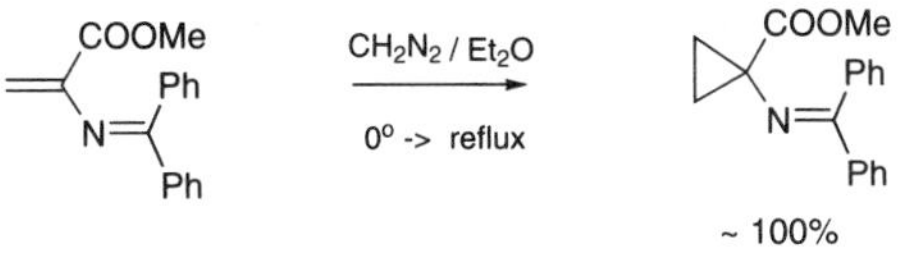

~ 100%

[1] J. Vallgarda, U. Appelberg, I. Csoregh, and U. Hacksell, *JCS(P1)* 461 (1994).
[2] C. Cativiela, M. D. Diaz de Villegas, and A. I. Jiminez, *T* **50**, 9157 (1994).

Dibenzoyl peroxide.

O-Benzoylhydroxylamines.[1] Rapid reaction takes place with amines at room temperature. The products can be converted to *N*-alkylhydroxamic acids.

Free radical cyclization.[2] Dibenzoyl peroxide promotes the isomerization of β-iodoalkyl propynoates to α-(iodomethylene)-γ-lactones in refluxing benzene.

44%

[1] L. Grierson and M. J. Perkins, *TL* **34**, 7463 (1993).
[2] G. Haaima, J.-J. Lynch, A. Routledge, and R. T. Weavers, *T* **49**, 4229 (1993).

Dibromodifluoromethane.

Ramberg–Bäcklund reaction.[1] A sulfone is converted to the alkene by treatment with Br_2CF_2, KOH supported on alumina, and *t*-BuOH. The one-pot reaction avoids isolation of the halogenation product.

gem-Difluorides.[2] Instigated by activated zinc, the transfer of fluorine atoms to carbonyl compounds from the reagent is probably the method of choice. The innocuousness of the system is of great advantage.

Difluoromethylenation.[3] Using $(Me_2N)_3P$ and Zn, the Wittig reagent $(Me_2N)_3P=CF_2$ is prepared, which is capable of reacting with substrates such as protected ribonolactones.

[1] T.-L. Chan, S. Fong, Y. Li, T.-O. Man, and C.-D. Poon, *CC* 1771 (1993).
[2] C.-M. Hu, F.-L. Quing, and C.-X. Shen, *JCS(P1)* 335 (1993).
[3] J. S. Houlton, W. B. Motherwell, B. C. Ross, M. J. Tozer, D. J. Williams, and A. M. Z. Slawin, *T* **49**, 8087 (1993).

1,2-Dibromoethane.

Bromine source.[1] Some ester enolates form α-bromoesters instead of the bromoethylated products on reaction with 1,2-dibromoethane.

[1] J. Ibarzo and R. M. Ortuno, *T* **50**, 9825 (1994).

Dibromomethane.

α-Methylenation of aldehydes.[1] The treatment of aldehydes with CH_2Br_2 and diethylamine leads to α-substituted acroleins. The same results are obtained directly from the ozonides derived from the homologous 1-alkenes.

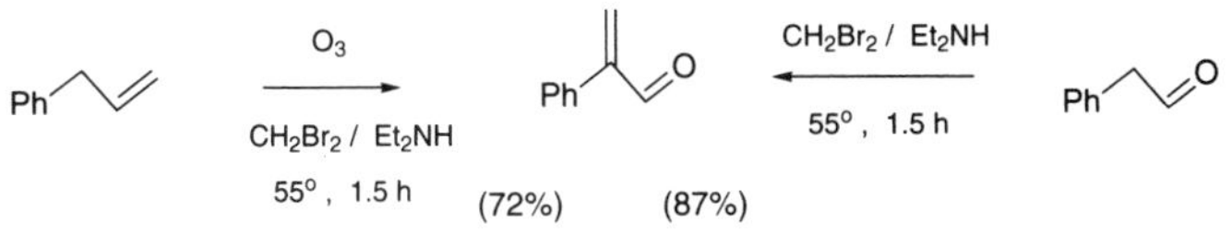

Ph⌒\\/   —→   Ph⌒\\\\₌O   ←—   Ph⌒\\\\₌O

O_3

CH_2Br_2 / Et_2NH

55°, 1.5 h   (72%)   (87%)

CH_2Br_2 / Et_2NH

55°, 1.5 h

N-Acyloxazolidines.[2]   Under the phase-transfer conditions smooth cyclization of 1,2-amido alcohols is observed.

[1] Y.-S. Hon, F.-J. Chang, and L. Lu, *CC* 2041 (1994).
[2] J. M. Vega-Perez, J. L. Espartero, and F. Alcudia, *JCC* **12**, 477 (1993).

1-(N,N'-Di-t-butoxycarbonylformamidinyl)pyrazole (1).

Di-Boc guanidines.[1]   The reagent is a donor of protected formamidine. Reaction with amines furnishes guanidine derivatives.

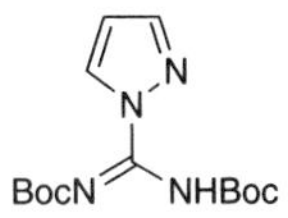

BocN⌒NHBoc

(1)

[1] B. Drake, M. Patek, and M. Lebl, *S* 579 (1994).

N,N'-Di-t-butoxycarbonylthiourea.

Guanidines.[1]   Guanidines are synthesized by reaction of the reagent with amines in the presence of $HgCl_2$ and Et_3N. The method is suitable for derivatization of electron-deficient amines such as trifluoroethylamine.

[1] K. S. Kim and L. Qian, *TL* **34**, 7677 (1993).

3,5-Di-t-butyl-1,2-benzoquinone. 15, 113

Cleavage of 1,2-amino alcohols.[1]   The oxidative cleavage to give aldehydes occurs at room temperature. However, the scope of this method has yet to be determined.

[1] V. Horak, Y. Mermersky, and D. B. Guirguis, *CCCC* **59**, 227 (1994).

Dibutyl disulfide.

[3+2]Cycloadditions.[1]   Methylenecyclopropane reacts with electron-deficient alkenes photochemically and in the presence of Bu_2S_2. This free radical reaction leads to methylenecyclopentane products, in which the electron-withdrawing substituent is at an allylic position.

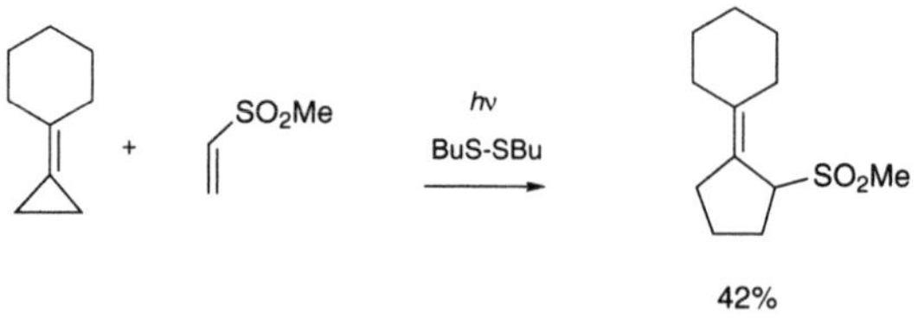

42%

[1] C. C. Huval, K. M. Church, and D. A. Singleton, *SL* 273 (1994).

Di-*t*-butyl peroxide.

Homolytic additions.[1]   Xanthates in which the *S*-bearing carbon site can support a free radical readily undergo homolysis by heating with *t*-Bu$_2$O$_2$. The two radical fragments can be taken up by an unsaturated compound. Group transfer to a remote double bond concurrent with cyclization of certain unsaturated xanthates is an intramolecular version. An oxocane precursor of lauthisan can be acquired (35%) in this manner.[2]

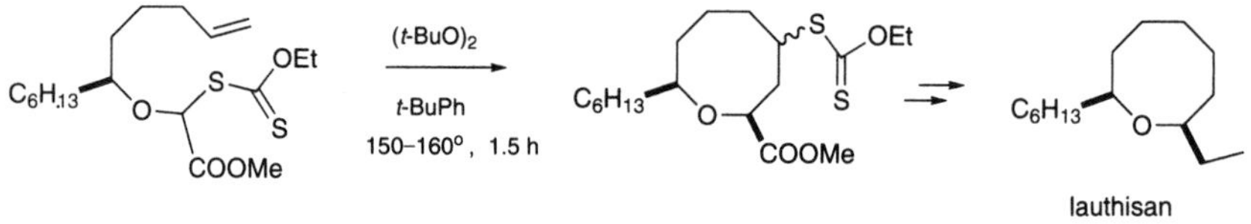

[1] J. H. Udding, H. Hiemstra, and W. N. Speckamp, *JOC* **59**, 3721 (1994).
[2] J. H. Udding, J. P. M. Giesselink, H. Hiemstra, and W. N. Speckamp, *JOC* **59**, 6671 (1994).

Dibutyltin bistriflate. 16, 111–112

1,3-Dithianes.[1]   The tin triflate serves as a Lewis acid catalyst.

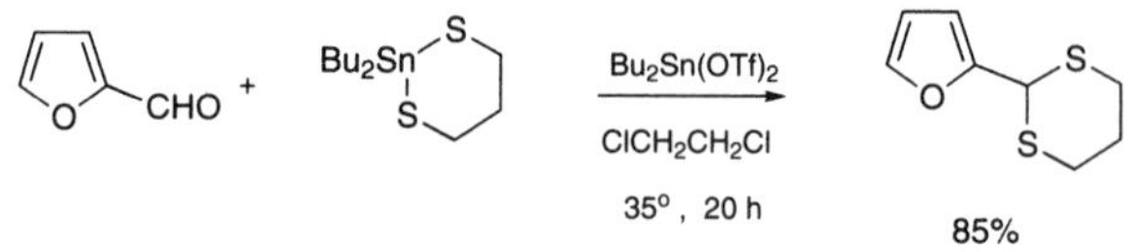

85%

[1] T. Sato, J. Otera, and H. Nozaki, *JOC* **58**, 4971 (1993).

Dibutyltin dialkoxides.

vic-Diol monofunctionalization.   Formation of stannylene derivatives permits selective cleavage and hence functionalization, including alkylation[1] and acylation.[2]

Dimethyl carbonate.[3] A direct preparation from methanol and carbon dioxide in the presence of $Bu_2Sn(OBu)_2$ is possible at elevated temperature and pressure. The addition of a water scavenger such as DCC increases the yield.

[1] G.-J. Boons, G. H. Castle, J. A. Clase, P. Grice, S. V. Ley, and C. Pinel, *SL* 913 (1993).
[2] D. J. Jenkins and B. V. L. Potter, *CR* **265**, 145 (1994).
[3] J. Kizlink and I. Pastucha, *CCCC* **59**, 2116 (1994).

Dibutyltin dichloride–silver perchlorate.

Activation of glycosyl fluorides.[1] The mixed reagents are a suitable catalyst system for the displacement of the fluorides by other nucleophiles, including *t*-BuOH, in the presence of 4A molecular sieves.

[1] H. Maeta, T. Matsumoto, and K. Suzuki, *CR* **249**, 49 (1993).

Dibutyltin oxide. 13, 95–96; 15, 116–117; 16, 112

Diol monofunctionalization. Formation of stannylene acetals and then treatment with electrophilic reagents achieve the purpose of selective diol derivatization. A primary hydroxyl group is preferentially protected, and in the case of a secondary–secondary diol their respective steric environments have important influence on the regiochemistry, as shown in several carbohydrates.[1-3]

93 -97%

Rapid stannyleneacetalization is achieved by microwave heating.[4]

[1] A. Glen, D. A. Leigh, R. P. Martin, J. P. Smart, and A. M. Truscello, *CR* **248**, 365 (1993).
[2] H. Qin and T. B. Grindley, *JCC* **13**, 475 (1994).
[3] B. Guilbert, N. J. Davis, M. Pearce, R. T. Aplin, and S. L. Flitsch, *TA* **5**, 2163 (1994).
[4] A. Morcuende, S. Valverde, and B. Herradon, *SL* 89 (1994).

Dicarbonyl(cyclopentadienyl)cobalt. 14, 116; 16, 112–113; 17, 102

Cycloisomerization.[1,2] The photoinduced formation of methylenecyclopentane derivatives from α-(4-pentynyl)acetoacetic esters represents a new way to prepare these substances.

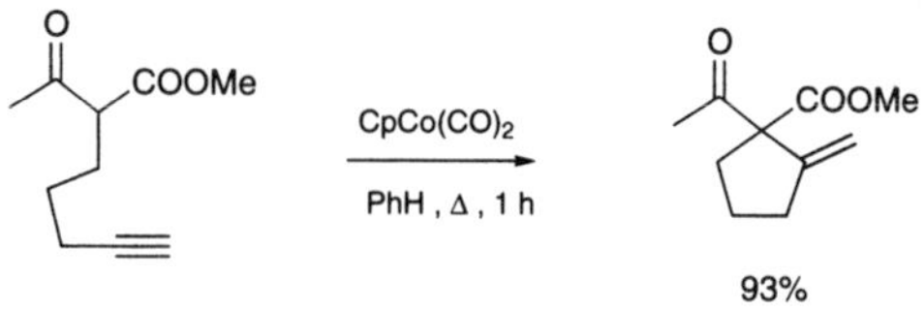

93%

[1] R. Stammler and M. Malacria, *SL* **92** (1994).
[2] P. Cruciani, C. Aubert, and M. Malacria, *TL* **35**, 6677 (1994).

Dicarbonyl(cyclopentadienyl)[(dimethylsulfonium)methyl]iron tetrafluoroborate. 13, 98; 15, 117

Cyclopropanation.[1] The reagent is an electrophilic carbene equivalent. Thus it behaves differently from dimethyloxosulfonium methylide toward 2-benzenesulfonyl-1,3-butadiene.

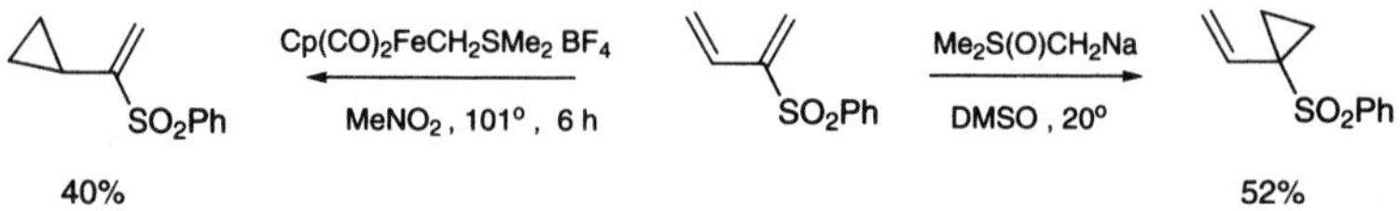

40%   52%

[1] J.-E. Bäckvall, C. Lofstrom, S. K. Kenfunen, and M. Mattson, *TL* **34**, 2007 (1993).

Dicarbonyl(cyclopentadienyl)(tetrahydrofuran)iron tetrafluoroborate.

Cyclopropanation.[1] The reagent is a catalyst for cyclopropanation of alkenes with diazoalkanes. High selectivities have been found for the formation of the *cis*-isomers.

Diels–Alder reactions.[2] A polymer-bound (to the cyclopentadiene unit) salt is a useful Lewis acid catalyst.

[1] W. J. Seitz and M. M. Hossain, *TL* **35**, 7561 (1994).
[2] A. K. Saha and M. M. Hossain, *TL* **34**, 3833 (1993).

(*E*)-1,2-Dicatecholboranylethene.

trans-Cyclohex-4-ene-1,2-diols.[1] The reagent is a reactive dienophile at moderate temperatures. The *vic*-diboranylcyclohexenes thus generated are readily oxidized to the diols with alkaline H_2O_2.

[1] D. A. Singleton and A. M. Redman, *TL* **35**, 509 (1994).

Dichlorine oxide.

Chlorination of quinones.[1] Cl_2O, which is generated in situ by passing Cl_2 through suspended yellow HgO in CCl_4, converts quinones to chloroquinones.

[1] P. C. Thapliyal, K. P. Singh, and R. N. Khanna, *SC* **24**, 1079 (1994).

Dichlorobis(triphenylphosphine)cobalt.
Carbonylation of benzyl halides.[1]

$$PhCH_2Cl \xrightarrow[\substack{Bu_4NOH,\ CO \\ PhH,\ 8\ h}]{(Ph_3P)_2CoCl_2}} PhCH_2COOH$$
$$88\%$$

[1] Y. Hu, J.-X. Wang, and W. Cui, *SC* **24**, 1743 (1994).

Dichlorobis(triphenylphosphine)palladium. **13**, 103–104; **15**, 123–124; **16**, 118–119; **17**, 111–112

Alkenes. Alk-3-en-1-ynes are obtained from dehydration of propynyl alcohols in the presence of $SnCl_2$ and $(Ph_3P)_2PdCl_2$.[1] With the same Pd complex as a catalyst carboxylic acids are degraded by one carbon.[2]

Reductive silylation.[3] α-Diketones are converted into 1,2-(bissiloxy)ethenes, while methyl benzoylformate undergoes reductive coupling to give dimethyl 2,3-diphenyl-2,3-bis(O-trimethylsilyl)tartrate with $Me_3SiSiMe_3$ [catalyzed by $(Me_3P)_2PdCl_2$].

Heck reaction. The coupling of iodoarenes with tertiary allylic amines gives the normal β-aminomethylstyrenes (in EtOH or PhMe) or enamines (in DMF).[4]

In the presence of CuI-Et_3N the Pd-catalyzed coupling can be employed in a synthesis of γ-(Z)-alkylidene butenolides from (Z)-3-bromoacrylic acid,[5] and the coupling of 1-penten-4-yn-3-ol with ArI occurs at the acetylenic terminus.[6]

An interesting chemoselectivity is shown for coupling with one of two enol triflates[7] as a result of slight structural variation.

Seven- and eight-membered cycloalkenes are available by intramolecular coupling of allenes.[8] The reductive cyclizations (requiring $SnCl_2$)[9] of o-nitrostyrenes and o-nitrobenzylidenimines afford indoles and 2H-indazoles, respectively.

Stille coupling. The effect of some additives on the coupling has been examined.[10] With Cu(I) salt as cocatalyst the process has proven valuable for the synthesis of hindered biphenyls and terphenyls.[11] Other reports concern the preparation of 2-substituted indoles,[12] 3,4-diarylfurans,[13] and carbazole precursors.[14] Aromatic amination is accomplished by reaction with in situ generated aminostannanes (from $RNHR'$ and Bu_3SnNEt_2).[15]

exo,exo-5-Alkynyl-6-arylnorbornenes are available from a three-component coupling.[16] Pyrolysis of the products gives (Z)-enynes.

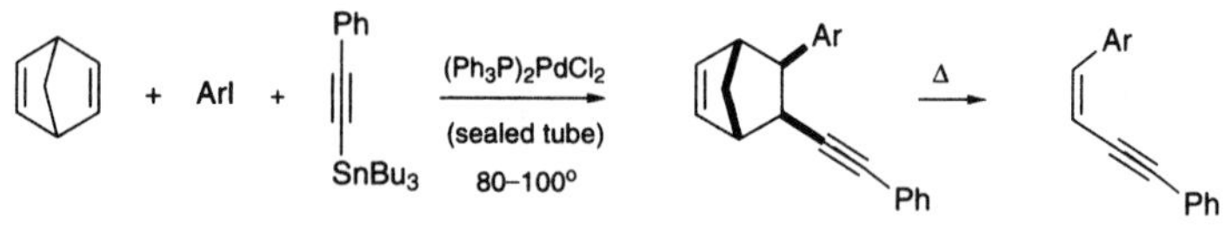

Coupling of alkenyl- or alkynyl-stannanes with alkenyl(phenyl)iodonium triflates leads to stereodefined dienes or enynes.[17]

The stannane–acyl chloride cross-coupling has been directed towards access to 3-acylcyclobutenones[18] and α-heterosubstituted ketones.[19]

Organozinc–aryl iodide coupling. The method serves for the preparation of unsymmetrical biaryls[20] and o-acylbenzylboronates and thence o-quinodimethanes.[21]

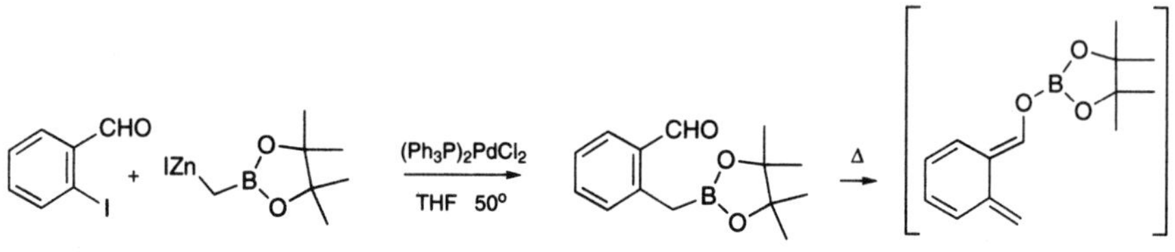

An intramolecular carbozincation product may be subjected to a metal–metal exchange in the presence of a slightly different Pd-complex, forming a cuprate reagent which is useful for assembly of ring structures bearing a functionalized side chain.[22]

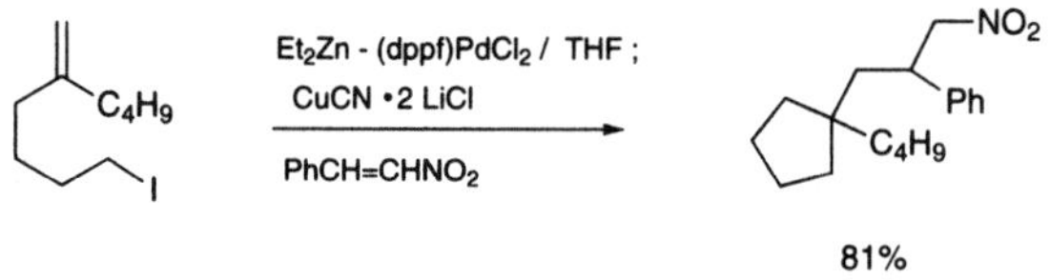

81%

Carbonylation and carboxylation. Araldehydes are derived from the Pd-catalyzed reaction of ArX with CO and HCOONa.[23] The Suzuki coupling under CO leads to diaryl ketones.[24] Chromones and quinolones are similarly acquired by carbonylation of o-iodophenols and anilines in the presence of alkynes.[25] 2-Aryl-benzimidazoles[26] and -benzothiazoles are produced from o-arenediamines and o-mercaptoanilines,[27] respectively.

Carbonylation of organohalides is achievable by using $CHCl_3$ and KOH,[28] whereas reductive alkoxycarbonylation[29,30] requires also $SnCl_2$.

A route to aryl aroates consists of catalyzed reaction from a mixture of ArX, ArOH, CO, and a base.[31] Phenol derivatives are formed in a cyclocarbonylation reaction of 2,4-pentadienyl acetates.[32]

$$Ph\diagup\!\!\diagdown\!\!\diagup\!\!\diagdown OAc \quad \xrightarrow[\substack{CO / PhH \\ Ac_2O - Et_3N \\ 140^\circ}]{(Ph_3P)_2PdCl_2} \quad Ph\text{—}\langle\text{—}\rangle$$

69%

Other couplings. 1-Alkylthio-1,3-butadienes are assembled by a one-pot method[33] involving hydroboration of thioalkynes and Suzuki coupling with alkenyl halides, both reactions catalyzed by Pd(II). Ullmann coupling using a Cu-$(Ph_3P)_2PdCl_2/DMSO$ system delivers nitrophenylpyridines.[34]

[1] Y. Masuyama, J. Takahara, K. Hashimoto, and Y. Kurusu, *CC* 1219 (1993).

[2] J. A. Miller, J. A. Nelson, and M. P. Byrne, *JOC* **58**, 18 (1993).

[3] H. Yamashita, N. P. Reddy, and M. Tanaka, *CL* 315 (1993).

[4] L. Filippini, M. Gusmeroli, and R. Riva, *TL* **34**, 1643 (1993).

[5] X. Lu, X. Huang, and S. Ma, *TL* **34**, 5963 (1993).

[6] N. G. Kundu, M. Pal, and C. Chowdhury, *JCR(S)* 4 (1995).

[7] M. Moniatte, M. Eckhardt, K. Brickmann, R. Brückner, and J. Suffert, *TL* **35**, 1965 (1994).

[8] S. Ma and E. Negishi, *JOC* **59**, 4730 (1994).

[9] M. Akazome, T. Kondo, and Y. Watanabe, *JOC* **59**, 3375 (1994).

[10] S. Gronowitz, P. Bjork, J. Malm, and A.-B. Hornfeldt, *JOMC* **460**, 127 (1993).

[11] J. M. Saa and G. Martorell, *JOC* **58**, 1963 (1993).

[12] S. S. Labadie and E. Teng, *JOC* **59**, 4250 (1994).

[13] Y. Yang and H. N. C. Wong, *T* **50**, 9538 (1994).

[14] M. Iwao, H. Takehara, S. Furukawa, and M. Watanabe, *H* **36**, 1483 (1993).

[15] A. S. Guram and S. L. Buchwald, *JACS* **116**, 7901 (1994).

[16] M. Kosugi, T. Kimura, H. Oda, and T. Migita, *BCSJ* **66**, 3522 (1993).

[17] R. J. Hinkle, G. T. Poulter, and P. J. Stang, *JACS* **115**, 11626 (1993).

[18] L. S. Liebeskind, G. B. Stone, and S. Zhang, *JOC* **59**, 7919 (1994).

[19] J. Ye, R. K. Bhatt, and J. R. Falck, *JACS* **116**, 1 (1994).

[20] Sibille, V. Ratovelomanana, J. Y. Nedelec, and J. Perichon, *SL* 425 (1993).

[21] G. Kanai, N. Miyaura, and A. Suzuki, *CL* 845 (1993).

[22] H. Stadtmuller, R. Lentz, C. E. Tucker, T. Studemann, W. Dorner, and P. Knochel, *JACS* **115**, 7027 (1993).

[23] T. Okano, N. Harada, and J. Kiji, *BCSJ* **67**, 2329 (1994).

[24] T. Ishiyama, H. Kizaki, N. Miyaura, and A. Suzuki, *TL* **34**, 7595 (1993).

[25] S. Torii, H. Okumoto, L. H. Xu, M. Sadakane, M. V. Shostakovsky, A. B. Ponomaryov, and V. N. Kalinin, *T* **49**, 6773 (1993).

[26] R. J. Perry and B. D. Wilson, *JOC* **58**, 7016 (1993).

[27] R. J. Perry and B. D. Wilson, *JOC* **13**, 3346 (1994).

[28] V. V. Grushin and H. Alper, *OM* **12**, 3846 (1993).

[29] R. Takeuchi and M. Sugiura, *JCS(P1)* 1031 (1993).

[30] R. Naigre, T. Chenal, I. Cipres, P. Kalck, J.-C. Daran, and J. Vaissermann, *JOMC* **480**, 91 (1994).

[31] Y. Kubota, T. Hanaoka, K. Takeuchi, and Y. Sugi, *SL* 515 (1994).

[32] Y. Ishii, C. Gao, W.-X. Xu, M. Iwasaki, and M. Hidai, *JOC* **58**, 6818 (1993).

[33] I. D. Gridnev, N. Miyaura, and A. Suzuki, *JOC* **58**, 5351 (1993).

[34] N. Shimizu, T. Kitamura, K. Watanabe, T. Yamaguchi, H. Shigyo, and T. Ohta, *TL* **34**, 3421 (1993).

2,3-Dichloro-5,6-dicyano-1,4-benzoquinone. **13**, 104–105; **14**, 126–127; **15**, 125–126; **16**, 120

Deacetalization and dethioacetalization.[1,2]   Dimethyl acetals, 4-p-methoxyphenyl-1,3-dioxolanes are cleaved with DDQ in the presence of water. For deprotection of thioacetals[3] under mild conditions photochemical assistance seems advantageous.[4]

The same reaction principle of deacetalization can be extended to ether exchange.[5] Thus alcohol protection is possible by mixing with 2,2-dimethoxypropane in the presence of DDQ, and proximal diols are converted to acetonides.[6] Replacement of anomeric arylmethoxyl groups by this method[7] complements other glycosylation procedures.

gem-Dialkoxycyclopropane fission.[8]   These compounds are transformed into α,β-unsaturated esters (33–86% yield).

Functionalization of benzyl ethers.   Dehydrogenation with interception of the electrophilic species generated from isochromans[9,10] extend the synthetic utility of such compounds.

Other C–C bond forming processes.   Oxidation of imines induces C–C bond formation at the α-carbon.[11] With alkynes present the products are the quinolines. C-Allylation of glycals by allyltrimethylsilane is induced by DDQ.[12]

[1] A. Oku, M. Kinugasa, and T. Kamada, *CL* 165 (1993).
[2] C. E. McDonald, L. E. Nice, and K. E. Kennedy, *TL* **35**, 57 (1994).
[3] K. Tanemura, H. Dohya, M. Imamura, T. Suzuki, and T. Horaguchi, *CL* 965 (1994).
[4] L. Mathew and S. Sankararaman, *JOC* **58**, 7576 (1993).
[5] O. Kjolberg and K. Neumann, *ACS* **48**, 80 (1994).
[6] O. Kjolberg and K. Neumann, *ACS* **47**, 843 (1993).
[7] J. Inaga, Y. Yokoyama, and T. Hanamoto, *CL* 85 (1993).
[8] M. Abe and A. Oku, *TL* 1994, 35, 3551.
[9] Y.-C. Xu, E. Lebeau, J.W. Gillard, and G. Attardo, *TL* **34**, 3841 (1993).
[10] Y.-C. Xu, C. Roy, and E. Lebeau, *TL* **34**, 8189 (1993).
[11] B. Bortolotti, R. Leardini, D. Nanni, and G. Zanardi, *T* **49**, 10157 (1993).
[12] K. Toshima, T. Ishizuka, G. Matsuo, and M. Nakata, *CL* 2013 (1993).

4,5-Dichloro-1,2,3-dithiazolium chloride (Appel's salt).

Condensation.[1]   This salt (**1**) is a dehydrating agent, capable of uniting a carboxylic acid and an alcohol to give the ester. Usually 2,6-lutidine is added to scavenge HCl.

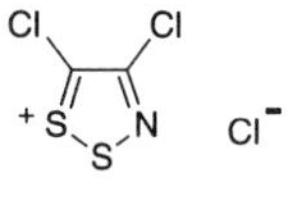

(1)

[1] J. J. Folmer and S. M. Weinreb, *TL* **34**, 2737 (1993).

Dichloro(ethoxy)oxovanadium(V).

Coupling of allyl and benzyl silanes.   Desilylative homo- and cross-coupling are operative,[1] but it is possible to achieve selective coupling of an allylsilane and a benzylsilane to give the 4-arylbutene predominantly.[2]

Cross-coupling of silyl enol ethers and allylic silanes.[3]   The formation of γ,δ-unsaturated carbonyl compounds involves one-electron oxidation.

Oxidative cleavage of 4-membered ketones.[4]   The oxovanadium species, being a Lewis acid, catalyzes aldolization with cyclobutanone as the acceptor, and furthermore a ring cleavage.

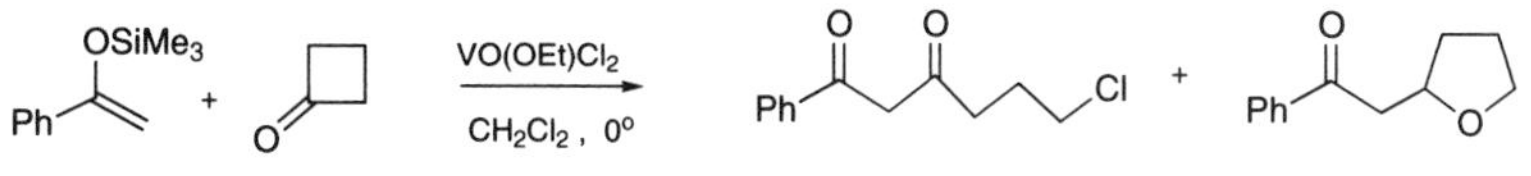

46%   (85:15)

Cyclobutenones give ethyl 2-chloro-3-alkenoates,[5] probably via the ketene tautomers and the confluent action of $VO(OEt)Cl_2$ and $CuCl_2$ on them.

[1] T. Fujii, T. Hirao, and Y. Ohshiro, *TL* **34**, 5601 (1993).
[2] T. Hirao, T. Fujii, and Y. Ohshiro, *TL* **35**, 8005 (1994).
[3] T. Hirao, T. Fujii, and Y. Ohshiro, *T* **50**, 10207 (1994).
[4] T. Hirao, T. Fujii, T. Tanaka, and Y. Ohshiro, *SL* 845 (1994).
[5] T. Hirao, T. Fujii, T. Tanaka, and Y. Ohshiro, *JCS(P1)* 3 (1994).

Dichloromethane.

Dialkylaminomethylation.[1]   The key intermediate in this reaction is the Mannich adduct $(R_2N=CH_2)^+$ formed upon treatment of CH_2Cl_2 with secondary amines.

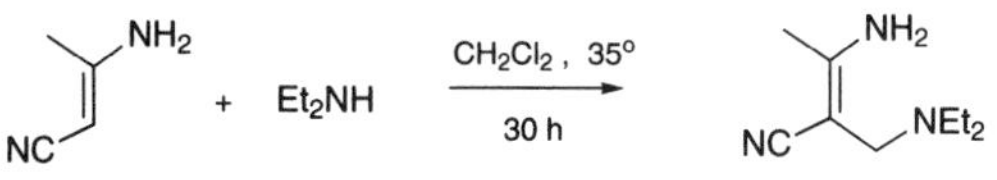

90%

[1] F. Souquet, T. Martens, and M.-B. Fleury, *SC* **23**, 817 (1993).

Dichloromethyl chloroformate.

Carbamates.[1] Alcoholysis of the chloroformate leads to mixed carbonates, which are reactive toward amines. Reaction with 2-amino alcohols directly furnishes oxazolones.

[1] T. Patonay, L. Hegedus, F. Mogyorodi, and L. Zolnai, *SC* **24**, 2507 (1994).

Dichloromethyl phenyl sulfoxide.

Homologation of carbonyl compounds.[1] The adducts of the derived sulfinyl α-carbanion with carbonyl compounds have been subjected to a ligand-exchange reaction of the sulfinyl group with EtMgBr. After rearrangement of the carbenoid, the ring-enlarged product is formed.

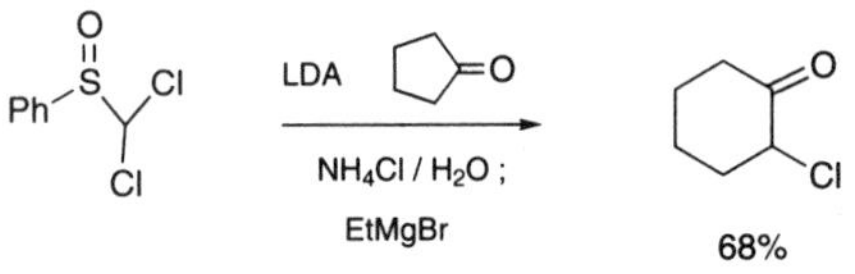

[1] T. Satoh, Y. Mizu, T. Kawashima, and K. Yamakawa, *T* **51**, 703 (1995).

Dicobalt octacarbonyl. 13, 99–101; 14, 117–119; 15, 117–118; 16, 113–115, 17, 102–105

Pauson–Khand reaction. Numerous examples of this reaction have been reported. Worthy of mentioning are those involving electron-deficient alkynones,[1] a methylenecyclopropane as terminator,[2] and a catalytic version in the presence of triphenyl phosphite and CO.[3]

Isomerization of 1-alkynylcyclopropanols.[4] 2-Cyclopentenones are the rearrangement products.

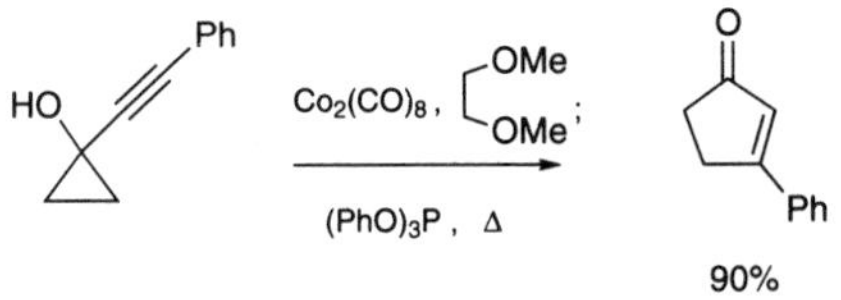

[1] T. R. Hoye and J. A. Suriano, *JOC* **58**, 1659 (1993).
[2] A. Stolle, H. Becker, J. Salaün, and A. de Meijere, *TL* **35**, 3517 (1994).
[3] N. Jeong, S. H. Hwang, Y. Lee, and Y. K. Chung, *JACS* **116**, 3159 (1994).
[4] N. Iwasawa and T. Matsuo, *CL* 997 (1993).

2,6-Dicyanato-9,10-anthraquinone.

Cyanation.[1]   The anthraquinone derivative readily transfers a cyano group to enamines and enolates of 1,3-dicarbonyl compounds.

[1] K. Buttke and H.-J. Niclas, *SC* **24**, 3241 (1994).

Dicyanoketene acetals.

Monothioacetalization.[1]   One alkoxy group of an acetal can be replaced by a thio group using RSH or RSSiMe$_3$ when catalyzed by a dicyanoketene acetal or tetracyanoethylene.

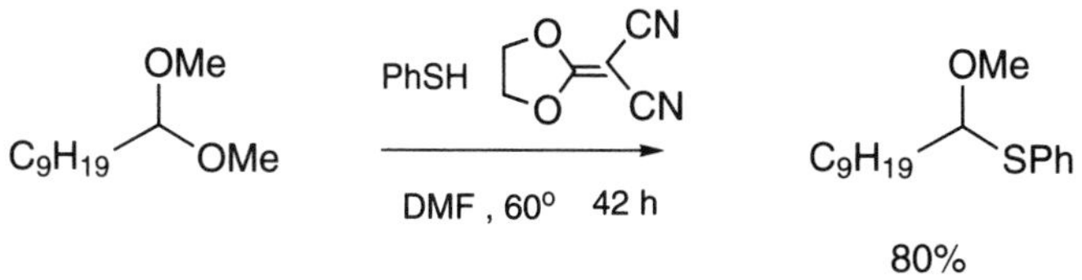

[1] T. Miura and Y. Masaki, *TL* **35**, 7961 (1994).

Dicyclohexylboron halides.

Aldolization.[1,2]   Using Et$_3$N as a base, the enolboration of carbonyl compounds is stereoselective; therefore, aldolization can be controlled. The substitution pattern at the β-carbon of an ester influences the configuration of the resulting ketene boryl ethers. Generally only a relatively small group can be accomodated by a (Z)-related dicyclohexylboroxy residue.

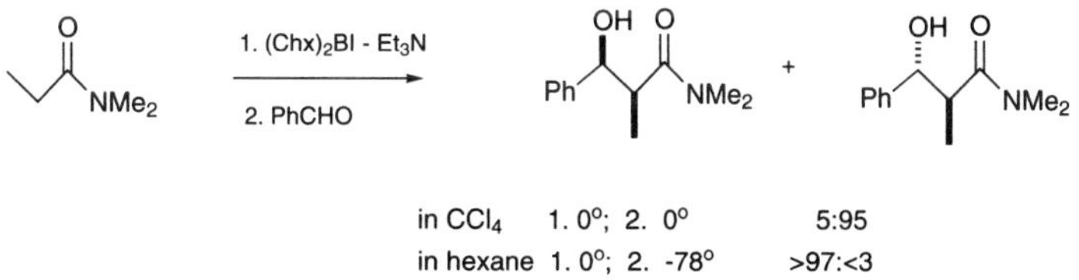

in CCl$_4$ 1. 0°; 2. 0°	5:95
in hexane 1. 0°; 2. -78°	>97:<3

[1] K. Ganesan and H. C. Brown, *JOC* **59**, 7346 (1994).
[2] K. Ganesan and H. C. Brown, *JOC* **59**, 2336 (1994).

N,N'-Dicyclohexylcarbodiimide. 14, 131–132; 16, 128

Imide formation.[1]   *N*-Carboxy-α-dehydroamino acid anhydrides couple with *N*-protected amino acids. Such anhydrides usually serve as acylating agents for amines.

C-Acylation of 1,3-dicarbonyl compounds.[2]   This transformation, which proceeds in CH_2Cl_2 at room temperature, is catalyzed by 4-dimethylaminopyridine.

Ar-Cyclohexylation.[3]   In the cyclohexylation DCC behaves as a typical alkylating agent. Promoters include H_2SO_4 and $AlCl_3$.

[1] C.-G. Shin, S. Honda, K. Morooka, and Y. Yonezawa, *BCSJ* **66**, 1844 (1993).
[2] H. Tabuchi, T. Hamamoto, and A. Ichihara, *SL* 651 (1993).
[3] J. N. Kim, K. H. Chung, and E. K. Ryu, *TL* **35**, 903 (1994).

Diethyl allylphosphonate.

Alkylation.[1]   Lithiation and reaction with aldehydes provide α-vinyl-β-hydroxyphosphonates, which are precursors of 2-diethylphosphonyl-1,3-dienes. The dehydration is easily accomplished with DCC-CuCl$_2$ in refluxing CH_2Cl_2.

[1] H. Al-Badri, E. About-Jaudet, and N. Collington, *S* 1072 (1994).

Diethylaluminum azide.

Acyl azides.[1]   Formation of $RCON_3$ from esters using this reagent proceeds in a nonpolar solvent (e.g., hexane) at room temperature. The obvious advantage is that inconveniences and limitations imposed by the formation of acyl chlorides and the workup are eliminated.

[1] V. H. Rawal and H. M. Zhong, *TL* **35**, 4947 (1994).

Diethylaluminum cyanide.

Strecker synthesis.[1]   *N*-Sulfimines add a cyanide ion from Et_2AlCN preferentially from a coordination complex. Thus 1,3-asymmetric induction arises from the chiral sulfinyl unit.

72% (*SS: SR* = 70:30)

[1] F. A. Davis, R. E. Reddy, and P. S. Portonovo, *TL* **35**, 9351 (1994).

N-Diethylalumino-2,2,6,6-tetramethylpiperidine.

Fischer indole synthesis.[1]   Phenylhydrazones are converted to indoles under mild conditions when they are treated with the hindered aluminum amide. Notably the regiochemistry follows the initial configuration of the hydrazones.

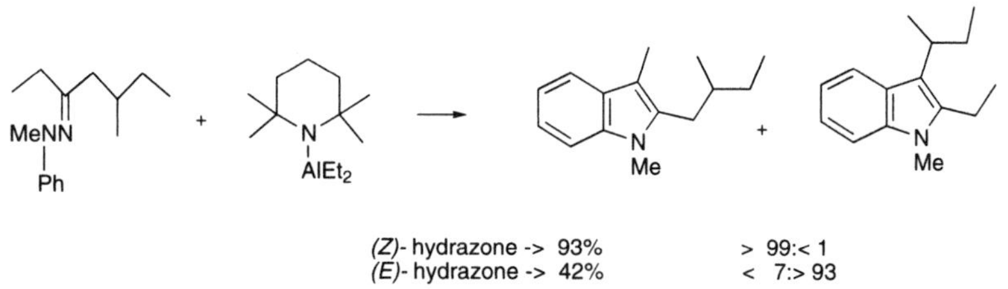

(Z)- hydrazone -> 93%   > 99:< 1
(E)- hydrazone -> 42%   < 7:> 93

[1] K. Maruoka, M. Oishi, and H. Yamamoto, *JOC* **58**, 7638 (1993).

Diethylaminosulfur trifluoride (DAST). 13, 110–112; 16, 128–129

Replacement of oxygen functionality by fluorine.   2-Fluoronitriles are produced from α-cyanohydrins or their silyl ethers,[1] while acyl cyanides give α,α-difluoronitriles.[2]

Terminally difluorinated allylic alcohols are converted to trifluoromethylalkenes.[3]

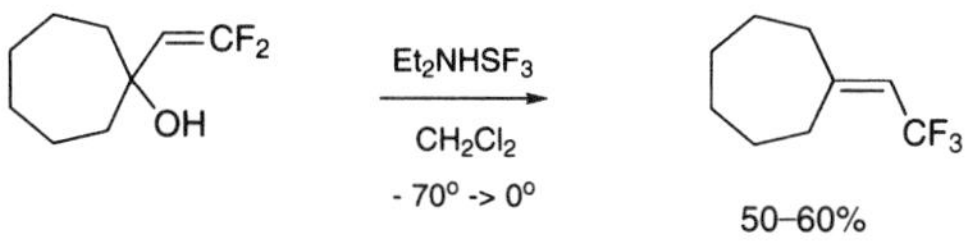

50–60%

α-Fluorination of thioethers.[4]   This transformation may be performed with DAST alone or in combination with $SbCl_3$.

[1] U. Stelzer and F. Effenberger, *TA* **4**, 161 (1993).
[2] E. Bartmann and J. Krause, *JFC* **61**, 117 (1993).
[3] F. Tellier and R. Sauvetre, *JFC* **62**, 183 (1993).
[4] M. J. Robbins and S. F. Wnuk, *JOC* **58**, 3800 (1993).

Diethyl (1,3-butadien-2-yl) phosphate.

Diels–Alder reaction.[1]   The substituted diene reacts with dienophiles in the presence of a Lewis acid. The stability towards acids makes it easier to handle than the corresponding alkyl and silyl ethers.

[1] H.-J. Liu, W. M. Feng, J. B. Kim, and E. N. C. Browne, *CJC* **72**, 2163 (1994).

Diethyl carboxymethylphosphonate.

1-(Dialkylaminomethyl)vinylphosphonates.[1]   The phosphonoacetic acid undergoes twofold Mannich reaction to give the vinylphosphonates. Elimination of the secondary amine reactant and CO_2 from the primary adducts occurs in situ.

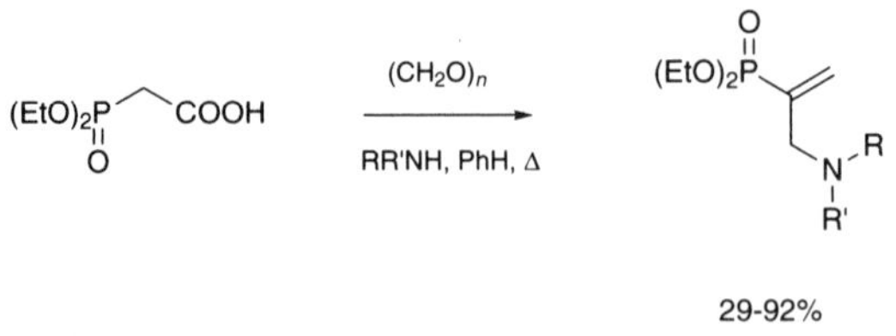

$(EtO)_2P(O)CH_2COOH$ + $(CH_2O)_n$ $\xrightarrow{RR'NH, PhH, \Delta}$ $(EtO)_2P(O)$ ... $N(R)(R')$

29-92%

[1] H. Krawczyk, *SC* **24**, 2263 (1994).

Diethyl chlorophosphate.

Claisen rearrangement.[1]   The quenching of the lithium enolates derived from allylic acetates results in enol phosphates, which show a higher propensity for Claisen rearrangement than ketene silyl ethers. The mixed anhydride products can be used directly to generate carboxylic acid derivatives.

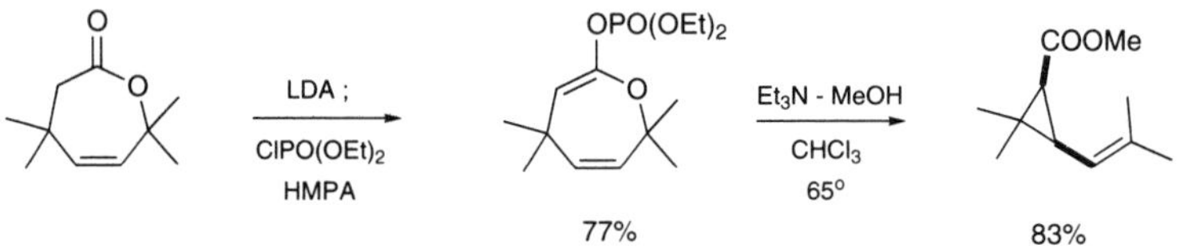

$\xrightarrow[\text{HMPA}]{\text{LDA ;} \atop \text{ClPO(OEt)}_2}$ OPO(OEt)$_2$ 77% $\xrightarrow[\text{CHCl}_3 \atop 65°]{\text{Et}_3\text{N - MeOH}}$ COOMe   83%

[1] R. L. Funk, J. B. Stallman, and J. A. Wos, *JACS* **115**, 8847 (1993).

Diethyl methylphosphonate.

1,1-Diiodoalkenes.[1]   Diethyl diiodo(lithio)methylphosphonate is formed in situ from MePO(OEt)$_2$, LiHMDS, and iodine. The Horner condensation products can also be dehydroiodinated to give iodoacetylenes.

(E)-Allylic amines.[2]   Reaction of the lithiated methylphosphonate ester with a nitrile, followed by addition of an aldehyde, furnishes a conjugated imine with an (*E*)-configuration. Immediate reduction (e.g., with NaBH$_4$) gives the allylic amine.

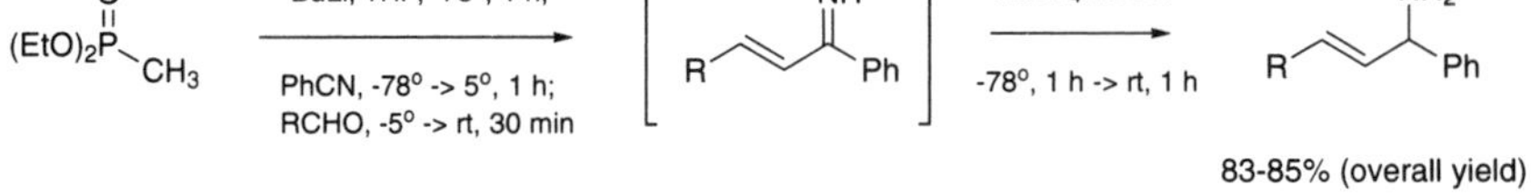

$(EtO)_2P(O)CH_3$ $\xrightarrow[\substack{\text{PhCN, -78° -> 5°, 1 h;} \\ \text{RCHO, -5° -> rt, 30 min}}]{\text{BuLi, THF, -78°, 1 h;}}$ [R ... NH ... Ph] $\xrightarrow[\text{-78°, 1 h -> rt, 1 h}]{\text{NaBH}_4\text{/MeOH}}$ R ... NH$_2$ ... Ph

83-85% (overall yield)

Functionalized β-keto phosphonates.[3]   (*Note:* Dimethyl methylphosphonate, which should have exactly the same reactivity, is described in the cited report.) After reaction of the α-lithio derivative with succinic anhydride and esterification of the product, a triply functionalized compound is obtained. Various synthetic uses of this phosphonate are conceivable.

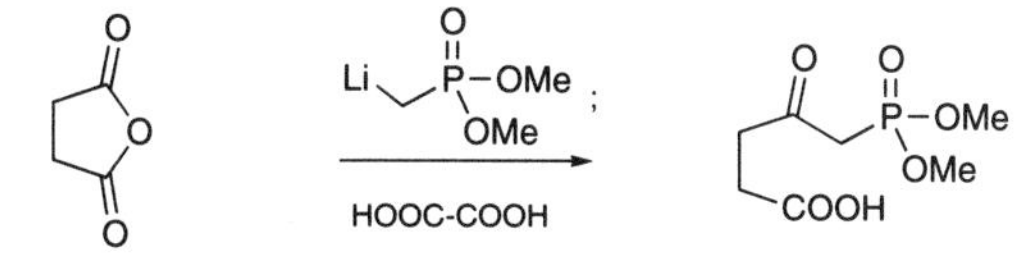

[1] B. Bonnet, Y. Le Gallic, G. Ple, and L. Duhamel, *S* 1071 (1993).
[2] W. S. Shin, K. Lee, and D. Y. Oh, *TL* **36**, 281 (1995).
[3] I. Delamarche and P. Mosset, *JOC* **59**, 5453 (1994).

Diethyl methylsulfonylmethylphosphonate, (EtO)₂P(O)CH₂S(O)₂CH₃.

$(EtO)_2P(O)CH_2S(O)_2CH_3$ heading rendering: **Diethyl methylsulfonylmethylphosphonate, $(EtO)_2P(O)CH_2S(O)_2CH_3$.**

Vinyl sulfones.[1]   Both lithiation and condensation with ketones are promoted by ultrasound.

[1] H. El Fakih, F. Pautet, D. Peters, H. Fillion, and J. L. Luche, *SC* **24**, 3225 (1994).

Diethyl oxalate.

Isatins.[1]   These compounds are obtained by interception of the *o*-lithiated *N*-Boc anilines with diethyl oxalate followed by acid treatment.

[1] P. Hewawasam and N. A. Meanwell, *TL* **35**, 7303 (1994).

Diethyl phosphite.

Aryl phosphonates.[1]   Phosphite anions react with diaryliodonium salts in hot DMF. Arylphosphonate esters are obtained in 79–93% yield.

Cyclic β-keto phosphonates.[2]   A new synthesis involves the treatment of α-nitro epoxides with diethyl phosphite and a base at room temperature.

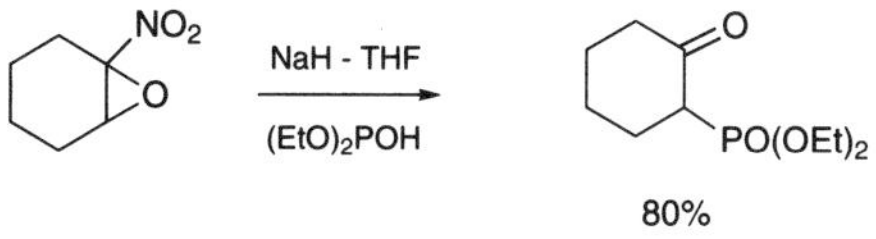

α-Amino phosphonic esters.[3]   A very simple procedure for the preparation of these phosphonopeptide precursors involves admixture of the phosphite with aldehydes at 60°C in an ethanolic medium containing NH₄OAc and 3A molecular sieves.

Reductive deconjugation of 2-bromo-2-alkenoates.[4]   The process probably involves Michael addition, debromination, and phosphite elimination prior to the kinetic protonation of the ester enolates. Triethylamine is required as base to promote the reaction.

2-Azadienes and isoquinolines.[5] Diethyl [(triphenylphosphoranylidene)-amino]phosphonate is generated from 1-(triphenylphosphoranylidene)amino-methyl-benzotriazole by reaction with diethyl phosphite anion. It undergoes typical Horner reactions.

[1] Z.-D. Liu and Z.-C. Chen, *S* 373 (1993).
[2] D. Y. Kim and M. S. Kong, *JCS(PI)* 3359 (1994).
[3] H. Takahashi, M. Yoshioka, N. Imai, K. Onimura, and S. Kobayashi, *S* 763 (1994).
[4] T. Hirao, K. Hirano, and Y. Ohshiro, *BCSJ* **66**, 2781 (1993).
[5] A. R. Katritzky, G. Zhang, and J. Jiang, *JOC* **59**, 4556 (1994).

Diethyl phosphoramidate.

β-Amino esters.[1]   The reagent is a Michael donor that adds to acrylic esters in the presence of K_2CO_3 and Bu_4NBr in refluxing toluene. The free amines are liberated on treating the Michael adducts with HCl in benzene at room temperature.

[1] K. Osowska-Pacewicka, S. Zawadzki, and A. Zwierzak, *PSS* **82**, 49 (1993).

Diethyl (trichloromethyl)phosphonate.

Alkylphosphonate esters.[1]   The carbanion generated from dechlorination with BuLi condenses with aldehydes (and some ketones). Catalytic hydrogenation of the (chlorovinyl)phosphonate products affords phosphonates.

[1] G. T. Lowen and M. R. Almond, *JOC* **59**, 4548 (1994).

Difluoroiodomethane.

1,1-Difluoro-2-iodoalkanes.[1]   A practical preparation of HCF_2I is by reaction of FSO_2CF_2COF with KI in MeCN between 30 and 40°C for 30 min. The iododifluoromethylation of alkenes is initiated by sodium dithionite.

[1] P. Cao, J.-X. Duan, and Q.-Y. Chen, *CC* 737 (1994).

2,2-Difluorovinyllithium.

α,β-Unsaturated acids and derivatives.[1]   The adducts with carbonyl compounds are unstable. On treatment with H_2SO_4 , H_2SO_4-MeOH, and Et_2NLi-THF they are transformed into the unsaturated acids, methyl esters, and diethylamides, respectively.

[1] F. Tellier and R. Sauvetre, *TL* **34**, 5433 (1993).

Diiodomethane. 13, 113–115, 275–276; **16**, 184–185; **17**, 155

Ring expansion of silacyclobutanes.[1]   The lithium carbenoid generated from LDA-CH$_2$I$_2$ inserts into the cyclic C–Si bond to give 2-iodosilacyclopentanes, which have found some interesting synthetic applications.

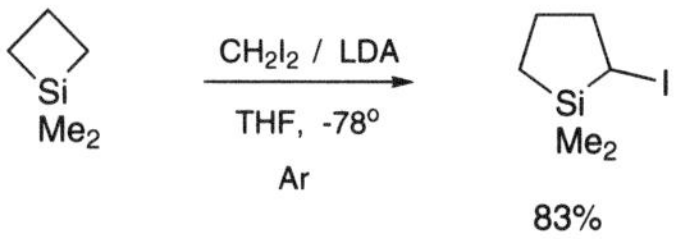

Simmons–Smith reaction.   Asymmetric methylene transfer to allyl glycosides containing a free hydroxyl group at C-2 arises from its directing effect. Both β-L-glucopyranosides[2] and their α-D- analogs[3] have been exploited for the access of chiral cyclopropanes.

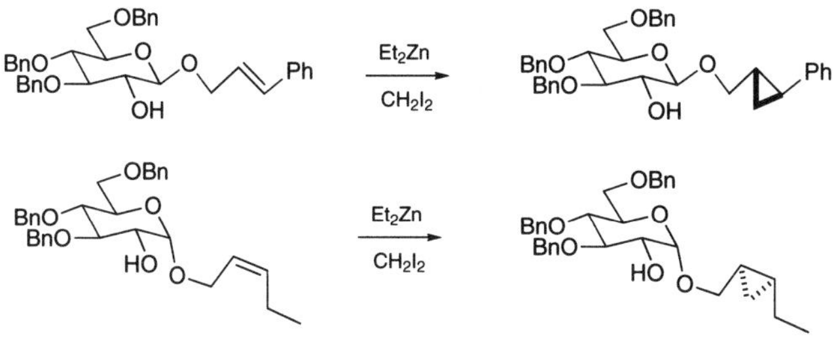

An allylic alcohol may undergo an asymmetric cyclopropanation[4] by attachment of a chiral ligand (e.g., a tartrate ester) to the derived zinc alkoxide.

Trimethylaluminum can serve the same role as Zn-Cu or Et$_2$Zn to activate diiodomethane for cyclopropanation.[5]

Homologation of allylic alcohols.[6]   The Simmons–Smith reaction of certain allylic alcohols becomes a minor pathway when they are treated with Et$_3$Al/Et$_2$AlOEt while using CH$_2$I$_2$-Et$_2$AlCl as reagents. Homoallylic iodides are obtained.

Methylenation of ketones.[7]   The combination Zn-TiCl$_4$ transforms CH$_2$I$_2$ into a methylenating agent. The Wittig-like reaction is accelerated by PbCl$_2$. Traces of lead are found to suppress the reactivity of Zn toward CH$_2$I$_2$, but the reactivity (for the Simmons–Smith reaction) is recovered by adding Me$_3$SiCl.

Methylene homologation of vinylcoppers.[8]   Allylmetal species are formed, which undergo transpositional alkylation.

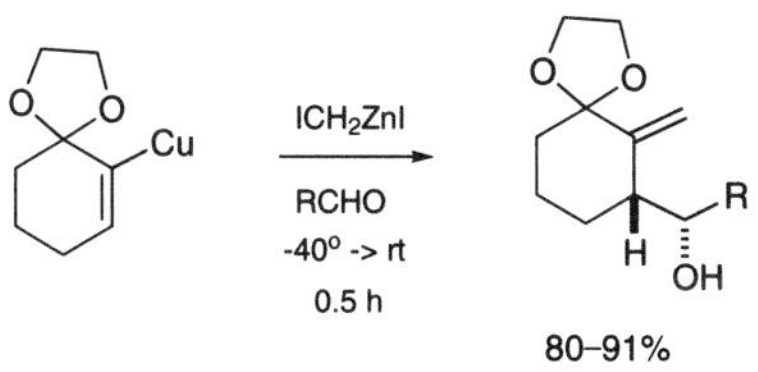

[1] K. Matsumoto, Y. Aoki, K. Oshima, K. Utimoto, and N. A. Rahman, *T* **49**, 8487 (1993).
[2] A. B. Charette and B. Cote, *JOC* **58**, 933 (1993).
[3] A. B. Charette, N. Turcotte, and J.-F. Marcoux, *TL* **35**, 513 (1994).
[4] Y. Ukaji, K. Sada, and K. Inomata, *CL* 1227 (1993).
[5] J. M. Russo and W. A. Price, *JOC* **58**, 3589 (1993).
[6] Y. Ukaji and K. Inomata, *CL* 2353 (1992).
[7] K. Takai, T. Kakiuchi, Y. Takaoka, and K. Utimoto, *JOC* **59**, 2668, 2671 (1994).
[8] A. Sidduri, M. J. Rozema, and P. Knochel, *JOC* **58**, 2694 (1993).

Diiron nonacarbonyl. 13, 320–321; 15, 334; 16, 351–353

Activation of allylic ethers. An allylic ether γ to an electron-withdrawing group is activated by forming the $Fe(CO)_4$ complex. On acid treatment ionization occurs to generate the allyl cation (still complexed to iron), which is reactive towards nucleophiles such as silyl enol ethers, malonate ester enolates, etc.[1] The substitution is stereoselective.[2]

[1] T. Zhou and J. R. Green, *TL* **34**, 4497 (1993).
[2] D. Enders, B. Jandeleit, and G. Raabe, *ACIEE* **33**, 1949 (1994).

Diisobutylaluminum hydride. 13, 115–116; 15, 137–138; 16, 134–135; 17, 123–125

Monosilyl acetals and amines. The DIBALH reduction products of esters can be converted into monosilyl acetals by treatment with TMSOTf-pyridine[1] and those of nitriles into amines by the addition of organolithium reagents.[2]

Opening of oxabicycles.[3] The S_N2' reduction opens the 2,5-dihydrofuran unit. Usually the presence of an *endo*-OH is required to assist the transformation. This method is applicable to the elaboration of the C_{17}–C_{23} segment of ionomycin.

83%

Special preparations. α-Trialkylsiloxy aldehydes[4] and 1-tosyloxy-2-alkanols[5] have been synthesized using DIBALH reduction of cyanohydrin silyl ethers and epoxy tosylates, respectively. Indoles and (Z)-allylic alcohols are acquired after simple manipulations of the primary reduction products of nitriles and α-hydroxy esters, respectively.

66%

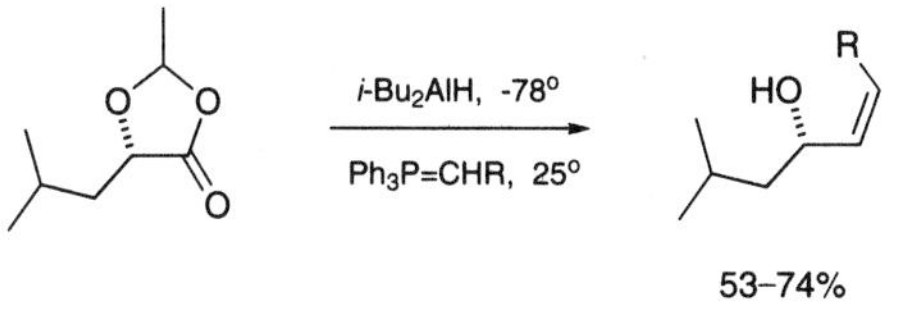

53–74%

[1] S. Kikooka, M. Shirouchi, and Y. Kaneko, *TL* **34**, 1491 (1993).
[2] G. Cainelli, M. Panunzio, M. Contento, D. Giacomini, E. Mezzina, and D. Giovagnoli, *T* **49**, 3809 (1993).
[3] M. Lautens, P. Chiu, and J.T. Colucci, *ACIEE* **32**, 281 (1993).
[4] M. Hayashi, T. Yoshiga, K. Nakatani, K. Ono, and N. Oguni, *T* **50**, 2821 (1994).
[5] J.M. Chong and J. Johansen, *TL* **35**, 7197 (1994).
[6] J.P. Marino and C.R. Hurt, *SC* **24**, 839 (1994).
[7] D.J. Krysan, *SC* **24**, 1589 (1994).

Diisobutylaluminum hydride–zinc chloride.

Reduction of α-sulfinyl ketones.   Chelation control operates when ZnCl₂ is added to the medium. The relative configuration of the resultant alcohol is often opposite to that obtained without the metal salt.[1,2] Magnesium bromide can be used instead of ZnCl₂.[3]

[1] J.L.G. Ruano, A. Fuerte, and M.C. Maestro, *TA* **5**, 1443 (1994).
[2] A.B. Bueno, M.C. Carreno, J.L.G. Ruano, B. Pefia, A. Rubio, and M.A. Hoyos, *T* **50**, 9355 (1994).
[3] G. Guanti, L. Banfi, R. Riva, and M.T. Zannetti, *TL* **34**, 5483 (1993).

Diisopinocampheylborane, (Ipc)₂BH, and B derivatives. 13, 117–118

Chiral 1,4-diols.[1]   Reaction of allyl ketones with this organoborane proceeds by a rapid hydroboration, which is followed by an intramolecular reduction.

An alternative method[2] comprises enantioselective allylation of aldehydes, which results in cyclic oxaboranes, transboration with BH₃ · SMe₂, and then oxidation. Prior to the oxidation the carbon chain can be elongated by reaction with BrCH₂Cl, BuLi. The diols have been converted to optically active 2-alkylpyrrolidines and -piperidines.

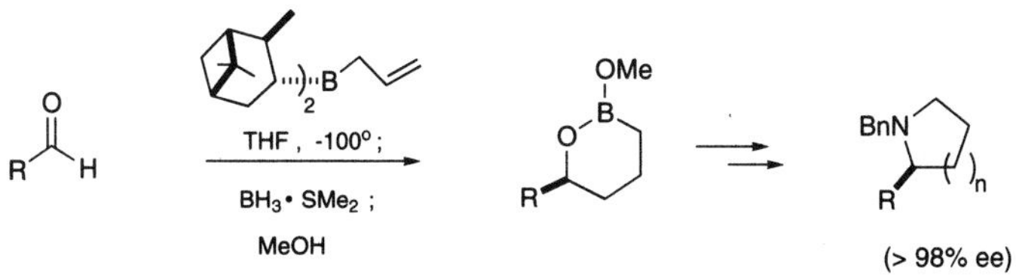

(> 98% ee)

Functionalized 1-alkenylboronates. The useful synthetic reagents are prepared by hydroboration of substituted 1-alkynes with (Ipc)$_2$BH and treatment of the alkenylboranes with acetaldehyde and tetramethylethylene glycol.[3]

Carbinols. *B*-Chlorodiisopinocampheylborane is a reducing agent that furnishes a hydride by severing the C–H bond geminal to the methyl group. Carbinols in good ee are formed from the trifluoromethyl ketones[4,5] and *o*-hydroxyacetophenones.[6] Interestingly, the configuration of the latter series is opposite to that of the reduction products of *o*-methoxyacetophenones.

3-Hetero allylboranes.[7,8] The 3-amino- and 3-silylallyl derivatives are very valuable reagents for the synthesis of *anti*-3-amino-1-alken-4-ols and *anti*-1-alkene-3,4-diols, respectively, by condensation with aldehydes. In the latter cases, an oxidative desilylation is required.

[1] G. A. Molander and K. L. Bobbitt, *JOC* **59**, 2676 (1994).
[2] T. Nguyen, D. Sherman, D. Ball, M. Solow, and B. Singaram, *TA* **4**, 189 (1993).
[3] A. Kamabuchi, T. Moriya, N. Miyaura, and A. Suzuki, *SC* **23**, 2851 (1993).
[4] P. V. Ramachandran, A. V. Teodorovic, and H. C. Brown, *T* **49**, 1725 (1993).
[5] P. V. Ramachandran, B. Gong, A. V. Teodorovic, and H. C. Brown, *TA* **5**, 1061, 1075 (1994).
[6] P. V. Ramachandran, B. Gong, and H. C. Brown, *TL* **35**, 2141 (1994).
[7] A. G. M. Barrett and M. A. Seefeld, *T* **49**, 7857 (1993).
[8] A. G. M. Barrett and J. W. Malecha, *JCS(P1)* 1901 (1994).

Diisopropenyl oxalate.

Oxalylation.[1]   Selective reaction with nucleophiles is accomplishable. For example, mixed oxamides are formed in a stepwise process with two different amines.

[1] M. Neveux, C. Bruneau, S. Lecolier, and P. H. Dixneuf, *T* **49**, 2629 (1993).

Diisopropoxytitanium(III) borohydride.

1,2-Reduction.[1]   This reducing agent is prepared in situ from (*i*-PrO)$_2$TiCl$_2$ and (PhNEt$_3$)BH$_4$. It shows chemoselectivity in the reduction of α,β-unsaturated carbonyl compounds.

[1] K. S. Ravikumar, S. Baskaran, and S. Chandrasekaran, *JOC* **58**, 5981 (1993).

Dilithium tetramethylcobaltate.

Substitution of alkenyl and alkynyl halides.[1]   Replacement of the chlorides and fluorides with a methyl group in excellent yields is achieved with the cobaltate or ferrate reagent.

[1] T. Kauffmann, R. Salker, and K.-U. Vob, *CB* **126**, 1447 (1993).

Dimesityl(alkyl)borane. 14, 6

Homologation of alkyl halides.[1]   A new way for the conversion of an alkyl halide to the homologous alcohol exploits the facile formation of dimesitylboryl-(alkyl)lithium by deprotonation and the excellent nucleophilicity of the lithiated species. Conventional oxidation after the alkylation completes the process. Epoxides can also be used as electrophiles.

[1] A. Pelter, L. Warren, J.W. Wilson, G. F. Vaughan-Williams, and R. M. Rosser, *T* **49**, 2988, 3007 (1993).

Dimethylaluminyl phenyl sulfide.

2-Alkyl-2-cyclohexenones.[1]   The alkylation of 2-cyclohexenone can be initiated by conjugate addition and trapping of the aluminum enolate with an aldehyde. Regeneration of the double bond and removal of the hydroxyl group are accomplished by mesylation. Actually, the liberated thiolate displaces the allylic mesylate in the process; thus the final operation involves desulfurization with Raney nickel.

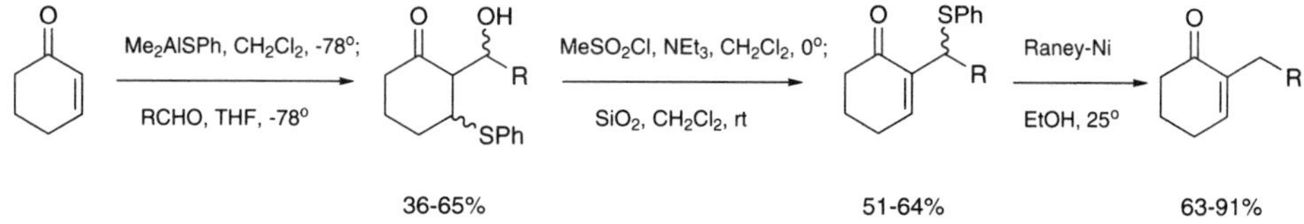

[1] M. A. Armitage, D. C. Lathbury, and M. B. Mitchell, *JCS(PI)* 1551 (1994).

Dimethylamino(dimethyl)[*o*-hydroxy-(*E*)-styryl]silane.

Alcohol protection.[1]   An alcohol displaces the dimethylamino group from silicon at room temperature (70–95%). Such silyl ethers are photolabile (254 nm); therefore, the deprotection avoids the addition of reagents.

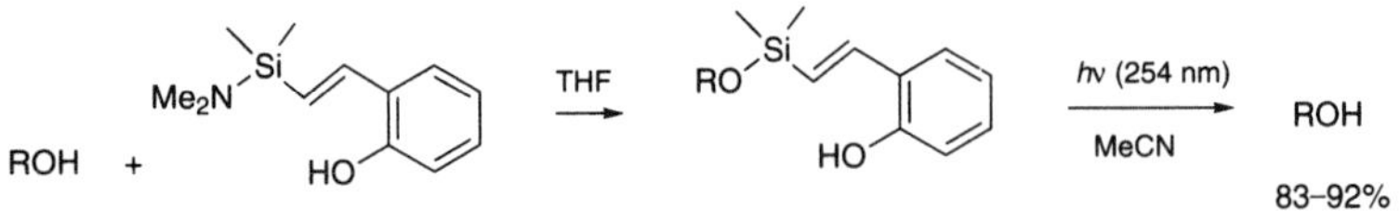

ROH + [structure with Me₂N-Si and HO-aryl] → THF → [structure with RO-Si and HO-aryl] → hν (254 nm) / MeCN → ROH

83–92%

[1] M. C. Pirrung and Y. R. Lee, *JOC* **58**, 6961 (1993).

Dimethyl carbonate.

Ketone–ester exchange.[1] Carbomethoxylation of ketones may be followed by deacylation in situ when the reaction is carried out at high temperatures.

Methylation of activated toluenes.[2] Arylacetonitriles and methyl arylacetates are monomethylated at the benzylic position on heating with $(MeO)_2CO$ and K_2CO_3 in an autoclave at 180°C.

3-Methyloxazolin-2-ones.[3] Under similar conditions ketoximes undergo the very unusual heterocyclization in 22–48% yield.

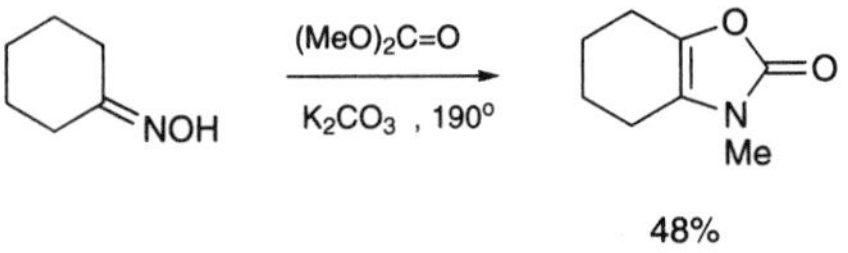

48%

[1] M. Selva, C. A. Marques, and P. Tundo, *G* **123**, 515 (1993).
[2] M. Selva, C. A. Marques, and P. Tundo, *JCS(PI)* 1323 (1994).
[3] C. A. Marques, M. Selva, P. Tundo, and F. Montanari, *JOC* **58**, 5765 (1993).

Dimethyldioxirane. 12, 413; 13, 120; 14, 148; 15, 143–144; 16, 142–144

Generated from Caro's acid and used in situ.[1]

Epoxidation. The epoxidation products of allenic compounds are prone to secondary reactions; thus an aldehyde group in proper distance may induce isomerization.[2]

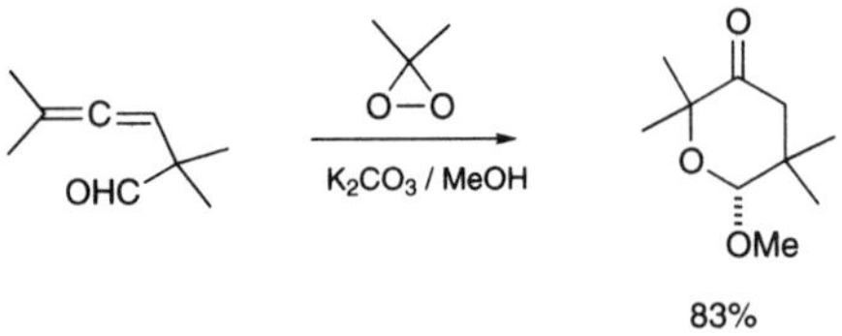

83%

Oxidations. Secondary alcohols are selectively oxidized to ketones, including *vic*-diols (without C–C bond cleavage).[3,4] *sec,sec*-1,2-Diols, which are C_2-symmetrical, give chiral ketols in high optical purity.[5]

> 96%   (94% ee)

Sulfides are oxidized to sulfoxides.[6] One application of this reaction is the removal of a benzylthioethyl group[7] from protected oligosaccharides. The oxidation makes the group base-labile. When the sulfide is coordinated to a metal that also bears chiral ligands, the oxidation becomes enantioselective.[8]

Cyclic thioamides are desulfurized to afford imines.[9]

C=N → C=O.   Ketoximes,[10] hydrazones,[11] and diazo compounds[12,13] give carbonyl products.

Hydroxylations.   Titanium enolates undergo diastereoselective hydroxylation to furnish α-ketols $(60-97\%)$[14] on exposure to dimethyldioxirane. The reaction is very rapid $(-78°C, 1 \text{ min})$.

De-O-benzylation.[15]   The oxidative cleavage occurs without affecting many other functional groups, including secondary alcohol, OTBS.

Oxidation of aromatic and heteroaromatic compounds.   The oxidation of methoxyarenes to afford *p*-quinones[16] requires acid catalysts.

Benzofurans and substituted indoles undergo epoxidation, but the products rearrange readily to give benzannulated lactones[17] and lactams.[18] Furan itself is converted to malealdehyde,[19] which can be trapped by Wittig reagents.

62%

[1] C. W. Jones, J. P. Sankey, W. R. Sanderson, M. C. Rocca, and S. L. Wilson, *JCR(S)* 114 (1994).

[2] J. K. Crandall and E. Rambo, *TL* **35**, 1489 (1994).

[3] R. Curci, L. D'Accolti, A. Detomaso, C. Fusco, K. Takeuchi, Y. Ohga, P. E. Eaton, and Y. C. Yip, *TL* **34**, 4559 (1993).

[4] P. Bovicelli, P. Lupattelli, A. Sanetti, and E. Mincione, *TL* **35**, 8477 (1994).

[5] L. D'Accolti, A. Detomaso, C. Fusco, A. Rosa, and R. Curci, *JOC* **58**, 3600 (1993).

[6] R. S. Glass and Y. Liu, *TL* **35**, 3887 (1994).

[7] T.-H. Chan and C.-P. Fei, *CC* 825 (1993).

[8] W. A. Schenk, J. Frisch, W. Adam, and F. Prechtl, *ACIEE* **33**, 1609 (1994).

[9] C. Claudia, E. Mincione, R. Saladino, and R. Nicoletti, *T* **50**, 3259 (1994).

[10] G. A. Olah, Q. Liao, C.-S. Lee, G. K. S. Prakash, *SL* 427 (1993).

[11] A. Altamura, R. Curci, and J. O. Edwards, *JOC* **58**, 7289 (1993).

[12] A. Saba, *SC* **24**, 695 (1994).

[13] P. Darkins, N. McCarthy, M. A. McKervey, and T. Ye, *CC* 1222 (1993).
[14] W. Adam, M. Müller, and F. Prechtl, *JOC* **59**, 2358 (1994).
[15] R. Csuk and P. Dörr, *T* **50**, 9983 (1994).
[16] W. Adam and M. Shimizu, *S* 560 (1994).
[17] W. Adam and M. Sauter, *T* **50**, 11441 (1994).
[18] Z. Zhang and C. S. Foote, *JACS* **115**, 8867 (1993).
[19] B. J. Adger, C. Barrett, J. Brennan, P. McGuigan, M. A. McKervey, and B. Tarbit, *CC* 1220 (1993).

N,N-Dimethylformamide–phosphoryl chloride.

Cyclodehydration.[1] A mild and efficient cyclodehydration of *o*-hydroxyalkyl-phenols calls for the use of $ClCH=N^+Me_2\ Cl^-$. The method is adaptable to large-scale operations.

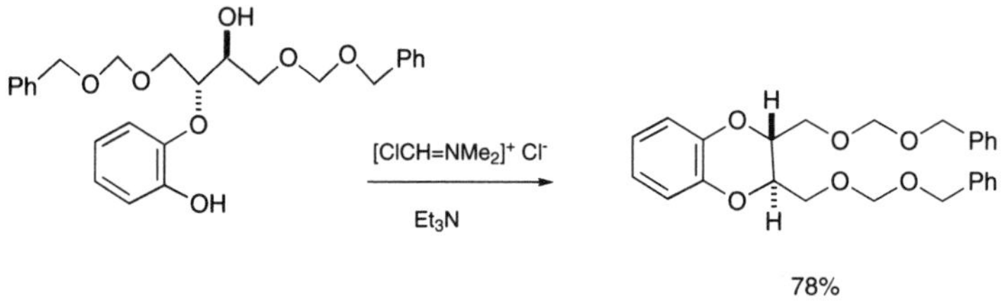

78%

[1] P. A. Procopiou, A. C. Brodie, M. J. Deal, and D. F. Hayman, *TL* **34**, 7483 (1993).

N,N-Dimethylformamide–thionyl chloride.

Sulfene.[1] Methanesulfonic acid gives sulfene ($CH_2=SO_2$) on reaction with the Vilsmeier reagent.

[1] D. Prajapati, S. P. Singh, A. R. Mahajan, and J. S. Sandhu, *S* 468 (1993).

2,2-Dimethylhydrazino(dimethyl)aluminum.

Hydrazones.[1] The reagent is prepared from *N,N*-dimethylhydrazine and trimethylaluminum in hot toluene (97% yield). It is useful for derivatization of unreactive ketones (e.g., ferrocenyl ketones) in refluxing toluene. The fully substituted hydrazones can undergo exchange on reaction with N_2H_4 in ethanol. Tropone hydrazone, which has previously been unavailable, can be prepared by this method.

[1] B. Bilstein and P. Denifl, *S* 158 (1994).

1,3-Dimethylimidazolium iodide.

α-Diketones.[1] *N*-Phenylimidoyl chlorides react with aldehydes in the presence of the imidazolium iodide and a base (ylide generation). Acid hydrolysis of the resulting α-keto imines leads to α-diketones.

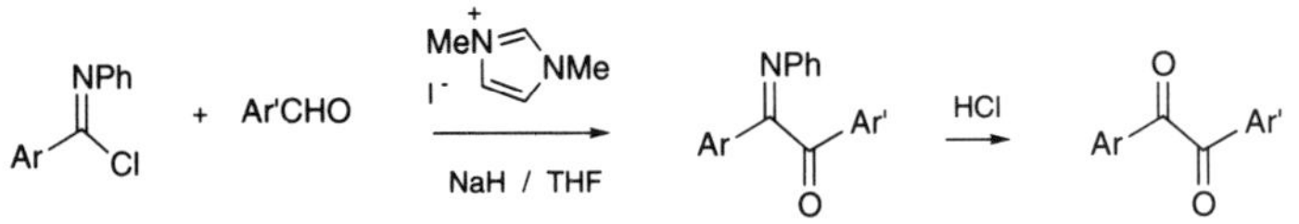

[1] A. Miyashita, H. Matsuda, and T. Higashino, *CPB* **40**, 2627 (1992).

Dimethyl methylphosphonate. 16, 145

Cycloalkenones.   The nucleophilic attack of lithiomethylphosphonates on 2,2-disubstituted 1,3-cycloalkanediones results in fragmentation and Emmons–Wadsworth reaction to give 3-substituted 2-cycloalkenones.[1]

The reaction of the same reagent with α,ω-diesters furnishes bis(β-ketophosphonates), which may undergo cyclization.[2,3] The glutarate-derived species undergoes rapid aldolization, permitting subsequent addition of aldehydes to generate 3-alkenyl-2-cyclohexenones.

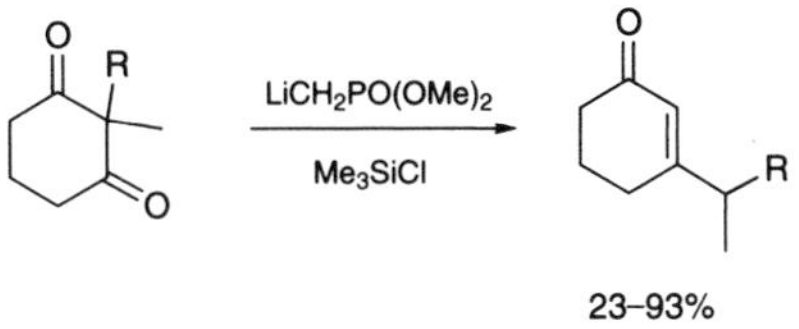

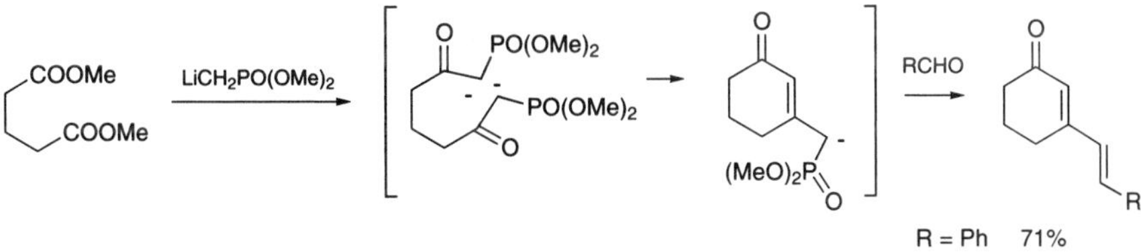

[1] T. Furuta, E. Oshima, and Y. Yamamoto, *HC* **3**, 471 (1992).
[2] E. Wenkert and M. K. Schorp, *JOC* **59**, 1943 (1994).
[3] M. Mikolajczyk and M. Mikina, *JOC* **59**, 6760 (1994).

Dimethyl(methylthio)sulfonium salts.

Glycosylation.[1]   The triflate is effective in the activation of an anomeric *N*-allylcarbamoyl group for reaction with glycosyl acceptors.

Disulfide bond formation.[2]   Coupling of cysteine units in peptides is easily achieved on treatment with the tetrafluoroborate salt.

[1] H. Kunz and J. Zimmer, *TL* **34**, 2907 (1993).
[2] P. Bishop, C. Jones, and J. Chmielewski, *TL* **34**, 4469 (1993).

Dimethyloxosulfonium methylide. 14, 152; **15**, 147; **16**, 146; **17**, 126

Methylenation.[1]   Acylsilanes give silyl enol ethers.

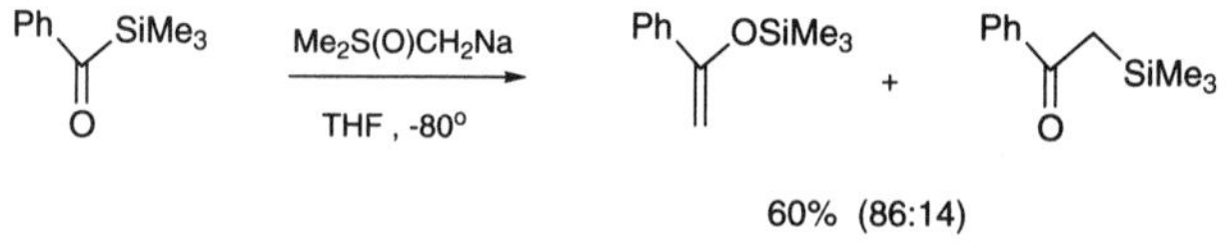

60% (86:14)

Cyclopropanation.   The reaction of the ylide with enones can be performed in the solid state (yields 79–91%) using KOH as base.[2] Under the same conditions epoxides and aziridines are obtained from saturated ketones and imines, respectively.

[1] T. Nakajima, M. Segi, F. Sugimoto, R. Hioki, S. Yokota, and K. Miyashita, *T* **49**, 8343 (1993).
[2] F. Toda and N. Imai, *JCS(P1)* 2673 (1994).

1,4-Dimethylpiperidine-2,3-dione.

α-Diketones.[1]   Both symmetrical and unsymmetrical α-diketones can be made from organometallic reaction with the dilactam.

[1] U. T. Mueller-Westerhoff and M. Zhou, *JOC* **59**, 4988 (1994).

N,N′-**Dimethylpropyleneurea (DMPU). 13**, 122; **16**, 146–147

Solvent for alkylation.[1]   *N*-Alkylation of chiral amines in DMPU gives secondary amines in excellent enantiopurity.

[1] E. Juaristi, P. Murer, and D. Seebach, *S* 1243 (1994).

Dimethylsilyl bistriflate.

Enolsilylation.[1]   Silyl enolates derived from this reagent (in the presence of *i*-PrNEt₂) undergo aldol and Michael reactions without additional catalysts.

[1] S. Kobayshi and K. Nishio, *JOC* **58**, 2647 (1993).

Dimethylsilyl dichloride. 17, 113–114

Tethering through dialkoxysilanes.[1]   The linking of hydroxyl groups from two molecules favor their subsequent reaction. More important, the tactic affords stereocontrol in these reactions, as shown in an intramolecularized glycosylation.

[1] M. Bols, *T* **49**, 10049 (1993).

Dimethyl sulfide–*N*-chlorosuccinimide.

Methylthiomethylation. In the presence of a tertiary amine, phenols[1] and indole derivatives[2] undergo *C*-alkylation with the sulfonium salt derived from the two reagents.

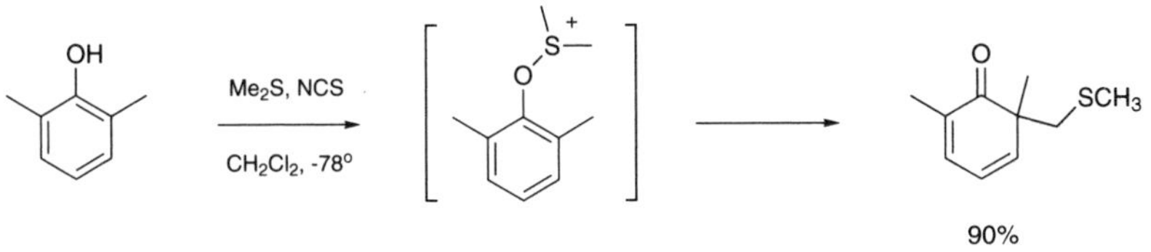

90%

[1] S. Katayama, T. Watanabe, and M. Yamauchi, *CPB* **41**, 439 (1993).
[2] S. Katayama, T. Watanabe, and M. Yamauchi, *CPB* **40**, 2836 (1992).

Dimethylsulfonium methylide.

Vinylation.[1] Carbonyl compounds react with two equivalents of the ylide to give allylic alcohols via epoxide intermediates.

Oxidative methylenation.[2] Alkyl halides and mesylates are homologated to alkenes. Similarly, epoxides give allylic alcohols with one more carbon atom.

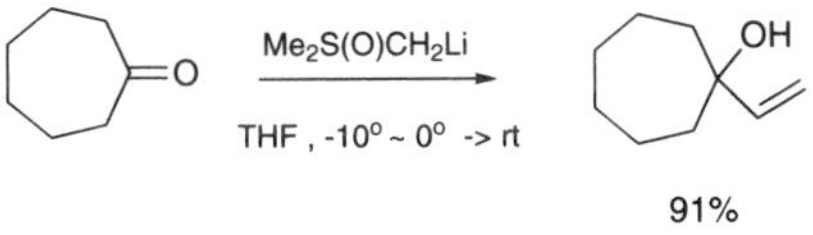

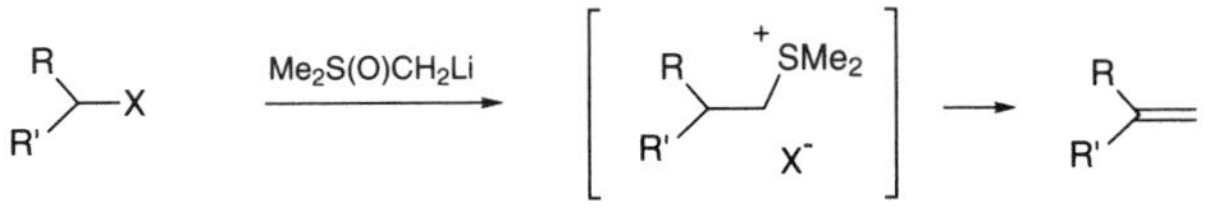

91%

[1] J. Harnett, L. Alcaraz, C. Mioskowski, J. P. Martel, T. L. Gall, D.-S. Shin, and J. R. Falck, *TL* **35**, 2009 (1994).
[2] L. Alcaraz, J. J. Harnett, C. Mioskowski, J. P. Martel, T. L. Gall, D.-S. Shin, and J. R. Falck, *TL* **35**, 5449, 5453 (1994).

Dimethyl sulfoxide. 13, 124; 16, 149

α,β-Unsaturated aldehydes.[1] Heating β-hydroxydithioacetals in DMSO at 160°C gives the conjugated aldehydes in 63–92% yield.

[1] C. S. Rao, M. Chandrasekharam, B. Patro, H. Ila, and H. Junjappa, *T* **50**, 5783 (1994).

Dimethyl sulfoxide–trimethylsilyl chloride. 15, 146

(Pyrrol-2-yl)dimethylsulfonium chlorides.[1] Substitution at C-2 of pyrroles is accomplished in MeCN at 0°C.

[1] F. Bellesia, F. Ghelfi, R. Grandi, U. M. Pagnoni, and A. Pinetti, *JHC* **30**, 617 (1993).

3,3-Dimethyl-1,2,5-thiadiazolidine 1,1-dioxide.

Esters.[1] The derived zwitterionic species (1) can be used to condense acid and alcohol pairs.

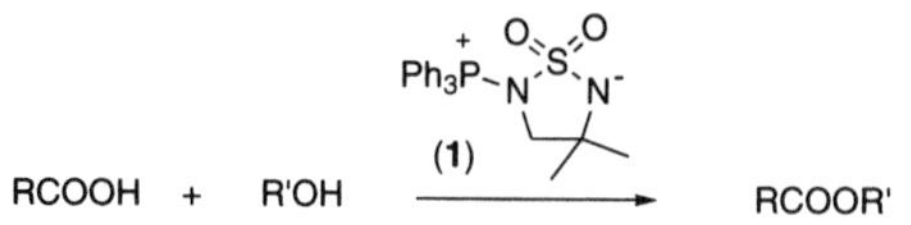

[1] J. L. Castro, V. G. Matassa, and R. G. Ball, *JOC* **59**, 2289 (1994).

Dinitrogen pentoxide.

Opening of small heterocycles.[1] Epoxides, oxetanes, and aziridines are opened to form dinitrates or nitroaminoethyl nitrates.

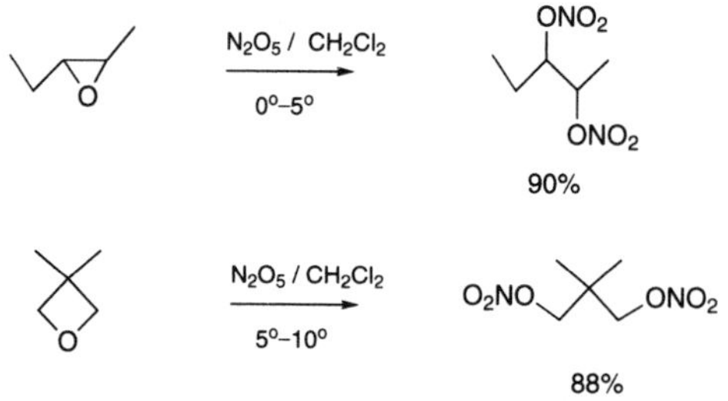

Nitration.[2] A new nitration system consisting of N_2O_5 and SO_2 has the obvious advantage in the isolation of the products, since any excess of reagents is very easily removed. Thus 3-nitropyridine is readily formed at low temperatures.

[1] P. Golding, R. W. Millar, N. C. Paul, and D. H. Richards, *T* **49**, 7037, 7051, 7063 (1993).
[2] J. M. Bakke and I. Hegbom, *ACS* **48**, 181 (1994).

Diorganyl tellurides.

(Z)-Alkenylcuprates.[1] (Z)-Dialkenyl tellurides, prepared from alkynes, are readily transformed into cuprate reagents by mixing with (2-thienyl)CuBu(CN)Li$_2$. With these species all the typical cuprate reactions can be carried out.

Olefination of aldehydes.[2] Diazo compounds are converted to stabilized telluronium ylides by diorganyl tellurides (typically Bu$_2$Te) in the presence of a catalytic amount of CuI. The olefination of aldehydes with such reagents is feasible.

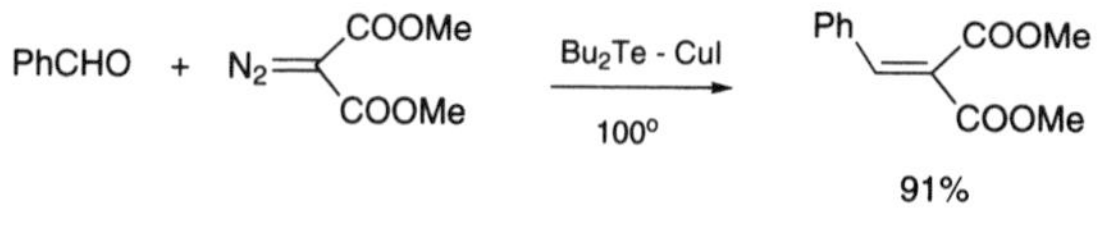

[1] J. P. Marino, F. Tucci, and J.V. Comasseto, *SL* 761 (1993).
[2] Z.-L. Zhou, Y.-Z. Huang, and L.-L. Shi, *T* **49**, 6821 (1993).

Diphenyl azidophosphonate (= diphenyl phosphorazidate).

Isocyanates.[1] This reagent has similar properties to diphenylphosphinic azide. The reaction on conjugated acids produces isocyanates which can be used in cycloadditions.

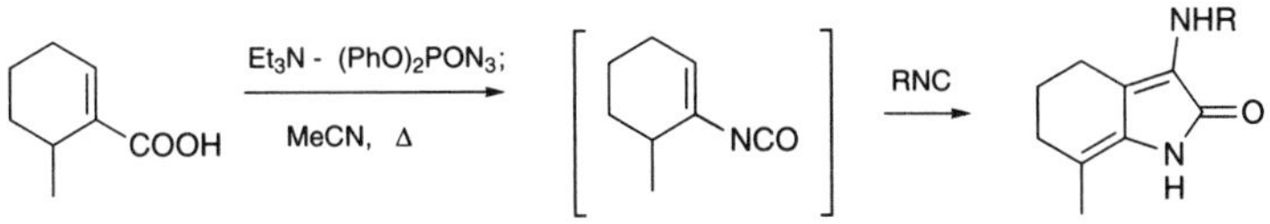

Alkyl azides.[2] The substitution of an alcohol to give the azide is an alternative to the Mitsunobu reaction.

[1] J. H. Rigby, M. Qabar, G. Ahmed, and R. C. Hughes, *T* **49**, 10219 (1993).
[2] A. S. Thompson, G. R. Humphrey, A. M. DeMarco, D. J. Mathre, and E. J. J. Grabowski, *JOC* **58**, 5886 (1993).

Diphenyl chlorophosphonate (= diphenyl phosphorochloridate).

Azirenes from amides.[1] Ketene iminium salts are generated from amide enolates. Further reaction with NaN_3 leads to 2-aminoazirenes.

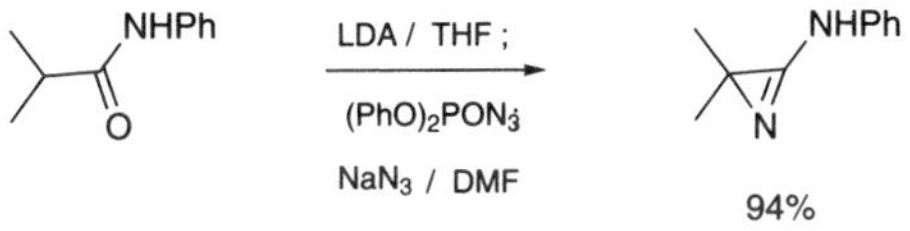

[1] J. M. Villalgordo and H. Heimgartner, *HCA* **76**, 2830 (1993).

Diphenyl diselenide. 13, 125

A new preparative method1 uses diphenyliodonium chloride and $(EtO)_2P(O)SeNa$, PhOH in DMF. Air oxidation of the product in aqueous NaOH is needed.

Oxidative functionalization of alkenes. Activated by cerium(IV) ammonium nitrate in methanol Ph2Se2 initiates addition to an alkene. Some unsaturated alcohols undergo phenylselenocycloetherification,[3] and in case that the activation is by a photoinduced electron transfer process the phenylseleno group of the products can be replaced.

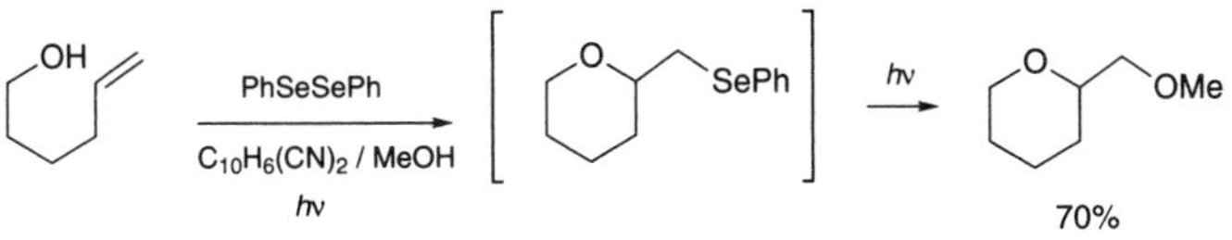

β-Hydroxy ketones from epoxy ketones.[4] Using thiol/diselenide exchange to generate the benzeneselenolate anion, catalytic reductive ring opening of epoxy ketones is achieved. N-Acetylcycteine may be employed as the supplementary reagent.

Peptide synthesis.[5]   Coupling reaction using Ph_2Se_2–Bu_3P is a replica of the previous method in which a diaryl disulfide is employed.

[1] Z.-D. Liu and Z.-C. Chen, *SC* **23**, 2673 (1993).
[2] C. Bosman, A. D'Annibale, S. Resta, and C. Trogolo, *TL* **35**, 6525 (1994).
[3] G. Pandey and B. B. V. S. Sekhar, *JOC* **59**, 7367 (1994).
[3] L. Engman and D. Stern, *JOC* **59**, 5179 (1994).
[5] S. K. Ghosh, U. Singh, M. S. Chadha, and V. R. Mamdapur, *BCSJ* **66**, 1566 (1993).

Diphenyl diselenide–ammonium persulfate.

γ-Oxy α,β-unsaturated esters.[1]   Transposed oxygenation of β,γ-unsaturated esters by the reagent combination in hydroxylic solvents is initiated by oxidation of Ph_2Se_2. The addition to the double bond is followed by another oxidation of the phenyl selenide and selenoxide elimination. The elimination seems to occur only when a conjugate system is formed. Note that β,γ-unsaturated acids are converted to butenolides,[2] and α-alkenyl β-keto esters to furans.[3]

Heterocyclization.   3-Butenyl ketoximes give either 1,2-oxazines or cyclic nitrones, depending on the relative configuration of the OH and the unsaturated side chain.[4] Oxazine formation is possible only when the oxygen atom of the oximes is on the same side as the double bond.

β,γ-Unsaturated hydroxamic acids do not have the same geometrical restriction. However, a kinetic process leads to cyclic N-hydroxy imidates, and the thermodynamic reaction products are the N-hydroxylactams.[5]

Isoxazolidines are also available from O-allyl hydroxylamines[6] and O-allyl oximes.[7]

[1] M. Tiecco, L. Testaferri, M. Tingoli, L. Bagnoli, and C. Santi, *CC* 637 (1993).
[2] M. Tiecco, L. Testaferri, M. Tingoli, L. Bagnoli, and C. Santi, *SL* 798 (1993).
[3] M. Tiecco, L. Testaferri, M. Tingoli, and F. Marini, *SL* 373 (1994).
[4] M. Tiecco, L. Testaferri, M. Tingoli, and F. Marini, *JCS(P1)* 1989 (1993).
[5] M. Tiecco, L. Testaferri, M. Tingoli, and F. Marini, *CC* 221 (1994).
[6] M. Tiecco, L. Testaferri, M. Tingoli, and C. Santi, *TL* **36**, 163 (1995).
[7] M. Tiecco, L. Testaferri, M. Tingoli, L. Bagnoli, and F. Marini, *CC* **235**, 237 (1995).

Diphenyl diselenide–phenyliodine(III) dicarboxylate.

Selenosulfonates.[1]   Oxidation of Ph_2Se_2 with $PhI(OCOCF_3)_2$ in the presence of an arenesulfinate salt gives the Se-sulfonate in one step (68–81% yield).

Azido-selenenylation.[2]   The introduction of a terminal azido group to an unsaturated ketone allows azacycles to be prepared. Access to pyridine derivatives from 3-butenyl ketones is particularly convenient based on azido-selenenylation because of the facile elimination of the selenenyl group after oxidation.

[1] D.-W. Chen and Z.-C. Chen, *TL* **35**, 7637 (1994).
[2] M. Tingoli, M. Tiecco, and L. Testaferri, *JOC* **58**, 6097 (1993).

Diphenyl disulfide. 14, 155–156

Stilbene isomerization.[1]   The (Z)- to-(E) isomerization is effected thermally.

(E)-Selective Wittig reactions.[2]   The reaction is conducted in the presence of Ph_2S_2 under visible light to maximize the (E)-isomer.

Oxidative sulfenylation of fluorinated amines.[3]   Anodic conditions convert the amines into sulfenylated imines, which can be used to react with organometallic reagents and enolates.

γ-Phenylthio ketones.[4]   These compounds are available from a three-component condensation involving α-mercurio ketones, alkenes, and Ph_2S_2 under uv irradiation.

63%

[1] M. A. Ali and Y. Tsuda, *CPB* **40**, 2842 (1992).
[2] J. K. Matilainen, S. Kaltia, and T. Hase, *SL* 817 (1994).
[3] T. Fuchigami, S. Ichikawa, and A. Konno, *CL* 2405 (1992).
[4] G. A. Russell and S. V. Kulkarni, *JOC* **58**, 2678 (1993).

2-Diphenylmethylsilylethanol.

Phosphorus protection.[1]   The protecting group is stable to acid and is removed by fragmentation using aqueous NH_4OH. It is introduced by the phosphoramidite approach and is useful for internucleotidic bonds in oligodeoxynucleotide synthesis.

[1] V. T. Ravikumar, T. K. Wyrzkiewicz, and D. L. Cole, *T* **50**, 9255 (1994).

Diphenylphosphine.

Free radical cyclization.[1]

X = O, NCOOMe

[1] J. E. Brumwell, N. S. Simpkins, and N. K. Terrett, *TL* **34**, 1215, 1219 (1993).

Diphenylphosphinic azide, $Ph_2P(O)N_3$.

Isocyanates.[1] The azide in combination with 1,8-bis(dimethylamino)-naphthalene effects the direct conversion of carboxylic acids to isocyanates.

[1] J.W. Gilman and Y. A. Otonari, *SC* **23**, 335 (1993).

Diphenylphosphinic chloride, $Ph_2P(O)Cl$. **15**, 150

N-Diphenylphosphinoylaziridines.[1] 1,2-Amino alcohols are doubly derivatized. As the phosphonate anion is sufficiently electrofugal, the treatment of the derivatives with NaH causes cyclization.

Allylic diphenylphosphinates.[2] The esterification is mediated by imidazole. The salt is probably the active reagent.

[1] H. M. I. Osborn, A. A. Cantrill, J. B. Sweeney, and W. Howson, *TL* **35**, 3159 (1994).
[2] J. S. McCallum, and L. S. Liebeskind, *S* 819 (1993).

Diphenylsilane.

Defunctionalization.[1] Removal of halogen atoms from bromo and iodo compounds, deoxygenation of alcohols, and deamination can be achieved by free radical chain processes. The reaction employs diphenylsilane as the hydrogen source to effect the hydrogenolysis; alcohols and amines must be derivatized as xanthates and isocyanides, respectively.

[1] D. H. R. Barton, D. O. Jang, and J. C. Jaszberenyi, *T* **49**, 7193 (1993).

Diphenyltin sulfide–silver perchlorate.

Aldol condensation.[1] Silyl enol ethers and silyl ketene acetals react with aldehydes smoothly in the presence of Ph_2SnS or Lawessons' reagent in combination with the very weak Lewis acid $AgClO_4$ in CH_2Cl_2 at $-78°C$.

[1] T. Mukaiyama, K. Saito, H. Kitagawa, and N. Shimomura, *CL* 789 (1994).

Diphosgene.

Esters and thiol esters.[1] Treatment of acids with diphosgene and pyridine at $-40°C$ and then with alcohols or thiols produces esters and thiol esters.

[1] D. Ravi, N. R. Rao, G. S. Reddy, K. Sucheta, and V. J. Rao, *SL* 856 (1994).

Dipotassium tetracarbonylferrate.

This compound is nonpyrophoric and easier to handle than commercial Collman's reagent. Its preparation[1] in essentially quantitative yield from $Fe(CO)_5$ involves treatment with KOH in MeOH for 0.5 h, and with Bu_3P in refluxing THF. The pure product can be stored under argon for months.

[1] A. Baby, J.-J. Brunet, F. B. Kindela, and D. Neibecker, *SC* **24**, 2827 (1994).

Disodium tetracarbonylcobaltate.

NpOTs → NpCOOMe.[1] Monoesters and diesters are produced from the corresponding tosylates.

[1] G. Cometti, A. Du Vosel, F. Francalanci, R. Santi, W. Cabri, and M. Foa, *JOMC* **451**, C13 (1993).

Disodium tetracarbonylferrate. 15, 152–153

α-Diketones.[1] Reaction of an alkyl bromide with $Na_2Fe(CO)_4$ under CO and then with CuCl, followed by demetallation (CAN), completes a double carbonylation process.

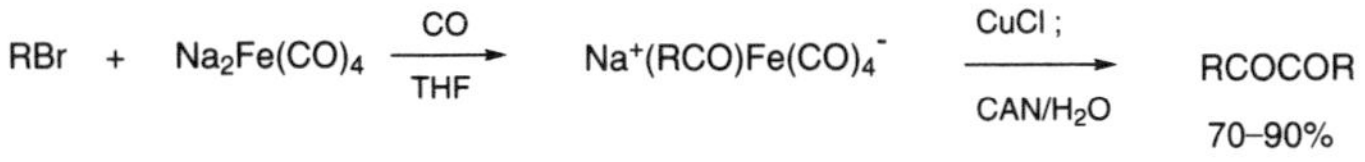

$$RBr + Na_2Fe(CO)_4 \xrightarrow[THF]{CO} Na^+(RCO)Fe(CO)_4^- \xrightarrow[CAN/H_2O]{CuCl;} RCOCOR$$

70–90%

Cyclohexanone-2-carboxylic esters.[2] The reagent acts as a condensation catalyst for twofold Michael additon of a donor to an acrylate ester and Dieckmann cyclization of the 1:2 adducts. However, NaOMe has the same capability; therefore, $Na_2Fe(CO)_4$ is not unique.

[1] M. Periasamy, A. Devasagayaraj, and U. Radhakrishnan, *OM* **12**, 1424 (1993).
[2] M. Periasamy, M. R. Reddy, U. Radhakrishnan, and A. Devasagayaraj, *JOC* **58**, 4997 (1993).

Disodium tetrachloropalladate.

Carbonylation of iodoxyarenes.[1] Aroic acids are obtained from $ArIO_2$ by reaction with CO in water in the presence of the palladium salt.

[1] V. V. Grushin and H. Alper, *JOC* **58**, 4794 (1993).

trans-**1,3-Dithiane 1,3-dioxide.**

α-Hydroxy acid derivatives.[1] Alkylation with aldehydes, Pummerer rearrangement, and ancillary manipulations render the acid derivatives in chiral form.

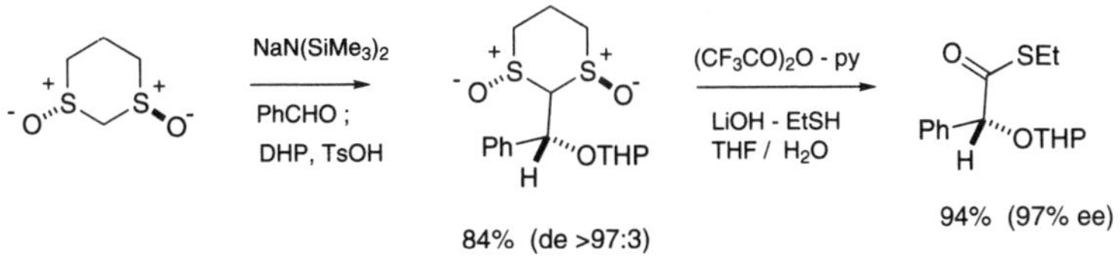

84% (de >97:3)

94% (97% ee)

[1] V. K. Aggarwal, A. Thomas, and R. J. Franklin, *CC* 1653 (1994).

1,3-Dithiolane 1,1,3,3-tetroxides.

Vinyl sulfones.[1] Ring cleavage of these heterocycles takes place on treatment with base (e.g., i-Pr$_2$NEt).

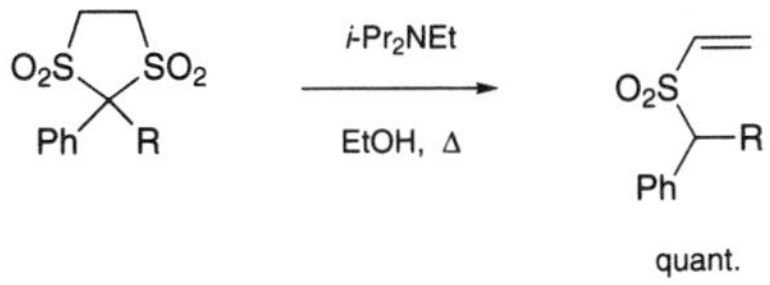

quant.

[1] B. E. Love and L. Chao, *SC* **23**, 3073 (1993).

E

α-Ethoxyalkenyllithium reagents.
An improved preparation of these reagents consists of deprotonating vinyl ethers with *t*-BuLi in tetrahydropyran.

α-Ethoxyvinyl ketones.[1] Reaction of the lithium reagent with *N*,*N*-dialkylcarboxamides in THF gives ketones.

[1] M. Shimano and A. I. Meyers, *TL* **35**, 7727 (1994).

Ethoxycarbonylhydrazine.
α-Methoxylation of enones.[1] In a one-pot reaction the hydrazone of an enone is converted to the α-methoxy enone on treatment with bromine, acidic methanol, and formaldehyde.

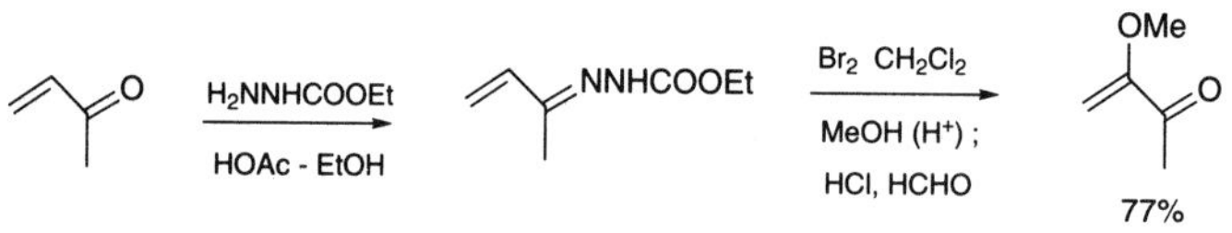

[1] A. Feuerer and T. Severin, *JOC* **59**, 6026 (1994).

α-Ethoxyvinyl acetate.
Pummerer reaction.[1] Upon reaction with α-ethoxyvinyl acetate chiral sulfoxides in which the α-carbon is substituted with an electron-withdrawing group are transformed into α-acetoxy sulfides. In most cases the enantiomer excess of the products amount to 70–80%.

[1] Y. Kita, N. Shibata, N. Kawano, S. Fukui, and C. Fujimori, *TL* **35**, 3575 (1994).

N-Ethyl-2-azidopyridinium tetrafluoroborate.
Transdiazotization.[1] Acylacetaldehydes are diazotized by the reagent (**1**) in nonbasic conditions. This method is distinguished by the retention of the formyl group in the products.

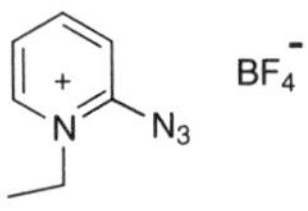

(1)

[1] O. Sezer and O. Anac, *HCA* **77**, 2323 (1994).

Ethylene glycol. 15, 156

Pyrrole formation.[1] 3,4-Diaroylpyrrolidines are converted to 2-aryl-3-methyl-4-acylpyrroles on heating in ethylene glycol at 130°C. No reaction occurs without the solvent.

[1] S. Mataka, H. Kitagawa, O. Misumi, M. Tashiro, and K. Kamata, *CC* 670 (1993).

Ethylene oxide.

Cyclopropanation.[1] Sodium enolates of β-ketophosphonate esters react with ethylene oxide in a sealed tube to form spiroannulated cyclopropyl ketones in moderate yields. Other epoxides give substituted cyclopropane derivatives.

40%

Acetalization.[2] Gentle and rapid acetalization is accomplished in the presence of a Lewis acid catalyst.

[1] T. E. Jacks, H. Nibbe, and D. F. Wiemer, *JOC* **58**, 4584 (1993).
[2] D. S. Torok, J. J. Figueroa, and W. J. Scott, *JOC* **58**, 7274 (1993).

Ethyl diazoacetate. 16, 163–164

Homologation of allyl sulfides.[1] Sulfonium ylides formed at low temperature are liable to undergo a [2,3]sigmatropic rearrangement on treatment with a mild base (e.g., DBU).

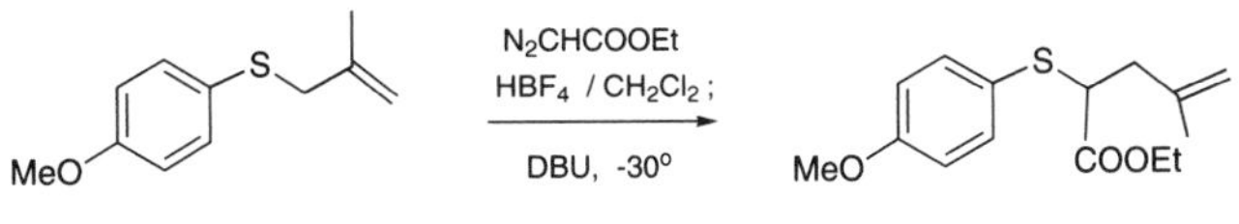

86%

[1] R. C. Hartley, S. Warren, and I. C. Richards, *JCS(PI)* 507 (1994).

1-Ethyl-3-(3-dimethylaminopropyl)carbodiimide, polymer supported.

Amide formation.[1]   Dehydrative coupling of an acid and an amine is readily achieved with this readily prepared reagent. No aqueous workup is necessary, and the product is easily separated from the polymer-bound urea by-product by simple filtration, rinse and evaporation.

[1] M. C. Desai and L. M. S. Stramiello, *TL* **34**, 7685 (1993).

Ethylene sulfate. **15**, 105–107

Cyclopropanation.[1]   This is a superior reagent for cyclopropanation of dibenzylaminoacetonitrile compared to 1,2-dibromoethane. In the presence of LDA extensive elimination occurs with the dibromide, and the latter also induces dimerization of the aminonitrile.

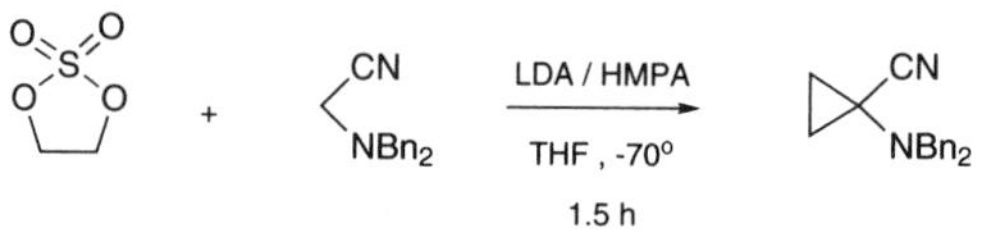

4-Hydroxyalkanonitriles.[2]   This cyclic sulfate and especially the substituted congeners are epoxide equivalents in reactivity toward carbon nucleophiles such as the acetonitrile carbanion.

[1] D. Guillaume, M. Brum-Bousquet, D. J. Aitken, and H.-P. Husson, *BSCF* **131**, 391 (1994).
[2] T. R. Hoye and K. B. Crawford, *JOC* **59**, 520 (1994).

Ethyl isocyanoacetate. **16**, 164–166

Pyrrole-2-carboxylates.[1]   The ester enolate reacts with vinyl sulfones by way of conjugate addition and displacement.

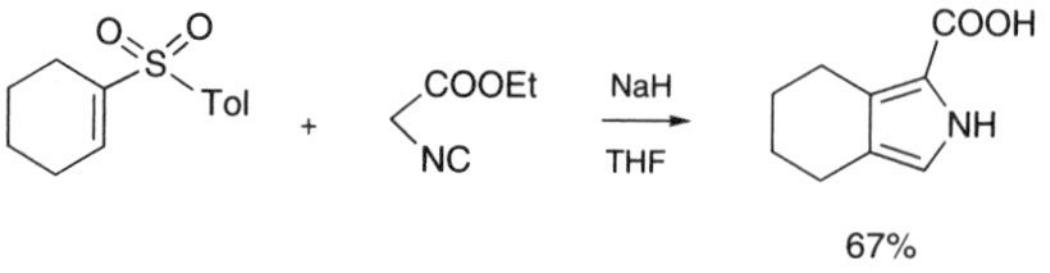

[1] D. P. Arnold, L. Burgess-Dean, J. Hubbard, and M. A. Rahman, *AJC* **47**, 969 (1994).

Ethyl *N*-(4-nitrophenylsulfonyloxy)carbamate.

Aziridination.[1] Ethoxycarbonylnitrene generated in the presence of solid K_2CO_3 or CaO adds to alkenes. The rapid reaction, requiring no catalyst, is useful for the aziridination of allylic and homoallylic acetals, which are quite unreactive toward other conditions. The reaction probably occurs at the solid–solid interface.

Azepine synthesis.[2] A similar reaction (vide supra) with aromatic compounds gives moderate yields of the heterocyclic derivatives.

Allylic carbamates.[3] Allylsilanes undergo desilylative 1,3-transpositional functionalization on exposure to the ethoxycarbonylnitrene. Yields up to 60% have been obtained.

[1] S. Fioravanti, M. A. Loreto, L. Pellacani, and P. A. Tardella, *TL* **34**, 4353 (1993).
[2] M. Barani, S. Fioravanti, M. A. Loreto, L. Pellacani, and P. A. Tardella, *T* **50**, 3829 (1994).
[3] S. Fioravanti, M. A. Loreto, L. Pellacani, S. Raimondi, and P. A. Tardella, *TL* **34**, 4101 (1993).

Ethyl trimethylsilylacetate.

Peterson olefination.[1] Stereocontrol of the reaction is achievable.

70 % (*Z*:*E* 88 : 12)

[1] N. Y. Grigoieva, O. A. Pinsker, and A. M. Moiseenkov, *MC* 129 (1994).

Europium(III) chloride.

Michael addition.[1] In the presence of $EuCl_3$ and molecular sieves 1,3-dicarbonyl compounds add to Michael acceptors at moderate temperatures.

[1] F. Bonadies, A. Lattanzi, L. R. Orelli, S. Pesci, and A. Scettri, *TL* **34**, 7649 (1993).

Europium tris[di(perfluoro-2-propoxypropionyl)]methanate.

Aldol condensation.[1] The Eu complex can recognize the difference of α- and β-alkoxy aldehydes and also the size of ketene silyl acetals. Thus, depending upon the particular substrates, it catalyzes the aldolization either in the chelation-controlled or in the non-chelation-controlled mode.

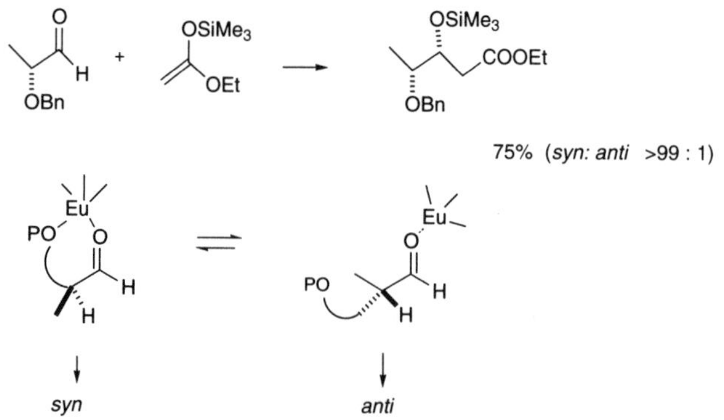

75% (*syn: anti* >99 : 1)

syn   anti

[1] K. Mikami, M. Terada, and T. Nakai, *CC* 343 (1993).

Europium tris[trifluoromethyl(hydroxymethylene)-*d*-camphorate].

Diels–Alder reaction.[1]   Eu(hfc)₃ is an efficient catalyst for cycloaddition involving electron-deficient α-pyrone systems.

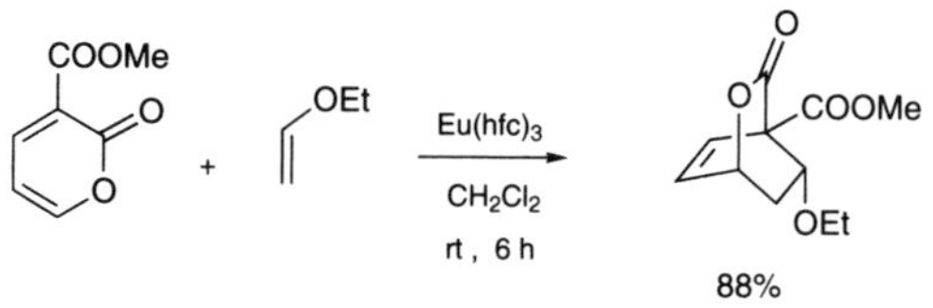

88%

[1] I. E. Marko and G. R. Evans, *SL* 431 (1994).

F

Fluorine. **13**, 135; **14**, 167; **15**, 160

Electrophilic fluorination.[1] Fluorination of aromatic compounds is effective in 98% formic acid or sulfuric acid with a 1:9 mixture of F_2-N_2 at room temperature. The yield for direct fluorination of 1,3-dicarbonyl compounds[2] by this system varies from 15% to 90%.

Fluorination of allenes.[3] Allene behaves differently toward fluorine in the presence or absence of dry NaF. The reaction of various allenes is explosion-prone and it must be carried out at low temperature. Cyanoallene requires even lower temperatures to avoid charring.

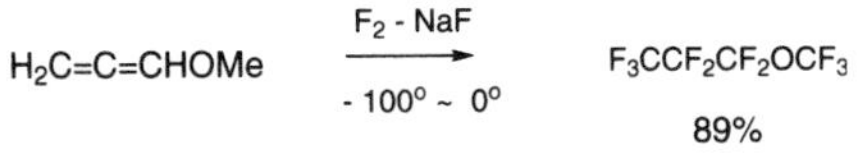

$$H_2C=C=CHOMe \quad \xrightarrow[-100° \sim 0°]{F_2 - NaF} \quad F_3CCF_2CF_2OCF_3$$

$$89\%$$

[1] R. D. Chambers, C. J. Skinner, J. Thomson, and J. Hutchinson, *CC* 17 (1995).
[2] R. D. Chambers, M. P. Greenhall, and J. Hutchinson, *CC* 21 (1995).
[3] T. Arimura, M. Shibakami, M. Tamura, S. Kurosawa, and A. Sekiya, *JCR(S)* 89, (1994).

N-Fluorobis(benzenesulfonyl)imide.

Fluorination of aromatics.[1] $(PhSO_2)_2NF$ is useful for delivering a fluorine atom to aryllithiums derived from directed *ortho*-metallation.

[1] V. Snieckus, F. Beaulieu, K. Mohri, W. Han, C. K. Murphy, and F. A. Davies, *TL* **35**, 3465 (1994).

(2-Fluoro-4'-carboxy)triphenylmethyl bromide.

Peptide synthesis.[1] The reagent (1) is useful for solid-phase peptide synthesis as an acid-labile handle. A Fmoc amino acid may be derivatized as the trityl ester, thereafter the Fmoc group is removable by treatment with 40% piperidine-DMF without affecting the ester linkage.

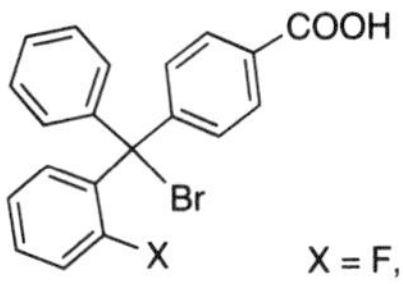

COOH

Br

X   X = F, Cl

(1)

[1] C. C. Zikos and N. G. Ferderigos, *TL* **35**, 1767 (1994).

Fluoroboric acid.

Allylic rearrangement.[1]   2-Alkenyl-2-alkylthio cyclic ethers are converted to enol ethers at low temperature.

Electrophilic substitutions.   In the presence of 4Å molecular sieves, tricarbonyl(dienyl)iron cations generated from the appropriate dienols[2] can be trapped with an alcohol that is too bulky to enter the cavities of the molecular sieves. The molecular sieves also play an important role in regiocontrol. When anhydrous $MgSO_4$ is used, some reaction occurs at the internal allylic position.

Enol silyl ethers undergo alkylation with dicobalt-complexed propargylic alcohols after ionization of the latter with fluoroboric acid.[3] A 1,1,1,3,3,3-hexafluoroisopropylphosphite ligand on the cobalt exerts its effect by virtue of a weakly σ-donating and strongly π-accepting behavior.

[1] J. P. Hagen, *JOC* **58**, 506 (1993).
[2] C. Quirosa-Guillou and J.-P. Lellouche, *JOC* **59**, 4693 (1994).
[3] A. J. M. Caffyn and K. M. Nicholas, *JACS* **115**, 6438 (1993).

N-Fluoro-2,10-(3,3-dichlorocamphorsultam).

Fluorination of enolates.[1]   Asymmetric fluorination is observed in moderate yields and enantiomer excesses with reagent (**1**).

(1)

[1] F. A. Davis, P. Zhou, and C. K. Murphy, *TL* **34**, 3971 (1993).

N-Fluoropyridinium salts. 16, 170–171

Functionalization of pyridine.   Regioselective reaction with oxygen, sulfur, and nitrogen nucleophiles[1] and with carbon nucleophiles[2] at C-2 occurs via an addition–elimination pathway.

[1] A. S. Kiselyov and L. Strekowski, *JHC* **30**, 1361 (1993).
[2] A. S. Kiselyov and L. Strekowski, *JOC* **58**, 4476 (1993).

N-Fluoro-2,4,6-trimethylpyridinium triflate.

Hydrolysis of dithioacetals.[1]   Generation of carbonyl compounds in moderate to good yields is accomplished in a mixed organic (e.g., CH_2Cl_2) and aqueous medium at room temperature. The reagent serves as an electron acceptor to initiate a single electron transfer process.

Dimeric alkenes from stabilized Wittig reagents.[2]   In the presence of traces of water, the reaction proceeds by a single electron transfer mechanism. The cation radical dimerizes, and the dimer loses two phosphine molecules.

[1] A. S. Kiselyov, L. Strekowski, and V. V. Semenov, *T* **49**, 2151 (1993).
[2] A. S. Kiselyov, *TL* **35**, 8951 (1994).

Fluorosilicic acid. 17, 139

Selective deprotection of trialkylsilyl ethers.[1]   A triisopropylsiloxy group can be retained when a *t*-butyldimethylsilyl group is removed on treatment with fluorosilicic acid at room temperature.

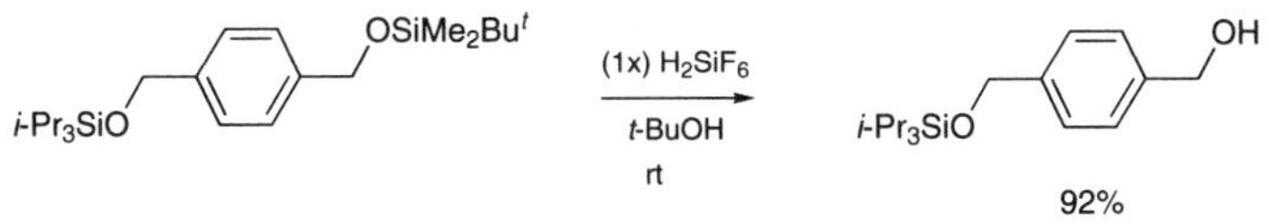

[1] A. S. Pilcher and P. DeShong, *JOC* **58**, 5130 (1993).

Formic acid. 13, 137

Deprotection of Adpoc-amino acids and peptides.[1]   The 1-adamantyl-1-methylethoxycarbonyl group from the *N*-protected amino acid or peptide is efficiently cleaved by treatment with formic acid in trifluoroethanol and chloroform.

Reductive dechlorination of chloroarenes.[2]   Catalyzed by Pd-C the dechlorination proceeds under mild conditions. PCBs are destroyed by this method.

Semihydrogenation of alkynes.[3]   Formic acid is a hydrogen source for the Pd(0)-catalyzed transfer reduction of the triple bond to afford the (Z)-alkene with a selectivity of 89–98%. The reducing system also contains triethylamine.

Reduction of carbonyl compounds.   The reduction is effected with 1,5-dihydro-5-deazaflavins.[4] Formic acid recycles the spent reagent.

α-Hydroxy acids are obtained, in the presence of triethylamine, by ruthenium-catalyzed cleavage of diallyl oxosuccinates.[5] Presumably via ester cleavage, decarboxylation, and reduction of the ketone.

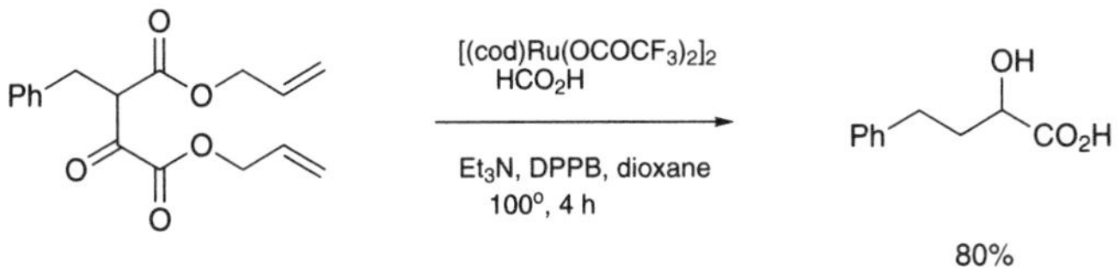

Hydration of terminal alkynes.[6]   In anhydrous media, formic acid acts as an equivalent of water in the conversion. Activation with $Ru_3(CO)_{12}$ is necessary for the reaction of functionalized alkynes, which are otherwise inert to the treatment.

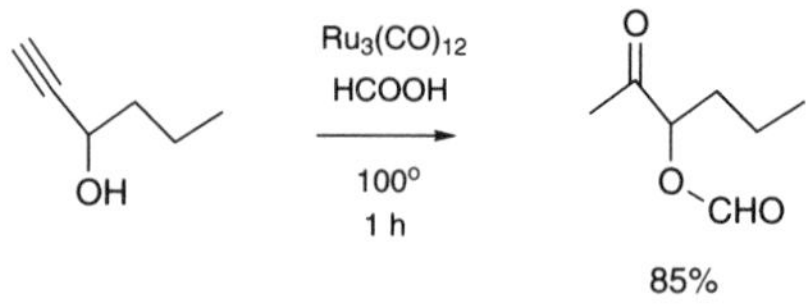

[1] W. Voelter and H. Kalbacher, *LA* 131 (1993).

[2] J. P. Barren, S. S. Baghel, and P. J. McCloskey, *SC* **23**, 1601 (1993).

[3] K. Tani, N. Ono, S. Okamoto, and F. Sato, *CC* 386 (1993).

[4] K. Kuroda, T. Nagamatsu, R. Yanada, and F. Yoneda, *JCS(PI)* 547 (1993).

[5] Y. Maruyama, T. Sezaki, M. Tekawa, T. Sakamoto, I. Shimizu, and A. Yamamoto, *JOMC* **473**, 257 (1994).

[6] N. Menashe and Y. Shvo, *JOC* **58**, 7434 (1993).

G

Gadolinium(III) isopropoxide.

Redox reactions.[1]   In the presence of cyclohexanone and isopropanol, respectively, oxidation of alcohols and reduction of ketones are catalyzed by the lanthanoid isopropoxide at room temperature or below.

[1] T. Okano, M. Matsuoka, M. Kinoshita, and J. Kiji, *NKK* 487 (1993).

Gallane.

Reduction of carbonyl compounds with Lewis base adducts.[1]   The highly efficient reduction with $GaH_3 \cdot NR_3$ or $GaH_3 \cdot PR_3$ also shows diastereoselectivity. For example, 4-*t*-butylcyclohexanone furnishes the alcohols with a *trans*:*cis* ratio of 87:13. As gallium is more electronegative than aluminum, gallane is less reactive than alane. However, the two reagents behave similarly in the reduction of enones, giving allylic alcohols.

[1] C. L. Raston, A. F.-H. Siu, C. J. Tranter, and D. J. Young, *TL* **35**, 5915 (1994).

Gallium.

Allyl- and propargylgallium reagents.[1]   Allylic and propargylic bromides are converted to allylating agents for carbonyl compounds. Regioselective reactions are observed with silyl-substituted bromides.

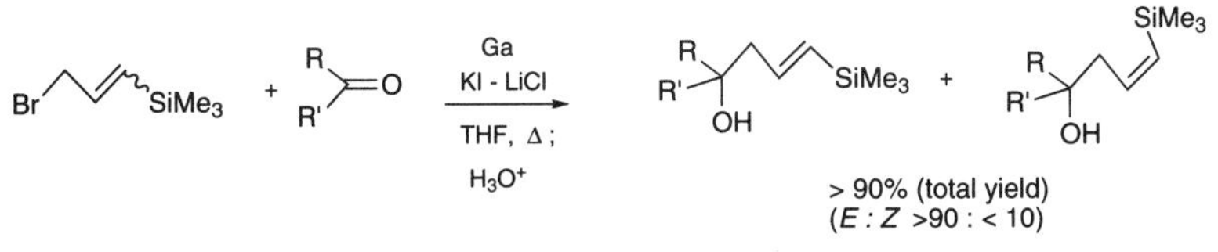

[1] Y. Han and Y.-Z. Huang, *TL* **35**, 9433 (1994).

Gallium(II) chloride. 17, 140

Reductive Friedel–Crafts alkylation.[1]   The reagent is actually a double salt, $GaCl \cdot GaCl_3$, and advantages can be taken of the reducing power of Ga(I) and the Lewis acidity of Ga(III). Thus the reaction of aldehydes, ketones, or acetals with arenes directly affords alkylarenes.

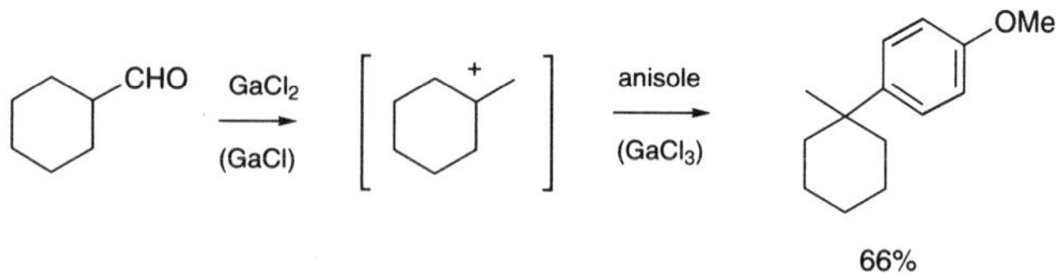

66%

[1] Y. Hashimoto, K. Hirata, H. Kagoshima, N. Kihara, M. Hasegawa, and K. Saigo, *T* **49**, 5969 (1993).

Germanium(II) iodide.

Allylation of carbonyl compounds.[1]   The reaction of allyl bromide with carbonyl compounds in the presence of zinc iodide is mediated by germanium(II) iodide. Ketones are less reactive than aldehydes and require an excess of the reagent and longer reaction times.

[1] Y. Hashimoto, H. Kagoshima, and K. Saigo, *TL* **35**, 4805 (1994).

Glyoxal.

α-Amino acid synthesis.[1]   The condensation products of glyoxal with *N*-homoallylic chiral amino ethanols are liable to undergo dehydration and cationic aza-Cope rearrangement. Further manipulation leads to amino acids.

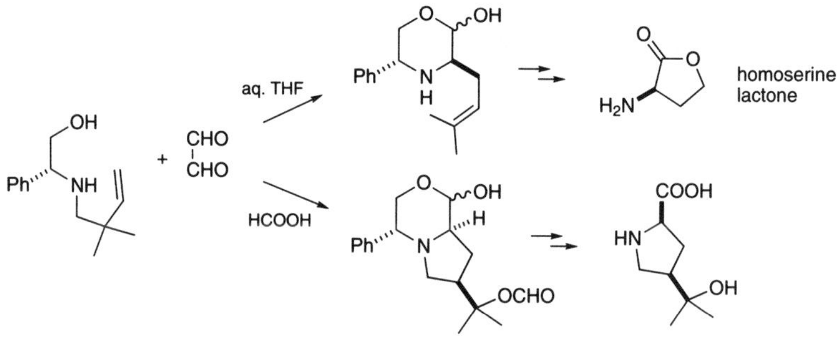

[1] C. Agami, F. Couty, J. Lin, A. Mikaeloff, and M. Poursoulis, *T* **49**, 7239 (1993).

Glyoxylic acid.

Arylmaleic acids.[1]   The base catalyzed reaction of arylacetonitriles with glyoxylic acid gives β-cyanocinnamic acids, which are readily transformed into arylmaleic anhydrides.

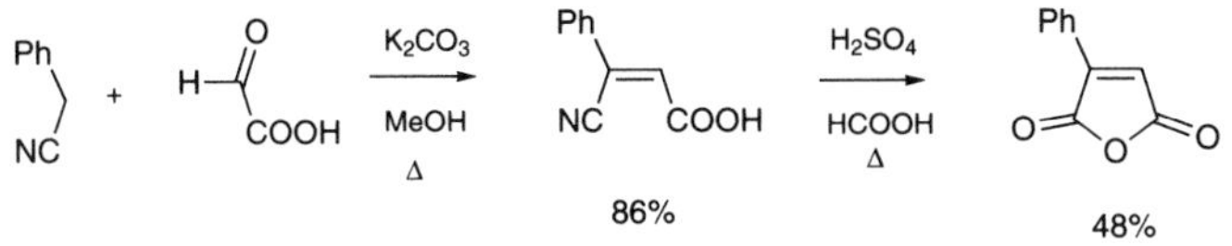

[1] W. D. Dean and D. M. Blum, *JOC* **58**, 7916 (1993).

Grignard reagents. 13, 138–140; 14, 171–172; 16, 172–173; 17, 141–142

A quantitative assay using both menthol and 1,10-phenanthroline in dry THF shows vivid endpoints in violet or burgundy.[1]

As bases. Formation of magnesium enolates from α-chloro-α-arenesulfinylcarboxylic acid derivatives involves desulfinylation with a Grignard reagent.[2] *t*-Butyl Grignard reagents are preferred in certain circumstances for the deprotonation of carbon acids. This method has been applied to a synthesis of chiral β-hydroxy esters from arenesulfinylacetic esters.[3]

Diastereoselective additions. The γ-substituents of allenyl carbonyl compounds[4] exert stereodirecting effects on the Grignard reaction. On the other hand, the diastereoselectivity observed in the reactions of the aldehydes derived from a serine orthoester[5] and 2-acyl-1,3-dithiane 1-oxides[6] can be attributed to chelation control.

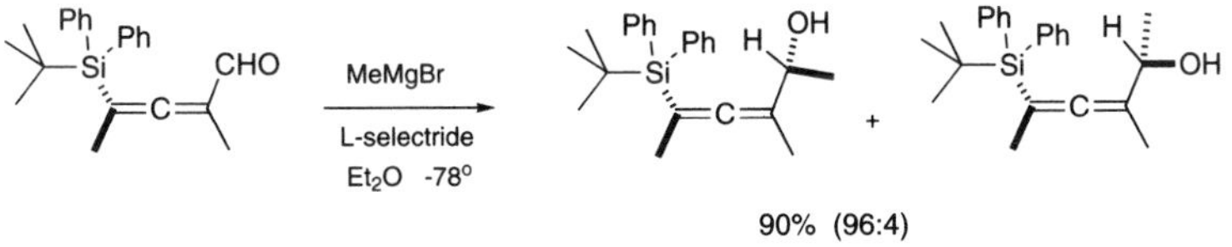

90% (96:4)

Addition to imines. In the presence of LiClO$_4$, ω-bromonitriles are attacked by Grignard reagents to form cyclic imines in a tandem addition–cyclization reaction.[7] Imines themselves are activated by 1-(trimethylsilyl)benzotriazole in the Grignard reaction.[8] Chiral *N*-(alkylthio)imines give optically active homoallylic amine derivatives on reaction with allylmagnesium bromide.[9]

The presence of additives can change the diastereoselectivity of Grignard reactions.[10]

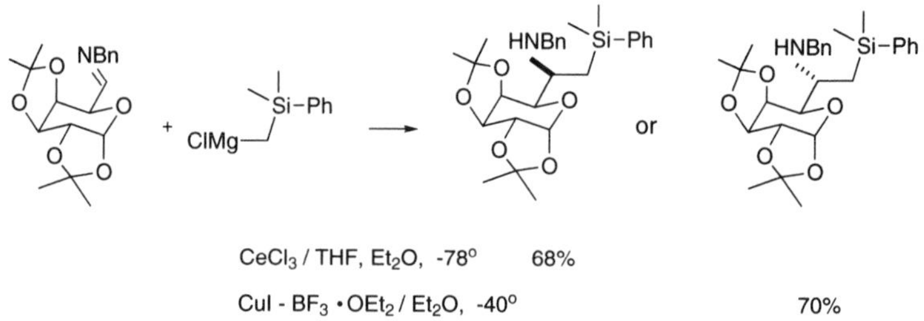

CeCl$_3$ / THF, Et$_2$O, -78°   68%

CuI - BF$_3$ ·OEt$_2$ / Et$_2$O, -40°   70%

Addition to chiral pyridinium salts. N-Alkyl-[11] or N-alkoxycarbonyl derivatives[12] are equally susceptible to asymmetric induction during attack at C-2.

Conjugate addition. Charge-directed addition to α-silylated α,β-unsaturated
amidate anions is observed.[13]-1,2-Addition is suppressed in such substrates.

Addition–fragmentation. α,β-Epoxy hydrazones react with Grignard reagents
to afford allylic alcohols.[14]

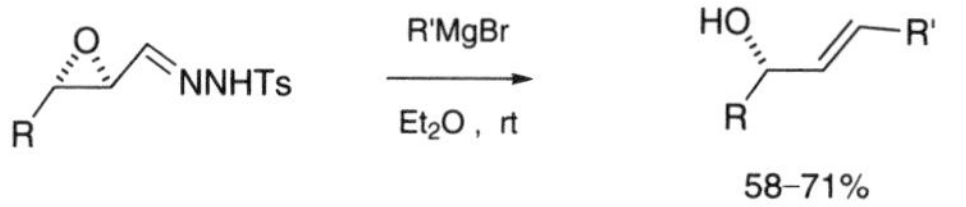

Ring cleavage of α-nitrocycloalkanones provides a synthesis of β-keto silanes.[15]

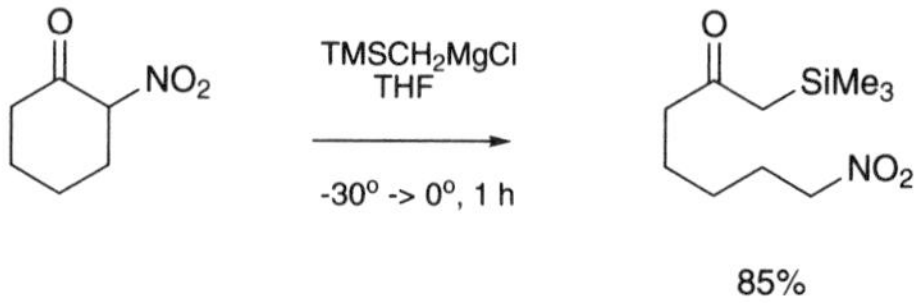

Addition–elimination. Axially chiral 1,1'-biphenyl-2-carboxylate esters are
obtained by the reaction of 2-menthoxybenzoates with aryl Grignard reagents.[16]

Conjugate addition followed by elimination of a malonic ester unit constitutes a
useful method for the access to (Z)-alkenes.[17] The reagents are 1,1-dimetalloalkanes.

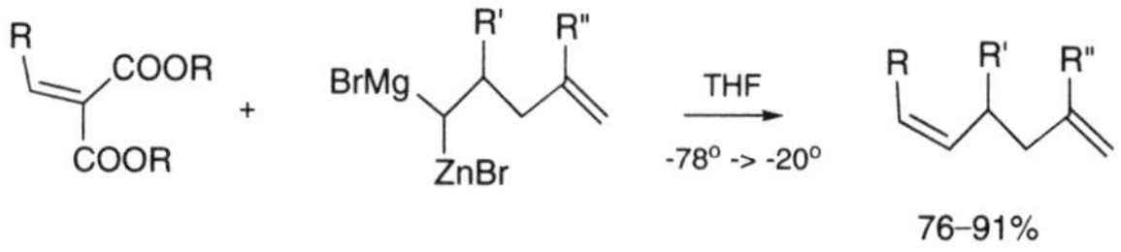

2,2-Difluoro enol silyl ethers are formed by the Grignard reaction of trifluoromethyl triphenylsilyl ketone.[18]

Some trifluoromethyl compounds such as 2-trifluoromethylaniline are converted to alkenes.[19] Trifluoroacetyl-stabilized Wittig reagents undergo Grignard reactions, but the products decompose on acid treatment to give alkenes.[20] Noteworthy is the dependency of the stereoselectivity on the acid strength.

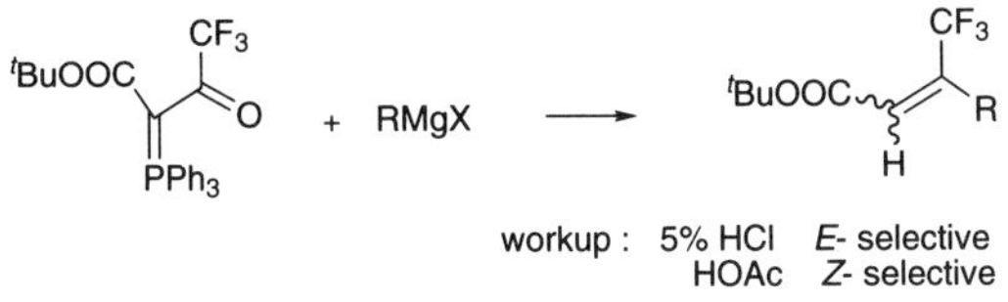

workup : 5% HCl *E*- selective
HOAc *Z*- selective

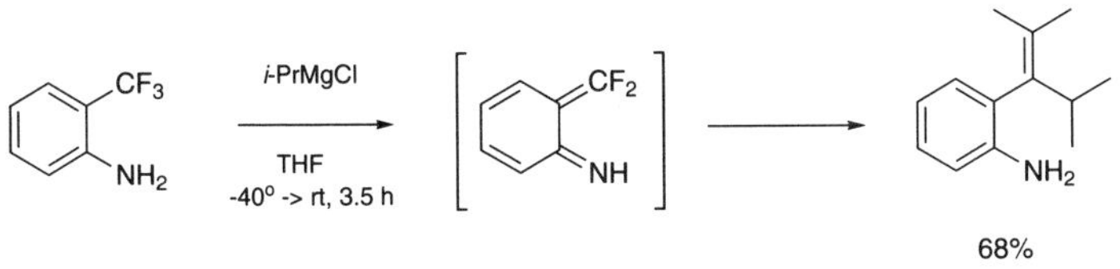

68%

The reaction of (*Z*)-1-halo-1-alkenyl-1,3,2-dioxaborolanes with allylmagnesium halides results in β,γ-unsaturated ketones.[21]

Displacement reactions.   1,3-Dibromo-1-trimethylsilylpropene is available from allyltrimethylsilane by reaction with NBS. The dibromo compound couples with Grignard reagents to effect chain elongation. With further displacement of the vinylic bromine and protodesilyation, it constitutes an intriguing approach to (*Z*)-alkenes.[22]

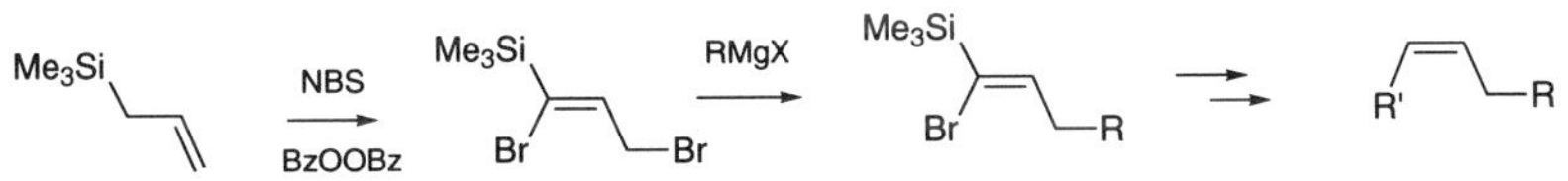

A method for the enantioselective synthesis of α-amino acetals (aldehydes)[23] is through ring opening of formyl-protected, chiral 2-formyl-1,3-oxazolidines with Grignard reagents. 2,2-Dimethyl-1,3-dioxolanes and dioxanes are subject to regioselective ring opening on reaction with methylmagnesium iodide.[24] The products contain a *t*-butoxy group on the more highly substituted carbon atom. Styrene derivatives are formed[25] when 2-aryl-1,3-dithiolane-*S*-oxides are similarly treated.

Grignard reagents are thiophilic towards sugar thiocyanates, forming thioglycosides[26] with retention of configuration.

α-Chloro-α-tolylsulfinylalkanoic esters react selectively with Grignard reagents, displacing the sulfinyl group.[27]

A benzotriazol-1-yl group attached to an activated benzylic position is nucleofugal. Its displacement by Grignard reagents[28] has been demonstrated.

Symmetrical α-diketones can be synthesized from 1,4-dimethylpiperazine-2,3-dione[29] or 1,1'-oxalyldiimidazole.[30]

In a general approach to phosphonodithioic acid derivatives,[31] the first step is the Grignard reaction of *P*-chloro-1,3,2-dithiaphospholane.

Displacement–rearrangement.[32] The reaction of allenesulfinate esters with alkenyl Grignard reagents proceeds by displacement on sulfur. [2,3]Sigmatropic rearrangement and dimerization follow.

S_N2' displacement. α,β-Disubstituted acrylic esters and nitriles are available from allylic displacement[33] of 3-acetoxy-2-methylenealkanoic acid derivatives, which are readily obtained from a Baylis–Hillman reaction.

Allylic phosphates are displaced, showing γ-selectivity with respect to allylic Grignard reagents.[34]

α-Trifluoromethylacrylic acid is transformed into α-substituted β,β-difluoro-acrylic acids[35] with various unsaturated Grignard reagents.

[1] H.-S. Lin and L. A. Paquette, *SC* **24**, 2503 (1994).
[2] T. Satoh, Y. Kitoh, K. Onda, and K. Yamakawa, *TL* **34**, 2331 (1993).
[3] R. J. Butlin, I. D. Linney, D. J. Critcher, M. F. Mahon, K. C. Molloy, and M. Wills, *JCS(P1)* 1581 (1993).
[4] J. A. Marshall and Y. Tang, *JOC* **58**, 3233 (1993).
[5] M. A. Blaskovich and G. Lajoie, *JACS* **115**, 5021 (1993).
[6] P. C. B. Page, J. C. Prodger, and D. Westwood, *T* **49**, 10355 (1993).
[7] D. F. Fry, C. B. Fowler, and R. K. Dieter, *SL* 836 (1994).
[8] A. R. Katritzky, Q. Hong, and Z. Yang, *JOC* **59**, 7947 (1994).
[9] T.-K. Yang, R.-Y. Chen, D.-S. Lee, W.-S. Peng, Y.-Z. Jiang, A.-Q. Mi, and T.-T. Jong, *JOC* **59**, 914 (1994).
[10] F. L. van Delft, M. De Kort, G. A. Van der Marel, and J. H. van Boom, *TA* **5**, 2261 (1994).
[11] Y. Genisson, C. Marazano, and B. C. Das, *JOC* **58**, 2052 (1993).
[12] D. L. Comins, S. P. Joseph, and R. R. Goehring, *JACS* **116**, 4719 (1994).
[13] M. P. Cooke, Jr. and C. M. Pollock, *JOC* **58**, 7474 (1993).
[14] S. Chandrasekar, M. Takhi, and J. S. Yadav, *TL* **36**, 307 (1995).
[15] R. Ballini, G. Bartoli, R. Giovannini, E. Marcantoni, and M. Petrini, *TL* **34**, 3301 (1993).

[16] T. Hattori, N. Koike, and S. Miyano, *JCS(P1)* 2273 (1994).

[17] C. E. Tucker and P. Knochel, *S* 530 (1993).

[18] F. Jin, Y. Xu, and W. Huang, *JCS(P1)* 795 (1993).

[19] M. Hojjat, A. S. Kiselyov, and L. Strekowski, *SC* **24**, 267 (1994).

[20] Y. Shen and S. Gao, *JOC* **58**, 4564 (1993).

[21] H. C. Brown and R. Soundararajan, *TL* **35**, 6963 (1994).

[22] R. Angell, P. J. Parsons, and A. Naylor, *SL* 189 (1993).

[23] K. R. Muralidharan, M. K. Mokhallalati, and L. N. Pridgen, *TL* **35**, 7489 (1994).

[24] W.-L. Cheng, S.-M. Yeh, and T.-Y. Luh, *JOC* **58**, 5576 (1993).

[25] W.-L. Cheng and T.-Y. Luh, *HC* **3**, 505 (1992).

[26] Z. Pakulski, D. Pierozynski, and A. Zamojski, *T* **50**, 2975 (1994).

[27] T. Satoh, Y. Kitoh, K.-I. Onda, K. Takano, and K. Yamakawa, *T* **50**, 4957 (1994).

[28] A. R. Katritzky, H. Lang, and X. Lau, *T* **49**, 7445 (1993).

[29] U. T. Mueller-Westerhoff and M. Zhou, *TL* **34**, 571 (1993).

[30] R. H. Mitchell and V. S. Iyer, *TL* **34**, 3683 (1993).

[31] S. F. Martin, A. S. Wagman, G. G. Zipp, and M. K. Gratchev, *JOC* **59**, 7957 (1994).

[32] J.-B. Baudin, M.-G. Commenil, S. A. Julia, L. Toupet, and Y. Wang, *SL* 839 (1993).

[33] D. Basavaiah, P. K. S. Sarma, and A. K. D. Bhavani, *CC* 1091 (1994).

[34] A. Yanagisawa, H. Hibino, N. Nomura, and H. Yamamoto, *JACS* **115**, 5879 (1993).

[35] S. Watanabe, K. Sugahara, T. Fujita, M. Sakamoto, and T. Kitazume, *JFC* **62**, 201 (1993).

Grignard reagents/cerium(III) chloride.

Cerium(III) chloride polarizes substrates to enhance their reactivities and regiose-lectivities in Grignard reactions. β-Enamino ketones and β-nitrostyrenes afford enones[1] and nitroethylarenes,[2] respectively, under the influence of $CeCl_3$.

65-85%

[1] G. Bartoli, C. Cimarelli, E. Marcantoni, G. Palmieri, and M. Petrini, *CC* 715 (1994).

[2] G. Bartoli, M. Bosco, L. Sambri, and E. Marcantoni, *TL* **35**, 8651 (1994).

Grignard reagents/copper salts.

Coupling. Dilithium tetrachlorocuprate can be used to couple allylmagnesium bromide with α,ω-dibromoalkanes to provide ω-bromoalkenes.[1] Vinylic tellurides have also been used as coupling partners.[2]

S_N2' displacements. The opening of allylic carbonates[3] and displacement reactions of chlorides are regio- and stereoselective processes.[4] The thio unit of function-alized allyl thiazolin-2-yl sulfides is selectively removed during the reaction.[5]

A convenient preparation of tributylstannylalkenes[6] involves the displacement of an allylic acetate. By using a bromoallene in the reaction the synthesis of alkynes[7] branching at the propargylic position is realized.

Opening of aziridines. Aziridines with an activating group on nitrogen undergo electrophilic reactions.[8,9]

Conjugate addition. Precursors of unusual amino acids are acquired by tandem addition and bromination of *N*-alkenoyloxazolidin-2-ones.[10] 2-(Oxazolidin-2-on-3-yl)acrylates[11] also act as acceptors in an alternative route to those compounds.

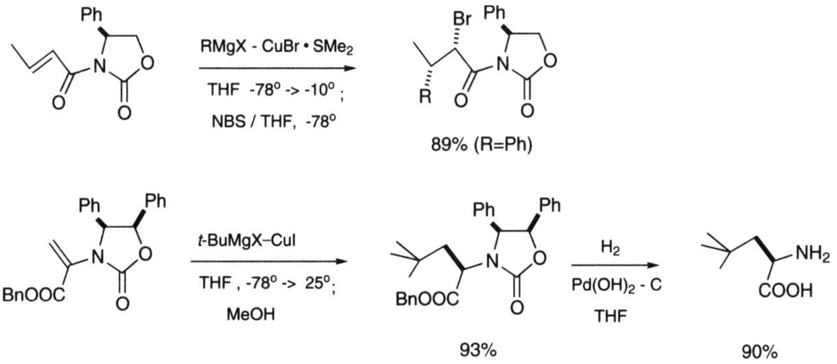

Alkynylphosphonates undergo stereoselective addition, making disubstituted alkenylphosphonates[12] readily available.

Sometimes vinylic Grignard reagents are prepared[13] indirectly, for example, from the readily available vinylstannanes via the vinyllithiums. The Sn/Li exchange is rapid at low temperature, and a primary chloride may be present.

[1] D. K. Johnson, J. Donohoe, and J. Kang, *SC* **24**, 1557 (1994).
[2] A. Chieffi and J.V. Comasseto, *TL* **35**, 4063 (1994).
[3] S.-K. Kang, D.-G. Cho, J.-U. Chung, and D.-Y. Kim, *TA* **5**, 21 (1994).
[4] J.-E. Bäckvall, E. S. M. Persson, and A. Bombrun, *JOC* **59**, 4126 (1994).
[5] V. Calo, V. Fiandanese, A. Nacci, and A. Scilimati, *T* **50**, 7283 (1994).
[6] F. Bellina, A. Carpita, M. De Santis, and R. Rossi, *T* **50**, 4853 (1994).
[7] F. D'Aniello, A. Mann, M. Taddei, and C.-G. Wermuth, *TL* **35**, 7775 (1994).
[8] J. E. Baldwin, A. C. Spivey, C. J. Schofield, and J. B. Sweeney, *T* **49**, 6309 (1993).
[9] H. M. I. Osborn, J. B. Sweeney, and W. Howson, *TL* **35**, 2739 (1994).
[10] G. Li, M. A. Jarosinski, and V. J. Hruby, *TL* **34**, 2561 (1993).
[11] P. A. Lander and L. S. Hegedus, *JACS* **116**, 8126 (1994).

[12] H.-J. Cristau, M.-B. Gasc, and X.Y. Mbianda, *JOMC* **474**, C14 (1994).
[13] E. Piers, B.W. A. Yeung, and F. F. Fleming, *CJC* **71**, 280 (1993).

Grignard reagents/nickel complexes.

Coupling with neopentyl iodides.[1] On treatment with $ZnCl_2$ and (dppf)NiCl$_2$, Grignard reagents couple with neopentyl iodides.

Reaction with enol ethers. 2-Substituted dihydropyrans and dihydrofurans are reduced to the (E)-alkenols[2] with isopropylmagnesium bromide in the presence of a nickel chloride complex. The reaction has been applied to a formal synthesis of recifeiolide.

Displacement of chalcogenides.

Displacement of chalcogenides. The catalyzed Grignard reaction is useful for C–C bond formation from vinylic chalcogenides[3,4] and cyclic dithioacetals.[5]

Conjugate addition-elimination. The displacement of the o-(t-butylsulfonyl) group of benzamides with aryl Grignard reagents furnishes unsymmetrical biphenyls.[6]

Thioamides. Good yields of the products are obtained in coupling reactions with chlorothioformamides.[7]

[1] K. Park, K. Yuan, and W.J. Scott, *JOC* **58**, 4866 (1993).
[2] J.-P. Ducoux, P. Le Menez, N. Kunesch, and E. Wenkert, *JOC* **58**, 1290 (1993).
[3] L. Hevesi, B. Hermans, and C. Allard, *TL* **35**, 6729 (1994).
[4] F. Babudri, V. Fiandanese, L. Mazzone, and F. Naso, *TL* **35**, 8847 (1994).
[5] L. L. Shiu, C.-C. Yu, K.-T. Wong, B.-L. Chen, W.-L. Cheng, T.-M. Yuan, and T.-Y. Luh, *OM* **12**, 1018 (1993).
[6] J. Clayden, J. J. A. Cooney, and M. Julia, *JCS(P1)* 7 (1995).
[7] F. Babudri, V. Fiandanese, G. Marchese, and A. Punzi, *SL* 719 (1994).

Grignard reagents/palladium chloride complexes.

Coupling with vinylic halides.[1,2] The C–C bond-forming reactions proceed with retention of configuration.

S$_N$2 Displacements.[3] Regio- and stereoselective methylation of γ-silylallyl phosphates has been observed.

Biphenyl synthesis.[4] Monoarylation of 1,4-dibromobenzene makes further functionalization of biphenyl derivatives possible.

[1] R.W. Hoffmann, V. Giesen, and M. Fuest, *LA* 629 (1993).
[2] Y. Sugihara and K. Ogasawara, *SL* 665 (1994).
[3] H. Urabe, H. Inami, and F. Sato, *CC* 1595 (1993).
[4] N. A. Bumagin, F. S. Safarov, and A. I. P. Beletskaya, *DC* **332**, 195 (1993).

Grignard reagents/titanium(IV) compounds. 14, 121–122

α-Hydroxystannanes.[1]   Ring opening of C_2-symmetrical 2-tributylstannyl-1,3-dioxanes followed by regioselective ether cleavage represents a general route to chiral hydroxystannanes.

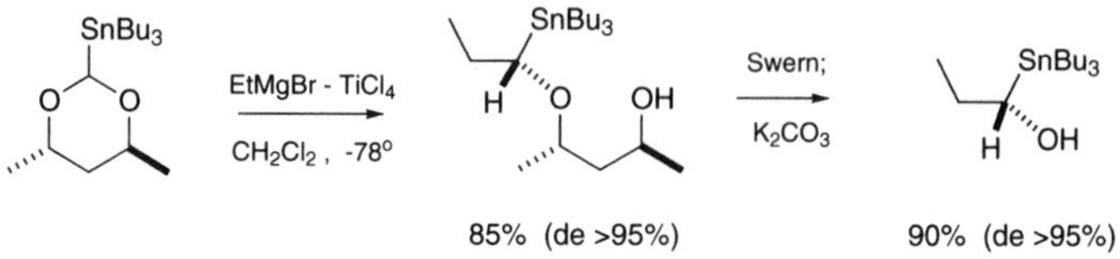

85% (de >95%)   90% (de >95%)

Conjugate addition to hindered enones.[2]   A methyltitanate complex generated from a methyl Grignard reagent is highly effective in delivering the methyl group to hindered enones in the presence of Ni(acac)$_2$.

[1] K. Tomooka, T. Igarashi, and T. Nakai, *TL* **35**, 1913 (1994).
[2] S. Flemming, J. Kabbara, K. Nickisch, H. Neh, and J. Westermann, *TL* **35**, 6075 (1994).

Grignard reagents/zinc chloride.

Reaction with α-ketoesters.[1]   In a regio- and diastereoselective addition reaction, chelation control provides optically active α-hydroxy esters.

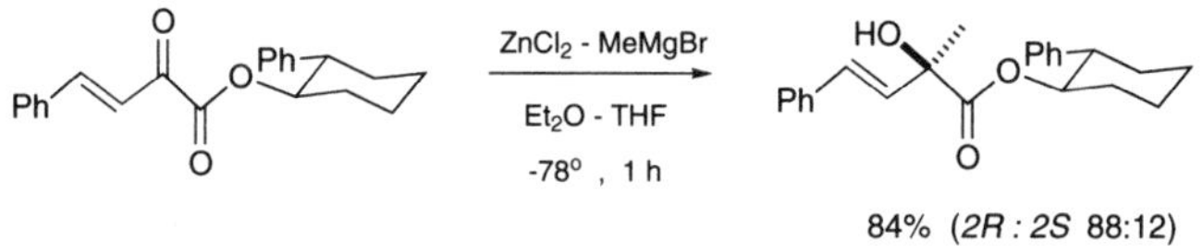

84% (*2R : 2S* 88:12)

[1] H. Sugimura and T. Watanabe, *SL* 175 (1994).

Grignard reagents/zirconocene dichloride.

S_N2' reactions.[1]   Allyl phenyl ethers are susceptible to transpositional attack with phenolate acting as a leaving group.

Addition to 1,3-diynes.[2]   Enynes are produced in moderate to good yields.

[1] N. Suzuki, D. Y. Kondakov, and T. Takahashi, *JACS* **115**, 8485 (1993).
[2] T. Takahashi, K. Aoyagi, V. Denisov, N. Suzuki, D. Choueiry, and E.-I. Negishi, *TL* **34**, 8301 (1993).

H

Hafnium(IV) triflate.

Friedel–Crafts acylation.[1] This salt in LiClO$_4$–nitromethane medium catalyzes the acylation of arenes at room temperature, usually in excellent yields (except benzene → acetophenone)

[1] I. Hachiya, M. Morikawi, and S. Kobayashi, *TL* **36**, 409 (1995).

Haloboranes. **13**, 47–48, 72, 73; **14**, 82–83; **15**, 163

Hydroboration.[1] Thexylhaloborane–dimethyl sulfide complexes are useful for monohydroboration at room temperature with superior regioselectivity. Its employment in converting terminal alkynes to aldehydes has been demonstrated.

Secondary amines from azides.[2] The reaction of alkyldichloroboranes with azides leads to secondary amines directly. This process has been incorporated into a polyamine synthesis. Note that the dichloroboranes can be prepared from organobis (diisopropylamino)boranes.[3]

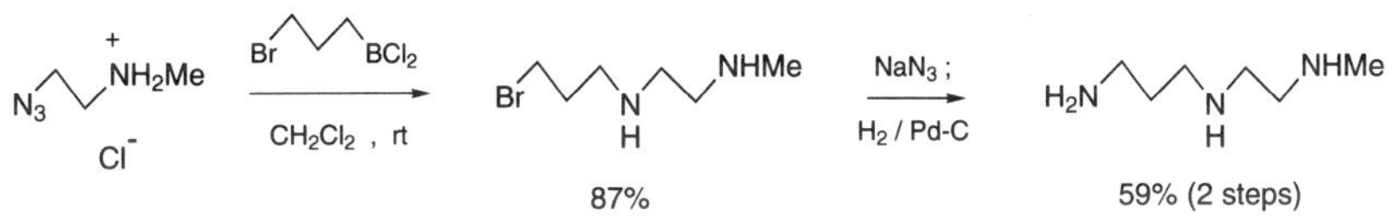

[1] J. S. Cha, S. J. Min, J. M. Kim, and O. O. Kwon, *TL* **34**, 5113 (1993).
[2] B. Carboni, A. Benalil, and M. Vaultier, *JOC* **58**, 3736 (1993).
[3] P.-Y. Chavant, F. Lhermitte, and M. Vaultier, *SL* 519 (1993).

Heteromethyl tris(o-methoxymethoxyphenyl)phosphonium salts.

(Z)-Alkenes.[1] The ylides from such phosphonium salts participate in Wittig reactions to give (Z)-alkenes selectively. The only unusual case is that involving benzaldehyde and the fluoromethylene ylide.

[1] X.-P. Zhang and M. Schlosser, *TL* **34**, 1925 (1993).

Hexaalkylditin. **13**, 142; **14**, 173–174; **16**, 174; **17**, 143–144

Aryltrimethyltins.[1] These aryltin compounds can be obtained from aryl bromides or triflates and (R$_3$Sn)$_2$ with a Pd(0) catalyst.

***vic*-Bistrimethylstannylation of alkynes.**[2]   The metathetic transformation is mediated by a Pd(0) catalyst in THF at room temperature. On warming to 75–95°C, the (Z)-alkenes undergo isomerization, presumably via the allenyl *O*-stannyl acetals.

Free radical substitutions.[3]   The cross-coupling of an alkyl halide and an allyl or vinyl halide is mediated by hexabutylditin under uv irradiation.

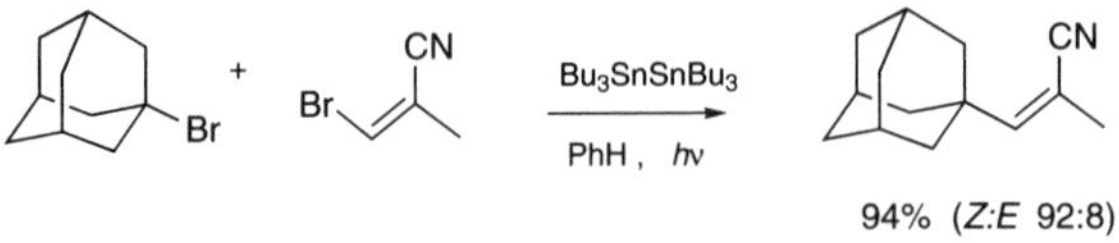

94% (*Z:E* 92:8)

Free radical addition to double bonds.   Photodissociation of hexabutylditin generates a tin radical, which can be used to abstract a halogen atom or thio group. Intramolecular or intermolecular processes can be designed to trap the resulting carbon radical. Synthetic applications include vicinal functionalization of an enone double bond[4] and the construction of a fused cyclopropane.[5]

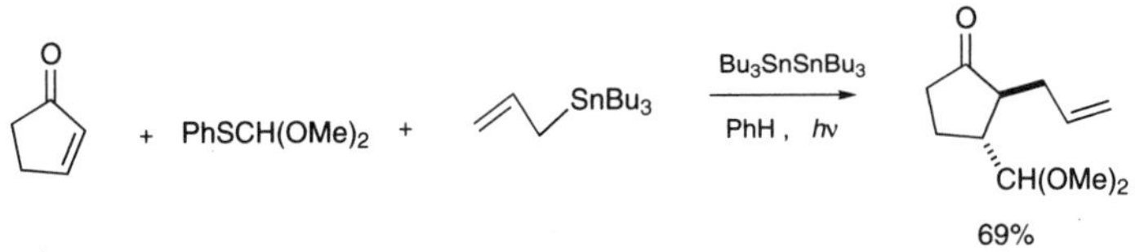

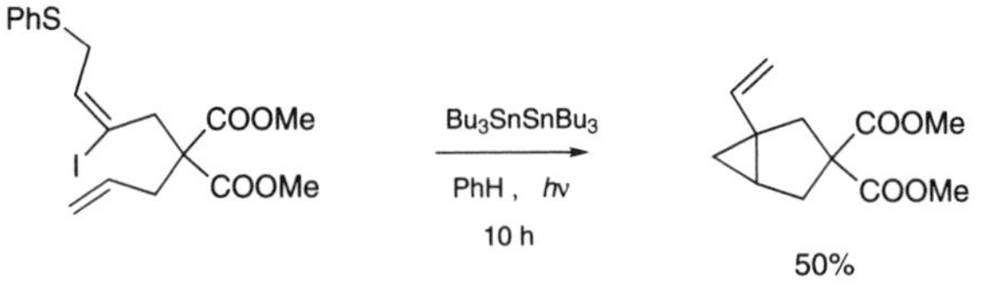

69%

50%

Reformatsky-type reaction.[6]   The condensation of α-iodoketones with aldehydes is promoted by $(Bu_3Sn)_2$.

Oximes.[7]   Alkyl radicals generated from halides by means of reaction with $(Bu_3Sn)_2$ attack tributylstannyl nitrite. Alkylnitroso compounds formed tautomerize to the oximes.

[1] M. P. Maguire, K. R. Sheets, K. McVety, A. P. Spada, and A. Zilberstein, *JMC* **37**, 2129 (1994).
[2] E. Piers and R. Skerlj, *CJC* **72**, 2468 (1994).
[3] C. C. Huval and D. A. Singleton, *TL* **34**, 3041 (1993).
[4] G. E. Keck and C. P. Kordik, *TL* **34**, 6875 (1993).
[5] R. C. Denis and D. Gravel, *TL* **35**, 4531 (1994).
[6] I. Shibata, T. Yamaguchi, A. Baba, and H. Matsuda, *CL* 97 (1993).
[7] R. J. Fletcher, M. Kizil, and J. A. Murphy, *TL* **36**, 323 (1995).

Hexaalkylguanidinium chlorides, silica supported.
Esterification.[1]

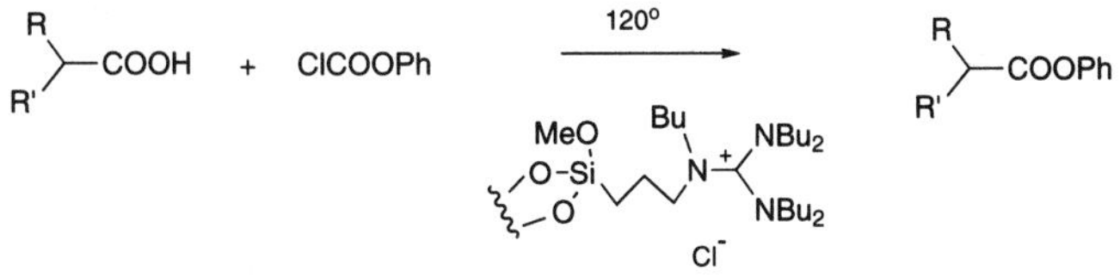

β-Chloroalkyl esters.[2] These reagents enable transformation of epoxides to β-chloroalkyl esters without simultaneous generation of HCl; therefore, potentially sensitive compounds (acrylate, etc.) are prepared cleanly. The silica-linked hexabutylguanidinium chloride is particularly effective.

[1] P. Gros, P. Le Perchec, P. Gauthier, and J. P. Senet, *SC* **23**, 1835 (1993).
[2] P. Gros, P. Le Perchec, and J. P. Senet, *JOC* **59**, 4925 (1994).

Hexafluorodimethyl disulfide.
Trifluoromethyl sulfides.[1] This reagent converts the diaminophosphite of an alcohol into a trifluoromethyl sulfide. An initial attack at the phosphorus atom (P–S bond formation) weakens the C–O bond for substitution by the CF_3S^- ion.

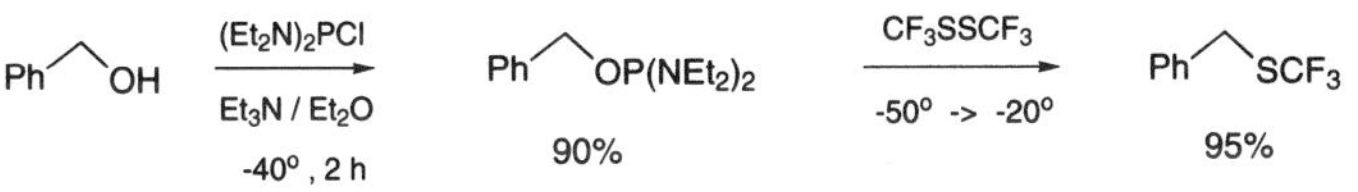

[1] A. A. Kolomeitsev, K. Y. Chabanenko, G.-V. Röschenthaler, and Y. L. Yagupolskii, *S* 145 (1994).

1,1,1,3,3,3-Hexafluoro-2-phenylisopropyl alcohol.
Primary alcohol protection.[1] The ether formation is mediated by Ph_3P-DEAD in 84–99% yield. The protecting group is remarkably robust; its removal requires treatment with Li naphthalenide in THF at −78°C for several hours (recovered yields: 73–89%).

[1] H.-S. Cho, J. Yu, and J. R. Falck, *JACS* **116**, 8354 (1994).

Hexamethyldisilazane. 13, 141
Silyl ethers.[1] By $ZnCl_2$ catalysis alcohols and phenols react with this reagent to give trimethylsilyl ethers, but amines and thiols remain unchanged.

Aromatic Claisen rearrangement.[2] The presence of the silylating agent prevents the so-called abnormal Claisen rearrangement from occurring.

Bis(trimethylsilyl) phosphonite. The reaction[3] with ammonium phosphinate gives $(Me_3SiO)_2PH$, which is reactive toward unactivated alkyl halides and imines. Accordingly, alkylphosphinic acids and α-aminophosphinic acids are more readily accessible.[4]

[1] H. Firouzabadi and B. Karimi, *SC* **23**, 1633 (1993).
[2] T. Fukuyama, T. Li, and G. Peng, *TL* **35**, 2145 (1994).
[3] E. A. Boyd, A. C. Regan, and K. James, *TL* **35**, 4223 (1994).
[4] X.-Y. Jiao, C. Verbruggen, M. Borloo, W. Bollaert, A. De Groot, R. Dommisse, and A. Haemers, *S* 23 (1994).

Hexamethylenetetramine.

Tröger's bases.[1] Hexamethylenetetramine is a source of formaldehyde. Thus its admixture with an aniline in trifluoroacetic acid affords Tröger's base.

Bromine carrier.[2] The hydrotribromide of hexamethylenetetramine is a mild and selective brominating agent for many arenes.

[1] R. A. Johnson, R. R. Gorman, R. J. Wnuk, N. J. Crittenden, and J. W. Aiken, *JMC* **36**, 3202 (1993).
[2] S. C. Bisarya and R. Rao, *SC* **23**, 779 (1993).

Hexamethylphosphoric triamide. 13, 142–143; 14, 176; 15, 165–166; 16, 174–175

Solvent effects.[1] The stereochemical outcome of the alkylation of α-alkyl β-keto ester enamines prepared from (S)-valine *t*-butyl ester is subject to solvent control. It seems that the bulky HMPA coordinates to Li from the opposite face of the isopropyl group, preventing the approach of the electrophile from that side. On the other hand, a smaller and weaker ligand allows the electrophile to attack from the opposite side of the isopropyl moiety.

[1] K. Ando, Y. Takemasa, K. Tomioka, and K. Koga, *T* **49**, 1579 (1993).

High-pressure reactions.

Aminolysis of epoxides.[1]   The reaction of epoxides with glycine *t*-butyl ester is catalyzed by silica gel and promoted by high pressure.

[1] H. Kotsuki, T. Shimanouchi, M. Teraguchi, M. Kataoka, A. Tatsukawa, and H. Nishizawa, *CL* 2159 (1994).

Hydrazine hydrate. 13, 144

Selective deoxygenation of isatins.[1]   A facile conversion of isatins to oxindoles involves brief heating with N_2H_4.

Reduction of 1,2-bis(trimethylstannyl)alkenes.[2]   Diimine generated in situ with hydrogen peroxide and $CuSO_4$ catalyst (1%) is useful for the reduction. *vic*-Silyl/stannylalkene analogues are also hydrogenated.

Liberation of amines.   A modification of the classical Gabriel synthesis to gain access to α-amino acids using chiral α-bromoalkanoic esters to react with potassium phthalimide catalyzed by a chiral phase transfer agent has been reported.[3] Unfortunately, the results are not very satisfactory.

Protected amino acids and peptides in the form of aminomethylene–dimedone derivatives[4] are formed and cleaved very readily. The cleavage with hydrazine in DMF proceeds at room temperature. This protection method is suitable for solid-phase peptide synthesis.

N-Aminoaziridines are released from a quinazolinone system[5] on reaction with hydrazine. 3-Acetoxyamino-2-trifluoromethylquinazolin-4(3*H*)-one is a source of the *N*-aminonitrene.

Preparation of bis(2-hydroxyethyl) diselenide.[6]   Reduction of selenium by hydrazine in the presence of KOH and alkylation in situ provide the diselenide, which is a useful source of the nucleophilic 2-hydroxyethyl selenide anion or the electrophilic 2-hydroxyethylselenenyl halides. The preparation can be run on a large scale.

[1] C. Crestini and R. Saladino, *SC* **24**, 2835 (1994).
[2] T. N. Mitchell and B. Kowall, *JOMC* **481**, 137 (1994).
[3] S. Guifa and Y. Lingchong, *SC* **23**, 1229 (1993).
[4] B. W. Bycroft, W. C. Chan, S. R. Chhabra, P. H. Teesdale-Spittle, and P. M. Hardy, *CC* 776, 778 (1993).
[5] R. S. Atkinson, M. P. Coogan, and C. L. Cornell, *CC* 1215 (1993).
[6] O. M. Jakiwczyk, E. M. Kristoff, and D. J. McPhee, *SC* **23**, 195 (1993).

Hydrazoic acid.

Arylamines.[1]   Direct aromatic amination occurs when an arene is treated with hydrazoic acid in the presence of both triflic acid and trifluoroacetic acid.

[1] H. Takeuchi, T. Adachi, H. Nishiguchi, K. Itou, and K. Koyama, *JCS(PI)* 867 (1993).

Hydriodic acid. 15, 166

Dehalogenation of α-haloketones.[1]   The reduction can be conducted in 57% HI without solvent.

Opening of 2-iminooxetanes.[2]   The easy availability of 2-iminooxetanes from Lewis acid–catalyzed cycloaddition of ketene imines and aldehydes makes it desirable to find some synthetic uses. 3-Iodocarboxamides, which are generated by reaction with HI, are sources of β-lactams. Remarkably little hydroxy amides is formed.

95%

[1] M. Penso, S. Mottadelli, and D. Albanese, *SC* **23**, 1385 (1993).
[2] G. Barbaro, A. Battaglia, and P. Giorgianni, *JOC* **59**, 906 (1994).

Hydrobromic acid/hydrogen bromide.

An expedient and economical method for the preparation of anhydrous HBr is the thermal decomposition of Ph_3PHBr in xylene under nitrogen.[1] The salt can be made from Ph_3P and *t*-BuBr in the presence of Bu_4NBr.

Selective bromination.[2]   Primary alcohols are converted to the bromides with HBr/HOAc in dioxane, although cyclic ethers are also obtained as side products. Under such conditions secondary alcohols are acetylated.

ArOMe → ArOH.[3]   Demethylation of a methyl ether without disturbing the chiral amino acid moiety attached to the aryl ring is accomplished with $HBr–NaI–H_2O$ in a sealed tube.

[1] X.-H. Wang and M. Schlosser, *S* 479 (1994).
[2] A. El Anzi, M. Benazza, C. Frechou, and G. Demailly, *TL* **34**, 3741 (1993).
[3] G. Li, D. Patel, and V. J. Hruby, *TL* **34**, 5393 (1993).

Hydrogen fluoride.

Halogen exchange.[1]   Aromatic halides undergo exchange to fluorides in moderate yields.

Formylation.[2] HF-SbF$_5$ catalyzes formylation of polynuclear aromatic compounds with carbon monoxide. Thus naphthalene affords 1,5-naphthalene-dialdehyde in 53% yield.

[1] T. Fukuhara and N. Yoneda, *CL* 509 (1993).
[2] M. Tanaka, M. Fujiwara, H. Ando, and Y. Souma, *JOC* **58**, 3213 (1993).

Hydrogen fluoride–amine. 16, 286–287

The two most frequently used complexes for organic synthesis are HF–pyridine and HF–Et$_3$N, both of which contain more than one equivalent of HF.

Aromatic fluorides.[1] Aromatic diazonium fluoroborates are readily transformed into the fluorides by treatment with HF–pyridine, induced either thermally or photochemically. The method is particularly useful for accessing those fluorides having polar substituents (e.g., OH, OMe, CF$_3$, halogens).

Epoxide opening.[2-4] Regioselective formation of fluorohydrins is observed.

Substitutions. Glycosyl fluorides are formed from glycosyl bromides with inversion of configuration.[5] When a vicinal *trans* dialkylamino group to a mesylate is present, the displacement of the mesylate group by fluorine proceeds with neighboring group participation.[6]

Another method for the synthesis of glycosyl fluorides consists of 1,3-dipolar cycloaddition of glycosyl azides with di-*t*-butyl acetylenedicarboxylate and subsequent treatment of the triazole derivatives with HF–pyridine.[7]

Electrochemical fluorination. The α-fluorination of β-lactams,[8] the conversion of aldehydes to acyl fluorides,[9] and of phenyltellurides (PhTeR) to difluorotelluranes (PhTeF$_2$R)[10] are some of the transformations that can be performed electrochemically in MeCN in the presence of nHF–Et$_3$N (n = 3 or 5).

[1] T. Fukuhara, M. Sekiguchi, and N. Yoneda, *CL* 1011 (1994).
[2] F. Ammadi, M. M. Chaabouni, H. Amri, and A. Baklouti, *SC* **23**, 2389 (1993).
[3] A. Hedhli and A. Baklouti, *JFC* **70**, 141 (1995).
[4] J. Umezawa, O. Takahashi, K. Furuhashi, and H. Nohira, *TA* **4**, 2053 (1993).
[5] R. Miethchen and G. Kolp, *JFC* **60**, 49 (1993).
[6] M.-B. Giudicelli, M.-A. Thome, D. Picq, and D. Anker, *CR* **249**, 19 (1993).
[7] W. Broder and H. Kunz, *CR* **249**, 221 (1993).
[8] S. Narizuka and T. Fuchigami, *JOC* **58**, 4200 (1993).
[9] N. Yoneda, S.-Q. Chen, T. Hatakeyama, S. Hara, and T. Fukuhara, *CL* 849 (1994).
[10] T. Fuchigami, T. Fujita, and A. Konno, *TL* **35**, 4153 (1994).

Hydrogen fluoride–pyridine–*N*-bromosuccinimide.

Bromofluorination.   The addition to vinyloxiranes[1] proceeds without affecting the epoxide. β-Phenylthio α,β-unsaturated ketones[2] undergo desulfenylating bromofluorination.

Fluorodesulfurization.   Replacement of a C–S bond with the C–F bond is most convenient with this reagent combination. Accordingly, dithioacetals are transformed into *gem*-difluorides[3] and thioalkanoic *O*-esters into α,α-difluoroalkyl ethers.[4]

[1] A. Hedhli and A. Baklouti, *JOC* **59**, 5277 (1994).
[2] R. Bohlmann, *TL* **35**, 85 (1994).
[3] M. Kuroboshi and T. Hiyama, *JFC* **69**, 127 (1994).
[4] M. Kuroboshi and T. Hiyama, *SL* 251 (1994).

Hydrogen fluoride–pyridine–nitrosonium tetrafluoroborate.

Perfluorination.   Ketoximes[1] are converted to *gem*-difluorides and diarylacetylenes[2] to 1,2-diaryltetrafluoroethanes with this combination of reagents. The latter transformation is actually initiated by nitrosofluorination to generate fluorinated oxime intermediates.

[1] C. York, G. K. S. Prakash, Q. Wang, and G. A. Olah, *SL* 425 (1994).
[2] C. York, G. K. S. Prakash, and G. A. Olah, *JOC* **59**, 6493 (1994).

Hydrogen fluoride–pyridine–phenyliodine(III) bis(trifluoroacetate).

4-Fluorocyclohexadienones.[1]   Oxidation of 4-alkylphenols in the presence of HF–pyridine leads to the introduction of a fluorine atom to C-4. The yields are moderate (6 examples, 42–77%).

[1] O. Karam, J.-C. Jacquesy, and M.-P. Jouannetaud, *TL* **35**, 2541 (1994).

Hydrogen peroxide, acidic. **14**, 176; **15**, 167–168; **16**, 177–178; **17**, 145

ArCHO → ArCOOH.[1]   Hydrogen peroxide in formic acid effects this transformation. However, methoxybenzaldehydes are reported to give phenols.[2]

Baeyer–Villiger oxidation.   30% Hydrogen peroxide in acetic acid is able to oxidize cyclobutanones to the γ-lactones.[3] In the presence of myristic acid and immobilized *Candida antarctica* lipase[4] other ketones also undergo Baeyer–Villiger oxidation with H_2O_2.

Acyl phosphonates behave similarly to α-diketones towards hydrogen peroxide, affording mixed anhydrides.[5]

Epoxidation.[6]   Treated with chloroperoxidase in a citrate buffer (pH 5), alkenes are epoxidized by H_2O_2 to afford chiral products.

[1] R. H. Dodd and M. Le Hyaric, *S* 295 (1993).
[2] R. N. Baruah, *IJC(B)* **33B**, 1103 (1994).

[3] T. Honda and N. Kimura, *CC* 77 (1994).
[4] S. C. Lemoult, P. F. Richardson, and S. M. Roberts, *JCS(P1)* 89 (1995).
[5] N. J. Gordon and S. A. Evans, *JOC* **58**, 4516 (1993).
[6] E. J. Allain, L. P. Hager, L. Deng, and E. N. Jacobsen, *JACS* **115**, 4415 (1993).

Hydrogen peroxide, basic. **13**, 145; **14**, 156; **15**, 167

Epoxidation. 2-(Benzenesulfonyl) 1,3-dienes are epoxidized selectively at the conjugated double bond with H_2O_2 in the presence of methanolic NaOH,[1] in a useful complementary manner to their reaction with peracids.

Active epoxidizing agents can be generated in situ using basic hydrogen peroxide in combination with N-arenesulfonylimidazoles[2] and with phosphonic anhydrides.

Sulfoxides. A convenient and simple method for oxidation of sulfides[4] involves the use of basic H_2O_2 and acetonitrile in methanol.

Oxidative desilylation.[5] From silanes bearing an electron-withdrawing substituent (e.g., Ph, OR), the method provides access to alcohols with retention of configuration. A double bond in the same molecule is retained. Vinylsilanes furnish ketones.[6]

Cleavage of nitronate anions. Conversion of conjugated nitroalkenes to ketones[7] can be accomplished through borohydride reduction and oxidation of the nitronate anions with 30% H_2O_2.

Oxidative cleavage of α-diketones.[8] Sodium percarbonate ($Na_2CO_3 \cdot 3/2\ H_2O_2$) is a convenient reagent for the cleavage to carboxylic acids.

[1] J.-E. Bäckvall, A. M. Ericsson, S. K. Juntunen, C. Najera, and M. Yus, *JOC* **58**, 5221 (1993).
[2] M. Schulz, R. Kluge, and M. Lipke, *SL* 915 (1993).
[3] A. S. Kende, P. Delair, and B. E. Blass, *TL* **35**, 8123 (1994).
[4] P. C. B. Page, A. E. Graham, D. Bethell, and B. K. Park, *SC* **23**, 1507 (1993).
[5] D. F. Taber, L. Yet, and R. S. Bhamidipati, *TL* **36**, 351 (1995).
[6] L. H. Li, D. Wang, and T. H. Chan, *OM* **13**, 1757 (1994).

[7] R. Ballini and G. Bosica, *S* 723 (1994).
[8] D. T. C. Yang, T. T. Evans, F. Yamazaki, C. Narayanna, and G. W. Kabalka, *SC* **23**, 1183 (1993).

Hydrogen peroxide–metal catalysts. 13, 145; 14, 177; 15, 294; 17, 146–148

Dehydrogenation. Pyrazolines, available from the reaction of hydrazones and conjugated esters, undergo dehydrogenation[1] readily on exposure to H_2O_2 in the presence of iron(II) chloride.

Oxidation of sulfur compounds. Aromatic amides are obtained from thioamides[2] on treatment with the H_2O_2–$CoCl_2$ system. Aliphatic thioamides do not behave in the same way.

Methyltrioxorhenium(VII) ($MeReO_3$) is an excellent catalyst for the oxidation of sulfides to sulfoxides.[3]

Phenols → p-quinones.[4] Hydrogen peroxide in the presence of $MeReO_3$ is capable of the transformation. However, the yields vary from 4 to 74% in 20 examples studied.

Aromatic bromination.[5] Addition of KBr generates positive bromine species that can brominate aromatic compounds.

Amine oxidation. The treatment of primary aliphatic amines with H_2O_2 and a catalytic amount of titanium silicate molecular sieves leads to oximes.[6]

Arylamines with various substituents can be converted to the corresponding nitroso compounds[7] by hydrogen peroxide and peroxomolybdenum complexes. The cetylammonium heteropolyoxometalate (e.g., peroxotungstophosphate) -catalyzed oxidation of aromatic amines leads to ArNO or $ArNO_2$, depending on the reaction conditions.[8] At room temperature the reaction can be stopped at the lower oxidation state, but $ArNO_2$ is produced in refluxing $CHCl_3$. When the $ArNH_2$ is cooxidized with primary aliphatic amines, arylazoxyalkanes are formed.

Oxidation of allenes and alkynes.[9] The oxidation promoted by cetylpyridinium peroxotungstophosphate in an alcoholic solvent produces α-alkoxy ketones from allenes and α,β-epoxy ketones from alkynes (along with α,β-unsaturated ketones). Diphenylacetylene gives benzil (93% yield at 45% conversion).

Wacker-type oxidation.[10]   Methyl ketone formation from $RCH=CH_2$ is effected with H_2O_2 in the presence of $(Ph_3P)_4Pd$.

Epoxidation.   (Salen)Mn complexes have been developed for asymmetric epoxidation.[11] Usually an imidazole is added as an additional ligand for the metal center in these oxidations.[12,13] A great variety of other oxidants have been used instead of H_2O_2 in this process, including oxygen, *t*-BuOOH, NaOCl, periodates, and high-valent organoiodine compounds.

Esters from alkanes.[14]   Various Rh salts catalyze the oxidation of alkanes. Thus cyclohexane gives the trifluoroacetate on reaction with $H_2O_2–CF_3COOH$.

Free radical reactions.   N-(ω-Iodoalkyl)pyrroles and indoles cyclize in the presence of $H_2O_2–FeSO_4$ in DMSO under ultrasonic irradiation. The method is useful for the synthesis of mitomycin skeleton and (-)-monomorine.[15]

78%   (-)-monomorine

[1] Y. Kamitori, M. Hojo, R. Masuda, M. Fujishiro, and M. Wada, *H* **38**, 21 (1994).

[2] N. Borthakur and A. Goswami, *IJC(B)* **32B**, 800 (1993).

[3] W. Adam, C. M. Mitchell, and C. R. Saha-Moller, *T* **50**, 13121 (1994).

[4] W. Adam, W. A. Herrmann, J. Lin, and C. R. Saha-Moller, *JOC* **59**, 8281 (1994).

[5] B. M. Choudary, Y. Sudha, and P. N. Reddy, *SL* 450 (1994).

[6] J. S. Reddy and P. A. Jacobs, *JCS(P1)* 2665 (1993).

[7] S. Tollari, M. Cuscela, and F. Porta, *CC* 1520 (1993).

[8] S. Sakaue, T. Tsubakino, Y. Nishiyama, and Y. Ishii, *JOC* **58**, 3633 (1993).

[9] S. Sakaguchi, S. Watase, Y. Katayama, Y. Sakata, Y. Nishiyama, and Y. Ishii, *JOC* **59**, 5681 (1994).

[10] M. Ioele, G. Ortaggi, M. Scarsella, and G. Sleiter, *G* **122**, 531 (1992).

[11] T. Schwenkreis and A. Berkessel, *TL* **34**, 4785 (1993).

[12] P. Pietikainen, *TL* **35**, 941 (1994).

[13] R. Irie, N. Hosoya, and T. Katsuki, *SL* 255 (1994).

[14] K. Nomura and S. Uemura, *CC* 129 (1994).

[15] D. R. Artis, I.-S. Cho, S. Jaime-Figueroa, and J. M. Muchowski, *JOC* **59**, 2456 (1994).

Hydrogen peroxide–urea. 16, 379; 17, 148

Heterocycle N-oxides.[1]   This stable source of H_2O_2 (with addition of CF_3COOH) effects oxidation of pyridine derivatives and other substrates. The reagent has considerable advantages over others.

Hydration of nitriles.[2]   In the presence of a weak base (aq. K_2CO_3), nitriles are converted to amides in acetone.

Epoxidation and Baeyer–Villiger oxidation.[3]   The complex together with maleic anhydride is equivalent to many ordinary peracids in its effectiveness.

[1] B. Ocana, M. Espada, and C. Avendano, *T* **50**, 9505 (1994).
[2] R. Balicki and L. Kaczmarek, *SC* **23**, 3149 (1993).
[3] L. Astudillo, A. Galindo, A. G. Gonzalez, and H. Mansilla, *H* **36**, 1075 (1993).

Hydrogen sulfide.

O-Alkyl thiocarboxylates.[1] Exposure of ketene silyl acetals to hydrogen sulfide at $-78°C$ leads to thiocarboxylic esters. Thus an ordinary ester is converted to the thioester in two steps.

3H-1,2-Dithiole-3-thiones.[2] The treatment of β-oxodithioic acids with H_2S and bromine leads to the heterocyclic compounds. Bromine serves as an agent for the oxidative cyclization.

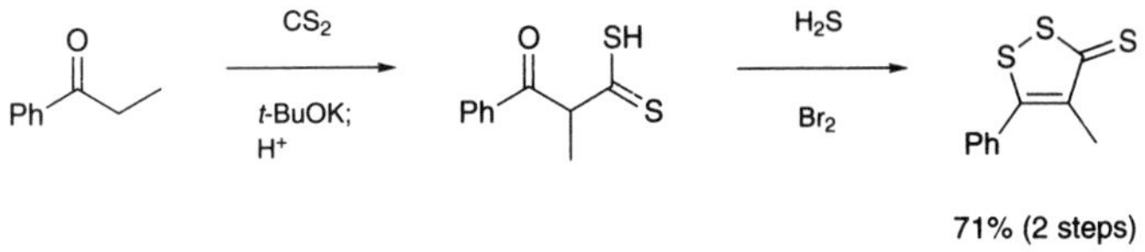

71% (2 steps)

[1] S. W. Wright, *TL* **35**, 1331 (1994).
[2] T. J. Curphey and H. H. Joyner, *TL* **34**, 3703 (1993).

Hydroxy(diphenylphosphoryloxy)iodobenzene.

Diphenyl β-oxoalkyl phosphates.[1] The reagent converts terminal alkynes into the α-ketol derivatives in aqueous acetonitrile.

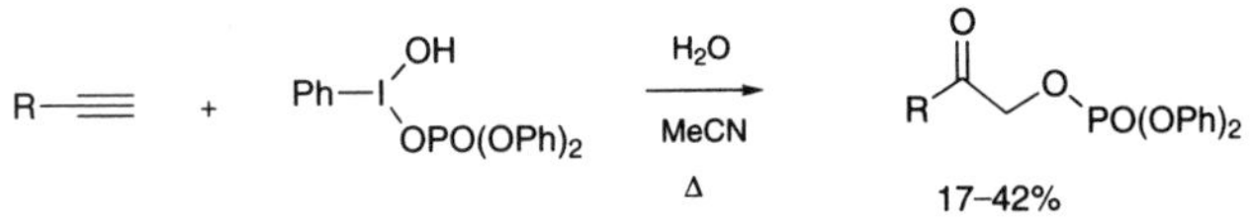

17–42%

[1] G. F. Koser, X. Chen, K. Chen, and G. Sun, *TL* **34**, 779 (1993).

Hydroxylamine. 15, 170

Pyridine formation.[1] On treatment with hydroxylamine and $AlCl_3$ in acetic acid 6,8-dioxabicyclo[3.2.1]octanes are transformed into pyridine derivatives (7 examples, 52–99% yield).

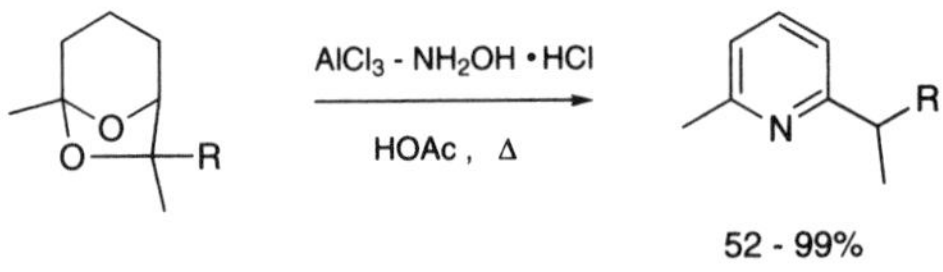

52 - 99%

Nitriles from acetals.[2] The direct conversion is carried out in refluxing ethanol.

[1] J.-G. Jun, H. S. Shin, and S. H. Kim, *JCS(P1)* 1815 (1993).
[2] M. Yamauchi, *CPB* **41**, 2042 (1993).

1-Hydroxypyridine-2-(1*H*)-thione.

S-Aryl thiosulfonates.[1] This novel preparation involves reaction of arenesulfinyl chlorides with the reagent in pyridine. The products are powerful sulfenylating agents.

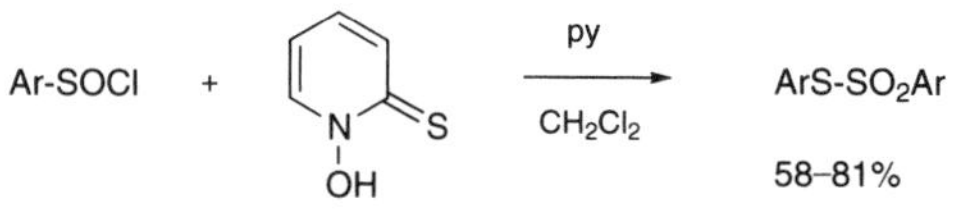

[1] W. Sas, *JCR(S)* 160 (1993).

Hydroxy(tosyloxy)iodobenzene. 14, 179–180; 16, 179; 17, 150

Oxidation of chromanones.[1] Formation of chromones is achieved under reflux or with ultrasound irradiation. 2-Spiroannulated chromanones give fused-ring systems.

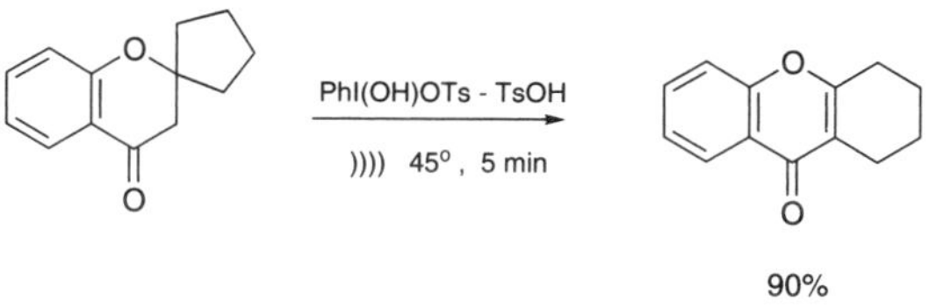

90%

Allylic oxidation.[2] Enones are obtained from glycal derivatives.

C-Tosyloxylation. Acetophenones,[3] 2-picolines,[4] and related compounds are subjected to the functionalization. Thus the original nucleophilic site is transformed into an electrophilic center.

[1] D. Kumar, O. V. Singh, O. Prakash, and S. P. Singh, *SC* **24**, 2637 (1994).
[2] A. Kirschning, G. Drager, and J. Harders, *SL* 289 (1993).
[3] O. Prakash and N. Saini, *SC* **23**, 1455 (1993).
[4] I. P. Andrews, N. J. Lewis, A. McKillop, and A. S. Wells, *H* **38**, 713 (1994).

Hydroxy(tosyloxy)iodoperfluoroalkanes.

These promising reagents are prepared[1] from perfluoroalkyl iodides on successive treatment with trifluoroperacetic acid in trifluoroacetic acid and tosic acid monohydrate. If the tosic acid is replaced with Me₃SiOTf, the corresponding triflates are formed.

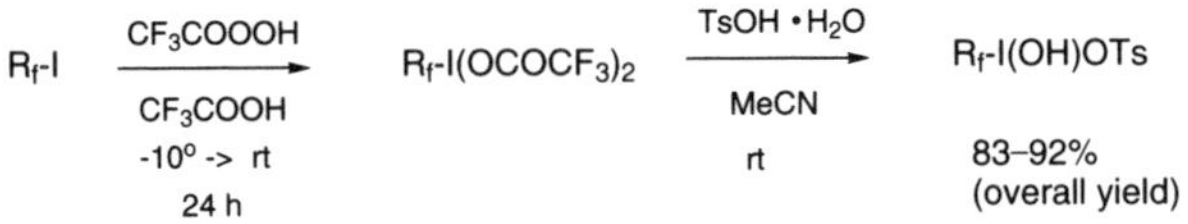

$$R_f\text{-}I \xrightarrow[\substack{CF_3COOH \\ -10° \rightarrow rt \\ 24\ h}]{CF_3COOOH} R_f\text{-}I(OCOCF_3)_2 \xrightarrow[\substack{MeCN \\ rt}]{TsOH \cdot H_2O} R_f\text{-}I(OH)OTs$$

83–92%
(overall yield)

[1] V. V. Zhdankin and C. Kuehl, *TL* **35**, 1809 (1994).

Hypofluorous acid–acetonitrile.

Oxidation.   This complex is made by passing gaseous fluorine into aqueous acetonitrile. It converts secondary alcohols to ketones (note that unsaturated alcohols are selectively oxidized to epoxy alcohols) and eventually to lactones.[1] Phenols and polycyclic arenes are rapidly oxidized to give quinones.[2]

Other oxidations achieved with this reagent include sulfides to sulfones[3,4] and amines to nitro compounds.[5] Actually, polyfunctional compounds show useful chemoselectivities. α-Nitroacetic esters are quantitatively prepared from glycine esters.

[1] S. Rozen, Y. Bareket, and M. Kol, *T* **49**, 8169 (1993).
[2] M. Kol and S. Rozen, *JOC* **58**, 1593 (1993).
[3] S. Rozen and Y. Bareket, *TL* **35**, 2099 (1994).
[4] S. Rozen and Y. Bareket, *CC* 1959 (1994).
[5] S. Rozen, A. Bar-Haim, and E. Mishani, *JOC* **59**, 1208 (1994).

I

1H-Indazole.

N-Methylation of primary amines.[1] Condensation of 1-hydroxymethyl-1H-indazole with an aromatic primary amine and reductive cleavage of the resulting aminal with lithium aluminum hydride afford the secondary amine.

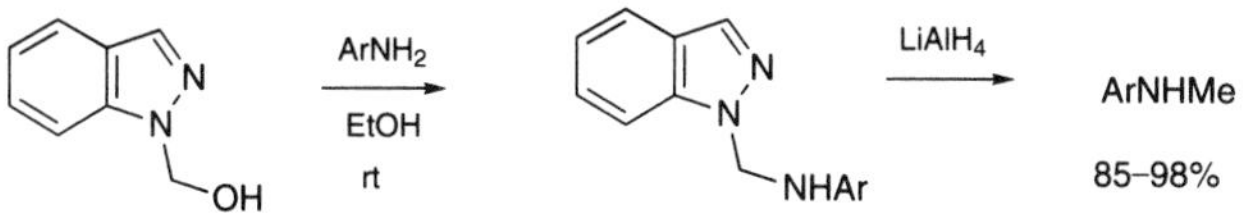

[1] R. Saladino, C. Crestini, and R. Nicoletti, *H* **38**, 567 (1994).

Indium. 14, 181; 16, 181–182

Reformatsky-type reaction.[1] Bromoacetonitrile reacts rapidly with indium and subsequent treatment with carbonyl compounds leads to β-hydroxy nitriles.

In an electrochemical version[2] of the modified Reformatsky reaction a sacrificial indium anode is used. Using this method, fully substituted β-lactones are obtained directly from the proper substrates.

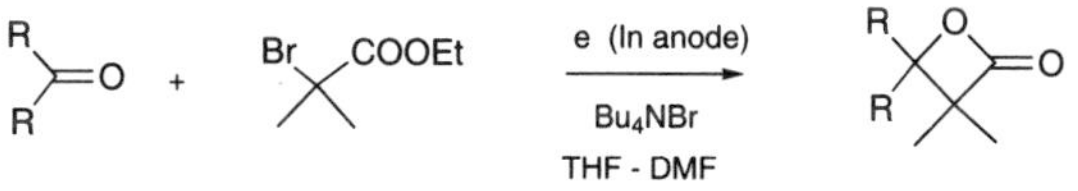

Reductive coupling of imines.[3] ArCH=NAr′ afford *vic*-diamines (*meso:dl* 1:1) on heating with indium rod in aqueous ethanol containing ammonium chloride, in 40–100% yield. Simple reduction side products are not formed.

[1] S. Araki, M. Yamada, and Y. Butsugan, *BCSJ* **67**, 1126 (1994).
[2] H. Schick, R. Ludwig, K.-H. Schwarz, K. Kleiner, and A. Kunath, *JOC* **59**, 3161 (1994).
[3] N. Kalyanam and G.V. Rao, *TL* **34**, 1647 (1993).

Iodine. 13, 148–149; 14, 181–182; 15, 172–173; 16, 182

Cleavage of carbon–metal bonds. When acyliron complexes such as those derived from carbonylation of $(\eta^3$-allyl)Fe(CO)$_2$NO are treated with an alcoholic iodine solution, esters are obtained.[1] Zirconacyclopentenes, which are formed by cross-coupling of alkynes and alkenes with Zr complexes, are stereoselectively trans-formed into iodoalkenes.[2] Cyclopropylthiocarbene–chromium complexes undergo ring scission[3] with the net extrusion of the metal moiety and introduction of two iodine atoms. 1,4-Diiodoalkenes are formed.

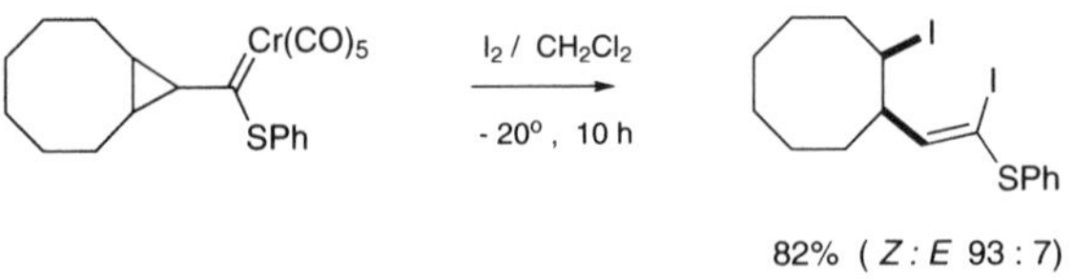

82% (Z : E 93 : 7)

Aromatic iodination.[4]   Activation of iodine is necessary for this reaction. For this purpose, mercuric nitrate appears to be quite effective.

α-Iodoenones.[5]   An iodine atom can be selectively introduced into the α,β-enone system with pyridine or PDC as catalyst.

Aromatization.[6]   α,β;γ,δ-Unsaturated ketones that can be deconjugated (or enolized) cyclize and dehydrate on heating with iodine. Thus β-ionone gives 1,1,6-trimethyltetralin in excellent yield.

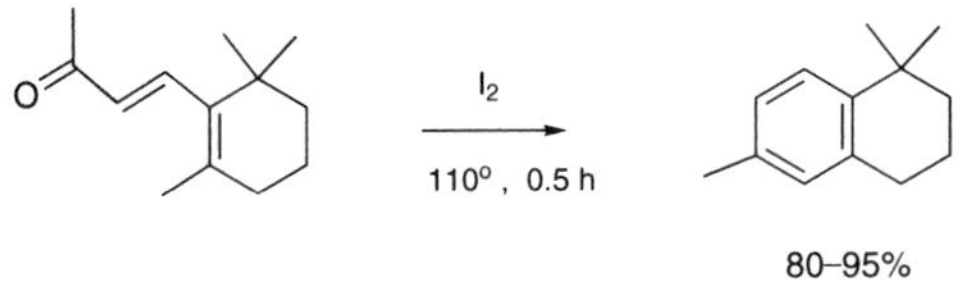

80–95%

Deesterification.[7]   3-Methyl-2-butenyl esters are cleaved by I_2 at room tempera-ture; therefore, such esters may serve in selective protection of acids. Simple allyl esters are unreactive.

Intramolecular cyclopropanation.[8]   Malonic esters of certain allylic alcohols form bicyclic lactones by way of iodination and generation of carbene intermediates. Thus when the system containing the malonic ester, iodine, solid K_2CO_3 and tri-caprylmethylammonium chloride in toluene is heated, the reaction occurs.

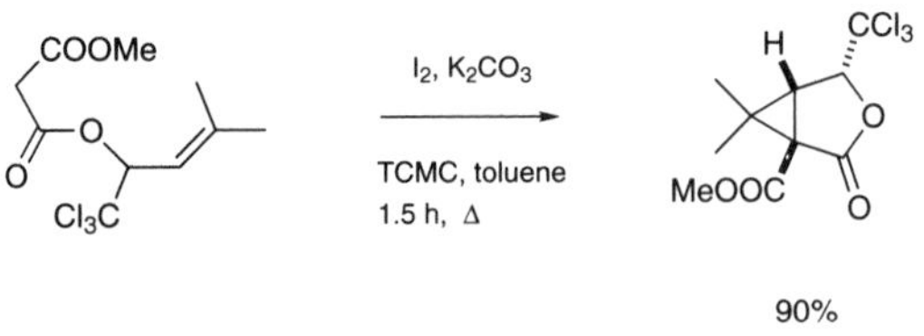

90%

[1] S. Nakanishi, T. Yamamoto, N. Furukawa, and Y. Otsuji, *S* 609 (1994).
[2] T. Takahashi, K. Aoyagi, R. Hara, and N. Suzuki, *CC* 1042 (1993).
[3] J. W. Herndon and M. D. Reid, *JACS* **116**, 383 (1994).
[4] A. Bachki, F. Foubelo, and M. Yus, *T* **50**, 5139 (1994).
[5] P. Bovonsombat, G. A. Angara, and E. McNelis, *T* **35**, 6787 (1994).
[6] J. J. Parlow, *T* **49**, 2577 (1993).
[7] J. Cossy, A. Albouy, M. Scheloske, and D. G. Pardo, *TL* **35**, 1539 (1994).
[8] L. Töke, Z. Hell, G. T. Szabo, G. Toth, M. Bihari, and A. Rockenbauer, *T* **49**, 5133 (1993).

Iodine–cerium(IV) ammonium nitrate. 15, 70

Iodination of chromones.[1]   Chromones, flavones, and thio analogues thereof form 3-iodo derivatives.

α,α′-Diiodination of ketones.[2]   The reaction is carried out in hot MeCN.

Alkoxyiodination and hydroxyiodination of enones.[3]   The reaction in an alcohol or 1:1 aqueous MeCN gives the adducts. In MeCN the products are α-iodo-β-nitrato ketones.

[1] F. J. Zhang and Y. L. Li, *S* 565 (1993).
[2] C. A. Horiuchi and E. Takahashi, *BCSJ* **67**, 271 (1994).
[3] C. A. Horiuchi, K. Ochiai, and H. Fukunishi, *CL* 185 (1994).

Iodine–hydroxy(tosyloxy)iodobenzene.

Ring expansion.[1]   1-Alkynylcyclopentanols give 2-(iodomethylene)-cyclohexanones on exposure to the reagent combination in MeCN at room temperature. The reaction proceeds in good yields (75–86%) in a limited number of examples. Substrates of other ring sizes have not been studied.

[1] P. Bovonsombat and E. McNelis, *T* **49**, 1525 (1993).

Iodine–lead(II) acetate.

trans-Iodoacetoxylation.[1]   Alkenes give *vic*-iodoacetates by treatment with iodine in the presence of lead(II) acetate in acetic acid at room temperature.

[1] A. V. Bedekar, K. B. Nair, and R. Soman, *SC* **24**, 2299 (1994).

Iodine–silver oxide.

Iodohydrins.[1] In aqueous THF alkenes undergo addition, and the resulting iodohydrins eliminate HI on exposure to base. Accordingly, ^{18}O-epoxides are available in two steps if isotopically labeled water is present in the medium.

[1] B. Borhan, S. Nazarian, E. M. Stocking, B. D. Hammock, and M. J. Kurth, *JOC* **59**, 4316 (1994).

Iodine–silver trifluoroacetate. 17, 154

Iododesilylation.[1] Controlled replacement of a silyl group in the furan ring by iodine has synthetic significance. It is possible to obtain 3-iodo-4-trimethylsilyl-furan by this method. Interestingly, trisfuranylboroxines are converted to the iodo-furans by I_2–$AgBF_4$.

[1] Z. Z. Song and H. N. C. Wong, *LA* 29 (1994).

Iodine–titanium tetra-*t*-butoxide.

Cyclization. 4-Pentenylmalonic esters are converted to cyclopentane derivatives[1] under the influence of the combined reagents. Apparently $Ti(OBu^t)_4$ acts as a base in the process. When CuO is also present, bicyclic lactones are formed.[2]

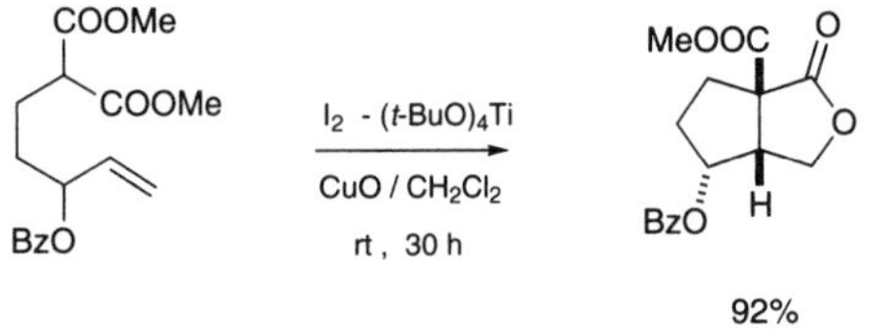

[1] O. Kitagawa, T. Inoue, K. Hirano, and T. Taguchi, *JOC* **58**, 3106 (1993).
[2] O. Kitagawa, T. Inoue, and T. Taguchi, *TL* **35**, 1059 (1994).

Iodine–triphenylbismuth dibromide.

Dehydration.[1] This reactive combination of reagents cannot be replaced with Ph_3Bi–I_2, $PhBiI_2$, or Ph_2BiI. With Ph_3BiBr_2 alone much longer reaction time is required for dehydration of alcohols.

[1] R. L. Dorta, E. Suarez, and C. Betancor, *TL* **35**, 5035 (1994).

Iodine–triphenylphosphine–imidazole.

Iodination. This reagent system is useful for the preparation of iodides[1] from acid-sensitive alcohols. The phosphine moiety may be incorporated in a polymer matrix.[2]

¹ Y. Wu and P. Ahlberg, *S* 463 (1994).
² R. Caputo, E. Cassano, L. Longobardo, and G. Palumbo, *TL* **36**, 167 (1995).

Iodine bromide. **17**, 152–153

*Cyclization.*¹   It is sometimes advantageous to use this reagent for the halogen-induced alkene functionalization with participation of an internal nucleophile. At low temperatures homoallylic carbonates react with enhanced diastereoselectivity.

89% (10:1)

¹ J. J.-W. Duan and A. B. Smith, *JOC* **58**, 3703 (1993).

Iodine chloride.

*Iododesilylation.*¹   Arylsilanes can be used as surrogates of aryl iodides by virtue of this facile conversion. A silver salt may² or may not be required to facilitate the reaction.

Iodination.   Different molecular complexes of ICl have been employed in iodination of organic molecules: ICl · pyridine for aromatic iodination,³ Ph₃P · ICl for conversion of fluorinated alkyl trimethylsilyl ethers to iodides.⁴

¹ M. Takahashi, K. Hatano, M. Kimura, T. Watanabe, T. Oriyama, and G. Koga, *TL* **35**, 579 (1994).
² L. A. Jacob, B.-L. Chen, and D. Stec, *S* 611 (1993).
³ H. A. Muathen, *JCR(S)* 405 (1994).
⁴ V. Montanari, S. Quici, and G. Resnati, *TL* **35**, 1941 (1994).

***N*-Iodosuccinimide.** **16**, 185–186

*Disulfide bridging.*¹   Cysteine units in peptides can be linked by reaction with NIS. Applications of this method in the synthesis of (Arg⁸)-vasopressin, oxytocin, and apamin have been demonstrated.

*Iodination.*²   Deactivated arenes can be iodinated with NIS in triflic acid at room temperature. Presumably, the actual reagent is the superelectrophilic iodine(I) triflate, which is formed in situ.

*Benzyne generation.*³   Several reagents [e.g., lead(IV) acetate, etc.] have been used to generate benzyne from 1-aminobenzotriazole. *N*-Iodosuccinimide is now recognized as a superior reagent for this purpose.

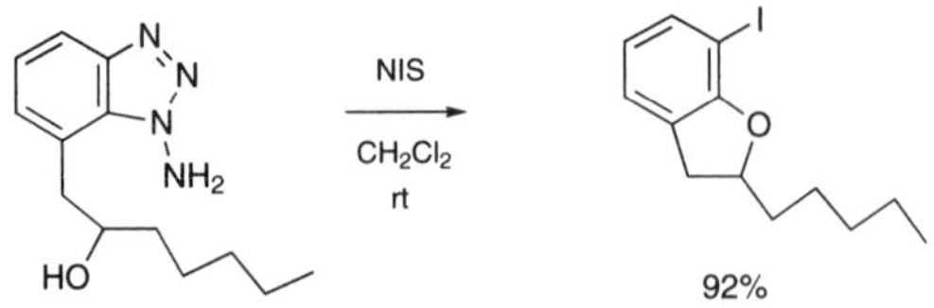

Isomerization–iodination of propargylic alcohols.[4]   1-Phenylethynylcyclopen-tanols do not undergo ring expansion on reaction with NIS–PhI(OH)OTs. The phenyl group apparently changes the polarization of the alkyne unit, which favors formation of α-iodo enones.

[1] H. Shih, *JOC* **58**, 3003 (1993).
[2] G. A. Olah, Q. Wang, G. Sandford, and G. K. S. Prakash, *JOC* **58**, 3194 (1993).
[3] M. A. Birkett, D. W. Knight, and M. B. Mitchell, *SL* 253 (1994).
[4] P. Bovonsombat and E. McNelis, *TL* **34**, 8205 (1993).

Iodosylbenzene. 13, 151; 16, 186

Oxidation of sulfides.[1]   *p*-Toluenesulfonic acid has a catalytic effect in the con-version to sulfoxides. The actual reagent is PhI(OH)OTs.

Asymmetric epoxidation.[2]   With a chiral (salen)Mn(III) complex to mediate the epoxidation of alkenes, iodosylbenzene supplies the active oxygen atom.

[1] R.-Y. Yang and L.-X. Dai, *SC* **24**, 2229 (1994).
[2] N. Hosoya, R. Irie, and T. Katsuki, *SL* 261 (1993).

Iodosyl fluorosulfate.

Diaryliodonium salts.[1]   This is a powerful electrophile to effect a Friedel–Crafts-type reaction of aromatic compounds (alkyl and haloarenes). Nitrobenzene reacts to give bis(*m*-nitrophenyl) iodonium bisulfate in 60% yield.

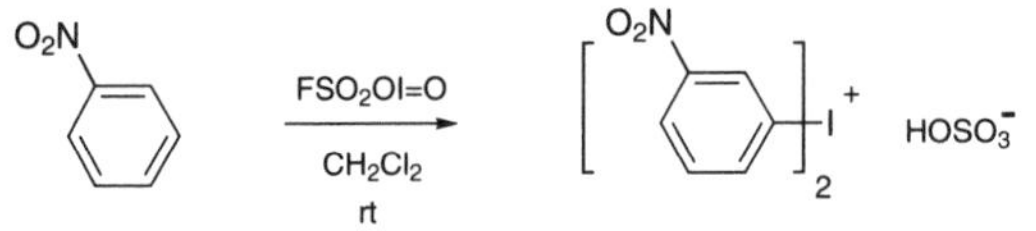

[1] N. S. Zefirov, T. M. Kasumov, A. S. Koz'min, V. D. Sorokin, P. J. Stang, and V. V. Zhdankin, *S* 1209 (1993).

Ion-exchange resins. 13, 152; 15, 178

Cyclization.[1]   Ring formation involving an enone and an allylsilane is usually initiated by a Lewis acid catalyst. However, Amberlyst-15 serves the same purpose yet entails a simpler procedure.

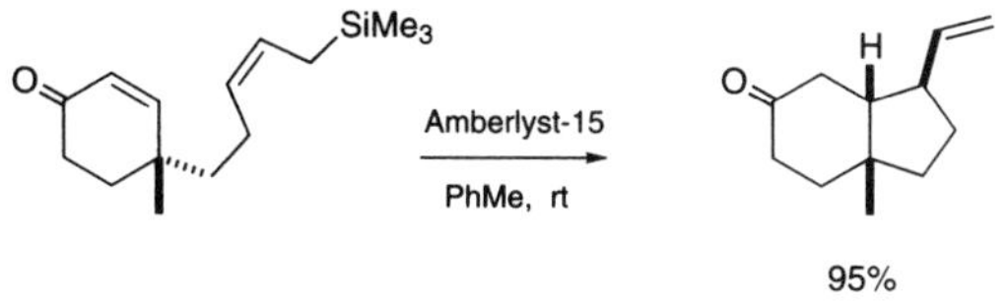

95%

Henry reaction.[2]   The condensation of a nitroalkane with an aldehyde in the presence of Amberlyst-A21 can be accomplished in good yields in a short time. Solvent is not needed.

Deacetalization and dethioacetalization.   With Dowex 50W-X8 as catalyst, a terminal acetonide can be hydrolyzed while keeping internal units intact. This accomplishment makes possible an easy access to 2,5-dideoxy-2,5-imino-D-mannitol.[3]

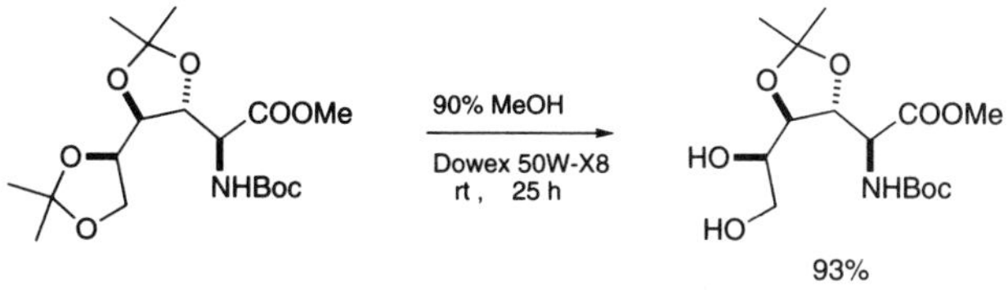

93%

Dithioacetals are hydrolyzed in acetone. Paraformaldehyde is added to trap the released mercaptan.[4]

Monoacylation of 1,ω-diols.[5]   By a transesterification with a simple ester (e.g., ethyl propionate), symmetrical 1,ω-diols can be derivatized at one site in the presence of Dowex 50W-X2.

[1] D. Schinzer and K. Ringe, *SL* 463 (1994).
[2] R. Ballini and G. Bosica, *JOC* **59**, 5466 (1994).
[3] K. H. Park, Y. J. Yoon, and S. G. Lee, *TL* **35**, 9737 (1994).
[4] V. S. Giri and P. J. Sankar, *SC* **23**, 1795 (1993).
[5] T. Nishiguchi, S. Fujisaki, Y. Ishii, Y. Yano, and A. Nishida, *JOC* **59**, 1191 (1994).

Ion-exchange resins, modified.

Oxidation.[1]   A supported ammonium perborate resin has been used to oxidize aromatic aldehydes to carboxylic acids.

Triazenes.[2]   Association of diazonium ions with a cation exchanger such as Amberlite IR-120 provides reagents for coupling with amines.

[1] J.-W. Chen, J.-X. Huang, and J. Xu, *YH* **13**, 537 (1993).
[2] P. J. Das and S. Khound, *T* **50**, 9107 (1994).

Iridium complexes. 13, 88–89

Reduction.   Dichlorohydridobis(triisopropylphosphine)iridium forms a catalyst with NaBH$_4$ that effects reduction[1] of ketones, alkenes, and alkynes.

Another catalyst for asymmetric hydrogenation of cycloalkanones is based on the BINAP–Ir(I)–aminophosphine system.[2]

[1] H. Werner, M. Schulz, M. A. Esteruelas, and L. A. Oro, *JOMC* **445**, 261 (1993).
[2] X. Zhang, T. Taketomi, T. Yoshizumi, H. Kumobayashi, A. Akutagawa, K. Mashima, and H. Takaya, *JACS* **115**, 3318 (1993).

Iron(III) acetylacetonate.

Coupling.[1]   Grignard reagents couple with allylic and propargylic halides in the presence of Fe(acac)$_3$.

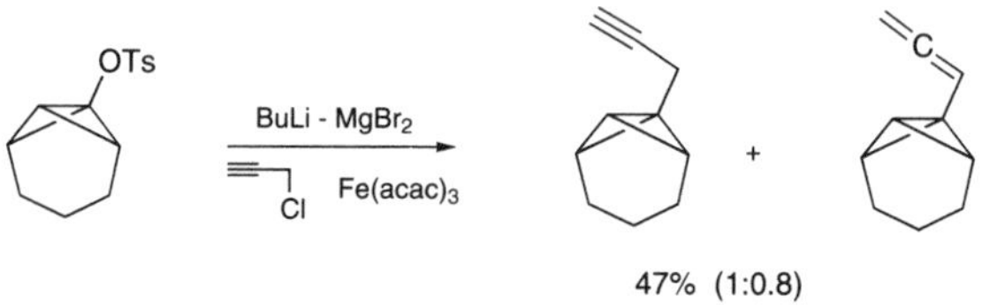

47%   (1:0.8)

α-Sulfonyl carbanions couple and form alkenes[2] in a dimerization reaction mediated by this catalyst.

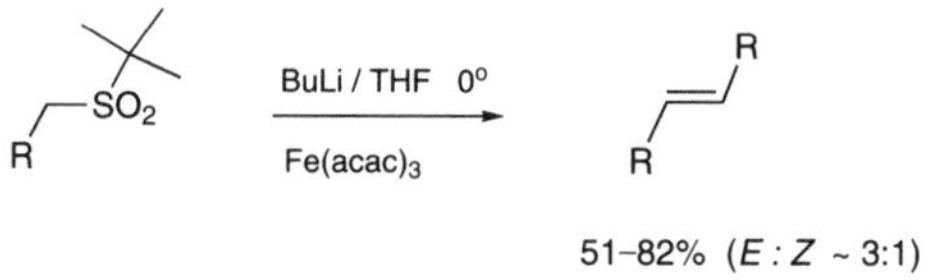

51–82% ($E:Z \sim 3{:}1$)

[1] A. S. K. Hashmi and G. Szeimies, *CB* **127**, 1075 (1994).
[2] L. Jin, M. Julia, and J. N. Verpeaux, *SL* 215 (1994).

Iron pentacarbonyl. 13, 152–153

Cyclopentenones.[1]   The carbonylative stitching method for cyclopentenone synthesis, typified by the Pauson–Khand reaction, includes the variant of using Fe(CO)$_5$ under carbon monoxide at relatively high temperatures.

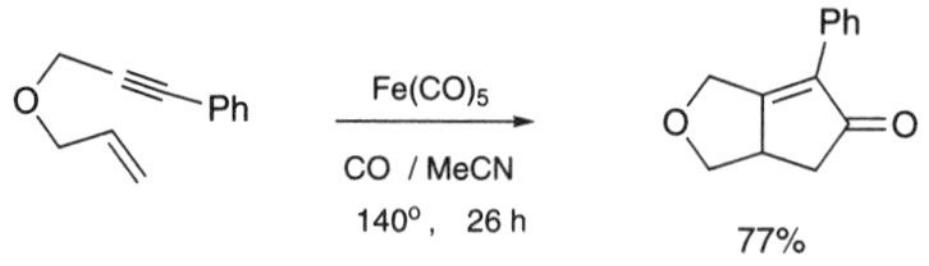

77%

[1] A. J. Pearson and R. A. Dubbert, *OM* **13**, 1656 (1994).

Iron(III) chloride. **13**, 133–134; **14**, 164–165; **15**, 158–159; **16**, 167–169, 190–191; **17**, 138–139

Alcoholysis of epoxides.[1] $FeCl_3$ is an effective catalyst for the transformation.

Oxidative cleavage of trimethylsiloxycyclopropanes.[2] This method of enone generation, in combination with conjugate addition, enolate trapping, and Simmons–Smith reaction, completes the sequence of ring expansion that also incorporates a substituent at the γ-position of a lower homolog. Thus 4-alkyl-2-cyclohexenones are readily prepared from cyclopentenone.

45-71% (last step)

Oxidative coupling.[3] 2,2'-Binaphthols are produced in excellent yields by heating 2-naphthols with $FeCl_3 \cdot 6H_2O$ in the neat with a microwave source.

Oxidative cyanation.[4] The methyl group of substituted $ArNMe_2$ is subjected to catalyzed autooxidation, and the intermediate radicals can be intercepted with PhCOCN, affording the *N*-cyanomethyl derivatives.

BnX → BnNHAc.[5] This transformation is related to the Ritter reaction. Activation of benzyl halides by $FeCl_3$ in refluxing MeCN is sufficient to bring about the displacement (and in situ hydration).

gem-Dichlorides.[6] Certain polychlorinated ketones undergo deoxygenative chlorination. The role of $FeCl_3$ in these cases is twofold: as Lewis acid to activate the carbonyl group and as chlorine donor.

Allylic amination.[7] Using a mixture of Fe(II) and Fe(III) chlorides as catalysts, alkenes undergo ene-type amination with PhNHOH.

48%

[1] N. Iranpoor and P. Salehi, *S* 1152 (1994).
[2] K. I. Booker-Milburn and D. F. Thompson, *TL* **34**, 7291 (1993).
[3] D. Villemin and F. Sauvaget, *SL* 435 (1994).
[4] S. Murata, K. Teramoto, M. Miura, and M. Nomura, *BCSJ* **66**, 1297 (1993).
[5] M. Kacan and A. McKillop, *SC* **23**, 2185 (1993).
[6] D. M. Antonov, L. I. Belen'kii, A. A. Dudinov, and M. M. Krayushkin, *MC* 130 (1994).
[7] R. S. Srivastava and K. M. Nicholas, *TL* **35**, 8739 (1994).

Iron(III) chloride, clay-supported.

Dithioacetalization.[1] $FeCl_3$ dispersed in silica gel catalyzes the rapid condensation of ketones with *o*-benzenebismethanethiol.

β-Amino esters.[2]   Iron montmorillonite shows good catalytic activities for the synthesis of β-amino esters from silyl ketene acetals and aldimines.

[1] H. K. Patney, *SC* **23**, 1829 (1993).
[2] M. Onaka, R. Ohno, N. Yanagiya, and Y. Izumi, *SL* 141 (1993).

Iron(III) perchlorate.

Transesterification.[1]   Ethyl esters from carboxylic acids and acetic esters of benzylic alcohols are obtained in the catalyzed transesterification processes, using ethyl acetate as the donor of ethoxy or acetyl group.

[1] B. Kumar, H. Kumar, and A. Parmar, *IJC(B)* **32B**, 292 (1993).

Iron(III) phthalocyanine.

Allylic amination.[1]   This catalyst effects anilination of alkenes with PhNHOH in refluxing PhMe. Other Fe catalysts cause reduction of the reagent to aniline to a great extent.

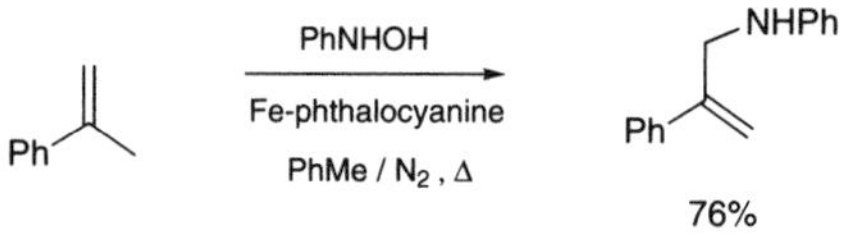

[1] M. Johannsen and K. A. Jorgensen, *JOC* **59**, 214 (1994).

Iron(II) sulfate.

$S_{RN}1$ reactions of haloarenes.[1]   The displacement of aromatic halides with enolates of acetic acid derivatives in liquid NH_3 to give arylacetic acid derivatives (esters, amides, etc.) has a broad scope. The reaction is initiated by electron transfer.

Reduction of benzofuroxans.[2]   The reduction produces *o*-nitroanilines. The iron salt may be present in catalytic amounts when PhSH is used as the reducing agent.

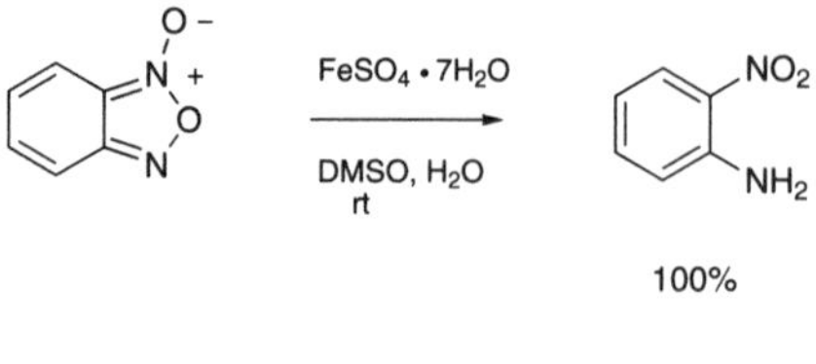

[1] M. van Leeuwen and A. McKillop, *JCS(P1)* 2433 (1993).
[2] A. M. Gasco, C. Medana, and A. Gasco, *SC* **24**, 2707 (1994).

Isocyanuric chloride.

α-Perchloro esters.[1]   Oxidative cleavage of cyclic acetals by this reagent leads to chloroalkyl esters that are also fully chlorinated at the α-position.

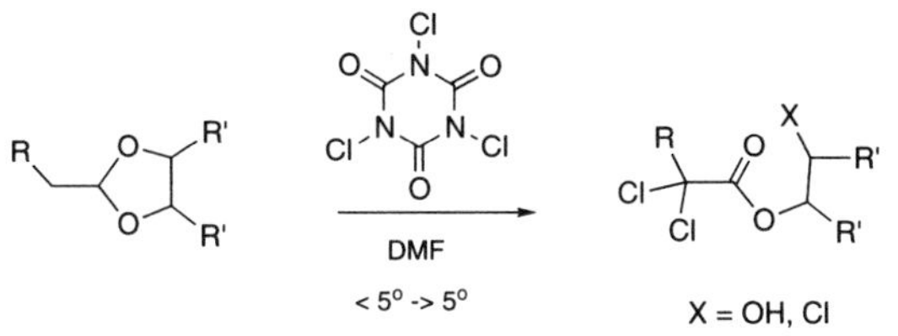

[1] F. Bellisia, M. Boni, F. Ghelfi, and U. M. Pagnoni, *TL* **35**, 2961 (1994).

N-Isopropylephedrine.

Enantioselective protonation.[1]   This chiral amino alcohol is an excellent proton source for thiol ester enolates.

[1] C. Fehr, I. Stempf, and J. Galindo, *ACIEE* **32**, 1042 (1993).

2-Isopropylapoisopinocampheylborane.

Enantioselective hydroboration.[1]   This bulkier chiral borane i-PraBH$_2$ achieves significantly better asymmetric induction than those realized by IpcBH$_2$ and related boranes.

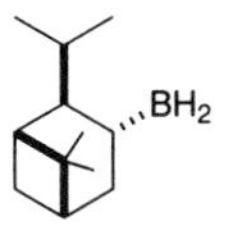

[1] U. P. Dhokte and H. C. Brown, *TL* **35**, 4715 (1994).

K

Ketene.

Acetylation.[1] This reagent effects rapid *C*-acetylation of β-diketone Cu(II) chelates at room temperature in $CHCl_3$ (75–95% yield). Contrarily, the use of acetyl chlorides (and other acyl chlorides) leads to a mixture of *C*- and *O*-acetyl derivatives.

[1] G. J. Matare, A. Bohac, and P. Hrnciar, *S* 381 (1994).

Ketene acetals.

A general preparative procedure[1] consists of passing HCl through a stoichiometric ethereal mixture of a nitrile and an alcohol (2 OH per CN) at 0°C and treatment of the adducts with NaOMe. Both acyclic and cyclic ketene acetals are obtainable.

[1] A. B. Argade and B. R. Joglekar, *SC* **23**, 1979 (1993).

Ketene *t*-butyldimethylsilyl methyl acetal.

Pummerer rearrangement. Sulfoxides are converted into α-siloxy sulfides,[1] in which the new O–C bond is *anti* to the original S–O bond. Accordingly, the rearrangement of chiral sulfoxides is enantioselective. The method is applicable to synthesis of β-lactam precursors.[2]

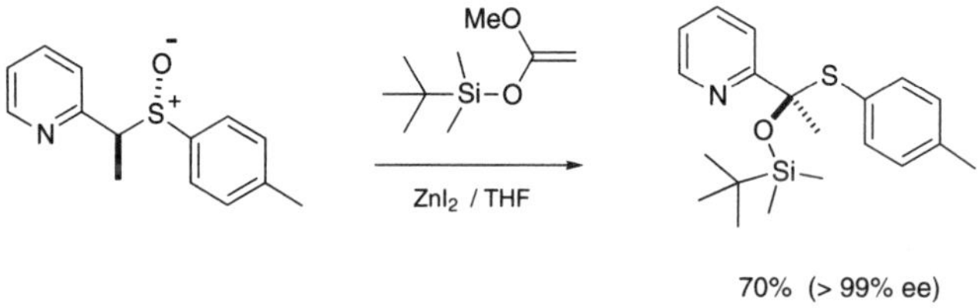

70% (> 99% ee)

[1] Y. Kita, N. Shibata, S. Fukui, and S. Fujita, *TL* **35**, 9733 (1994).
[2] Y. Kita, N. Shibata, N. Kawano, T. Tohjo, C. Fujimori, and K. Matsumoto, *TL* **36**, 115 (1995).

L

Lanthanum(III) bromide.

Benzyl esters.[1]   Benzyl ethers can be used to esterify carboxylic acids in the presence of LaBr$_3$. Other lanthanoid bromides (Nd, Sm, Dy, Er) are also effective.

[1] Y. Jiang and Y. Yuan, *SC* **24**, 1045 (1994).

Lanthanum(III) chloride.

β-Alkoxy carboxylic acids.[1]   The Michael addition of alcohols to unsaturated acids is catalyzed by La(III) salts.

[1] J. Huskens, J. A. Peters, and H. van Bekkum, *T* **49**, 3149 (1993).

Lanthanum(III) isopropoxide. 17, 160

β-Cyanohydrins.[1]   Group transfer from acetone cyanohydrin to epoxides occurs in the presence of Ln(III) alkoxides.

[1] H. Ohno, A. Mori, and S. Inoue, *CL* 975 (1993).

Lanthanum(III) tris-β-diketonates.

Selective alkylation.[1]   The reactivities of amines toward alkylating agents can be clearly differentiated in the presence of a lanthanide complex. Thus a tertiary amine undergo methylation (MeI or Me$_2$SO$_4$), while a primary amino group remains unchanged.

Hydrogenation.[2]   With lanthanum(III) tris-β-diketonates, the C=C bond in enones can be selectively reduced in a catalytic hydrogenation protocol.

[1] I. V. Komarov, V. E. Denisenko, and M. Yu. Kornilov, *T* **49**, 7593 (1993).
[2] I. V. Komarov, V. E. Denisenko, and M. Yu. Kornilov, *T* **50**, 6921 (1994).

Lead(IV) acetate. 13, 155–156; 14, 188; 16, 193–194

Cleavage of N-substituted phenylglycinols.[1]   The generation of primary amines from such derivatives bearing chiral substitutents proceeds without racemization.

4-Oxopentylation of heteroaromatics.[2]   The oxidation of 1-methylcyclobutanol (ring cleavage) in the presence of pyridine and related substances leads to useful yields of the products. The reaction involves radical substitution on protonated heteroaromatics. Manganese(III) acetate is a comparable reagent.

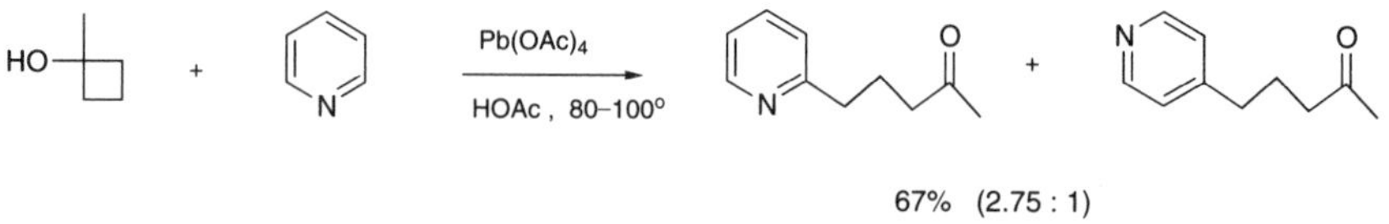

67% (2.75 : 1)

Dioxycarbenes.[3] A method for the carbene generation involves oxidation of methyl dimethylhydrazonocarboxylate, alcoholysis, and pyrolysis.

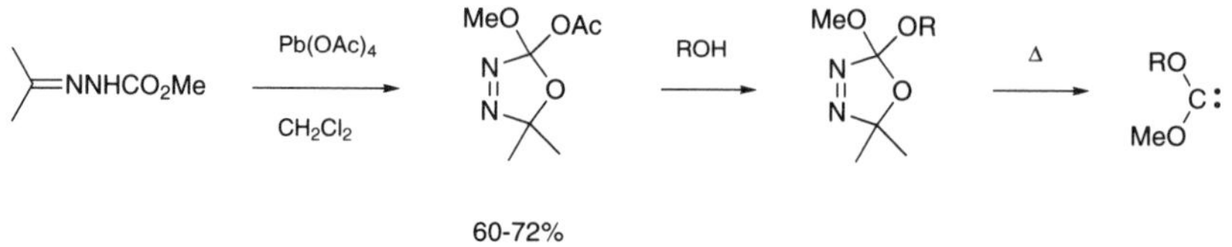

60-72%

Remote carbonylation.[4] Saturated alcohols form δ-lactones on reaction with Pb(OAc)$_4$ and carbon monoxide.

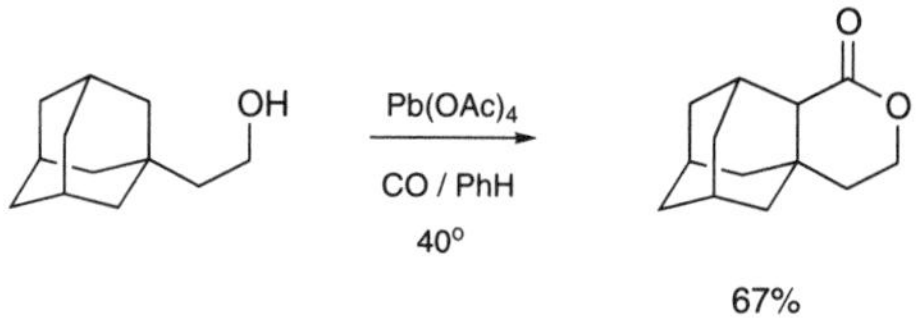

67%

[1] M. K. Mokhallalati and L. N. Pridgen, *SC* **23**, 2055 (1993).
[2] G. I. Nikishin, L. L. Sokova, and N. I. Kapustina, *DOK* **326**, 205 (1992).
[3] K. Kassam, D. L. Pole, M. El-Saidi, and J. Warkentin, *JACS* **116**, 1161 (1994).
[4] S. Tsunoi, I. Ryu, and N. Sonoda, *JACS* **116**, 5473 (1994).

Lead(IV) acetate–copper(II) acetate.

Dehydrogenation.[1] 2-(Alkoxycarbonyl)- and 2-(aminocarbonyl)cycloalkanones are converted to the cycloalkenones with the mixed acetates in refluxing benzene.

[1] A. G. Schultz and M. A. Holoboski, *TL* **34**, 3021 (1993).

Lipases.

Hydrolysis. The list includes acetates of primary alcohols adjacent to a stereocenter,[1] methyl 3-hydroxyalk-2-enoates,[2] dimethyl *trans*-aziridine-2,3-dicarboxylate,[3] acetates of secondary alcohols[4] including the homoallylic type,[5] 4-substituted oxazolin-5-ones,[6] and 1-chloroacetoxybicyclo[4.1.0]heptane.[7]

Resolution by transesterification.   Using vinylic acetates to esterify allyl alcohols,[8] propargyl alcohols,[9] 2-phenylthiocycloalkanols,[10] α-hydroxy esters,[11] methyl 5-hydroxy-2-hexenoates,[12] and 2-substituted 1,3-propanediols,[13] the enantioselective esterification provides a means of separation of optical isomers. Vinyl carbonates are also resolved by lipase-mediated enantioselective conversion to benzyl carbonates.[14]

Other esters that have also been used in the kinetic resolution include 2,2,2-trifluoroethyl propionate.[15,16] There is a report on a double enantioselective transesterification[17] of racemic trifluoroethyl esters and cyclic *meso*-diols by lipase catalysis.

Malonic esters are desymmetrized by exchange of one of the ester groups,[18] and chiral carboxylic esters are obtained from lipase-mediated alcoholysis of mixed carboxylic carbonic anhydrides.[19]

Chemoselective transesterification.   The esterification of the primary alcohol of carbohydrates,[20] the 2-OH of 4,6-*O*-benzylidene glycopyranosides,[21] and the 3-OH of 1,3-naphthalenediol[22] has been observed, although in the last case the selectivity is only moderate (3-OAc:1-OAc:1,3-$(OAc)_2$=80:16:4).

Amidation of esters.   Acrylic[23] and β-keto amides,[24] protected dipeptides,[25] as well as *N,N'*-diacyldiamines[26] are accessible by the enzymatic method. Succinic diesters are converted to *N*-aminosuccinimides[27] by hydrazines.

β-Hydroxy esters give equimolar mixtures of chiral 3-hydroxyamides and the resolved esters.[28] The former compounds are useful for the synthesis of 1,3-amino alcohols. Racemic amines, like alcohols, are resolved by enantioselective acetylation[29] or *N*-alkoxycarbonylation.[30]

[1] T. Matsumoto, Y. Takeda, E. Iwata, M. Sakamoto, and T. Ishida, *CPB* **42**, 1191 (1994).

[2] P. Allevi, M. Anastasia, P. Ciuffreda, and A. M. Sanvito, *TA* **4**, 1397 (1993).

[3] M. Bucciarelli, A. Forni, I. Moretti, F. Prati, and G. Torre, *JCS(P1)* 3041 (1993).

[4] Y. Naoshima, M. Kamezawa, H. Tachibana, Y. Munakata, T. Fujita, K. Kihara, and T. Raku, *JCS(P1)* 557 (1993).

[5] Y.-C. Pai, J.-M. Fang, and S.-H. Wu, *JOC* **59**, 6018 (1994).

[6] J. Z. Crich, R. Brieva, P. Marquart, R.-L. Gu, S. Flemming, and C. J. Sih, *JOC* **58**, 3252 (1993).

[7] J.-P. Barnier, L. Blanco, G. Rousseau, E. Guibe-Jampel, and I. Fresse, *JOC* **58**, 1570 (1993).

[8] N. W. Boaz and R. L. Zimmerman, *TA* **5**, 153 (1994).

[9] P. Allevi, M. Anastasia, F. Cajone, P. Ciuffreda, and A. M. Sanvito, *TA* **5**, 13 (1994).

[10] T. Fukazawa and T. Hashimoto, *TA* **4**, 2323 (1993).

[11] L. T. Kanerva and O. Sundholm, *JCS(P1)* 2407 (1993).

[12] H. Akita, I. Umezawa, M. Takano, and T. Oishi, *CPB* **41**, 680 (1993).

[13] T. Itoh, J. Chika, Y. Takagi, and S. Nishiyama, *JOC* **58**, 5717 (1993).

[14] M. Pozo and V. Gotor, *T* **49**, 10725 (1993).

[15] V. S. Parmar, R. Sinha, K. S. Bisht, S. Gupta, A. K. Prasad, and P. Taneja, *T* **49**, 4107 (1993).

[16] L. T. Kanerva and E. Vanttinen, *TA* **4**, 85 (1993).

[17] F. Theil, A. Kunath, M. Ramm, T. Reiher, and H. Schick, *JCS(P1)* 1509 (1994).

[18] M. Shapira and A. L. Gutman, *TA* **5**, 1689 (1994).

[19] E. Guibe-Jampel and M. Bassir, *TL* **35**, 421 (1994).

[20] R. Pulido and V. Gotor, *CR* **252**, 55 (1994); *JCS(P1)* 589 (1993).

[21] L. Panza, M. Luisetti, E. Crociati, and S. Riva, *JCC* **12**, 125 (1993).

[22] D. Lambusta, G. Nicolosi, M. Piattelli, and C. Sanfilippo, *IJC(B)* **32B**, 58 (1993).
[23] S. Puertas, R. Brieva, F. Rebolledo, and V. Gotor, *T* **49**, 4007 (1993).
[24] M. J. Garcia, F. Rebolledo, and V. Gotor, *T* **50**, 6935 (1994).
[25] K. Kawashiro, K. Kaiso, D. Minato, S. Sugiyama, and H. Hayashi, *T* **49**, 4541 (1993).
[26] C. Astorga, F. Rebolledo, and V. Gotor, *JCS(PI)* 829 (1994).
[27] C. Astorga, F. Rebolledo, and V. Gotor, *S* 287 (1993).
[28] M. J. Garcia, F. Rebolledo, and V. Gotor, *TA* **4**, 2199 (1993).
[29] M. T. Reetz and C. Dreisbach, *Chimia* **48**, 570 (1994).
[30] M. Pozo and V. Gotor, *T* **49**, 4321 (1993).

α-Lithioalkyl dimesitylborane. 13, 8

Alkenes.[1] The α-boryl carbanions react with carbonyl compounds to afford alkenes. The reagent is (E)-selective for ArCHO when the intermediates are trapped with Me_3SiCl and then treated with aq. HF–MeCN. Treatment with TFAA leads to (Z)-alkenes mainly. For generation of (E)-alkenes from aliphatic aldehydes they are added together with a protic acid (e.g., CF_3SO_3H) to the lithioalkylboranes.[2]

Ketones.[3] When the adducts from the lithioalkylboranes and aldehydes are treated with TFAA or NCS, ketones are formed, except that Mes_2BCH_2Li furnishes 1-alkenes.

[1] A. Pelter, D. Buss, E. Colclough, and B. Singaram, *T* **49**, 7077 (1993).
[2] A. Pelter, K. Smith, and S. M. A. Elgendy, *T* **49**, 7119 (1993).
[3] A. Pelter, K. Smith, S. M. A. Elgendy, and M. Rowlands, *T* **49**, 7104 (1993).

Lithiophosphine.

Alkylphosphines.[1] The reagent $LiPH_2$ is prepared from red P, Li, and t-BuOH in liquid ammonia, and used immediately. Thus the addition of an alkyl halide to the solution furnishes the primary alkylphosphine.[1] On the other hand, a one-pot synthesis of dialkylphosphines[2] needs $NaNH_2$ to promote the second alkylation.

[1] L. Brandsma, J. A. van Doorn, R.-J. de Lang, N. K. Gusarova, and B. A. Trofimov, *MC* 14 (1995).
[2] L. Brandsma, N. K. Gusarova, A. V. Gusarov, H. D. Verkruijsse, and B. A. Trofimov, *SC* **24**, 3219 (1994).

2-Lithiothiazole.

Formyl anion equivalent.[1]   2-Lithiothiazole reacts with various carbonyl compounds. Degradation of the heterocycle involves *S*-methylation with MeOTf, borohydride reduction, and hydrolysis ($HgCl_2$, aq. MeCN).

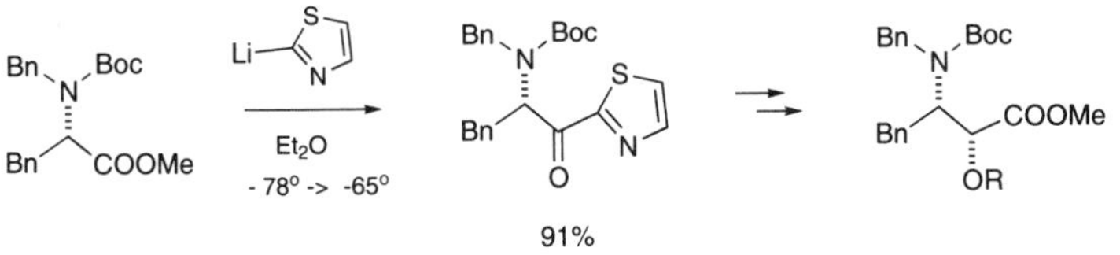

91%

[1] A. Dondoni and D. Perrone, *S* 1162 (1993).

5-Lithiomethyl-3-methylisoxazole.

Benzoisothiazoles.[1]   Annulation with β,β-bismethylthio α,β-unsaturated carbonyl compounds probably proceeds by a 1,4-addition, aldolization, and aromatization.

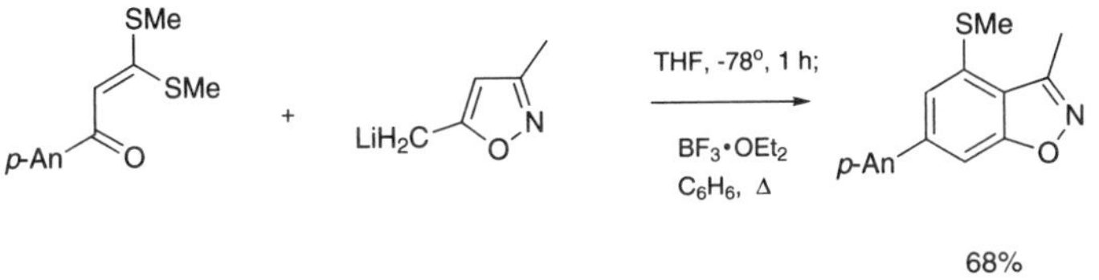

68%

[1] D. Pooranchand, J. Satyanarayana, H. Ila, and H. Junjappa, *S* 241 (1993).

3-Lithiothiophene.

3-Substituted thiophenes.[1]   3-Lithiothiophene is obtained by the Li/Br exchange method. It is stable at room temperature, and various derivatives can thus be prepared in a convenient manner.

[1] X. Wu, T.-A. Chen, L. Zhu, and R. D. Rieke, *TL* **35**, 3673 (1994).

Lithium. 13, 157–158; 15, 184

(E)-1,2-Dilithioalkenes.[1]   Carbon–carbon triple bonds add lithium to form dilithio derivatives. These may be alkylated.

5-Trimethylsilyl-3-cyclohexenone.[2]   This compound is obtained by treatment of anisole with Li–Me_3SiCl and hydrolysis of the enol ether. Further desilylative acylation and dehydrogenation result in *m*-acylphenols.

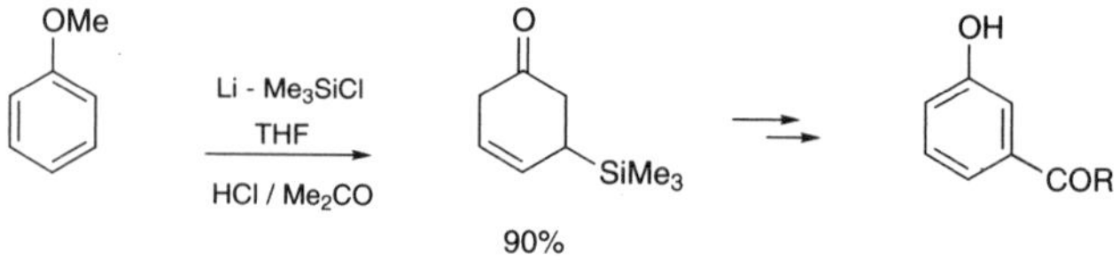

Reductive C–O bond cleavage. Benzylic methyl ethers give benzyl anions[3] which can be alkylated. The controlled stepwise cleavage of dimethyl acetals[4] makes the method synthetically useful.

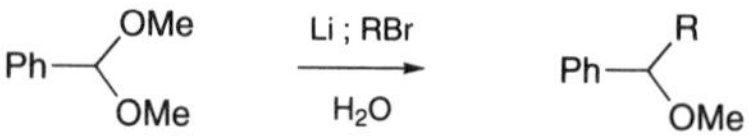

[1] A. Maercker and U. Girreser, *T* **50**, 8019 (1994).
[2] B. Bennetau, F. Rajarison, J. Dunogues, and P. Babin, *T* **50**, 1179 (1994).
[3] U. Azzena, G. Melloni, M. Fenude, C. Fina, M. Marchetti, and B. Sechi, *SC* **24**, 591 (1994).
[4] U. Azzena, G. Melloni, L. Pisano, and B. Sechi, *TL* **35**, 6759 (1994).

Lithium–ammonia. 13, 158; 17, 161

De-S-benzylation.[1] A synthesis of 1,2-dithiins from toluene-α-thiol and 1,3-alkanediynes requires debenzylation and oxidative coupling.

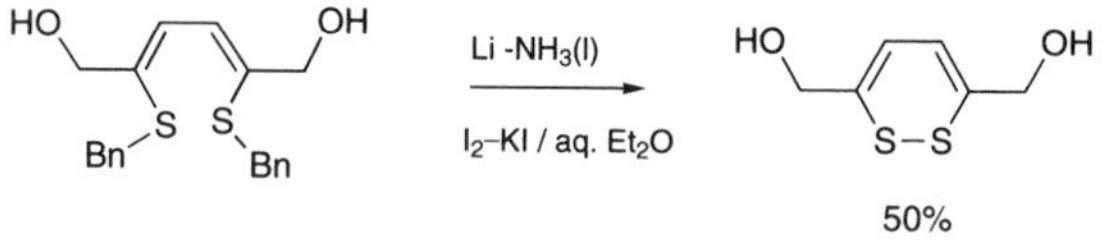

[1] M. Koreeda and W. Yang, *SL* 201 (1994).

Lithium cobaltbisdicarbollide $LiCo(B_9C_2H_{11})_2$.

Michael addition.[1] The reaction between silyl ketene acetals and very hindered enones is effectively catalyzed by this Lewis acid. In various considerations it is superior to the $LiClO_4$–Et_2O system.

[1] W. J. DuBay, P. A. Grieco, and L. J. Todd, *JOC* **59**, 6898 (1994).

Lithium–1,3-diaminopropane.

(E)-Alkenes.[1] Carbon–carbon triple bonds are reduced stereoselectively with this system.

[1] I. Kovarova and L. Streinz, *SC* **23**, 2397 (1993).

Lithium aluminum amides. 17, 162

Imines.[1] Efficient formation of imines from aldehydes and from unhindered ketones is carried out at room temperature.

[1] A. Solladie-Cavallo, M. Bencheqroun, and F. Bonne, *SC* **23**, 1683 (1993).

Lithium aluminum hydride. 14, 190–191

Selective desilylation.[1] Due to neighboring group participation, the selective removal of one *O*-silyl group has been observed.

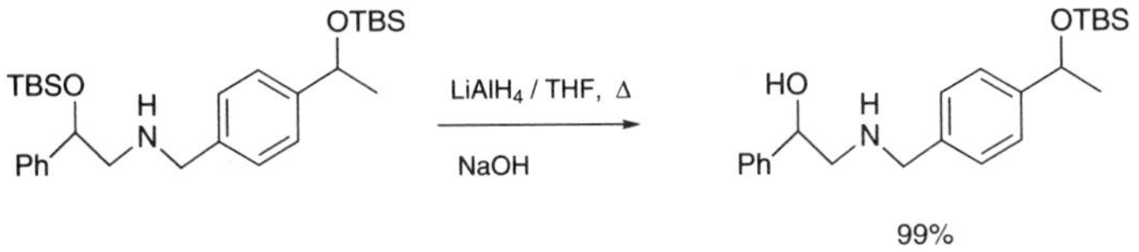

99%

Reduction of 2-arylidenecycloalkanones.[2] Formation of *trans*-2-arylmethyl-cycloalkanols is the result of a directed hydroalumination of the immediate allylic alcohols.

Heterocycle transformation.[3] 6*H*-1,2-Oxazines are converted to aziridines by way of N–O bond cleavage, C=N bond reduction, and cyclization.

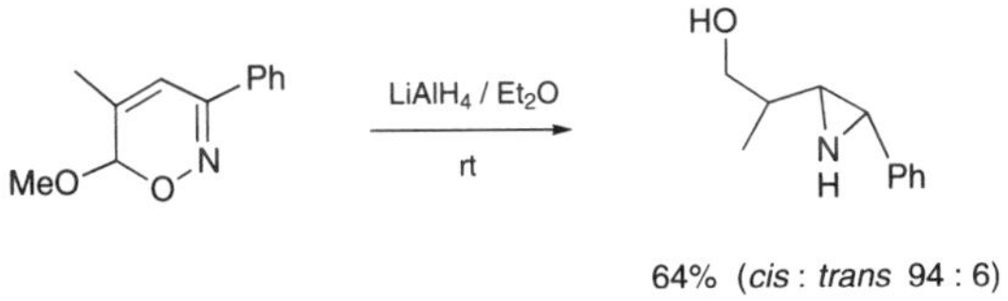

64% (*cis* : *trans* 94 : 6)

[1] E. F. J. de Vries, J. Brussee, and A. van der Gen, *JOC* **59**, 7133 (1994).
[2] K. Koch and J. H. Smitrovich, *TL* **35**, 1137 (1994).
[3] R. Zimmer, K. Homann, and H.-U. Reissig, *LA* 1155 (1993).

Lithium aluminum hydride–aluminum chloride.

Stereoselective cleavage of diphenylmethylene acetals.[1] Formation of axial diphenylmethyl ethers of pyranosides is noted. Preferential coordination to the more available equatorial oxygen atom prior to the hydride delivery may explain the results.

Hydroalumination of allenes.[2] Allylaluminum species are formed that on treatment with phenylboronic acid and a carbonyl compound furnish homoallylic alcohols.

$$C_6H_{13} \diagdown C \diagdown \quad \xrightarrow[\text{PhCHO, -78°}]{\substack{\text{LiAlH}_4\text{-AlCl}_3 \\ \text{PhB(OH)}_2}} \quad \begin{array}{c} \text{Ph} \diagdown \diagup \text{OH} \\ C_6H_{13} \diagup \diagdown \diagup \\ \text{62\%} \end{array}$$

[1] A. Borbas, J. Hajko, M. Kajtar-Peredy, and A. Liptak, *JCC* **12**, 191 (1993).
[2] S. Nagahara, K. Maruoka, and H. Yamamoto, *BCSJ* **66**, 3783 (1993).

Lithium aluminum hydride–copper(I) cyanide.

Reduction of allylic epoxides.[1]   Reduction by an S_N2' pathway leads to allylic alcohols in which the double bond is transposed.

[1] J. F. Genus, D. D. Peters, and T. A. Bryson, *SL* 759 (1993).

Lithium amides, chiral. 13, 159–160; 17, 163–164

Enantioselective deprotonation.[1]   Protected 4-hydroxycyclohexanones undergo regioselective enolization and can subsequently be converted to chiral compounds through *O*-derivatization (acetylation, silylation). The addition of LiCl seems to enhance the enantioselectivity[2] somewhat.

Michael additions.[3]   Chiral amides themselves may be used as the Michael addends for unsaturated esters. With an *anti*-selective alkylation in situ the overall process gives access to chiral β-amino esters.

The presence of chiral lithium amides has influence on the Michael addition involving achiral enolates.

[1] M. Majewski and J. MacKinnon, *CJC* **72**, 1699 (1994).
[2] B. J. Bunn, N. S. Simpkins, Z. Spavold, and N. J. Crimmin, *JCS(P1)* 3113 (1993).
[3] E. Juaristi, A. K. Beck, J. Hansen, T. Matt, T. Mukhopadhyay, M. Simon, and D. Seebach, *S* 1271 (1993).

Lithium aminoborohydrides. 17, 170

These nonpyrophoric reagents, $LiBH_3(NR_2)$, show virtually the same reducing power as $LiAlH_4$, yet react with protic solvents only slowly above pH 4. They can be prepared as solids or used in situ.[1] Advantageously, they can be handled in dry air as easily as $NaBH_4$.

Reduction of amides.[1]   The second stage of the reduction (i.e., C–O vs. C–N bond scission), can be controlled by using aminoborohydrides having different steric bulks. The cleavage of the C–N bond to give alcohols is favored by less hindered reducing agents.

Diastereoselective reduction of imines.[3]   Moderate 1,3-induction of chirality is observed.

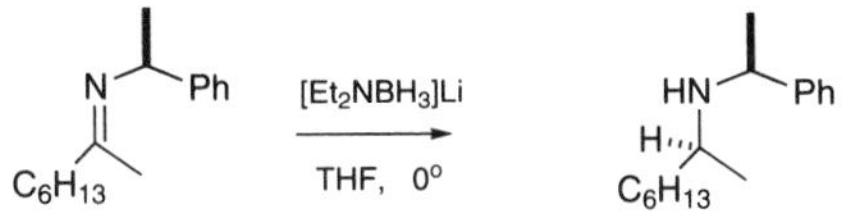

83% (34% de)

Chiral borohydrides.[4]   Transfer of [LiH] to chiral organoboranes leads to the chiral reducing agents.

[1] G. B. Fisher, J. C. Fuller, J. Harrison, S. G. Alvarez, E. R. Burkhardt, C. T. Goralski, and B. Singaram, *JOC* **59**, 6378 (1994).
[2] G. B. Fisher, J. C. Fuller, J. Harrison, C. T. Goralski, and B. Singaram, *TL* **34**, 1091 (1993).
[3] J. C. Fuller, C. M. Belise, C. T. Goralski, and B. Singaram, *TL* **35**, 5389 (1994).
[4] J. Harrison, S. G. Alvarez, G. Godjoian, and B. Singaram, *JOC* **59**, 7193 (1994).

Lithium bis(methylthio)trimethylsilylmethide.

Cascade reaction.[1]   With 2-oxiranylethyl tosylate and related bifunctional electrophiles that enable regeneration of a carbanionoid species by Si–C bond scission, ring formation is essentially accomplished in one step. For example, the preparation of 3,3-bismethylthiocyclopentanol in 80% yield has been reported.

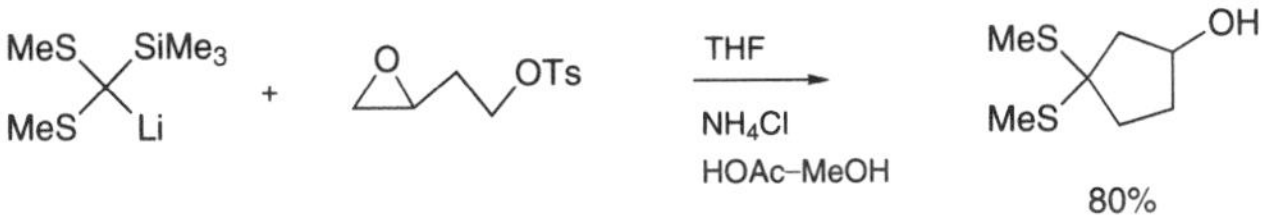

[1] M.-R. Fischer, A. Kirschning, T. Michel, and E. Schaumann, *ACIEE* **23**, 217 (1994).

Lithium borohydride–trimethylsilyl chloride. 15, 186

C=N Bond reduction.   Oximes[1] and various *O*-derivatives thereof[2] undergo reduction to give amines at room temperature.

[1] D. Green, G. Patel, S. Elgendy, J. A. Baban, G. Claeson, V. V. Kakkar, and J. Deadman, *TL* **34**, 6917 (1993).
[2] A. Banaszek and W. Karpiesiuk, *CR* **251**, 233 (1994).

Lithium bromide.

Finkelstein displacement.[1]   Primary alkyl chlorides are converted to the pure bromides on repeated treatment with LiBr in 2-butanone at 120°C. Typically, the yields range from 86% to 96%.

2-Alkylcycloalkanones.[2]   A useful procedure for the dealkoxycarbonylation of 2-alkyl-2-ethoxycarbonylcycloalkanones consists of microwave heating of the esters with LiBr, Bu$_4$NBr, and some water.

[1] X. Li, S. M. Singh, and F. Labrie, *SC* **24**, 733 (1994).
[2] J. P. Barnier, A. Loupy, P. Pigeon, M. Ramdani, and P. Jacquault, *JCS(P1)* 397 (1993).

Lithium *t*-butoxide.

1,4-Naphthalenediols from dimethyl phthalide-3-phosphonates.   The annulation by tandem Michael–Dieckmann reactions is promoted by *t*-BuOLi.

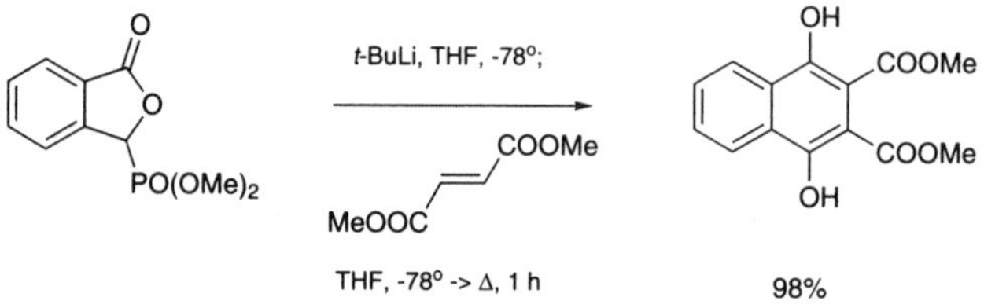

98%

Allylic alcohol dianions.[2]   2-Alkylpropenols are deprotonated by BuLi and *t*-BuOLi to provide the dianions. However, the reaction with electrophiles is not regioselective.

[1] M. Watanabe, H. Morimoto, K. Nogami, S. Ijichi, and S. Furukawa, *CPB* **41**, 968 (1993).
[2] T. Liu and R. M. Carlson, *SC* **23**, 1437 (1993).

Lithium 4,4'-di-*t*-butylbiphenylide (LDTBB). 13, 162–163; 16, 195–196; 17, 164

Lithium–halogen exchange.   The reagent has been proven particularly effective for chloro compounds, as shown in the Barbier-type reactions involving chloromethyl ethyl ether,[1] 1,4-dichloro-2-butyne[2] and -2-butenes.[3] Vinylic chlorides also readily undergo exchange,[4] and actually 2,3-dichloropropene is converted to the dilithio reagent.[5]

Reductive lithiation of allylic acetals.[6]   This method provides masked lithium homoenolates, which are used in reactions with carbonyl substrates.

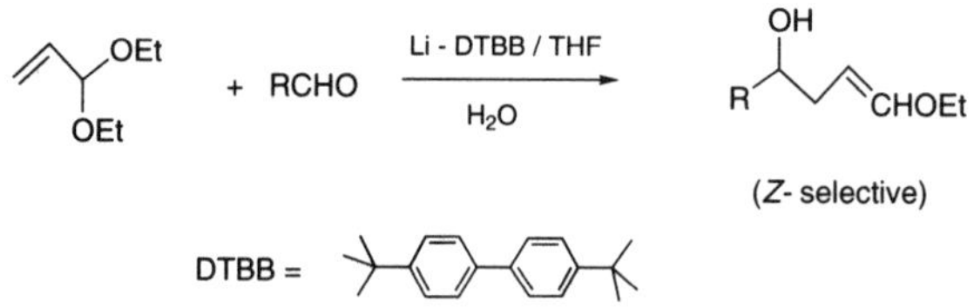

(*Z*- selective)

Ring opening of azetidines.[7]   3-Aminolithium reagents are obtained as nucleophilic intermediates.

Organolithiums.   Barbier-type condensation occurs when an alkyl nitrile and a carbonyl compound are treated with LDTBB.[8] Trialkyl phosphates are converted to alkyllithiums in the same way.[9]

Carbanions generated by reductive desulfurization react intermolecularly[10,11] or intramolecularly[12] according to their structural features. Thus the rearrangement of some species at higher temperatures allows access to novel skeletons.

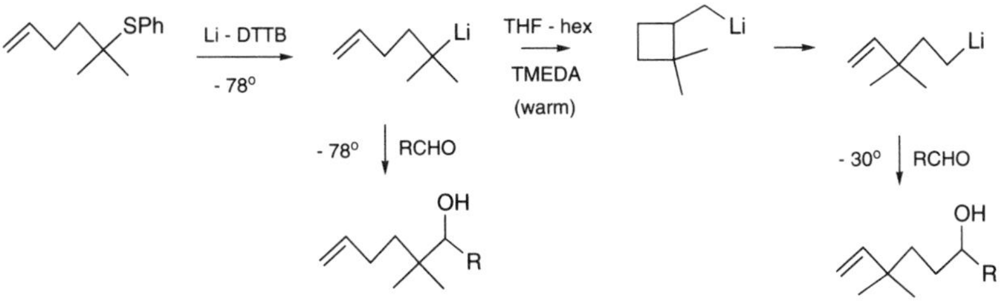

[1] D. Guijarro and M. Yus, *TL* **34**, 3487 (1993).
[2] D. Guijarro and M. Yus, *T* **51**, 231 (1995).
[3] D. Guijarro and M. Yus, *T* **50**, 7857 (1994).
[4] A. Bachki, F. Foubelo, and M. Yus, *TL* **35**, 7643 (1994).
[5] D. Guijarro and M. Yus, *TL* **34**, 2011 (1993).
[6] J. F. Gil, D. J. Ramon, and M. Yus, *T* **50**, 3437 (1994).
[7] J. Almena, F. Foubelo, and M. Yus, *T* **50**, 5775 (1994).
[8] D. Guijarro and M. Yus, *T* **50**, 3447 (1994).
[9] D. Guijarro, B. Mancheno, and M. Yus, *T* **50**, 8551 (1994).
[10] S. Marumoto and I. Kuwajima, *JACS* **115**, 9021 (1993).
[11] B. Mudryk and T. Cohen, *JACS* **115**, 3855 (1993).
[12] F. Chen, B. Mudryk, and T. Cohen, *T* **50**, 12793 (1994).

Lithium chloride.

β-Chlorohydrins from cyclic sulfites.[1]   Ring opening of cyclic sulfides by LiCl serves to differentiate the two hydroxyl groups of *vic*-diols.

Robinson annulation.[2]   Using the LiCl–HMPA system at high temperatures, 2-oxocycloalkanecarboxylic esters not only undergo annulation with enones, decarbalkoxylation also occurs.

[1] K. Nymann and J. S. Svendsen, *ACS* **48**, 183 (1994).
[2] Y. Ozaki, A. Kubo, and S.-W. Kim, *CL* 993 (1993).

Lithium cyanide.

α-Cyano phosphates.[1]   Lithium cyanide provides the nucleophile in the derivatization of carbonyl compounds. The reaction is performed in the presence of a chlorophosphate ester in DMF at room temperature.

[1] I. Mico and C. Najera, *T* **49**, 4327 (1993).

Lithium diethylamide. 15, 188

Diethylcarbamoyllithium.[1]   Reaction of Et$_2$NLi with CO gives a reagent that is reactive towards several sulfur compounds. Accordingly, it can be used to synthesize thiocarbamates.

γ,δ-Unsaturated anilides.[2]   Et$_2$NLi converts *N*-phenyl imidates to *N*-silyl ketene *N,O*-acetals, which undergo Claisen rearrangement at room temperature.

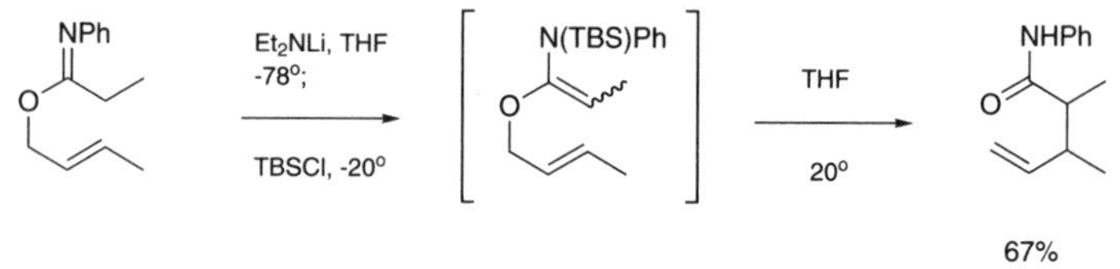

[1] T. Mizuno, I. Nishiguchi, and T. Hirashima, *T* **49**, 2403 (1993).
[2] P. Metz and C. Linz, *T* **50**, 3951 (1994).

Lithium diisopropylamide. 13, 163–164; 15, 188–189; 16, 196–197; 17, 165–167

Ester enolates.   Procedures for the preparation of (*E*)- and (*Z*)-ketene silyl acetals are well developed.[1] Enolates have been generated from conjugate esters by way of Michael addition, and when a remote halide is present, they are quenched by cyclization. Chiral Michael donors such as carbanions of the SAMP/RAMP hydrazones initiate formation of *trans*-2-(2'-oxoalkyl)cycloalkanecarboxylic esters[2] with excellent diastereomer excess and enantiomer excess.

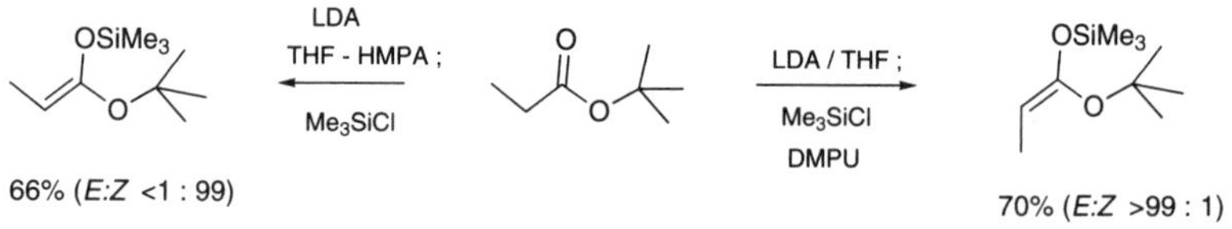

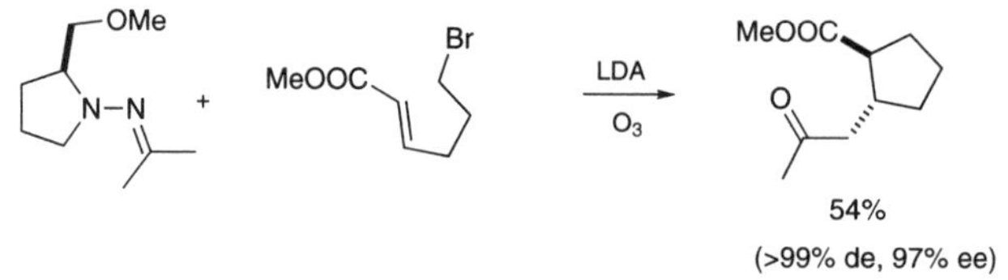

N-Protected iminodiacetic esters readily yield 1,3-dianions,[3] which are useful nucleophiles.

Directed lithiation.   2-Arenesulfonyl benzamides are deprotonated in a regioselective fashion, furnishing thioxanthen-9-one 10,10-dioxides[4] in good yields (57–96%). 3-Halopyridines, including the iodo compounds, manifest directed *o*-lithiation and regioselective reaction (at C-4) with electrophiles.[5] It should be noted that 3-chloro-4-iodopyridine generates 3,4-pyridyne on treatment with either BuLi or *t*-BuLi. Halogen dance is also observed.[6]

Vinyl esters are silylated at the trigonal α-carbon, making available the enol esters of silyl ketones.[7] The regioselective deprotonation of β-alkoxyacrylic esters and the ability of the lithio derivative to undergo a tandem Michael–Julia condensation with suitable acceptors are exploitable in the synthesis of highly functionalized cyclopentenones.[8]

55%
(ratio of diastereomers 2:1)

Alkenyl aryl sulfoxides are also deprotonated at the α-carbon, and the subsequent reaction with aldehydes is diastereoselective.[9]

Cyclopropanation.[10]   When a ketone or nitrile and 2-phenylsulfonyl-1,3-cyclohexadiene are mixed with LDA, a twofold reaction (Michael + S_N2) occurs.

62%

Assorted anions.   These are generated by deprotonation of allylic halides,[11] chloromethylphosphonic esters,[12] conjugated hydrazones,[13] chiral carbamates,[14] unsaturated α-aminonitriles,[15] phosphonamides,[16] and sulfonamides.[17] The dianions derived from ω-haloalkanecarboxylic acids cyclize, and this reaction forms the basis of a synthesis of *N*-Boc cyclic imino acids.[18] The conjugate bases of 2-(arylmethoxy)-methyl-2-oxazolines[19] are unstable as [2,3]sigmatropic rearrangement takes place even at $-75°C$.

70%

[1] J. Otera, Y. Fujita, and S. Fukuzumi, *SL* 213 (1994).
[2] D. Enders, H. J. Scherer, and J. Runsink, *CB* **126**, 1929 (1993).
[3] J. Einhorn, C. Einhorn, and J.-L. Pierre, *SL* 1023 (1994).
[4] F. Beaulieu and V. Snieckus, *JOC* **59**, 6508 (1994).
[5] G. W. Gribble and M. G. Saulnier, *H* **35**, 151 (1993).
[6] Rocca, C. Cochennec, F. Marsais, L. Thomas-dit-Dumont, M. Mallet, A. Godard, and G. Queguiner, *JOC* **58**, 7832 (1993).
[7] S. W. Wright, *TL* **35**, 1841 (1994).
[8] A. Datta and R. R. Schmidt, *TL* **34**, 4161 (1993).
[9] J. Fawcett, S. House, P. R. Jenkins, N. J. Lawrence, and D. R. Russell, *JCS(P1)* 67 (1993).
[10] A. M. Ericsson, N. A. Plobeck, and J.-E. Bäckvall, *ACS* **48**, 252 (1994).
[11] M. Julia, J.-N. Verpeaux, and T. Zahneisen, *BSCF* **131**, 539 (1994).
[12] S. Berte-Verrando, F. Nief, C. Patois, and P. Savignac, *JCS(P1)* 821 (1994).
[13] M. Yamashita, K. Matsumiya, and K. Nakano, *BCSJ* **66**, 1759 (1993).
[14] S. S. C. Koch and A. R. Chamberlin, *JOC* **58**, 2725 (1993).
[15] C.-J. Chang, J.-M. Fang, and L.-F. Liao, *JOC* **58**, 1754 (1993).
[16] V. J. Blazis, K. J. Koeller, and C. D. Spilling, *TA* **5**, 499 (1994).
[17] F. A. Davis, P. Zhou, and P. J. Carroll, *JOC* **58**, 4890 (1993).
[18] A. De Nicola, C. Einhorn, J. Einhorn, and J. L. Luche, *CC* 879 (1994).
[19] K. Kamata and M. Terashima, *CC* 2771 (1994).

Lithium diisopropylamide–hexamethylphosphoric triamide.

Enolization.[1] *N*-Aryl enaminones are enolized toward the α'-position despite the presence of HMPA. For alkylation at the γ-carbon it is necessary to use LiN-(SiMe$_3$)$_2$ as the base.

A general approach to itaconic esters[2] is via the α-dimethylaminomethyl-succinates, which are formed by alkylation of β-dimethylaminopropionic esters with a bromoacetate.

1,4-Asymmetric induction is evident in the alkylation[3] of chiral cyclic acetals of cycloalkanone-2-carboxylic esters.

Attack of a remote ester enolate onto a diene–Fe(CO)$_3$ complex under a CO atmosphere leads to a cyclic product. Fused and bridged ring systems can thus be constructed.[4]

79%

[1] D. Dugat, D. Gardette, J.-C. Gramain, and B. Perrin, *BSCF* 66 (1994).

[2] P. Dowd and B. K. Wilk, *SC* **23**, 2307 (1993).

[3] K. Kato, H. Suemune, and K. Sakai, *T* **50**, 3315 (1994).

[4] M.-C. P. Yeh, B. A. Sheu, H.-W. Fu, S.-I. Tau, and L.-W. Chuang, *JACS* **115**, 5941 (1993).

Lithium diisopropylamide–potassium *t*-butoxide. 13, 164

Epoxide isomerization.[1]   This strong base combination converts epoxides to allylic alcohols. The regiochemistry of the ring opening can be directed by a proximal heteroatom in the substrate.

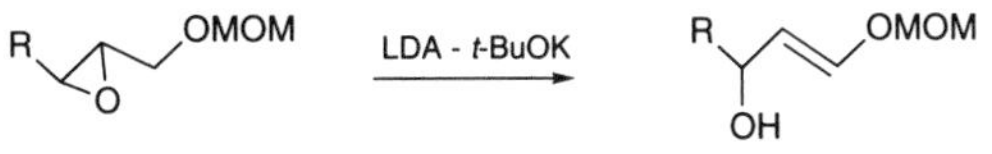

[1] A. Mordini, S. Pecchi, G. Capozzi, A. Capperucci, A. Degl'Innocenti, G. Reginato, and A. Ricci, *JOC* **59**, 4784 (1994).

Lithium diisopropylamide–silver cyanide.

Intramolecular Michael addition.[1]   The combination of reagents promotes formation of sulfonic lactones in good yields (60–93%) from γ-alkanesulfonyloxy-α,β-unsaturated esters due to preferential deprotonation at the α-carbon to the sulfur atom.

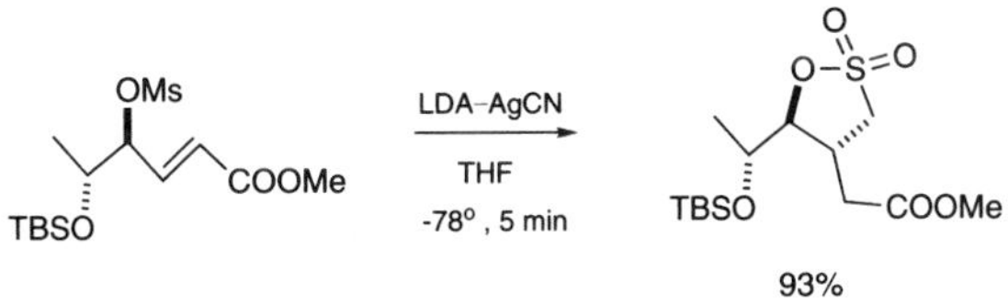

[1] N. Asao, M. Meguro, and Y. Yamamoto, *SL* 185 (1994).

Lithium hexamethyldisilazide. 13, 165; 14, 194

Amino acids.   Through enolization piperazine-2,5-diones[1] and perhydropyrimidin-4-ones[2] using LHMDS as the base for alkylation, α- and β-amino acids are synthesized, respectively. The stereoselectivity can be manipulated by varying chiral substituents on the nitrogen atoms.

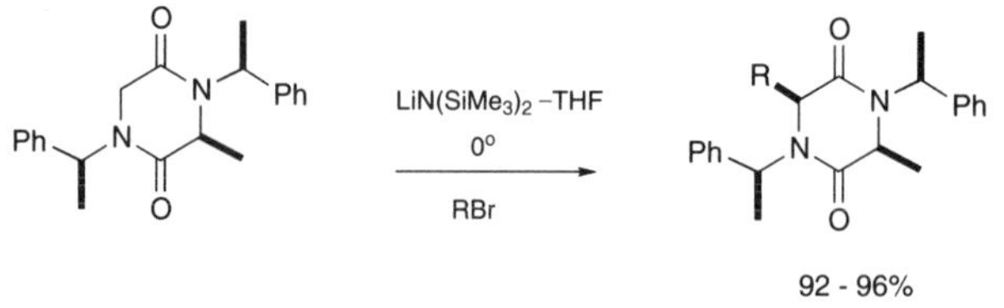

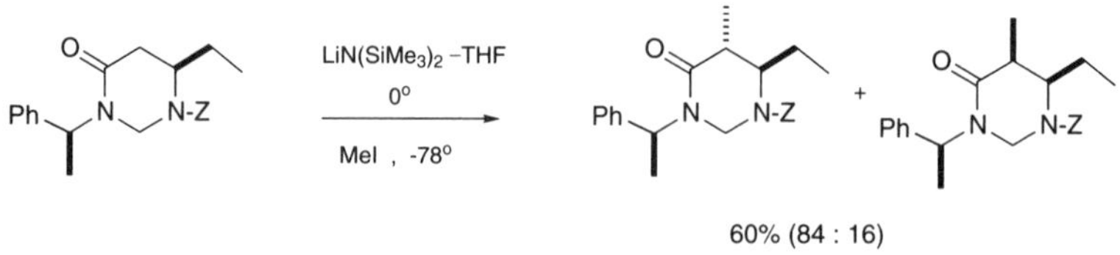

92 - 96%

60% (84 : 16)

Julia condensation.[3]   LHMDS is also an effective base for the generation of sulfonyl carbanions, which may undergo condensation with esters.

Horner olefination.[4]   With this stronger base a lower temperature (e.g., $-78°C$) is sufficient to achieve the transformation.

Elimination.[5] α,α-Dibromoketones undergo dehydrobromination and rearrangement to afford lithium alkynolates on consecutive exposure to LHMDS and BuLi. Such species are intercepted by acid chlorides.

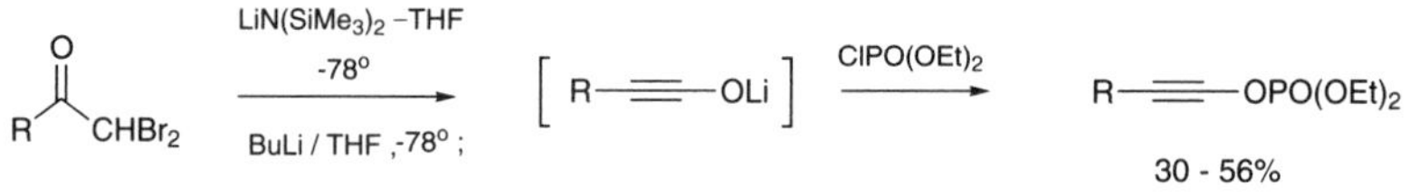

30 - 56%

[1] M. Orena, G. Porzi, and S. Sandri, *JCR(S)* 318 (1993).
[2] I. Braschi, G. Cardillo, C. Tomasini, and R. Venezia, *JOC* **59**, 7292 (1994).
[3] H. K. Jacobs and A. S. Gopalan, *JOC* **59**, 2014 (1994).
[4] M. Watanabe, S. Ijichi, H. Morimoto, K. Nogami, and S. Furukawa, *H* **36**, 553 (1993).
[5] V. V. Zhdankin and P. J. Stang, *TL* **34**, 1461 (1993).

Lithium hexamethyldisilazide–silver iodide.

Allylic and benzylic amines.   The substitution of benzylic[1] and allylic halides[2] proceeds in refluxing THF. The protecting groups on the N,N-bis(trimethylsilyl) amines are easily hydrolyzed.

[1] T. Murai, M. Ogami, F. Tsujimura, H. Ishihara, and S. Kato, *SC* **23**, 7 (1993).
[2] T. Murai, M. Yamamoto, S. Kondo, and S. Kato, *JOC* **58**, 7440 (1993).

Lithium hydride. 13, 165–166

Alkyl silyl ethers.[1] Activated commercial LiH promotes reductive silylation of carbonyl compounds in very high yields in the presence of Me_3SiCl and $Zn(OTf)_2$.

Reduction.[2] A wide scope of substrates reducible by activated LiH or LiD (derived from hydrogenation of *n*-BuLi in TMEDA) includes RCHO, R_2CO, RCOCl, RCOOR', and $(RO)_2CO$. Halides undergo elimination.

[1] T. Ohkuma, S. Hashiguchi, and R. Noyori, *JOC* **59**, 217 (1994).
[2] E. M. Zippi, *SC* **24**, 2515 (1994).

Lithium hydroxide.

Saponification.[1] The mildness of LiOH as a saponifying agent is demonstrated in a synthesis of chiral 2-alkyl-4-oxoalkanoic acids. In the first step, alkylation of β-keto esters with 2-triflyloxy esters provides the keto diesters.

Emmons–Wadsworth olefination.[2] Lithium hydroxide is a more convenient base to effect the olefination of aldehydes in THF at room temperature. For reaction with ketones, slow addition of the base to the reaction components in the presence of activated 4Å molecular sieves improves the yields significantly.

[1] R. V. Hoffman and H.-O. Kim, *TL* **34**, 2051 (1993).
[2] F. Bonadies, A. Cardilli, A. Lattanzi, L. R. Orelli, and A. Scettri, *TL* **35**, 3383 (1994).

Lithium naphthalenide (LN). 15, 190–191

Lithium–halogen exchange. The reagent is useful for preparation of organo-lithiums from acetal-containing substrates,[1] dihaloarenes (monofunctionalization),[2] and halopyridines.[3] Dechlorination of *N*-(β-chloroalkyl)amides apparently cannot be achieved with BuLi alone (stops at *N*-deprotonation?), and the use of LN is required.[4] Of course the *N,C*-dithio derivatives may be employed in C–C bond formation[5] (remarkably the coupling with aryl and vinyl halides). Barbier-type reactions have been achieved.[6]

Carbamoyl and thiocarbamoyl chlorides also undergo exchange to give the corresponding lithium reagents. These are readily trapped by carbonyl compounds to afford 2-hydroxy carboxamides.[7]

Although it is mentioned that cyclic acetals of saturated carbonyl compounds survive the LN treatment, those derived from phenones are susceptible, especially at higher temperatures (−40°C vs. −78°C). Actually both C–O bonds may be severed.[8]

The direct conversion of dialkyl sulfates into alkyllithium reagents[9] provides an alternative route to the R–OH → R–X → R–Li method.

Electrophilic substitution of phenones. The facile formation of *C,O*-dilithiated diarylmethanols from diaryl ketones and LN, coupled with subsequent reaction with various electrophiles,[10] solves the compatibility problem of the conventional addition reaction with functional nucleophiles to prepare diaryl *t*-carbinols.

1,2-Amino alcohols.[11]   The treatment of a mixture of an imine and a carbonyl compound with LN causes reductive cross-coupling.

β-Functional amines.[12]   β-Aminoethyllithium reagents are available by treatment of aziridines with LN, and thence the β-functionalized amines.

Reductive desulfonylation.[13]   Alkyl phenyl sulfones are cleaved unidirectionally, alkyllithium thus generated can be utilized accordingly.

[1] J. F. Gil, D. J. Ramon, and M. Yus, *T* **49**, 4923 (1993).
[2] M. S. Sell, M. V. Hanson, and R. D. Rieke, *SC* **24**, 2379 (1994).
[3] Y. Kondo, N. Murata, and T. Sakamoto, *H* **37**, 1467 (1994).
[4] F. Foubelo and M. Yus, *TL* **35**, 4831 (1994).
[5] J. Barluenga, J. M. Montserrat, and J. Florez, *JOC* **58**, 5976 (1993).
[6] C. Gomez, D. J. Ramon, and M. Yus, *T* **49**, 4117 (1993).
[7] D. J. Ramon and M. Yus, *TL* **34**, 7115 (1993).
[8] J. F. Gil, D. J. Ramon, and M. Yus, *T* **49**, 9535 (1993).
[9] D. Guijarro, G. Guillena, B. Mancheno, and M. Yus, *T* **50**, 3427 (1994).
[10] D. Guijarro, B. Mancheno, and M. Yus, *T* **49**, 1327 (1993).
[11] D. Guijarro and M. Yus, *T* **49**, 7761 (1993).
[12] J. Almena, F. Foubelo, and M. Yus, *JOC* **59**, 3210 (1994).
[13] D. Guijarro and M. Yus, *TL* **35**, 2965 (1994).

Lithium perchlorate. 17, 167–170

Aldol-type reactions.[1]   The condensation between an aldehyde and a ketene silyl acetal proceeds in CH_2Cl_2 with $LiClO_4$ as catalyst at $-30°C$. *syn*-Selectivity is observed when α-oxyaldehydes are used as reaction partners.

Diels–Alder reactions and 1,3-Claisen rearrangements.[2]   Lithium perchlorate in CH_2Cl_2 apparently mimics the $LiClO_4–Et_2O$ system in the catalysis of many reactions, including the Diels–Alder reactions and 1,3-Claisen rearrangements.

[1] M. T. Reetz and D. N. A. Fox, *TL* **34**, 1119 (1993).
[2] M. T. Reetz and A. Gansaur, *T* **49**, 6025 (1993).

Lithium perchlorate–diethyl ether.

Dithioacetalization of aldehydes.[1]   The reaction is performed at room temperature with high chemoselectivity.

β-Amino esters.[2]   A three-component condensation comprising an aldehyde, an *N*-silylamine, and a ketene silyl acetal is extremely efficient (requiring as short a reaction time as 10 min at room temperature).

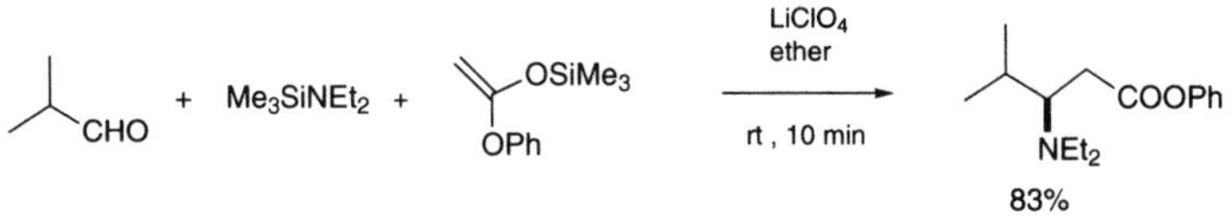

Electrophilic substitution of indole.[3]   This reaction strictly exploits the Lewis acidity of the $LiClO_4$–Et_2O system. It can be applied to a synthesis of yuehchukene.

Diels–Alder reactions and cyclocondensations.   Tricyclic skeletons containing up to 5 stereocenters are created by this method in an intramolecular Diels–Alder reaction.[4] Camphorsulfonic acid is also present as a catalyst. The Danishefsky cyclocondensation involving an aldehyde and a very reactive diene is also similarly catalyzed.[5]

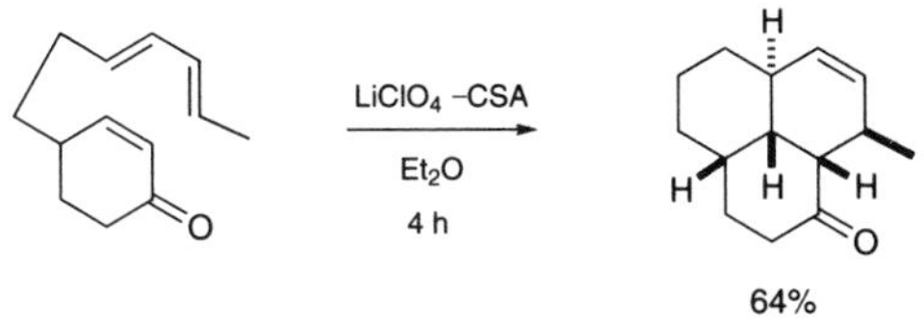

Ene reactions.[6]   Catalyzed reactions include alkenes and 1,2,4-triazole-3,5-diones.

Arylamines.   Electron-rich arenes react with bis(2,2,2-trichloroethyl) azodicarboxylate to give arylhydrazine derivatives[7] in good yields. Reduction with Zn–HOAc provides the amines in one operation. For example, *p*-methoxyaniline is prepared in about 70% yield.

[1] V. G. Saraswathy and S. Sankararaman, *JOC* **59**, 4665 (1994).
[2] M. R. Saidi, A. Heydari, and J. Ipaktschi, *CB* **127**, 1761 (1994).
[3] K. H. Henry, Jr., and P. A. Grieco, *CC* 510 (1993).
[4] P. A. Grieco, J. P. Beck, S. T. Handy, A. Saito, and J. F. Daeuble, *TL* **35**, 6783 (1994).
[5] P. A. Grieco and E. D. Moher, *TL* **34**, 5567 (1993).
[6] W. J. Kinart, *JCR(S)* 486 (1994).
[7] I. Zaltsgendler, Y. Leblanc, and M. A. Bernstein, *TL* **34**, 2441 (1993).

Lithium 2-(pyrrolidin-1-ylmethyl)pyrrolidide.

Chiral allylic alcohols.[1]   Opening of *meso*-epoxides by the chiral base furnishes products with moderate enantiomer excess.

[1] M. Asami, T. Ishizaki, and S. Inoue, *TA* **5**, 793 (1994).

Lithium tetrafluoroborate.

Activation of oxetanes.[1]   The ring opening by an amine in MeCN is made possible by the addition of $LiBF_4$. 1,3-Amino alcohols are obtained in 85–98% yields.

Cleavage of 4-(trimethylsilylmethyl)ethyleneacetals.[2]   The anion furnishes F^- to attack the silyl group, resulting in the fragmentation of the dioxolane system and emergence of the carbonyl group.

[1] M. Chini, P. Crotti, L. Favero, and F. Macchia, *TL* **35**, 761 (1994).
[2] B. M. Lillie and M. A. Avery, *TL* **35**, 969 (1994).

Lithium 2,2,6,6-tetramethylpiperidide (LTMP). 13, 167; 14, 194–195; 17, 171–172

Isomerization of epoxides.[1]   The conversion to carbonyl compounds is initiated by proton abstraction. With a highly hindered base the less substituted side of an epoxide is preferentially deprotonated; thus a monosubstituted epoxide tends to give an aldehyde product.

Ketene silyl acetals.[2]   The enolsilylation of esters is highly dependent on the bases used. Contrasting stereoselectivities in the silylation of methyl α-t-butyldimethylsiloxyacetate are found with respect to the reaction conditions: The (E)-isomer is obtained in the trimethylsilylation promoted by lithium tetramethylpiperidide, and the (Z)-isomer from the reaction with t-butyldimethylsilyl chloride in the presence of LHMDS and HMPA.

Directed lithiation.   The nitrogen atom of aromatic aldimines assists the deprotonation at a proximal benzylic position by LTMP.[3] The resulting lithium derivatives are nucleophilic and synthetically useful for elaboration of isoquinolines.[4]

Dianions of β-enamino ketones.[5]   The deprotonation is complete at room temperature, but the subsequent alkylation (e.g., with aldehydes) should be conducted at a lower temperature ($-70°C$).

[1] A. Yanagisawa, K. Yasue, and H. Yamamoto, *CC* 2103 (1994).
[2] K. Hattori and H. Yamamoto, *JOC* **58**, 5301 (1993); *T* **50**, 3099 (1994).
[3] L. A. Flippin, J. M. Muchowski, and D. S. Carter, *JOC* **58**, 2463 (1993).
[4] L. A. Flippin and J. M. Muchowski, *JOC* **58**, 2631 (1993).
[5] G. Bartoli, M. Bosco, C. Cimarelli, R. Dalpozzo, and G. Palmieri, *T* **49**, 2521 (1993).

Lithium tetramethylthallate. 17, 172

1,2-Addition to enones.[1] This methylating agent can discriminate an enone and a ketone, favoring the former. The reverse chemoselectivity may be due to the operation of a single-electron transfer mechanism.

[1] I. E. Marko and C. W. Leung, *JACS* **116**, 371 (1994).

Lithium trialkylstannate.

Allenyl stannanes.[1] The S_N2' displacement of propargyl tosylates in the presence of $CuBr \cdot SMe_2$ provides allenyl stannanes, which, by virtue of Pd(0)-catalyzed coupling with aryl iodides, are important precursors for arylallenes.

β-Trialkylstannylcyclohexanones.[2] Conjugated chiral (SAMP) hydrazones impose attack by organostannane reagents diastereoselectively and enantioselectively. The adducts are nucleophilic, thus allowing substitution at C-2. Entry of the electrophile *trans* to the tin residue is favored.

[1] I. S. Aidhen and R. Braslau, *SC* **24**, 789 (1994).
[2] D. Enders, K.-J. Heider, and G. Raabe, *ACIEE* **32**, 598 (1993).

Lithium tri-s-butylborohydride (L-selectride). 13, 167–168; 14, 195–196; 15, 192–193

Demethylation.[1] An unconventional application of the bulky reducing agent is the cleavage of aryl methyl ethers (68–100% yield).

[1] G. Majetich, Y. Zhang, and K. Wheless, *TL* **35**, 8727 (1994).

Lithium trimethylsilyldiazomethane.

Heterocycles. Both the formation of 5-trimethylsilyl-2,3-dihydrofurans[1] from β-siloxy ketones and of cycloheptapyrrolones[2] from *N*-methylanilides of α-keto acids involve condensation with the ketone group and subsequent S_N2 reaction or carbene insertion.

70%

[1] K. Miwa, T. Aoyama, and T. Shioiri, *SL* 461 (1994).
[2] H. Ogawa, T. Aoyama, and T. Shioiri, *SL* 757 (1994).

Lithium triorganozincates.

Halogen–zinc exchange. Haloarenes are converted by Me_3ZnLi to arylzinc nucleophiles.[1] Reaction of the latter with carbonyl compounds gives benzylic alcohols. Analogously, propargylic substrates give allenylzinc reagents that are sources of homopropargylic alcohols.[2]

Conjugate addition.[3] Silylzincate reagents are superior to silylcuprates in terms of their efficiency in the addition to the α,β-unsaturated carbonyl compounds.

[1] Y. Kondo, N. Takazawa, C. Yamazaki, and T. Sakamoto, *JOC* **59**, 4717 (1994).
[2] T. Katsuhira, T. Harada, K. Maejima, A. Osada, and A. Oku, *JOC* **58**, 6166 (1993).
[3] R. A. N. C. Crump, I. Fleming, and C. J. Urch, *JCS(P1)* 701 (1994).

Lithium tris(t-alkoxy)aluminum hydride.

α-Amino aldehydes. N-Boc α-amino phenyl esters[1] and carboxyanhydrides[2] are reduced to aldehydes in THF.

Chiral α-hydroxy acids.[3] α-Keto esters of cis-3-tosylaminoisoborneol are reduced stereoselectively. The chiral auxiliary can be removed by LiOH in aqueous THF at room temperature without racemization.

Stereoselective reduction of cyclic ketones.[4] Conformationally rigid ketones are reduced from the less hindered equatorial direction by the very bulky tris(t-butyldiethylmethoxy)aluminum hydride reagent. Thus 4-t-butylcyclohexanone furnishes a mixture of cis- and trans-alcohols in a 95:5 ratio.

[1] P. Zlatoidsky, *HCA* **77**, 150 (1994).
[2] J. A. Fehrentz, C. Pothion, J.-C. Califano, A. Loffet, and J. Martinez, *TL* **35**, 9031 (1994).
[3] Y. B. Xiang, K. Snow, and M. Belley, *JOC* **58**, 993 (1993).
[4] G. Boireau, A. Deberly, and R. Toneva, *SL* 585 (1993).

Lithium tris(methylthio)methide.

β-Hydroxy esters.[1] Epoxide ring opening with $LiC(SMe)_3$ with subsequent hydrolysis (mediated by $HgO–HgCl_2$) complements the conventional method involving formation of β-cyanohydrins. The overall yields of the new method are good; however, the reagents are much more expensive.

Homologous thioesters.[2] Besides complete hydrolysis to give esters, 1,1,1-tris-(methylthio)alkanes derived from alkyl halides can also be a source of methylthio esters. The transformation is accomplished by heating the tris(methylthio)alkanes with aqueous HBF_4 in DMSO at 130°C.

[1] N. A. Abood, *SC* **23**, 811 (1993).
[2] M. Barbero, S. Cadamuro, I. Degani, S. Dughera, and R. Fochi, *JCS(P1)* 2075 (1993).

Lithium tris(phenylthio)methide.

Ipso anion precursor.[1] Nucleophilic homologation of the reagent generates products that readily submit a PhS unit to another nucleophile (e.g., *sec*-BuLi). The resulting lithio compounds are acyl anion or methylene anion equivalents.

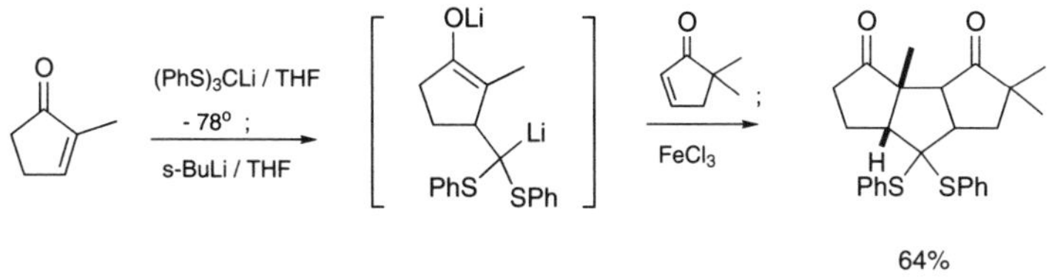

[1] T. Cohen, K. McNamara, M. A. Kuzemko, K. Ramig, J. L. Landi, and Y. Dong, *T* **49**, 7931 (1993).

M

Magnesium. 13, 170; **15**, 194; **16**, 198–199

1,2-Difunctionalization of conjugated dienes.[1]   Dienes are metallated by treatment with active Mg. A variety of compounds can be prepared by the subsequent reaction with electrophiles such as epoxides and CO_2.

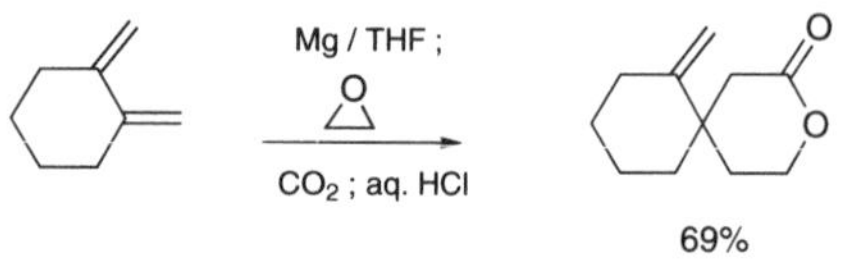

69%

Cleavage of esters.[2]   Many esters are cleaved on reaction with Mg in methanol at 0°C to room temperature. A special value of this method pertains to its selectivity: *p*-nitrobenzoate > acetate > benzoate > pivalate. To cleave one group from a polyester it requires the limitation of Mg to 1 equivalent. Otherwise an excess of Mg is used. In the case of an insoluble substrate, methanol can be diluted with THF.

Reduction of α,β-unsaturated esters.   Special features of a substrate may induce a tandem reaction.[3] A reductive cyclization of 7-oxo-2-alkenoates[4] is observed.

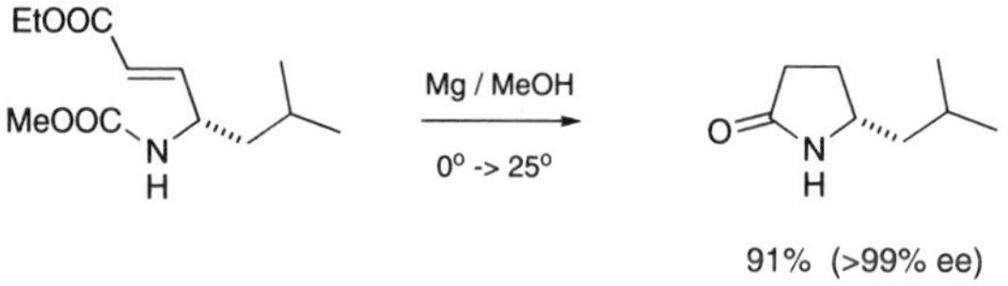

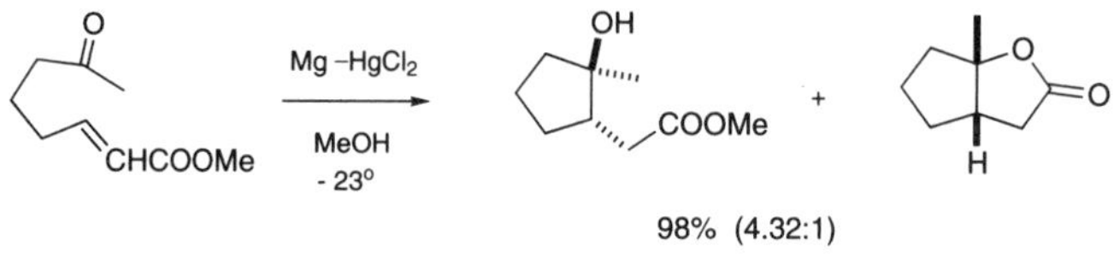

91% (>99% ee)

98% (4.32:1)

Note that a γ-substituent with sufficient leaving ability (e.g., epoxide) suffers reductive removal under these conditions.[5]

Reductive coupling. Using Mg as a sacrificial anode to perform electrolysis, aromatic esters are reduced to benzils[6] and isothiocyanate esters to dithiooxamides.[7]

Stannylation of organohalides.[8] Magnesium together with $PbBr_2$ promotes coupling of allyl, vinyl, propargyl, and aryl halides with Bu_3SnCl in THF at room temperature. The unsaturated stannanes are obtained in 66–99% yield.

[1] M.W. Sell, H. Xiong, and R. D. Rieke, *TL* **34**, 6007, 6011 (1993).
[2] Y.-C. Xu, E. Lebeau, and C. Walker, *TL* **35**, 6207 (1994).
[3] Z.-Y. Wei and E. E. Knaus, *TL* **34**, 4439 (1993).
[4] G. H. Lee, E. B. Choi, E. Lee, and C. S. Pak, *JOC* **59**, 1428 (1994).
[5] C. S. Pak, E. Lee, and G. H. Lee, *JOC* **58**, 1523 (1993).
[6] M. Heintz, M. Devaud, H. Hebri, E. Dunach, and M. Troupel, *T* **49**, 2249 (1993).
[7] Y.-P. Xiao and M.-Z. Bei, *YH* **13**, 84 (1993).
[8] H. Tanaka, A. K. M. abdul Hai, H. Ogawa, and S. Torii, *SL* 835 (1993).

Magnesium–cadmium chloride–water.

Reductions.[1] A number of functional groups are rapidly reduced by this system, which includes acid chlorides, benzyl halides, epoxides, and carbonyl compounds. THF is used as cosolvent.

Ketones from nitroalkenes.[2] Some selectives are observed.

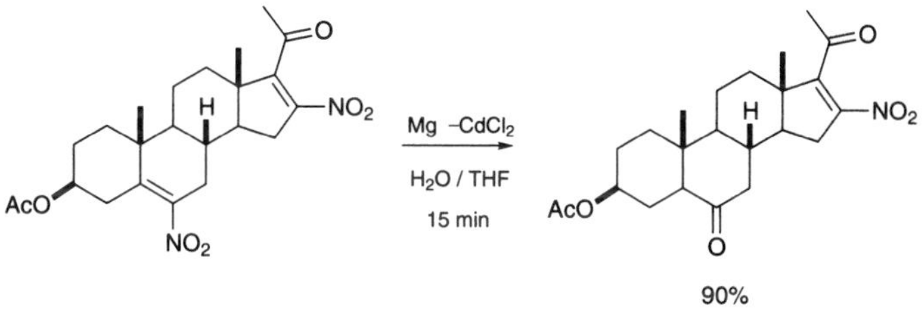

[1] M. Bordoloi, *TL* **34**, 1681 (1993).
[2] M. Bordoloi, *CC* 922 (1993).

Magnesium–mercury(II) chloride.
Desulfonylation.[1]

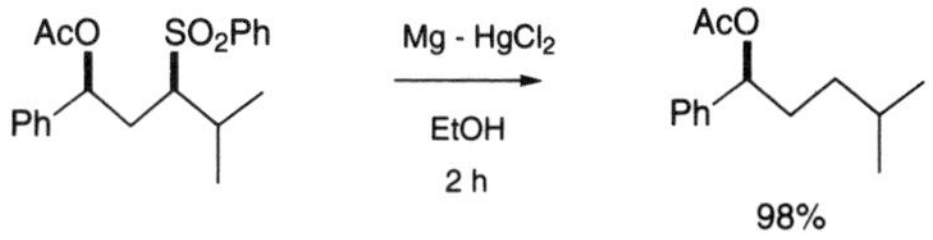

Phenyl sulfides from sulfoxides.[2]

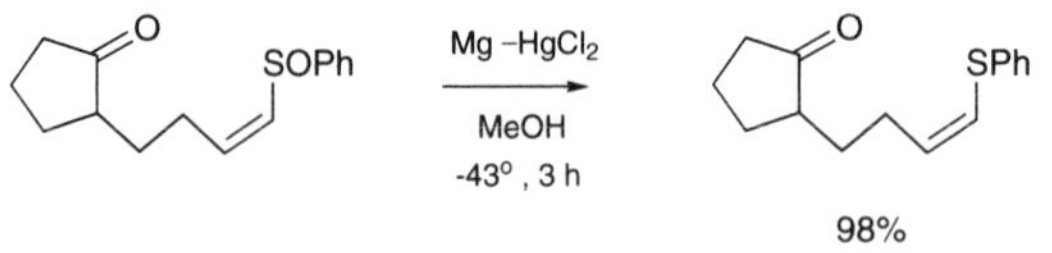

98%

[1] G. H. Lee, E. Lee, and C. S. Pak, *TL* **34**, 4541 (1993).
[2] G. H. Lee, E. B. Choi, E. Lee, and C. S. Pak, *TL* **35**, 2195 (1994).

Magnesium bromide. 15, 194–196; 16, 199; 17, 174

Carbonyl condensations.[1] The highly *syn*-selective addition of allylstannanes to protected amino aldehydes is mediated by MgBr$_2$.

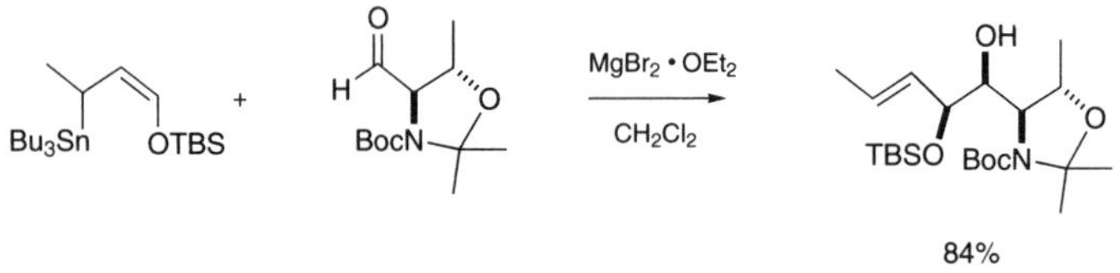

84%

α-Oxy o-quinodimethanes. MgBr$_2$ plays two roles in the synthesis of α-tetralols[2] from *o*-tributylstannylmethylbenzaldehydes. It promotes the conjugated enol generation, and it catalyzes the subsequent cycloaddition with dienophiles.

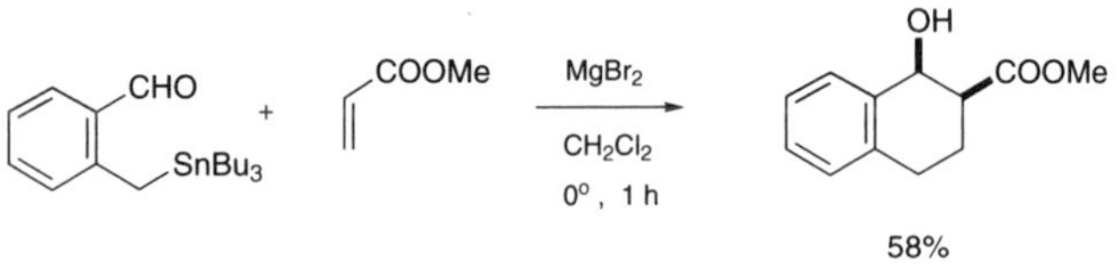

58%

Hydrolysis of methylthiomethyl esters.[3] The method has been applied to the synthesis of phosphoserine and phosphothreonine.

Bromohydrins from epoxy sulfoximines.[4] Chiral epoxides available from the unsaturated sulfoximines may be used as precursors of bromohydrins. Ring opening via the magnesium chelates affords α-bromoaldehydes, as a result of sulfinimine elimination. In the presence of Bu$_4$NBH$_4$ the aldehydes are immediately reduced without much racemization.

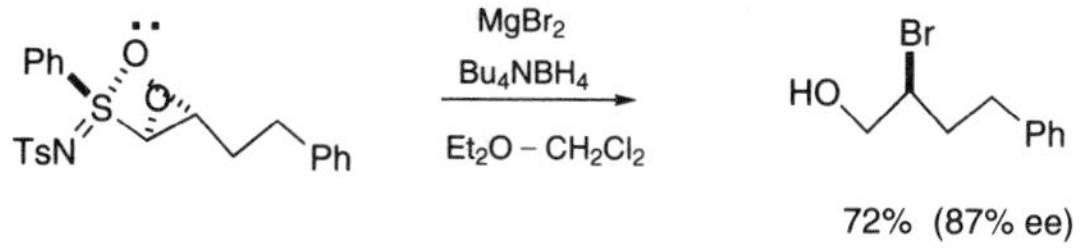

72% (87% ee)

[1] J. A. Marshall, B. M. Seletsky, and P. S. Coan, *JOC* **59**, 5139 (1994).
[2] S. H. Woo, *TL* **35**, 3975 (1994).
[3] N. Mora, J. M. Lacombe, and A. A. Pavia, *TL* **34**, 2461 (1993).
[4] P. L. Bailey, A. D. Briggs, R. F. W. Jackson, and J. Pietruzka, *TL* **34**, 6611 (1993).

Magnesium chloride.

β-Keto esters.[1] After conversion of potassium malonic monoesters to the Mg salts with MgCl$_2$–Et$_3$N, the reaction with acid chlorides proceeds smoothly and safely.

[1] R. J. Clay, T. A. Collom, G. L. Karrick, and J. Wemple, *S* 290 (1993).

Magnesium chlorochromate.

Ketones.[1] The oxidation of secondary alcohols with variable yields (17–100%) has been reported.

[1] P. H. J. Carlsen and K. Aasbo, *SC* **24**, 89 (1994).

Magnesium iodide–diethyl ether. 13, 171; 16, 199; 17, 174

Dithioacetalization.[1]

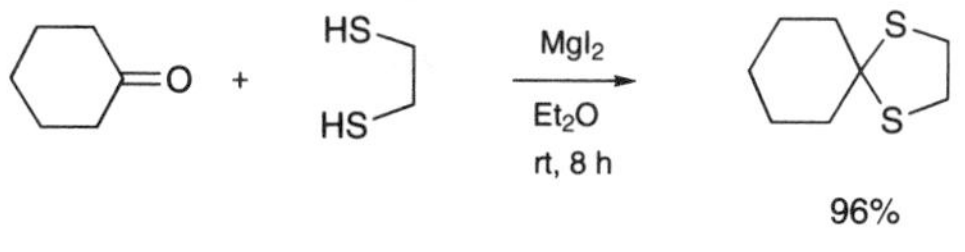

96%

[1] P. K. Chowdhury, *JCR(S)* 124 (1993).

Magnesium methoxide.

Decarbalkoxylation of carbamates.[1] Bidentate carbamates such as *N*-carbalkoxylactams react smoothly. Note that the use of NaOMe instead of Mg(OMe)$_2$ leads to lactam cleavage.

[1] Z.-Y. Wei and E. E. Knaus, *TL* **35**, 847 (1994).

Magnesium monoperoxyphthalate.

Oxidation of aldehyde hydrazones.[1]   The reaction produces nitriles, including chiral nitriles from SAMP-hydrazones in methanol at pH 7 and 0°C.

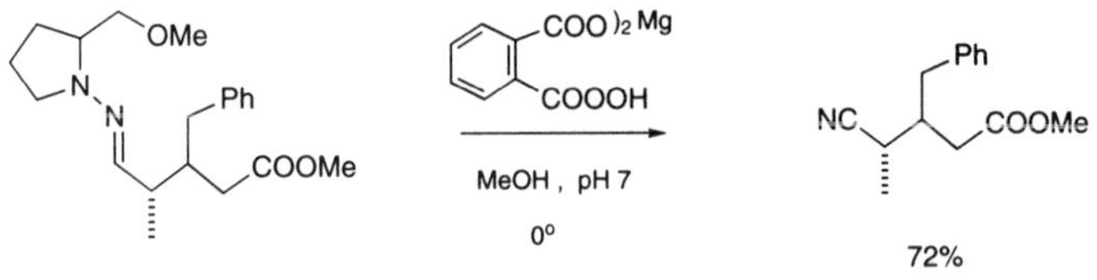

[1] D. Enders and A. Plant, *SL* 1054 (1994).

Magnesium oxide. 16, 200

Alkynylsilanes.[1]   A terminal alkyne and a hydrosilane undergo dehydrogenative coupling on solid bases. Magnesium oxide can be used to promote the reaction at room temperature.

[1] M. Itoh, M. Mitsuzuka, T. Utsumi, K. Iwata, and K. Inoue, *JOMC* **476**, C30 (1994).

Magnesium perchlorate.

Cleavage of N-Boc amides and carbamates.[1]   This Mg(ClO$_4$)$_2$-promoted cleavage does not affect *N*-Boc amines.

[1] J. A. Stafford, M. F. Brackeen, D. S. Karanewsky, and N. L. Valvano, *TL* **34**, 7873 (1993).

Malononitrile.

Polycyanocyclopropanes.[1]   Coelectrolysis with aldehydes leads to tetracyanocyclopropanes or bicyclic dicyanopyrrolines. At a higher current alcoholation of a cyano group and addition of the amine to the *cis*-nitrile result in the heterocyclization.

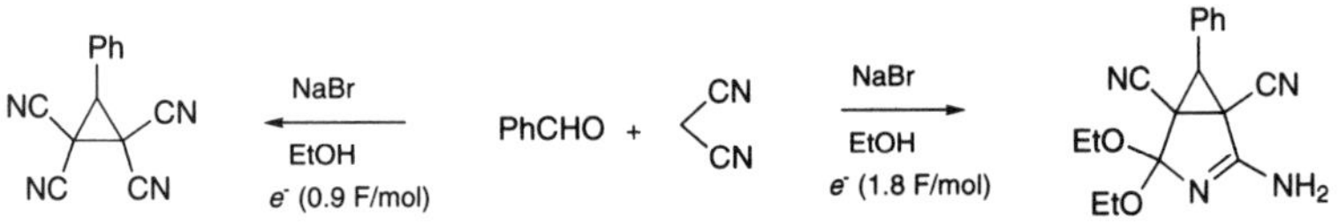

[1] M. N. Elinson, T. L. Lizunova, B. I. Ugrak, G. I. Nikishin, M. O. Kekaprilevich, and Y. T. Struchkov, *MC* 191 (1993).

Manganese(III) acetate. **13**, 171; **14**, 197–199; **16**, 200; **17**, 175–176

Cycloalkylation of β-dicarbonyl compounds. The radicals generated from Mn(III) oxidation of β-dicarbonyl compounds add to alkenes efficiently. Lactone formation from alkenes is improved by ultrasound (45–80% yield).[1]

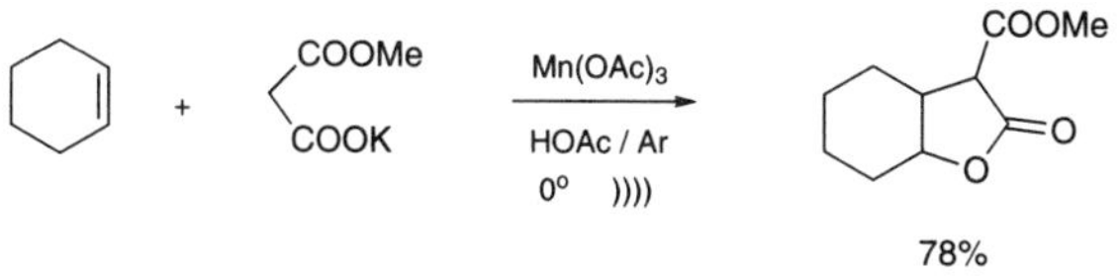

78%

Oxaspirolactones are products of formal cycloaddition of enols to exocyclic enol lactones.[2] The reaction with hexacarbonyldicobalt-coordinated enynes is regioselective.[3]

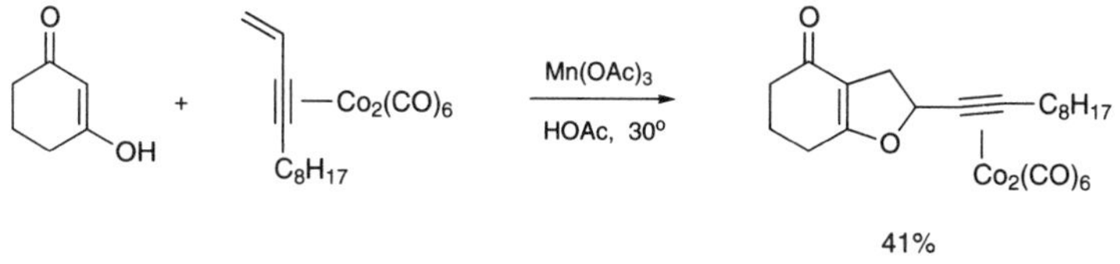

41%

When Mn(OAc)₃-mediated electrochemical oxidation of diethyl benzylmalonate is conducted in the presence of an alkyne or alkene, hydronaphthalene derivatives appear as products.[4]

Cyclic peroxides.[5] When a stream of dry air is admitted to the oxidation system [including Mn(OAc)₃ and Mn(OAc)₂], the first C–C bond formation in the reaction of alkenes and β-keto esters is followed by oxygen trapping and completed by ring closure toward the ketone or aldehyde group. The product yield can be as high as 90%.

α'-Acyloxylation of enones.[6]

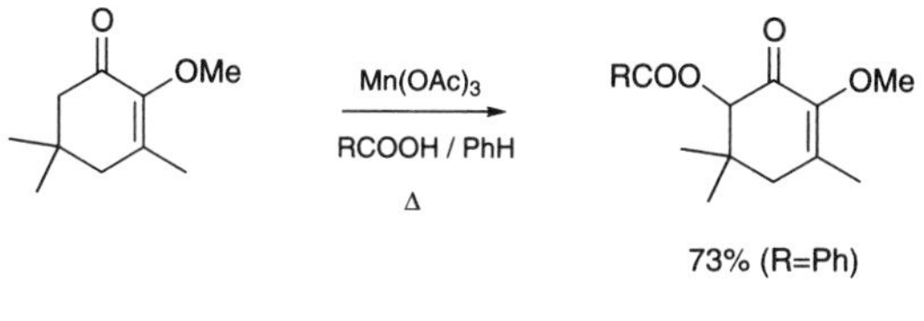

73% (R=Ph)

Oxidative cyclization of β-enamino ketones.[7] Radicals are produced that on intramolecular trapping lead to cyclic products.

38%

[1] M. Allegretti, A. D'Annibale, and C. Trogolo, *T* **49**, 10705 (1993).
[2] J. M. Mellor and S. Mohammed, *T* **49**, 7547 (1993).
[3] G. G. Melikyan, O. Vostrowsky, W. Bauer, H. J. Bestmann, M. Khan, and K. M. Nicholas, *JOC* **59**, 222 (1994).
[4] F. Bergamini, A. Citterio, N. Gatti, M. Nicolini, R. Santi, and R. Sebastiano, *JCR(S)* 364 (1993).
[5] T. Yamada, Y. Iwahara, H. Nishino, and K. Kurosawa, *JCS(P1)* 609 (1993).
[6] A. S. Demir and A. Saatcioglu, *SC* **23**, 571 (1993).
[7] J. Cossy, A. Bouzide, and C. Leblanc, *SL* 202 (1993).

Manganese(II) bromide.

Bromine–zinc exchange.[1] $MnBr_2$ together with CuI catalyzes the exchange, thus rendering the preparation of RZnBr from RBr and Et_2Zn quite readily in DMPU at room temperature.

[1] I. Klement, P. Knochel, K. Chan, and G. Cahiez, *TL* **35**, 1177 (1994).

Manganese dioxide. 14, 200–201; 15, 197–198

Oxidation of sulfides. The oxidation to sulfoxides is carried out in the presence of Me_3SiCl[1] or concentrated hydrochloric acid[2] in aqueous methanol. On the other hand, alkyl phenyl sulfides are converted to phenyl vinyl sulfides[3] with MnO_2–AcCl in DMF.

Oxidation of alcohols. 2,6-Bis(hydroxymethyl)phenols are oxidized by active MnO_2 to the salicylaldehydes at room temperature (86–90% yield).[4] In refluxing chloroform both benzylic alcohols are oxidized (79–82%).

Allylic and benzylic alcohols are oxidized by MnO_2 on bentonite with microwave irradiation.[5] No solvent is needed, and the reaction is rapid (1 min). Ultrasound is less effective in promoting the oxidation (yields 15–66% vs. 32–100%).

Unactivated alcohols can be oxidized with MnO_2 in the presence of $RuCl_2(p$-cymene)$_2$, K_2CO_3, and a free radical inhibitor.[6]

Hydration of hydroxy nitriles.[7] Manganese dioxide deposited on silica gel is prepared by adding a mixture of $KMnO_4$–SiO_2 to aqueous $MnSO_4$. This reagent effects hydration of hydroxy nitriles to amides (the OH group is important) at room temperature. However, the long reaction time makes the usefulness of the method somewhat questionable.

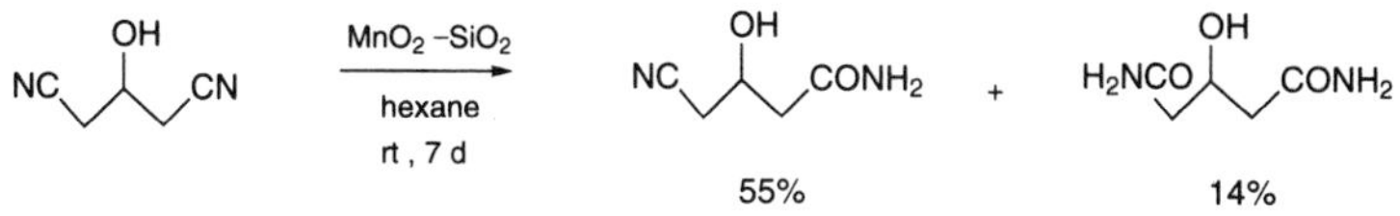

NC$\diagdown$/$\diagup$CN (OH) → MnO₂-SiO₂ / hexane, rt, 7d → NC$\diagdown$/$\diagup$CONH₂ (OH) 55% + H₂NCO$\diagdown$/$\diagup$CONH₂ (OH) 14%

Pyrroles from pyrrolidines.[8] Various *N*-substituted pyrrolidines undergo dehydrogenation on exposure to MnO₂ in refluxing THF. Yields range from 24% to 72% (14 examples).

[1] F. Bellesia, F. Ghelfi, U. M. Pagnoni, and A. Pinetti, *SC* **23**, 1759 (1993).
[2] A. Fabretti, F. Ghelfi, R. Grandi, and U. M. Pagnoni, *SC* **24**, 2393 (1994).
[3] F. Bellesia, F. Ghelfi, U. M. Pagnoni, and A. Pinetti, *G* **123**, 289 (1993).
[4] R.-G. Xie, Z.-J. Zhang, J.-M. Yan, and D.-Q. Yuan, *SC* **24**, 53 (1994).
[5] L. A. Martinez, O. Garcia, F. Delgardo, C. Alvarez, and R. Patino, *TL* **34**, 5293 (1993).
[6] U. Karlsson, G.-Z. Wang, and J.-E. Bäckvall, *JOC* **59**, 1196 (1994).
[7] P. Breuilles, R. Leclerc, and D. Uguen, *TL* **35**, 1401 (1994).
[8] B. Bonnaud and D. C. H. Bigg, *S* 465 (1994).

Manganese(III) tris(2-pyridinecarboxylate) [Mn(pic)₃].

β-Oxoalkyl radicals.[1] Cyclopropanol derivatives fragment, and the resultant radicals form adducts with alkenes (acrylonitrile, silyl enol ethers).

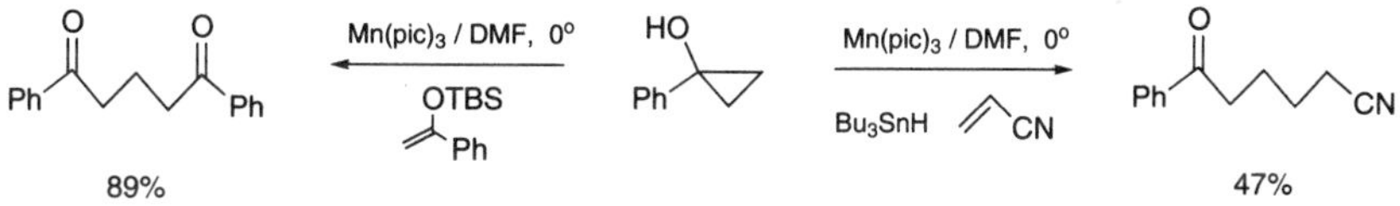

Sulfonyl radicals.[2] Generated from sodium arenesulfinates, the radicals react with alkenes.

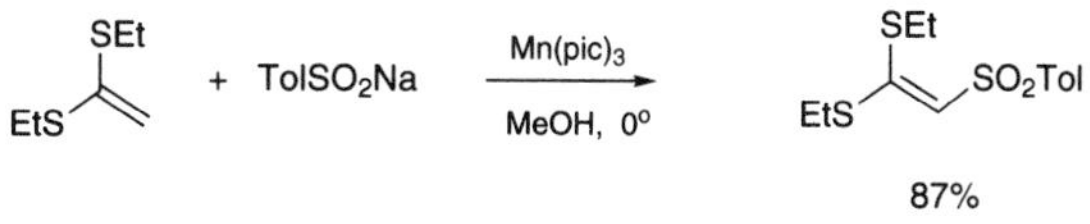

Oxidation of 1-oxidoalkylidenechromium(0) complexes.[3] C–C Bond cleavage by removal of Cr(CO)₆ from such a complex leaves the alkyl radical, which can be intercepted by silyl enol ethers.

[1] N. Iwasawa, S. Hayakawa, M. Funahashi, K. Isobe, and K. Narasaka, *BCSJ* **66**, 819 (1993).
[2] K. Narasaka, T. Mochizuki, and S. Hayakawa, *CL* 1705 (1994).
[3] K. Narasaka and H. Sakurai, *CL* 1269 (1993).

Mercury. **13**, 174; **15**, 198; **16**, 205–206

Allylmercury iodides.[1] These allylating agents are prepared from allylic halides (RCl or RBr requires NaI) and Hg in THF. They react with various acid chlorides in the presence of AlCl$_3$ to afford ketones. The more highly substituted allylic position participates in the C–C bond formation.

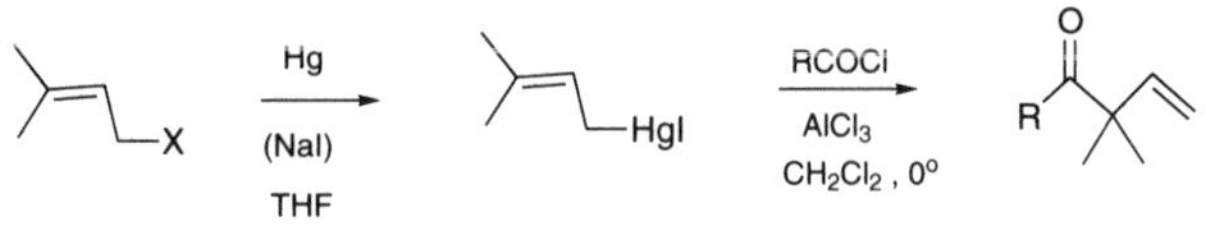

[1] R. C. Larock and Y.-d. Lu, *JOC* **58**, 2846 (1993).

Mercury(II) acetate. **15**, 198–199; **17**, 176–177

Hydroxymercuration.[1] Starting from olefins, an asymmetric synthesis of alcohols is achieved by the hydroxymercuration–demercuration method in the presence of β-cyclodextrin, albeit in relatively low optical yields.

[1] K. R. Rao and H. M. Sampathkumar, *SC* **23**, 1877 (1993).

Mercury(II) chloride. **13**, 175; **15**, 200

Cleavage of stannyl carboxylates.[1]

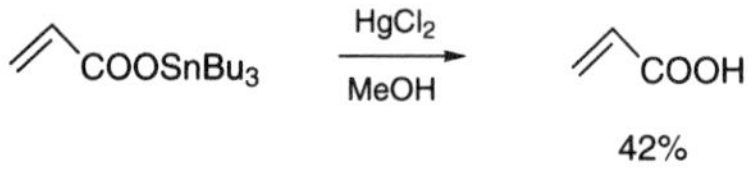

Desulfurization.[2] A synthesis of pyrroles from phenylthio dihydrofuran derivatives is mediated by HgCl$_2$.

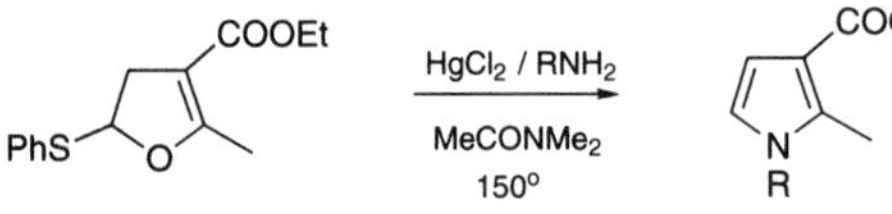

[1] C. Deb and B. Basu, *JOMC* **443**, C24 (1993).
[2] W. H. Chan, A. W. M. Lee, K. M. Lee, and T. Y. Lee, *JCS(P1)* 2355 (1994).

Mercury(II) iodide.

Condensations. HgI$_2$ is a mild Lewis acid, which, as a suspension in an aprotic solvent promotes condensation between silyl ketene acetals and carbonyl compounds.[1] The workup procedure comprises hexane dilution, filtration, and evaporation. Aldehydes and enones give 1,2- and 1,4-addition products.[2]

[1] I. B. Dicker, *JOC* **58**, 2324 (1993).
[2] J. Otera, Y. Fujita, S. Fukuzumi, K. Hirai, J.-H. Gu, and T. Nakai, *TL* **36**, 95 (1995).

Mercury(II) oxide–sulfuric acid.

Furans.[1] 1-Alkynyl-2,3-epoxy alcohols undergo Hg(II)-catalyzed isomerization. The epoxide ring suffers total destruction during its incorporation into a furan nucleus.

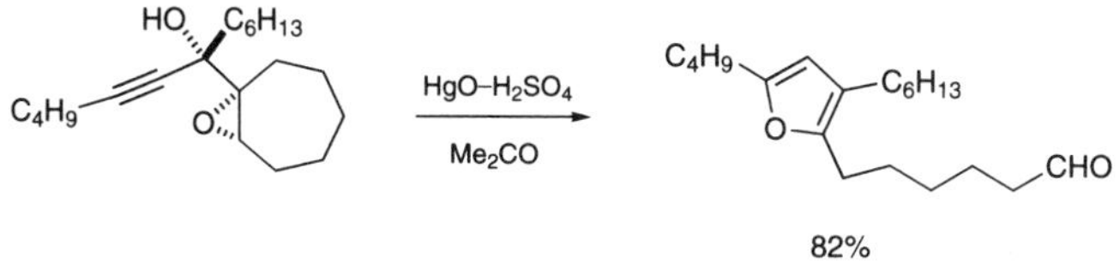

82%

[1] C. M. Marson, S. Harper, and R. Wrigglesworth, *CC* 1879 (1994).

Mercury(II) trifluoroacetate. 13, 175

3-Alkoxy-2-bromopropionate esters.[1] In comparison with Hg(OAc)$_2$ the trifluoroacetate has higher electrophilicity. Thus it reacts with acrylic esters, forming an adduct in conjunction with an alcohol. The trifluoroacetoxymercury group can then be replaced by sequential treatment with KBr and Br$_2$.

[1] P. L. Anelli, A. Beltrami, M. Lollo, and F. Uggeri, *SC* **23**, 2639 (1993).

Methanesulfonyl chloride. 13, 176

Dehydration of allylic alcohols.[1] The method involves treatment of the alcohols with MsCl and Et$_3$N (mesylation), and with *i*-Pr$_2$NEt in HMPA at 140°C for 10 min.

[1] T. Kitahara, T. Matsuoka, H. Kiyota, Y. Warita, H. Kurata, A. Horiguchi, and K. Mori, *S* 692 (1994).

α-Methoxyallenyltitanium triisopropoxide.

Ester homoenolate.[1] The reaction of this reagent with aldehydes gives γ-lactones.

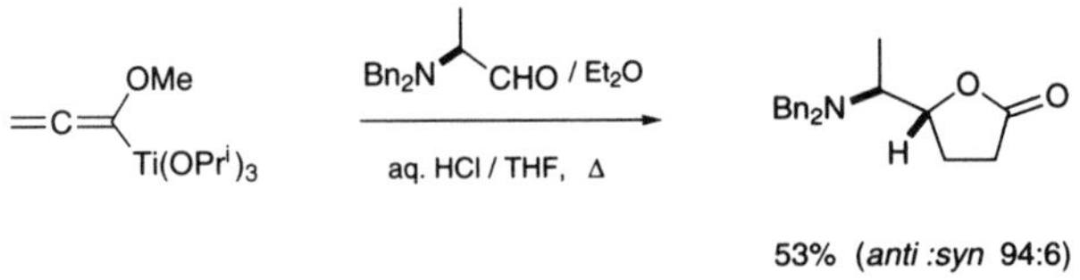

53% (*anti* :*syn* 94:6)

[1] S. Hormuth, H.-U. Reissig, and D. Dorsch, *ACIEE* **32**, 1449 (1993).

S-Methoxyl benzenesulfenylpyrrolidine tetraphenylborate.

Methylation of β-keto esters.[1]   This novel *C*-methylating agent (**1**) reacts with enolates to give products in 70–84% yield.

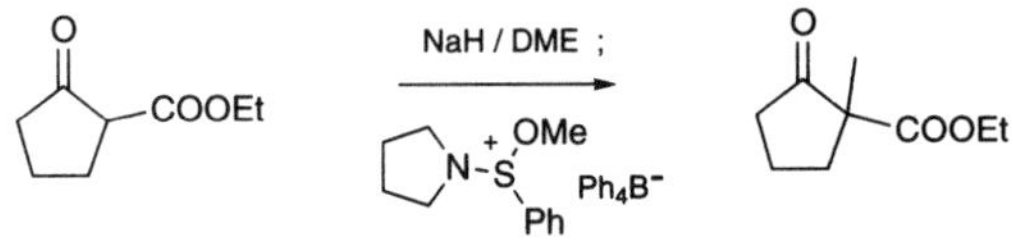

[1] I. F. Pickersgill, A. P. Marchington, and C. M. Rayner, *CC* 2597 (1994).

4-Methoxy-2-*t*-butyl-2,5-dihydroimidazole-1-carboxylic esters.

Amino acid synthesis.   The reagent is a chiral glycine synthon (**1**) with a *t*-butyl group as enantio-controller for alkylation. Amino acids are recovered from hydrolysis of the products.

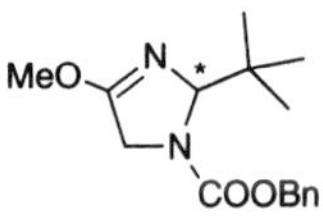

(**1**)

[1] S. Blank and D. Seebach, *ACIEE* **32**, 1765 (1993).

2-(2-Methoxyethoxy)prop-2-yl hydroperoxide.

Alkyl hydroperoxides.[1]   Alkylation of this easily handled hydroperoxide with alkyl halides furnishes mixed peroxides that are selectively hydrolyzed in aqueous acetic acid. The reagent is obtained in 53% yield by ozonolysis of 2,3-dimethyl-2-butene in neat 2-methoxyethanol, followed by dilution with water, extraction with EtOAc, and evaporation. It should be noted that the simpler 2-methoxyprop-2-yl hydroperoxide is more hazardous, as it can undergo exothermic decomposition on solvent removal at room temperature.

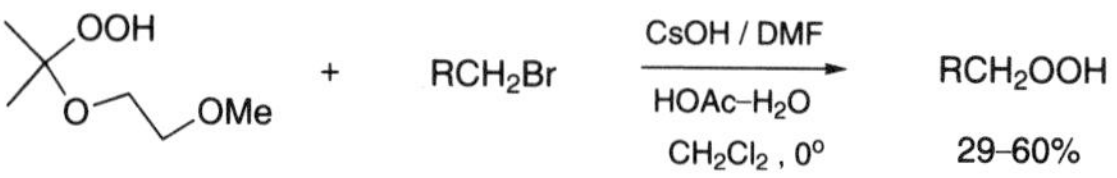

[1] P. H. Dussault, U. R. Zope, and T. A. Westermeyer, *JOC* **59**, 8267 (1994).

Methoxymethylenetriphenylphosphorane.

(E)-4-Hydroxyalk-2-enals.[1] The Wittig reaction with 2,3-epoxy aldehydes leads to unstable enol ethers, which are rapidly hydrolyzed. Since the epoxy aldehydes are available in chiral form from the allylic alcohols through Sharpless epoxidation and oxidation, 4-hydroxyalkenals with desired absolute configuration at the carbinol center are easily established.

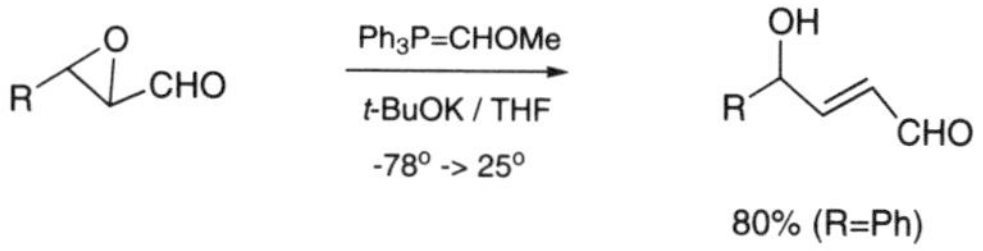

[1] L. Yu and Z. Wang, *CC* 232 (1993).

β-[N-(2-Methoxymethyl)pyrrolidinyl]nitroethene.

Chiral Michael acceptor. The chiral auxiliary of this nitroenamine (1) determines the stereochemical course of Michael reactions[1] such that enantioselective nitroolefination (e.g., with lactone enolates) becomes a reality.

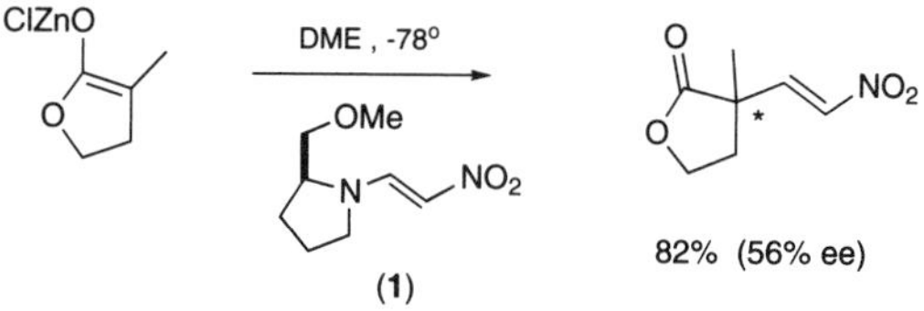

[1] M. Node, R. Kurosaki, K. Hosomi, T. Inoue, K. Nishide, T. Ohmori, and K. Fuji, *TL* **36**, 99 (1995).

2-(4-Methoxyphenyl)ethanol.

Protection of carboxylic acids.[1]   Esters are prepared using DCC-DMAP, and they are cleaved on contact with 1% trifluoroacetic acid in CH_2Cl_2 at room temperature for a short time. The deblocking conditions do not affect *t*-butyl esters and *t*-Boc amines.

[1] M. S. Bernatowicz, H.-G. Chao, and G. R. Matsueda, *TL* **35**, 1651 (1994).

1-Methoxy-3-trimethylsiloxy-1,3-butadiene.

Trichlorotropones.[1]   The Diels–Alder reaction with tetrachlorocyclopropene takes place at room temperature. 3,4,5-Trichlorotropone is isolated in 56% yield.

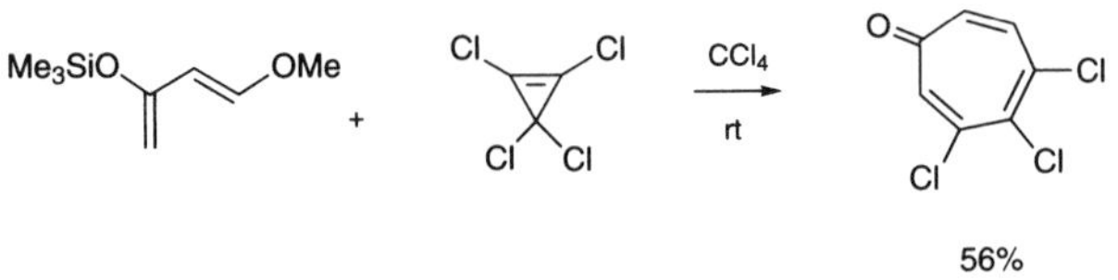

56%

[1] M. G. Banwell and J. H. Knight, *AJC* **46**, 1861 (1993).

Methylamine–potassium permanganate.

Amination of nitroisoquinolines.[1]

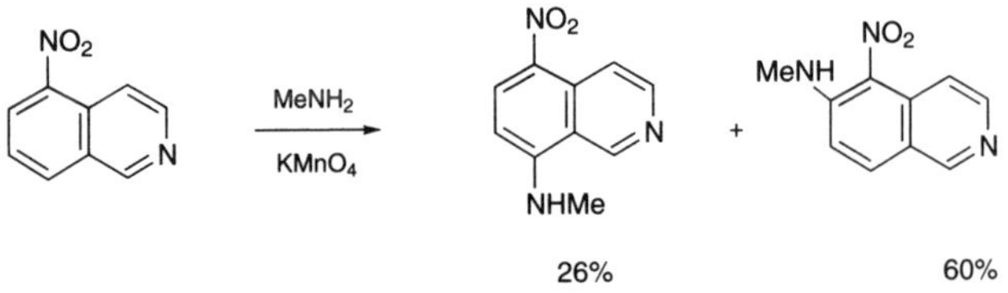

26%   60%

[1] M. Wozniak and K. Nowak, *LA* 355 (1994).

Methylaluminum 1,1'-bi(2,2'-naphthoxide).

β-Lactones.[1]   The [2+2]cycloaddition of ketene with aldehydes is subject to asymmetric induction (S) → (S).

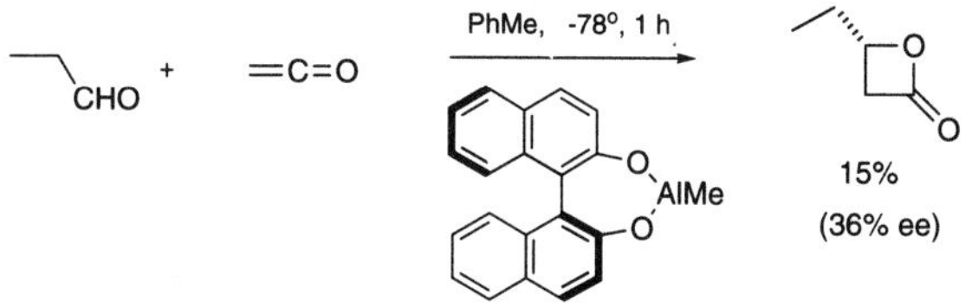

[1] Y. Tamai, M. Someya, J. Fukumoto, S. Miyano, *JCS(P1)* 1549 (1994).

Methylaluminum bis(2,6-di-*t*-butyl-4-X-phenoxide). **13**, 203; **14**, 206–207; **15**, 204; **16**, 209–212; **17**, 184–188

Homologation of carbonyl compounds. The very bulky organoaluminums promote reaction of aldehydes[1] and ketones[2] with diazoalkanes. Aldehydes afford ketones.

Isomerization of trisubstituted epoxides.[3] A reversal of relative migratory aptitudes is revealed in the rearrangement induced by SbF$_5$ and by the bulky aluminum phenoxide.

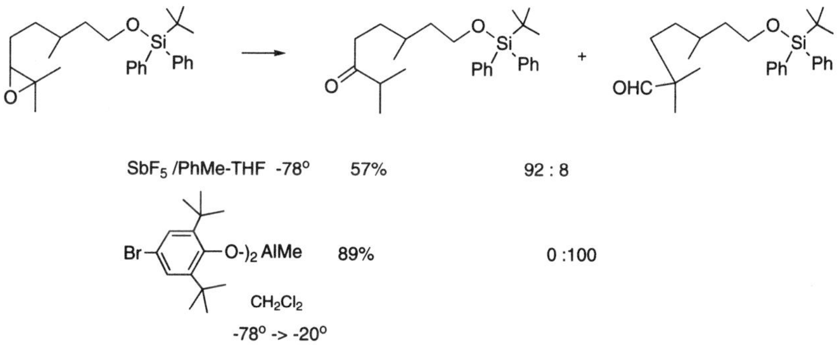

Chemoselective organolithium reactions.[4] The blocking of the equatorial plane of cyclohexanones by the bulky Lewis acid forces attack by nucleophiles from the axial direction. Also for steric reasons dialkyl ketones are less reactive than cyclohexanones in which C-2 and C-6 are not substituted.

Asymmetric Diels–Alder reactions.[5] Single-sited coordination of chiral dienophiles (e.g., the acrylate of D-pantolactone) confers diastereoselectivity to the cycloaddition.

[1] K. Maruoka, A. B. Concepcion, and H. Yamamoto, *SL* 521 (1994).
[2] K. Maruoka, A. B. Concepcion, and H. Yamamoto, *JOC* **59**, 4725 (1994).
[3] K. Maruoka, N. Murase, R. Bureau, T. Ooi, and H. Yamamoto, *T* **50**, 3663 (1994).
[4] K. Maruoka, H. Imoto, and H. Yamamoto, *SL* 441 (1994).
[5] K. Maruoka, M. Oishi, K. Shiohara, and H. Yamamoto, *T* **50**, 8983 (1994).

Methylaluminum bis(2,6-diphenylphenoxide). 15, 205; 16, 212–213

Organometallic reaction of aldehydes.[1]   Contrasteric functionalization of the more hindered aldehyde has been observed in a competing reaction, due to selective complexation of the less hindered substrates.

[1] K. Maruoka, S. Saito, A. B. Concepcion, and H. Yamamoto, *JACS* 115, 1183 (1993).

1-(*N*-Methylamido)-3-methylimidazolium iodides.

Ketones.   These imidazolium salts (1) are acyl transfer agents that react readily with organometallic compounds to give ketones.

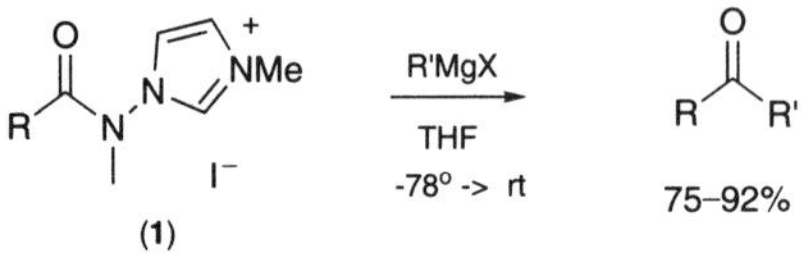

[1] M. A. de las Heras, A. Molina, J. J. Vaquero, J. L. G. Navio, and J. Alvarez-Builla, *JOC* 58, 5862 (1993).

O-Methyl benzenesulfenate. 16, 214

Alkyl phenyl sulfoxides.[1]   The methoxysulfonium salts obtained from reaction of the sulfenate ester with alkyl halides decompose in situ (elimination of MeX in analogy to the Arbusov reaction), leading to the sulfoxides.

[1] M. Kersten and E. Wenschuh, *PSS* 80, 81 (1993).

Methyl chlorodifluoroacetate.

Trifluoromethylarenes.[1,2]   The substitution of aryl halides to give $ArCF_3$ by heating with $ClCF_2COOMe$, KF, and CuI in DMF involves insertion of difluorocarbene into C–Cu bonds.

[1] J.-X. Duan, D.-B. Su, and Q.-Y. Chen, *JFC* 61, 279 (1993).
[2] J.-X. Duan, D.-B. Su, J.-P. Wu, and Q.-Y. Chen, *JFC* 66, 167 (1994).

Methylenecyclopropane.

1,3-Dipolar cycloadditions.[1]   The reaction of methylenecyclopropane with nitrones generates spirocyclic isoxazolidines that are prone to thermal rearrangement. Thus 4-piperidones can be prepared in a two-step process.

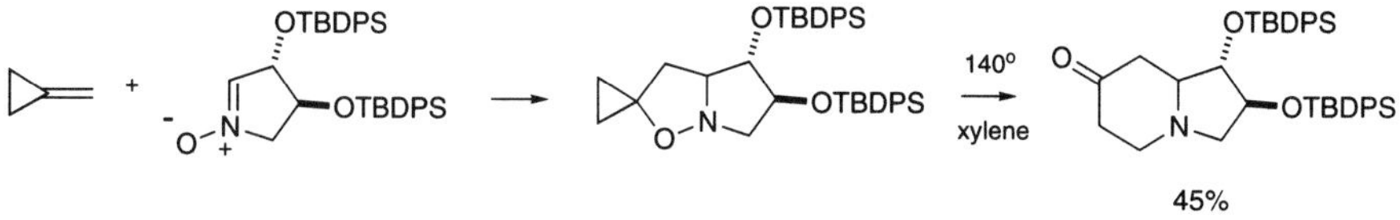

45%

[1] F. M. Cordero, S. Cicchi, A. Goti, and A. Brandi, *TL* **35**, 949 (1994).

Methylenetriphenylphosphorane.

Vinyltin compounds.[1]   Acyltins undergo normal Wittig reactions with this reagent. Accordingly, vinyltins are accessible.

[1] J.-B. Verlhac, H. Kwon, and M. Pereyre, *JCS(P1)* 1367 (1993).

Methyl N-ethyl-N-tributylstannylcarbamate.

Diacylcyclopropanes.[1]   The tin reagent promotes Michael reactions. Contrary to common tin enolates, which are not Michael donors, those derived from α-chloroketones react well, and cyclization following the conjugate addition results in cyclopropanes.

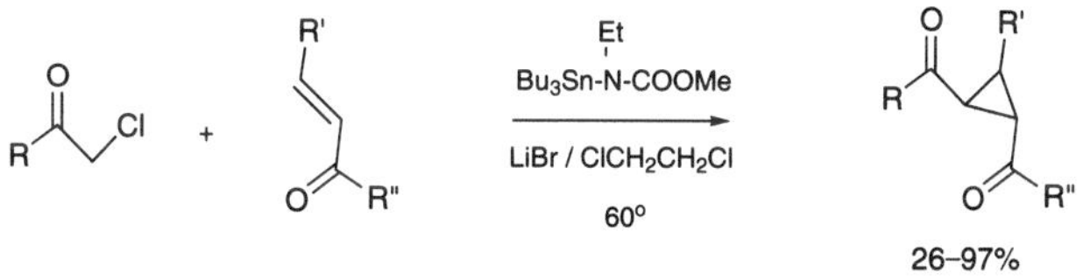

26–97%

[1] I. Shibata, Y. Mori, H. Yamasaki, A. Baba, and H. Matsuda, *TL* **34**, 6567 (1993).

N-Methylimidazole.

Reactions of π-allylnickels.[1]   As an additive, N-methylimidazole confers beneficial effects to allylation and coupling reactions.

[1] S. Knapp, J. Albaneze, and H. J. Schugar, *JOC* **58**, 997 (1993).

Methyl α-isocyanatoacrylate.

This dienophile is obtained in good yields from methyl α-azidopropionate by treatment with $NaReO_4$, $COCl_2$, and triflic acid.[1]

[1] F. Effenberger, J. Kuhlwein, and C. Baumgartner, *LA* 1069 (1994).

S-Methylisothiocarbonohydrazide salts.

sym-Tetrazines.[1] 6-Substituted 3-methylthio-1,2,4,5-tetrazines are formed when iminium chlorides are treated with the salts (1). The last step is an oxidative aromatization.

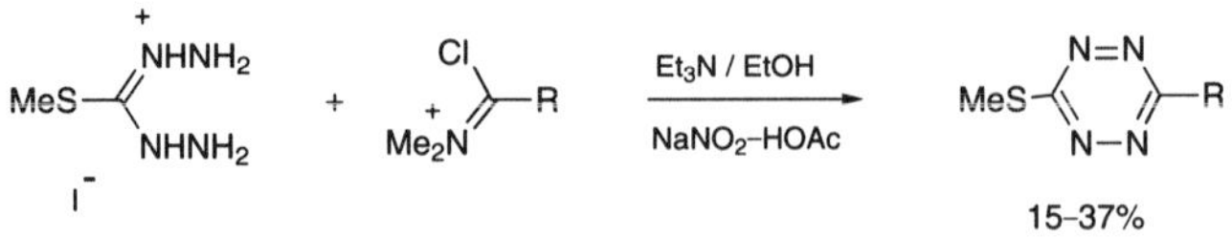

(1)

[1] S. C. Fields, M. H. Parker, and W. R. Erickson, *JOC* **59**, 8284 (1994).

Methyllithium. 13, 188–189; 14, 211; 15, 208

Deprotonation.   For alkylation of enamino ketones at the γ-position[1] and lithiation of allenes[2] the use of MeLi–HMPA is adequate.

Debromination.   Selective debromination of 1,1-dibromoalkenes affords predominantly (98:2) the (*E*)-isomers.[3] *gem*-Dibromocyclopropanes are debrominated. A preparation of 1-trialkylsilylcyclopropenes by this method takes advantage of a 1,2-silicon shift.[4]

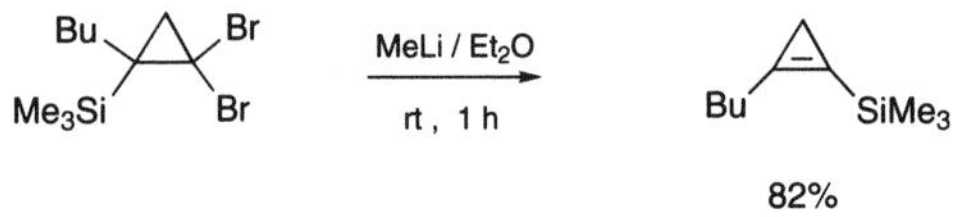

82%

Reaction with carbonyl compounds.   Methyl 2,3-epoxypropanoate derived from serine gives ketones on reaction with RLi or RMgX at low temperatures. It is important that Me₃SiCl is added before RM.[5]

The addition of organometallic reagents to α-nitroketones constitutes a straightforward route to β-nitro alcohols.[6]

[1] G. Bartoli, M. Bosco, C. Cimarelli, R. Dalpozzo, G. De Munno, and G. Palmieri, *TA* **4**, 1651 (1993).
[2] P. Audin, G. Drut-Grevoz, and J. Paris, *SC* **23**, 1139 (1993).
[3] D. Grandjean and P. Pale, *TL* **34**, 1155 (1993).
[4] M. S. Baird, C. M. Dale, and J. R. Al Dulayymi, *JCS(P1)* 1373 (1993).
[5] L. Pegorier, Y. Petit, A. Mambu, and M. Larcheveque, *S* 1403 (1994).
[6] R. Ballini, G. Bartoli, P. V. Gariboldi, E. Marcantoni, and M. Petrini, *JOC* **58**, 3368 (1993).

S-Methyl-*S*-neomenthyl-*N*-tosyl sulfoximine.

Chiral epoxides.[1]   The carbanion derived from the sulfoximine (1) reacts with aldehydes by methylene transfer.

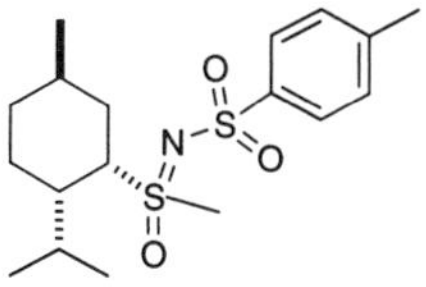

(1)

[1] S. S. Taj and R. Soman, *TA* **5**, 1513 (1994).

Methyl 2-pyridinesulfinate.

Prepared in 55% yield from 2-pyridinethiol in 1:1 MeOH–CH$_2$Cl$_2$ at 0°C by treatment with NBS in one portion.[1]

α,β-Unsaturated ketones.[1] The direct dehydrogenation of ketones involving the sulfinylation–dehydrosulfinylation sequence is achieved in a one-pot reaction using this reagent and without the need of a sulfenic acid trap. Sulfinylation of esters requires reflux in THF.

Ketones containing dithioacetal groups can be dehydrogenated, whereas conventional oxidative methods are incompatible. Addition of CuSO$_4$ (2 equiv.) before thermal decomposition of the α-sulfinylated mesocyclic ketones (7-, 8-, and 12-membered) is recommended.

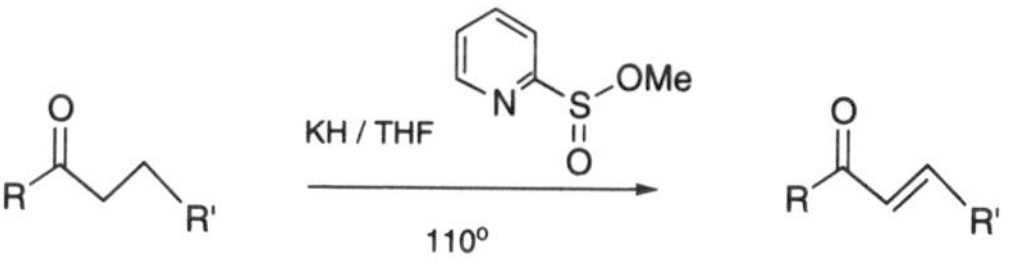

[1] B. M. Trost and J. R. Parquette, *JOC* **58**, 1579 (1993).

5-Methyl-1,5-tetramethylene-4-phenylcyclopentadiene.

Facially selective Diels–Alder reactions.[1] The reaction with various dienophiles gives adducts with which stereochemical manipulation is greatly simplified.

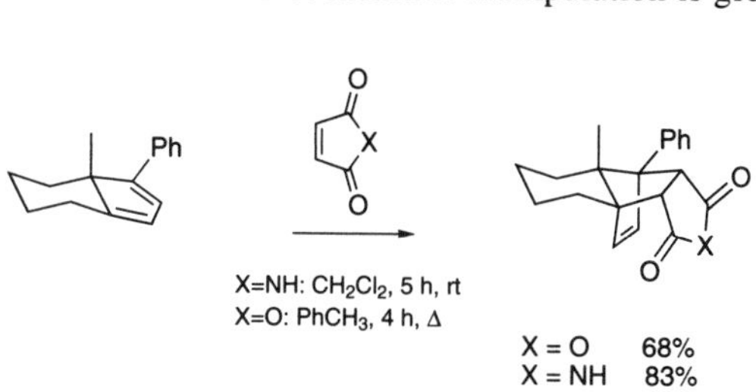

X=NH: CH$_2$Cl$_2$, 5 h, rt
X=O: PhCH$_3$, 4 h, Δ

X = O   68%
X = NH   83%

[1] M. Beckmann, T. Meyer, F. Schulz, and E. Winterfeldt, *CB* **127**, 2505 (1994).

Methyl(tributylstannyl)magnesium.

Addition to alk-3-en-1-ynes.[1] The method of vicinal dimetallation of the triple bond followed by sequential quenching with electrophiles is valuable for the synthesis of functionalized dienes with defined geometry. Thus treatment of the adducts of enynes and MeMgSnBu$_3$ with MeI and I$_2$ gives (*E*)-2-methyl-1-iododienes.

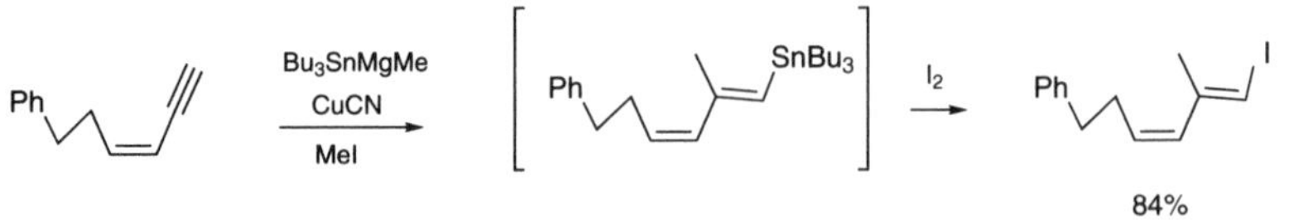

84%

[1] J. Uenishi, R. Kawahama, A. Tanio, and S. Wakabayashi, *CC* 1438 (1993).

Methyl(trifluoromethyl)dioxirane. 15, 212; 16, 224

Oxyfunctionalization.[1] Oxygen insertion into unactivated secondary and tertiary C–H bonds of protonated alkylamines can be very efficient with this dioxirane. Hydrogen bonding between the dioxirane and the NH$_3{}^+$ moiety is important in determining the regiochemistry of the functionalization.

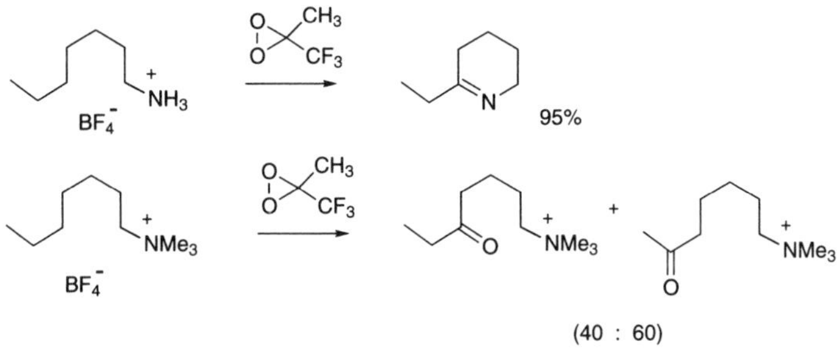

(40 : 60)

[1] G. Asensio, M. E. Gonzalez-Nunez, C. B. Bernardini, R. Mello, and W. Adam, *JACS* 115, 7250 (1993).

Molybdenum carbene complexes. 17, 194–195

1,4-Dialkoxy-1,3-dienes.[1] Propargyl ethers and α-alkoxyalkylidene–molybdenum pentacarbonyls undergo a novel transformation.

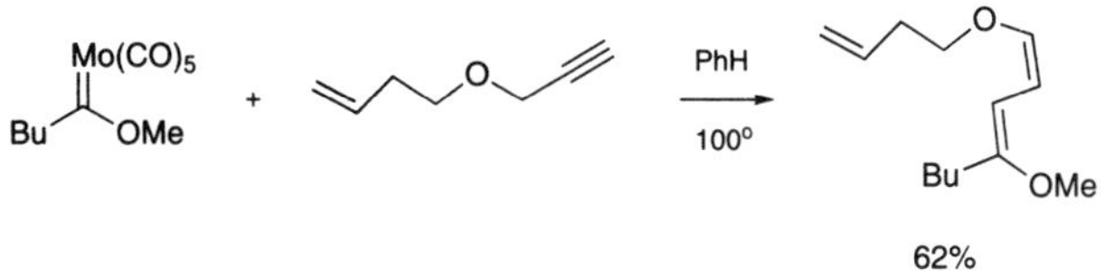

62%

Olefin metathesis. With catalyst (1) cycloalkenes including enol ethers[2] and unsaturated heterocycles[3] are effectively synthesized. Keto alkenes can also be used as substrates.[4]

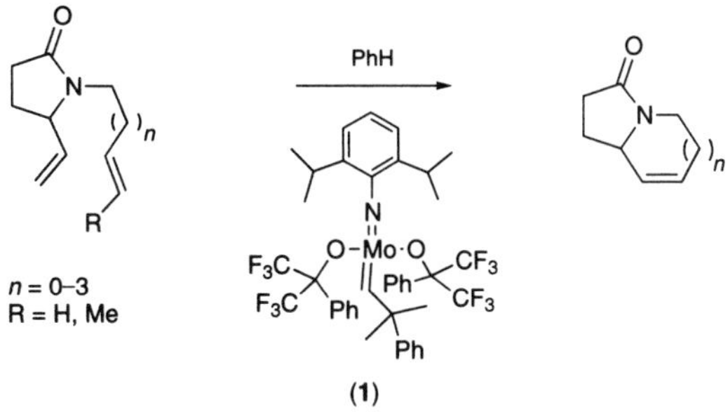

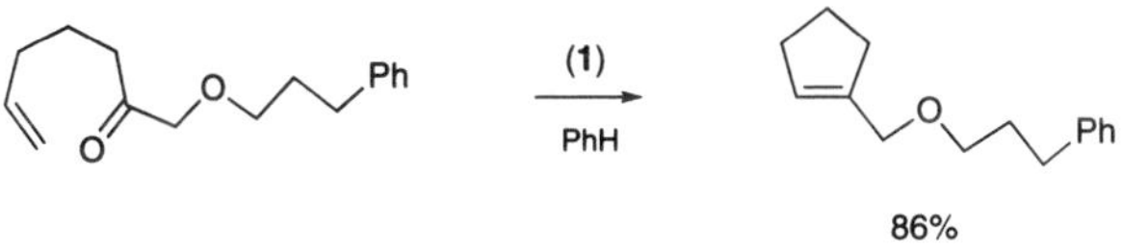

(1)

86%

[1] D. F. Harvey and D. A. Neil, *T* **49**, 2145 (1993).
[2] O. Fujimura, G. C. Fu, and R. H. Grubbs, *JOC* **59**, 4029 (1994).
[3] S. F. Martin, Y. Liao, H.-J. Chen, M. Pätzel, and M. N. Ramser, *TL* **35**, 6005 (1994).
[4] G. C. Fu and R. H. Grubbs, *JACS* **115**, 3800 (1993).

Molybdenum hexacarbonyl. 13, 194–195; 15, 212–213; 16, 225–226

Cleavage of isoxazoles.[1]   The mild conditions of ring cleavage generate enamino ketones without affecting alkene units present in the molecules.

Cycloisomerization.   Homopropargyl alcohols give, after mild oxidation, dihydrofurans.[2] The $(Et_3N)Mo(CO)_5$ complex, obtained by photochemically induced ligand exchange, promotes the isomerization of epoxyalkynes to furans.[3]

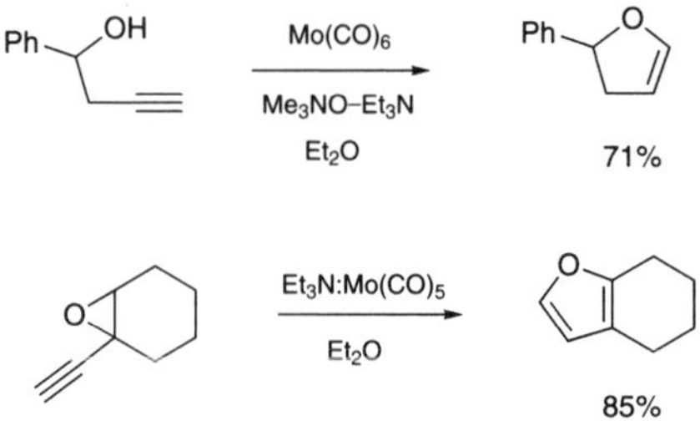

Cyclopentenones.[4]   The Pauson–Khand reaction is also mediated by Mo(CO)$_6$ in DMSO.

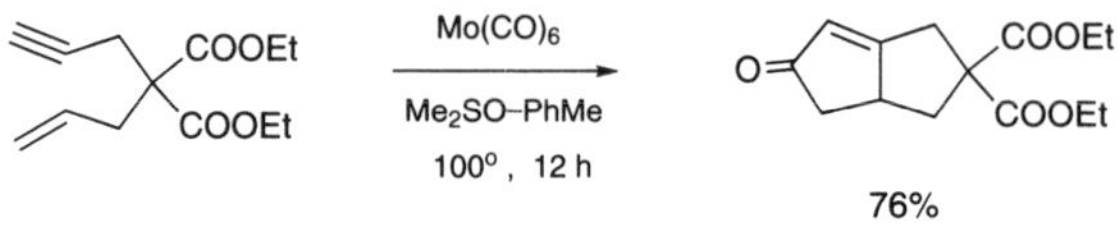

[1] R. C. F. Jones, G. Bhalay, and P. A. Carter, *JCS(P1)* 1715 (1993).
[2] F. E. McDonald, C. B. Connolly, M. M. Gleason, T. B. Towne, and K. D. Treiber, *JOC* **58**, 6952 (1993).
[3] F. E. McDonald and C. C. Schultz, *JACS* **116**, 9363 (1994).
[4] N. Jeong, S. J. Lee, B. Y. Lee, and Y. K. Chung, *TL* **34**, 4027 (1993).

Molybdenyl acetylacetonate.

Dehydration.[1]   Tertiary alcohols are dehydrated in refluxing dioxane in the presence of MoO$_2$(acac)$_2$. However, regioselectivity is not observed.

Etherification.   MoO$_2$(acac)$_2$ is a catalyst for the formation of tetrahydropyranyl ethers[2] and methoxymethyl ethers.[3]

[1] M. L. Kantam, A. D. Prasad, and A. L. Santhi, *SC* **23**, 45 (1993).
[2] M. L. Kantam and A. L. Santhi, *SC* **23**, 2225 (1993).
[3] M. L. Kantam and A. L. Santhi, *SL* 429 (1993).

Montmorillonite clays. 15, 213–214

Dehydration.   Many dehydration processes can be performed in the presence of montmorillonite. Besides generation of olefins[1] from alcohols, the formation of esters,[2] and anhydrides[3] has been effected (preferably with microwave irradiation). Other reactions include enamination,[4–7] acetalization,[8] dithioacetalization,[9] and Friedel–Crafts alkylation with allylic alcohols[10] and tertiary alcohols.[11]

β-Hydroxy sulfides.[12]   Epoxide opening with benzenethiol is catalyzed by clays.

Claisen rearrangement.[13]   The clay-catalyzed aromatic Claisen rearrangement is useful for access to 2-prenylphenols. C–C bond formation occurs at the less substituted site.

[1] M. L. Kantam, A. L. Santhi, and M. F. Siddiqui, *TL* **34**, 1185 (1993).
[2] A. Loupy, A. Petit, M. Ramdani, C. Yvanaeff, M. Majdoub, B. Labiad, and D. Villemin, *CJC* **71**, 90 (1993).
[3] D. Villemin, B. Labiad, and A. Loupy, *SC* **23**, 419 (1993).
[4] S. K. Dewan, U. Varma, and S. D. Malik, *JCR(S)* 21 (1995).
[5] M. E. F. Braibante, H. S. Braibante, L. Missio, and A. Andricopulo, *S* 898 (1994).
[6] B. Rechsteiner, F. Texier-Boullet, and J. Hamelin, *TL* **34**, 5071 (1994).
[7] P. Ruault, J.-F. Pilard, B. Touaux, F. Texier-Boullet, and J. Hamelin, *SL* 935 (1994).
[8] H. K. Patney, *SC* **23**, 1523 (1993).
[9] H. K. Patney, *SC* **23**, 2229 (1993).
[10] K. Smith and G. M. Pollaud, *JCS(P1)* 3519 (1994).
[11] O. Sieskind and P. Albrecht, *TL* **34**, 1197 (1993).
[12] A. K. Maiti, G. K. Biswas, and P. Bhattacharyya, *JCR(S)* 325 (1993).
[13] E. J. Corey and L. I. Wu, *JACS* **115**, 9327 (1993).

Montmorillonite clays, metal ion doped. 12, 231; 15, 101, 178–179

Cleavage of thiol acetates.[1]   Clay-supported $Fe(NO_3)_3$ takes off the acetyl group of thiol acetates and transforms them into disulfides.

Cyanosilylation of carbonyl compounds.[2]   The formation of cyanohydrin silyl ethers is catalyzed by the Fe-montmorillonite. Enones undergo 1,4-addition.

Alkylation of phenols and rearrangement of phenyl ethers.   4-Hydroxybutanone serves as an alkylating agent, giving 4-(4-hydroxyphenyl)-2-butanone in 28% yield[3] on heating with phenol in the presence of Zr(IV)-montmorillonite at 100°C for 48 h. The same compound is obtained in 34% yield when 3-oxobutyl phenyl ether is treated with Zn(II)-montmorillonite and phenol.[4]

Epoxidation.[5]   Montmorillonite impregnated with $Ni(acac)_2$ catalyzes the oxidation of alkenes (cooxidized with isobutyraldehyde) in compressed air (10 bar).

[1] H. M. Meshram, *TL* **34**, 2521 (1993).
[2] K. Higuchi, M. Onaka, and Y. Izumi, *BCSJ* **66**, 2016 (1993).
[3] J. Tateiwa, H. Horiuchi, K. Hashimoto, T. Yamauchi, and S. Uemura, *JOC* **59**, 5901 (1994).
[4] J. Tateiwa, T. Nishimura, H. Horiuchi, and S. Uemura, *JCS(P1)* 3367 (1994).
[5] E. Bouhlel, P. Laszlo, M. Levart, M.-T. Montaufier, and G. P. Singh, *TL* **34**, 1123 (1993).

N

Nafion-H. 14, 213

Oxindoles.[1] Cyclization accompanied by ester cleavage of α-carbomethoxy-α-diazoacetanilides takes place in variable yields. The presence of an *o*-substituent suppresses the cyclization due to steric interaction with the *N*-alkyl group.

68%

[1] A. G. Wee and B. Liu, *T* **50**, 609 (1994).

Nickel. 12, 335; **13**, 197; **14**, 213

Trisadamantylidene[3]radialene.[1] Nickel generated from NiI_2 with Li-DTTB in THF effects debrominative trimerization of dibromomethyleneadamantane.

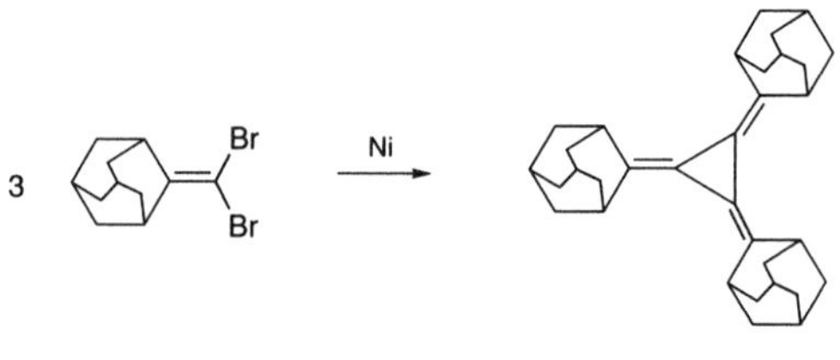

[1] K. Komatsu, H. Kamo, R. Tsuji, and K. Takeuchi, *JOC* **58**, 3219 (1993).

Nickel, Raney. 13, 265–266; **14**, 270; **15**, 278; **17**, 296

β-Ketoesters.[1] β,γ-Unsaturated β-nitro esters undergo reductive Nef reaction in the presence of Raney nickel and sodium hypophosphite in a buffered medium.

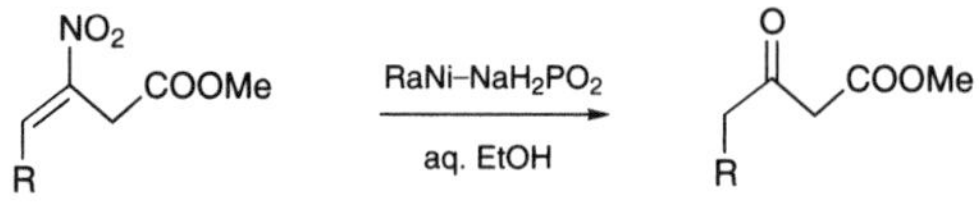

[1] R. Ballini and G. Bosica, *JCR(S)* 435 (1993).

Nickel–acetic acid.

Lactams. Trichloroacetamides bearing an unsaturated substituent on nitrogen are induced to undergo cyclization.[1] Trichloroacetanilides are converted to oxindoles[2] directly, as benzylic C-Cl bonds are susceptible to reduction.

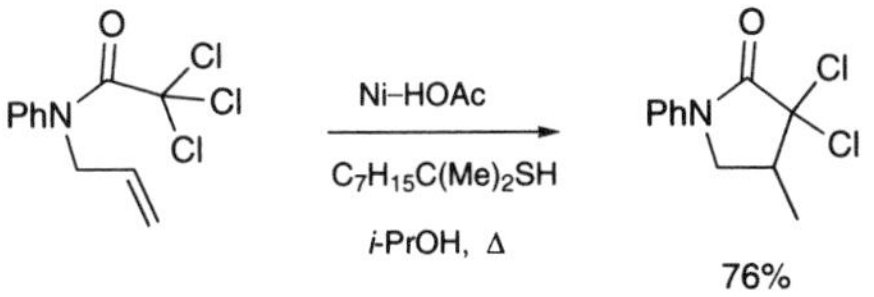

76%

[1] J. Boivin, M. Yousfi, and S. Z. Zard, *TL* **35**, 5629 (1994).
[2] J. Boivin, M. Yousfi, and S. Z. Zard, *TL* **35**, 9553 (1994).

Nickel(II) acetylacetonate. **17**, 201

β-Enamino lactones.[1] α-Acetyllactones condense with lactim ethers under the influence of Ni(acac)$_2$ to give β-enamino lactones in one step with low to moderate yields.

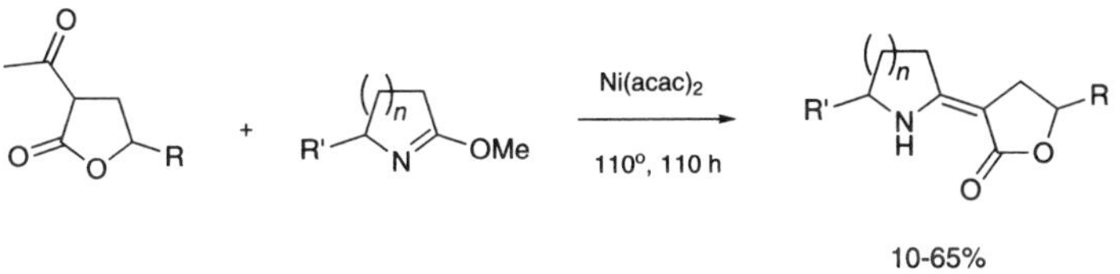

10-65%

Carbozincation.[2] The catalyzed reaction from an unsaturated iodo compound results in the cyclized organometallic species, which can be functionalized. Chain extension via organocopper intermediates is one possibility.

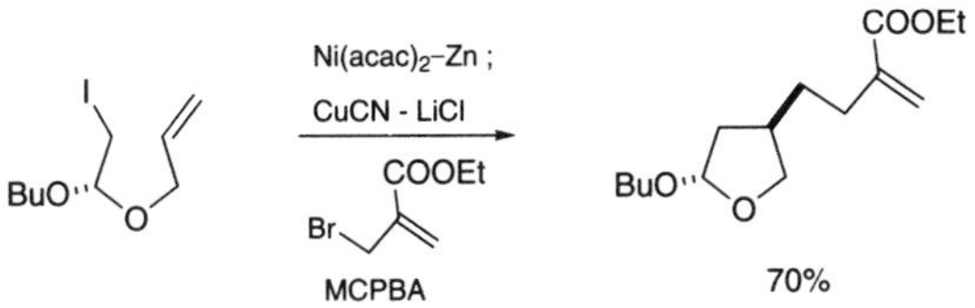

70%

Biphenyls.[3] The cross-coupling of arylzinc chloride reagents with aryl triflates constitutes a convenient method for assembling biphenyls. The zinc reagents are accessible from directed *o*-metallation and metal–metal exchange.

1,2-syn Addition to alkynes. The synthesis of enynes by a catalyzed coupling process features alkynyltin reagents and diisobutylaluminum hydride in the presence of acceptors (enones,[4] allyl halides[5]). The catalytic cycle involves formation of π-allylnickel species and vinylnickel intermediates.

80%

[1] O. Provot, J. P. Celerier, H. Petit, and G. Lhommet, *S* 69 (1993).
[2] A. Vaupel and P. Knochel, *TL* **35**, 8349 (1994).
[3] C. A. Quesnelle, O. B. Familoni, and V. Snieckus, *SL* 349 (1994).
[4] S. Ikeda and Y. Sato, *JACS* **116**, 5975 (1994).
[5] S. Ikeda, D.-M. Cui, and Y. Sato, *JOC* **59**, 6877 (1994).

Nickel–aluminum.

Arylamines from triazenes.[1] Arylamines can be protected as triazenes (formation by reaction of the diazonium salts with a secondary amine). Regeneration of $ArNH_2$ is accomplished by reductive cleavage using Al–Ni alloy in a basic solution.

[1] M. L. Gross, D. H. Blank, and W. M. Welch, *JOC* **58**, 2104 (1993).

Nickel boride. 13, 197–198; 16, 288

Desulfurization. The scope, selectivity, and stereochemistry of this method (C–S → C–H) has been studied using a variety of organosulfur compounds as substrates.[1] β-Lactams unsubstituted at C-4 are formed in a two-step process involving photocycloaddition of chromium carbene complexes and $RN=C(SMe)_2$ followed by desulfurization.[2]

Reduction of the nitro group and halides.[3] Amines are obtained in good yields. The reduction can also be performed by a reagent generated from borohydride resin.[4] The latter reagent converts alkyl bromides and iodides to the hydrocarbons,[5] whereas chlorides, tosylates, esters, and nitriles are not affected.

Reduction of anhydrides.[6] Semireduction of anhydrides releases an alcohol and an acid.

Homoallylic amines.[7] Isoxazolines derived from allylic phosphine oxides can be converted by nickel boride to the saturated 1,3-amino alcohols. On further reaction, elimination of phosphinic acid occurs.

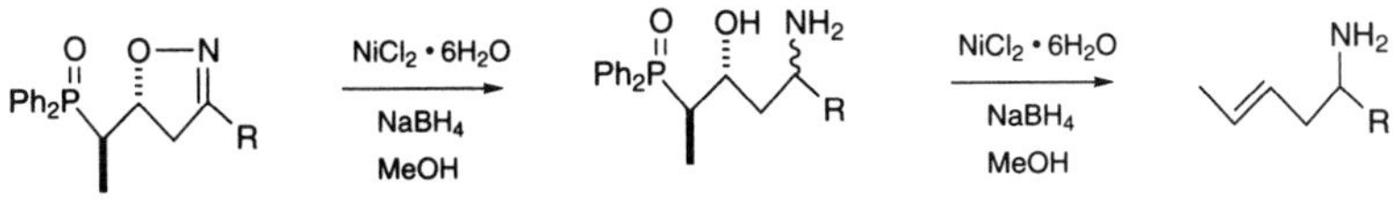

[1] T. G. Back, D. L. Baron, and K. Yang, *JOC* **58**, 2407 (1993).
[2] B. Alcaide, L. Casarrubios, G. Dominguez, and M. A. Sierra, *JOC* **59**, 7934 (1994).
[3] H. H. Seltzman and B. D. Berrang, *TL* **34**, 3088 (1993).
[4] N. M. Yoon and J. Choi, *SL* 135 (1993).
[5] N. M. Yoon, H. J. Lee, J. H. Ahn, and J. Choi, *JOC* **59**, 4687 (1994).
[6] R. H. Khan and R. C. Rastogi, *IJC(B)* **32B**, 898 (1993).
[7] S. K. Armstrong, E. W. Collington, J. G. Knight, A. Naylor, and S. Warren, *JCS(P1)* 1433 (1993).

Nickel carbonyl. 13, 198–199; 15, 216–217

Carbonylation. Vinylic bromides are carbonylated, and the primary products can be trapped intramolecularly. With proper functional groups as trapping agents, cyclopentenones[1] and lactones[2] can be generated.

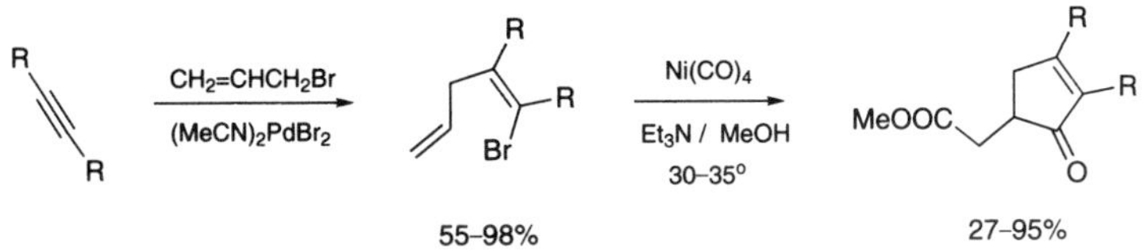

55–98%   27–95%

[1] A. Llebaria, F. Camps, and J. M. Moreto, *T* **49**, 1283 (1993).
[2] A. Llebaria, A. Delgado, F. Camps, and J. M. Moreto, *OM* **12**, 2825 (1993).

Nickel chloride.

Unsymmetrical carbodiimides.[1] Isocyanides react with primary amines under catalysis of $NiCl_2$. Air is the oxidant.

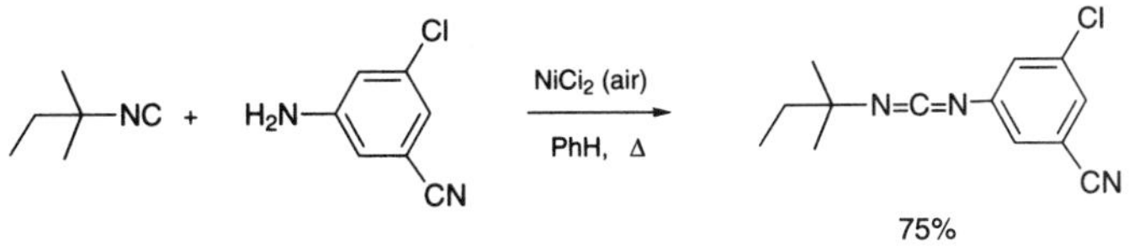

75%

Reduction of nitriles.[2] Primary amines are obtained in a nickel-catalyzed reduction of nitriles with hydrazine as the hydrogen source.

[1] T. Kiyoi, N. Seko, K. Yoshino, and Y. Ito, *JOC* **58**, 5118 (1993).
[2] Q.-D. You, H.-Y. Xhou, Q.-Z. Wang, and X.-H. Lei, *HX* **51**, 85 (1993).

Nickel chloride–phosphine complexes. 14, 125; **15**, 122; **16**, 124

Hydroboration.[1]   Regioselecive hydroboration of alkynyl sulfides with catecholborane to give bifunctional (*E*)-alkenes is catalyzed by NiCl₂(dppe).

Substitutions.   C–C bond formation to replace allylic[2,3] and aryl C–O bonds[4] is readily achieved. These catalysts also promote cross-coupling of vinyl Grignard reagents with vinyl bromides.[5]

57% (ratio 85:15)

Biaryls.[6]   Aryl triflates are self-coupled with the nickel catalysts in the presence of zinc powder. Excellent yields are obtained.

[1] I. D. Gridnev, N. Miyaura, and A. Suzuki, *OM* **12**, 589 (1993).
[2] L. Lardicci, C. Malanga, F. Balzano, and R. Menicagli, *T* **50**, 12953 (1994).
[3] Y. Kobayashi and E. Ikeda, *CC* 1789 (1994).
[4] A. F. Brigas and R. A. W. Johnstone, *CC* 1923 (1994).
[5] B. C. Fulcher, M. L. Hunter, and M. E. Welker, *SC* **23**, 217 (1993).
[6] A. Jutland and A. Mosleh, *SL* 568 (1993).

Nickel cyanide. 16, 288

Carboxylation.[1]   Under phase-transfer conditions, propargyl and allenyl halides in the presence of CO and Ni(CN)₂ are converted into allenic acids. Further reaction gives rise to 2-alkylidenesuccinic acids.

[1] H. Arzoumanian, F. Cochini, D. Nuel, and N. Rosas, *OM* **12**, 1871 (1993).

Nickel halide bis(triphenylphosphine). 16, 118

Halide coupling.   Vinylic bromides undergo homo-coupling[1] when treated with the nickel catalyst and zinc powder.

The cross-coupling of two aryl halides[2] in pyridine makes it possible to prepare a key intermediate for angiotensin II receptor antagonists.

Rearrangement.[3]   Effective catalysis of the vinylcyclopropane-to-cyclopentene rearrangement by a nickel complex in refluxing toluene renders siloxycyclopentenes readily available.

84%

[1] K. Sasaki, K. Nakao, Y. Kobayashi, M. Sakai, N. Uchino, Y. Sakakibara, and K. Takagi, *BCSJ* **66**, 2446 (1993).
[2] H. Kageyama, T. Miyazaki, and Y. Kimura, *SL* 371 (1994).
[3] I. Ryu, K. Ikura, Y. Tamura, J. Maenaka, A. Ogawa, and N. Sonoda, *SL* 941 (1994).

Niobium(III) chloride. **14**, 213–214; **16**, 229

Pinacol formation.[1] Reductive coupling of aliphatic aldehydes with this reagent proceeds in a highly stereoselective manner.

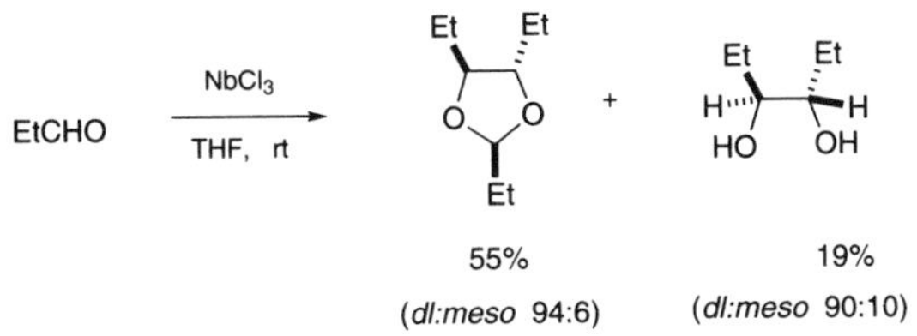

	55%	19%
	(*dl:meso* 94:6)	(*dl:meso* 90:10)

[1] J. Szymoniak, J. Besancon, and C. Moise, *T* **50**, 2841 (1994).

Nitric acid.

Nitration of activated methine.[1] β-Keto esters are nitrated with the HNO_3–H_2SO_4 system. Deacylation gives nitroacetic esters.

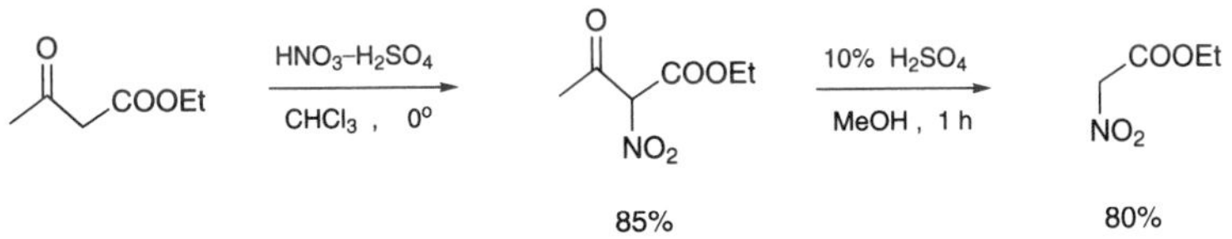

Oxidation of Hantzsch esters.[2] Nitric acid supported on bentonite oxidizes Hantzsch esters to the pyridines. The reaction is assisted by microwave irradiation.

Trifluorolactic acid.[3] This compound can be prepared from 1,2-epoxy-3,3,3-trifluoropropane in a Cu-catalyzed oxidation.

α-Disulfones.[4] Oxidation of *N,N'*-disulfonylhydrazines is one of the simplest ways to prepare these compounds.

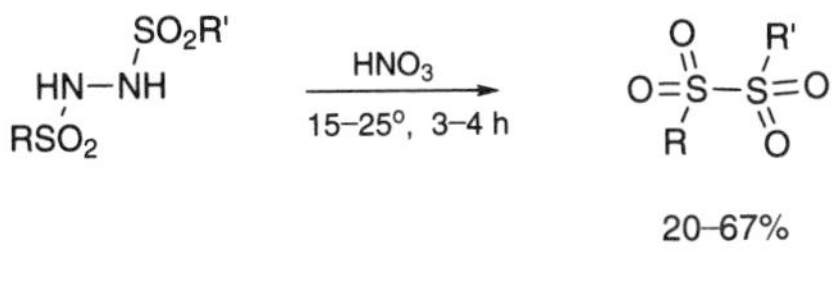

20–67%

[1] A. L. Laikhter, V. P. Isklyi, and V. V. Semenov, *MC* 20 (1993).
[2] O. Garcia, F. Delgado, A. C. Cano, and C. Alvarez, *TL* **34**, 623 (1993).
[3] T. Katagiri, F. Obara, S. Toda, and K. Furuhashi, *SL* 507 (1994).
[4] E. A. Bartmann, *S* 490 (1993).

Nitrobenzaldehyde, polymer supported.

Deprotection of oxathioacetals.[1]   Catalyzed by trimethylsilyl triflate the polymer acts as an efficient acceptor of 2-mercaptoethanol. Although *p*-nitrobenzaldehyde is equally effective, product separation is much less convenient.

[1] T. Ravindranathan, S. P. Chavan, and M. M. Awachat, *TL* **35**, 8835 (1994).

4-Nitrobenzenesulfenanilide.

Functionalization of multiple bonds.   The sulfenylating agent attacks alkenes and alkynes; thus in the presence of hydrobromic acid it results in bromosulfenylation.[1] Cyclic ethers with a phenylthioalkyl side chain are obtained from unsaturated alcohols.[2]

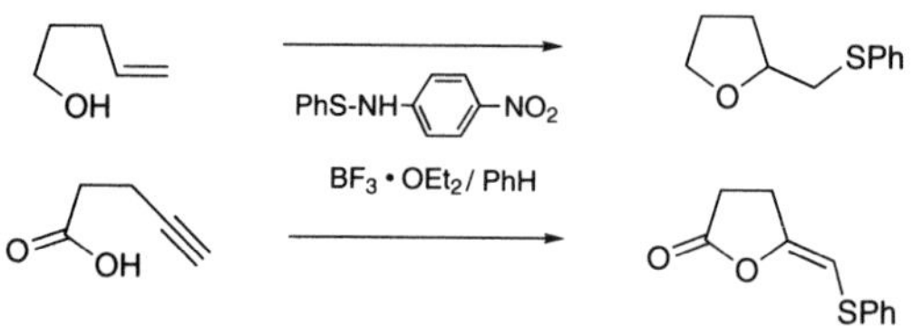

[1] L. Benati, P. C. Montevecchi, and P. Spagnolo, *T* **49**, 5365 (1993).
[2] L. Benati, L. Capella, P. C. Montevecchi, and P. Spagnolo, *T* **50**, 12395 (1994).

2-(4-Nitrobenzene)sulfonylethyl chloroformate.

Amino protection.[1]   The derived carbamates are readily cleaved by organic bases in aprotic solvents. The protected amino acids are useful for solid-phase peptide synthesis.

[1] V. Y. Samukov, A. N. Sabirov, and P. I. Pozdnyakov, *TL* **35**, 7821 (1994).

Nitrogen dioxide.

Catalyzed autoxidation.   Hydroquinones and benzoquinone dioximes are converted to quinones[1] and dinitrosobenzenes,[2] respectively, on autoxidation in the presence of NO$_2$.

Nitration. Mediated by ozone, the method is suitable for nitration of moderately activated to deactivated arenes.[3-5] Because of the nonacidic conditions, acetals can be preserved.[6]

[1] R. Rathore, E. Bosch, and J. K. Kochi, *TL* **35**, 1335 (1994).
[2] R. Rathore, J. S. Kim, and J. K. Kochi, *JCS(P1)* 2675 (1994).
[3] H. Suzuki, T. Mori, and K. Maeda, *S* 841 (1994).
[4] H. Suzuki, T. Murashima, I. Kozai, and T. Mori, *JCS(P1)* 1591 (1993).
[5] H. Suzuki and T. Murashima, *JCS(P1)* 903 (1994).
[6] H. Suzuki, S. Yonezawa, T. Mori, and K. Maeda, *JCS(P1)* 1367 (1994).

Nitronium triflate.

N-Nitration.[1] The salt, prepared from tetrabutylammonium nitrate and triflic anhydride in CH_2Cl_2 at 0°C, can be used directly to nitrate compounds such as lactams.

[1] C. M. Adams, C. M. Sharts, and S. A. Shackelford, *TL* **34**, 6069 (1993).

4-Nitrophthalic anhydride.

Amino protection.[1] Primary amines are liberated from the imides with methylhydrazine.

[1] H. Tsubouchi, K. Tsuji, and H. Ishikawa, *SL* 63 (1994).

N-Nitrosodiphenylamine.

Nitrosation.[1] This reagent is suitable for nitrosation of indoles.

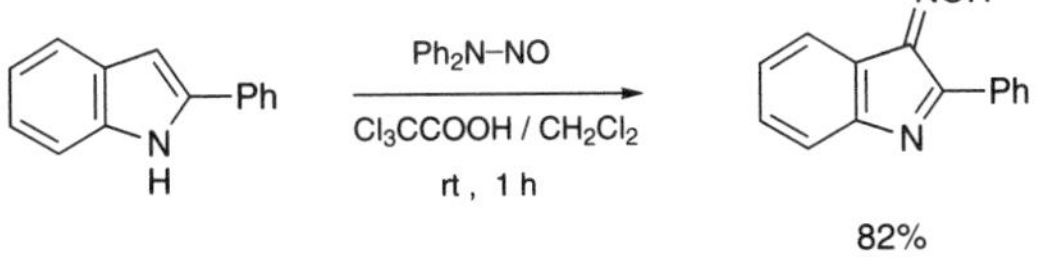

[1] L. Cardellini, L. Greci, and P. Stipa, *SC* **24**, 677 (1994).

Nitrosonium tetrafluoroborate. 14, 215

Nitrosation.[1] Aryl ethers and polymethylbenzenes give nitroso derivatives in reactions at room temperature.

Δ^2-*Isoxazolines.*[2] Allylsilanes undergo nitrosation and cyclization to form the heterocycles.

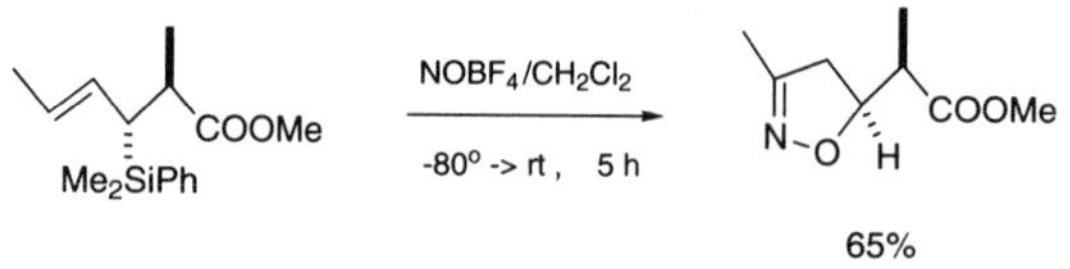

65%

[1] E. Bosch and J. K. Kochi, *JOC* **59**, 5573 (1994).
[2] J. S. Panek and R. T. Beresis, *JACS* **115**, 7898 (1993).

Nitrosyl benzenesulfonate.

Reductive deamination.[1] The reagent is prepared from benzenesulfinic acid and dinitrogen tetroxide in THF. Aromatic amines undergo deamination in the presence of $CuCl_2$.

[1] H. H. Shin, Y. J. Park, and Y. H. Kim, *HC* **4**, 259 (1993).

Nitrosyl chloride.

A convenient method for in situ generation[1] involves treatment of sodium nitrite with trimethylsilyl chloride in $CHCl_3$.

[1] P. K. Chowdhury, M. Barbaruah, and R. P. Sharma, *IJC(B)* **33B**, 71 (1994).

O

Organoantimony reagents. **17**, 204

Allenic and acetylenic alcohols. Propargyl- and allenylantimony(V) bromides react with main-group organometallics to form isomerized species; therefore, different alcohols are obtained on further reaction with aldehydes.[1] However, ω-trimethylsilylpropargylantimony bromides give regioisomers in ratios dependent on reaction conditions (solvents, additives, nature of the main-group organometallics used, etc.).[2]

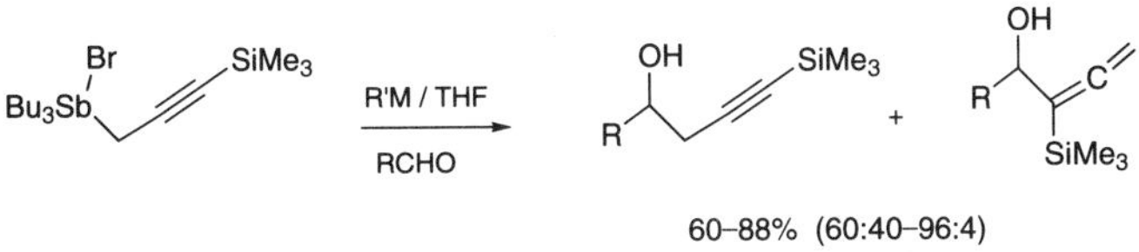

60–88% (60:40–96:4)

[1] L.-J. Zhang, Y.-Z. Huang, and Z.-H. Huang, *TL* **32**, 6579 (1991).
[2] L.-J. Zhang, X.-S. Mo, J.-L. Huang, and Y.-Z. Huang, *TL* **34**, 1621 (1993).

Organobarium reagents. **17**, 204–205

Allylation. Active barium is generated by reduction of BaI$_2$ with 2 equivalents of Li biphenylide in THF at room temperature. Allylic chlorides are converted to the allylbarium reagents, which can be used to effect allylation with excellent regio- and stereoselectivities. *threo*-Homoallylic alcohols from a γ-selective reaction with aldehydes are obtained.[1]

Siloxyallylbarium reagents (from allyl silyl ethers by consecutive treatment with *sec*-BuLi and BaI$_2$) also show a high γ-selectivity in alkylation reactions.[2]

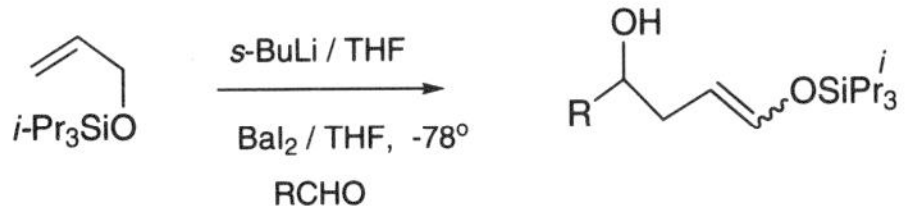

[1] A. Yanagisawa, S. Habaue, K. Yasue, and H. Yamamoto, *JACS* **116**, 6130 (1994).
[2] A. Yanagisawa, K. Yasue, and H. Yamamoto, *SL* 686 (1993).

Organocerium reagents. **13**, 206–207; **14**, 217–218; **15**, 221; **16**, 232; **17**, 205–207

Practically all these reagents are prepared[1] in situ from RLi or RMgX with $CeCl_3$ at $-78°C$ or below. The organolithiums may also be made by reductive cleavage of phenyl sulfides. The presence of some other anionic species such as thiolates does not affect the reactivity of $RCeCl_2$.

Reaction with carbonyl compounds. Such additions are highly diastereoselective.[2] Since the reagents are hard, they are well suited for addition to enolizable compounds, including β-diketones[3] and α-phosphonyl ketones.[4] In the ketone synthesis from lithium carboxylates and RLi the presence of $CeCl_3$ greatly increases the yields.[5]

Addition to the N=X bonds. Amine synthesis, particularly from chiral hydrazones, using $RCeCl_2$ as nucleophiles is well established.[6–8] The reagent formed from 1:1 RLi and $CeCl_3$ is superior to RLi alone or RMgX.[7] For conjugated aldimines, enantioselective 1,2-addition predominates. Allylcerium halide reagents can be prepared from allyl halides, $CeCl_3 \cdot 7H_2O$, and zinc.[9–11]

N,N-Disubstituted hydroxylamines are reaction products of nitroalkanes and organocerium reagents derived from RMgX and $CeCl_3$.[12]

[1] Y. Ahn and T. Cohen, *JOC* **59**, 3142 (1994).
[2] N. Greeves, L. Lyford, and J. E. Pease, *TL* **35**, 285 (1994).
[3] G. Bartoli, E. Marcantoni, and M. Petrini, *ACIEE* **32**, 1061 (1993).
[4] G. Bartoli, L. Sambri, E. Marcantoni, and M. Petrini, *TL* **35**, 8453 (1994).
[5] Y. Ahn and T. Cohen, *TL* **35**, 203 (1994).
[6] D. Enders, M. Klatt, and R. Funk, *SL* 226 (1993).
[7] S. E. Denmark and O. Nicaise, *JOC* **58**, 569 (1993).
[8] D. Enders, R. Funk, M. Klatt, G. Raabe, and E. R. Hovestreyadt, *ACIEE* **32**, 418 (1993).
[9] D. Enders, J. Schankat, and M. Klatt, *SL* 795 (1994).
[10] D. Enders and J. Schankat, *HCA* **76**, 402 (1993).
[11] T. Basile, A. Bocoum, D. Savoia, and A. Umani-Ronchi, *JOC* **59**, 7766 (1994).
[12] G. Bartoli, E. Marcantoni, and M. Petrini, *CC* 1373 (1993).

Organocopper reagents. **13**, 207–209; **14**, 218–219; **15**, 221–227; **16**, 232–238; **17**, 207–218

Organocopper reagents can be made directly from the bromides and active Cu, which is prepared by reduction of CuCN · nLiX with lithium naphthalenide.[1] Copper prepared from CuI · PR$_3$ (R=Bu, Ph) is most suitable for intramolecular reaction of bromoaryl epoxides to give dihydrobenzofurans.

Defunctionalization of heteroalkenes. The C–X bonds of alkenyliodonium salts,[2] 1,2-bis(phenyltelluro)alkenes,[3] and ketene bis(methylthio)acetals,[4] which carry an electron-deficient substituent, are selectively replaced. Only one group of the chalcogenides is affected, and the ketene dithioacetals undergo reductive cleavage. However, the alkenylcopper intermediates can be acylated.[5]

Epoxide opening. A stereoselective alkene synthesis[6] starts from reaction of triethylsilyloxirane with an organocuprate reagent, and it is concluded by oxidation of the β-silyl alcohol to the aldehyde, Grignard reaction and elimination of [Et$_3$Si/OH]. The elimination leads to either the (E)- or the (Z)-alkene by using different reagents.

Non-ate allylcopper is also useful in the synthesis of 5-hydroxyalkenes from allyl chlorides and epoxides.[7] 1,2-Amino alcohols are produced from reaction of epoxides with amide cuprate reagents.[8]

Chain extension. 1,1-Diiodo-[9] and 1,1-bis(triflyloxy)alkanes,[10] which can be made from aldehydes, undergo dialkylation or alkylative reduction. In using aspartic acid as a general building block for β-amino acids, the α-carboxyl group is converted to an iodomethyl group. Subsequent reaction with cuprate reagents completes the chain elaboration.[11]

A series of 4-substituted oxazolidin-2-ones and 2-aminoalkanols are formed by displacement reactions with organocoppers on tosyloxymethyl derivatives which are derived from serine[12] or glycidol.[13] Chiral stannanes have been acquired from α-mesyloxystannanes.[14]

Replacement of an α-methoxy group of an amide (or carbamate) with RCu in the presence of $BF_3 \cdot OEt_2$ probably involves an N-acyliminium intermediate.[15]

N-Boc amines are made from the metallated t-butyl N-tosyloxycarbamate.[16]

Allylic displacements. Applications of this process include vinyl- and allyltin compounds.[17,18] Homoallylic alcohols asymmetrically substituted at the allylic position are obtained from 4-bromo-2-alkenoyl derivatives of camphor sultam in two steps.[19]

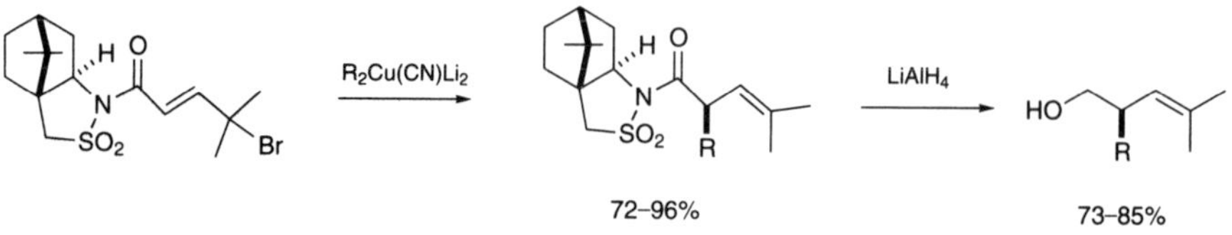

72–96%   73–85%

Propargyl acetates undergo reductive isomerization to give terminal allenes.[20]

Conjugate additions. Interesting preparations pertain to the synthesis of 3-trifluoromethyl-3-hydroxyalkanoic acids,[21] 3-aminomethylcycloalkanones,[22] and 4-vinyl-2-acetoxy-2-cyclobutenones.[23] Compounds of the last type give catechol derivatives on thermolysis.

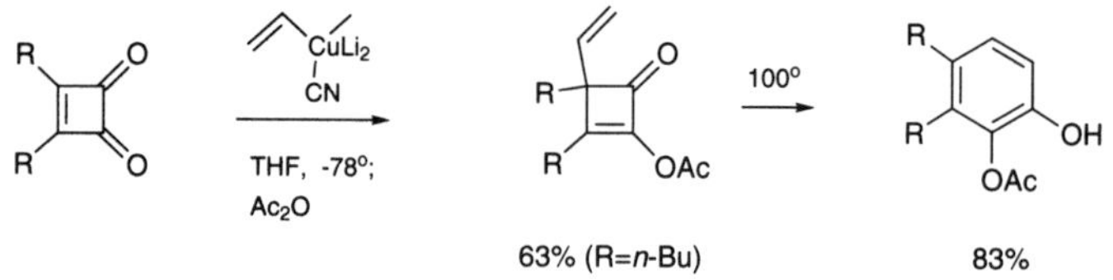

63% (R=n-Bu)   83%

The conjugate addition using silylcuprate reagents enables β-hydroxy carbonyl compounds to be synthesized.[24] Functional group transformation is relatively simple when the silicon atom of such reagents bears an electron-withdrawing substituent (e.g., Ph).

A homoallylic substituent (e.g., carbonate) imparts diastereoselectivity on the conjugate addition due to its interaction with the reagent.[25]

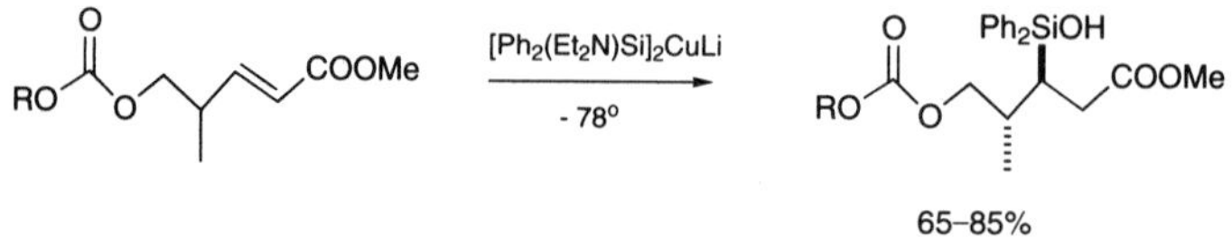

65–85%

The *syn*-1,3-dimethyl pattern is frequently found in carbon chains. Elaboration of the structural motif by an iterative process may employ the addition of Me_3CuLi_2 to γ-hydroxy α,β-unsaturated sulfones.[26] Potential precursors of substituted trimethyl-enemethanes are available from allenyl sulfones and bis(trimethylsilylmethyl)cuprate reagents.[27]

Differences in stereochemical approach of organocopper and cuprates are found in the addition on γ-bis(trimethylsilyl)amino-α-alkynoic esters, amides, and the alkynyl ketones.[28] 2-Alkynoic esters are converted into tetrasubstituted α,β-unsaturated esters by the addition-trapping technique.[29]

BuCu / Et$_2$O, -20°   R = Bu 90% (*E:Z* 83:17)

Me(hexynyl)CuLi / Et$_2$O, -50°   R = Me 95% (*E:Z* 10:90)

1,6-Addition to alk-4-yn-2-enoic esters gives deconjugated allenyl esters.[30] The extended enolate ions can be methylated at C-2. Interestingly, allylation occurs at C-4, and the products contain a conjugate diene unit.

A synthesis of α-substituted γ,δ-unsaturated acids exploits the facile Claisen rearrangement of allyl ester enolates, which can be generated by conjugate addition of organocopper reagents to allyl 2-alkenoates.[31]

Ketone synthesis. δ-Keto-α-amino acid derivatives[32] and 2-oxathianyl ketones[33] are accessible from the functionalized organocoppers and relatively simple acid chlorides. Thioesters can be used instead of acid chlorides for ketone synthesis.[34]

Addition to the carbonyl group. Reaction at the aldehyde function of β-formyl esters is under chelation control.[35] The entering group becomes *trans* to the original chain in the γ-lactones thus generated. More readily understood is the diastereoselectivity in the reactions of protected α-amino aldehydes.[36]

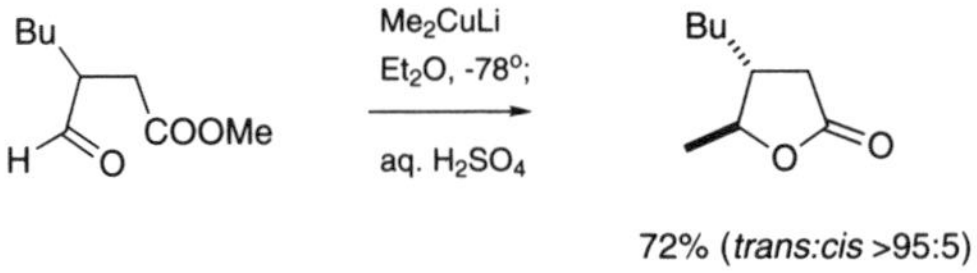

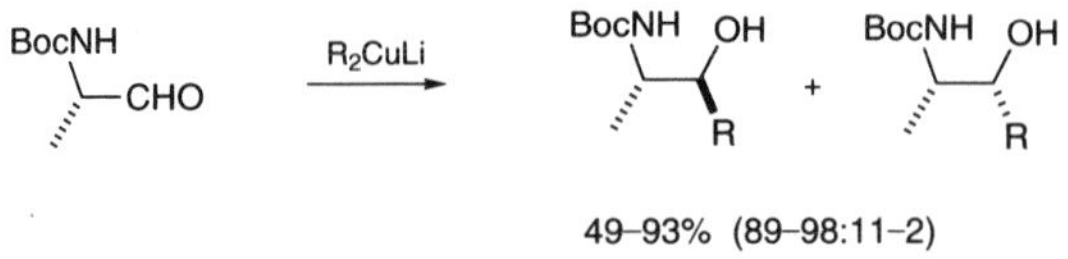

72% (*trans:cis* >95:5)

49–93% (89–98:11–2)

Other additions. 4-Substituted dihydropyridines are usually synthesized by addition reactions on activated pyridines. Derivatization of 3-pyridinecarbalde-hyde into a C_2-symmetrical imidazolidine enables the addition enantioselective[37] by using soft nucleophiles, the attack of which is preceded by coordination to the chiral auxiliary.

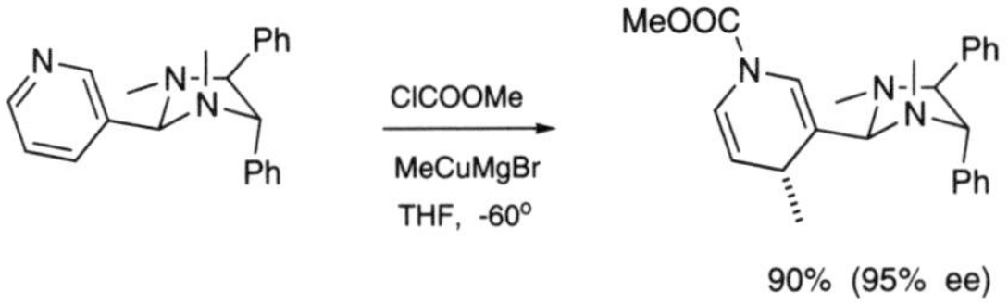

90% (95% ee)

Alkynes can be transformed into alkenylstannanes[38,39] by reaction with stannyl-cuprates. It is possible to trap the 1,2-dimetallic alkene species with various elec-trophiles. The analogous vicinal difunctionalization of alkynyl selenides[40] has also been reported. A route to trisubstituted alkenes from phenylthioacetylene[41] starts with cuprate addition, but a 1,2-metal rearrangment is involved. Enamines are ob-tained from *N*-ethynyldiphenylamine.[42] The alkenylcopper intermediate is also reac-tive toward many electrophiles. Silylcupration of functionalized alkynes may lead to cyclic products by virtue of intramolecular alkylation.[43]

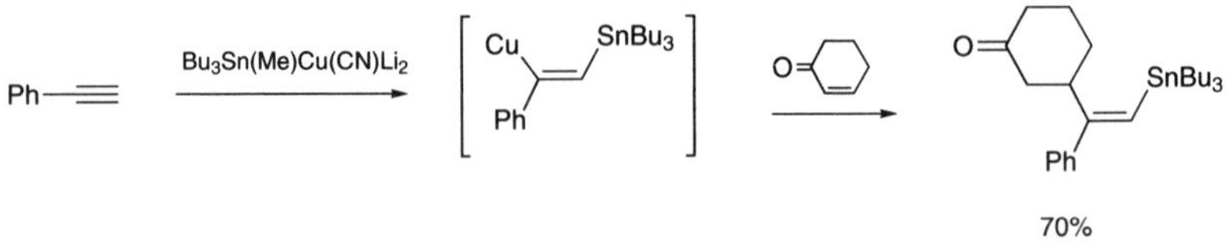

70%

$$\text{Ph}_2\text{N}\text{———} \xrightarrow[\text{THF - HMPA}]{\text{(Me}_3\text{Si)}_2\text{Cu} \cdot \text{LiCN}} \left[\begin{array}{c} \text{Cu} \diagup \text{SiMe}_3 \\ \diagdown \diagup \\ \text{Ph}_2\text{N} \end{array} \right] \xrightarrow[\text{MeOH}]{\text{MgBr}_2} \begin{array}{c} \diagup \text{SiMe}_3 \\ \text{Ph}_2\text{N} \end{array}$$

85%

[1] R. D. Rieke, D. E. Stack, B. T. Dawson, T.-C. Wu, and W. R. Klein, *JOC* **58**, 2483 (1993).

[2] P. J. Stang, T. Blume, and V. V. Zhdankin, *S* 35 (1993).

[3] A. Ogawa, Y. Tsuboi, R. Obayashi, K. Yokohoma, I. Ryu, and N. Sonoda, *JOC* **59**, 1600 (1994).

[4] M. Hojo, H. Harada, C. Watanabe, and A. Hosomi, *BCSJ* **67**, 1495 (1994).

[5] M. Hojo, H. Harada, and A. Hosomi, *CL* 437 (1994).

[6] D. C. Chauret and J. M. Chong, *TL* **34**, 3695 (1993).

[7] D. E. Stack, W. R. Klein, and R. D. Rieke, *TL* **34**, 3063 (1993).

[8] Y. Yamamoto, N. Asao, M. Meguro, N. Tsukada, H. Nemoto, N. Sadayori, J. G. Wilson, and H. Nakamura, *CC* 1201 (1993).

[9] A. G. Martinez, A. H. Fernandez, R. M. Alvarez, J. O. Barcina, C. G. Gomez, and L. R. Subramanian, *S* 1063 (1993).

[10] A. G. Martinez, J. O. Barcina, B. R. Diez, and L. R. Subramanian, *T* **50**, 13231 (1994).

[11] C. W. Jefford and J. Wang, *TL* **34**, 1111 (1993).

[12] M. P. Sibi, D. Rutherford, and R. Sharma, *JCS(P1)* 1675 (1994).

[13] S. Iwama and S. Katsumura, *BCSJ* **67**, 3363 (1994).

[14] J. Ye, D.-S. Shin, R. K. Bhatt, P. A. Swain, and J. R. Falck, *SL* 205 (1993).

[15] C. Ludwig and L.-G. Wistrand, *ACS* **48**, 367 (1994).

[16] C. Greck, L. Bischoff, A. Girard, J. Hajicek, and J.-P. Genet, *BSCF* **131**, 429 (1994).

[17] F. Bellina, A. Carpita, M. De Santis, and R. Rossi, *T* **50**, 4853 (1994).

[18] S. Watrelot, J.-L. Parrain, and J.-P. Quintard, *JOC* **59**, 7959 (1994).

[19] C. Girard, G. Mandville, and R. Bloch, *TA* **4**, 613 (1993).

[20] M. H. Nantz, D. M. Bender, and S. Janaki, *S* 577 (1993).

[21] M. Gautschi, W. B. Schweizer, and D. Seebach, *CB* **127**, 565 (1994).

[22] R. K. Dieter and C. W. Alexander, *SL* 407 (1993).

[23] A. Gurski and L. S. Liebeskind, *JACS* **115**, 6101 (1993).

[24] I. Fleming and S. B. D. Winter, *TL* **34**, 7287 (1993).

[25] M. R. Hale and A. H. Hoveyda, *JOC* **59**, 4370 (1994).

[26] E. Dominguez and J. C. Carretero, *T* **50**, 7557 (1994).

[27] M. Harmata and B. F. Herron, *S* 202 (1993).

[28] R. J. P. Corriu, G. Bolin, J. Iqbal, J. J. E. Moreau, and C. Vernhet, *T* **49**, 4603 (1993).

[29] D. G. Hall, D. Chapdelaine, P. Preville, and P. Deslongchamps, *SL* 660 (1994).

[30] S. Arndt, G. Handke, and N. Krause, *CB* **126**, 251 (1993).

[31] M. Eriksson, M. Nilsson, and T. Olsson, *SL* 271 (1994).

[32] R. F. W. Jackson, M. Wishart, and M. J. Wythes, *SL* 219 (1993).

[33] J. Wei, R. O. Hutchins, and J. Prol, *JOC* **58**, 2920 (1993).

[34] B. F. Bonini, A. Capperuci, M. Comes-Franchini, A. Degl'Innocenti, G. Mazzanti, A. Ricci, and P. Zani, *SL* 937 (1993).

[35] H.-U. Reissig, H. Angert, T. Kunz, A. Janowitz, G. Handke, and E. Bruce-Adjei, *JOC* **58**, 6280 (1993).

[36] M. T. Reetz, K. Rölfing, and N. Griebenow, *TL* **35**, 1969 (1994).

[37] P. Mangeney, R. Gosmini, S. Raussou, M. Commercon, and A. Alexakis, *JOC* **59**, 1877 (1994).

[38] A. Barbero, P. Cuadrado, I. Fleming, A. M. Gonzalez, F. J. Pulido, and R. Rubio, *JCS(P1)* 1657 (1993).

[39] I. Beaudet, V. Launay, J.-L. Parrain, and J.-P. Quintard, *TL* **36**, 389 (1995).

[40] A. L. Braga, A. Reckziegel, C. C. Silveira, and J. V. Comasseto, *SC* **24**, 1165 (1994).

[41] E. Creton, I. Marek, D. Brasseur, J.-L. Jestin, and J. F. Normant, *TL* **35**, 6873 (1994).

[42] L. Cappella, A. Capperucci, G. Curotto, D. Lazzari, P. Dembech, G. Reginato, and A. Ricci, *TL* **34**, 3311 (1993).

[43] I. Fleming and E. Martinez de Marigorta, *TL* **34**, 1201 (1993).

Organocopper/zinc reagents. 15, 229–230; **16**, 238–241; **17**, 218–223

Alkylation. The tolerance of functional groups in these organometallics makes them very valuable in synthesis. Thus their formation and reactivity are not affected by the presence of a pivalate ester at the α-position.[1]

S_N2' *Reactions.* Methyl α-(*t*-butyl)acrylate is obtained from reaction of the 2-chloromethyl-3-methyl-2-butenoic ester with a methylcopper/zinc reagent.[2] Higher-order zinc cuprates or vinyl–ZnCl (with catalytic Cu salt) effect displacement reactions on γ-mesyloxy α,β-enoates, whereas reduction is observed with vinyl–Cu(CN)M or (vinyl)$_2$Cu(CN)M$_2$.[3]

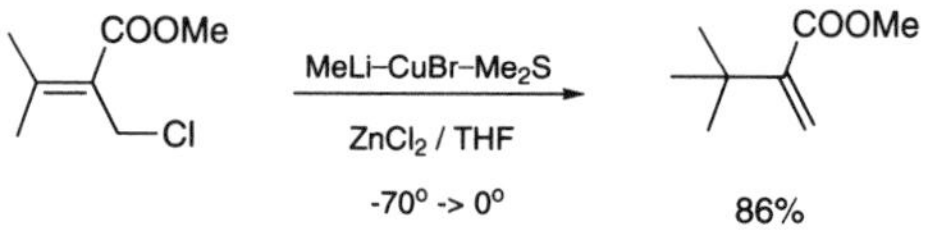

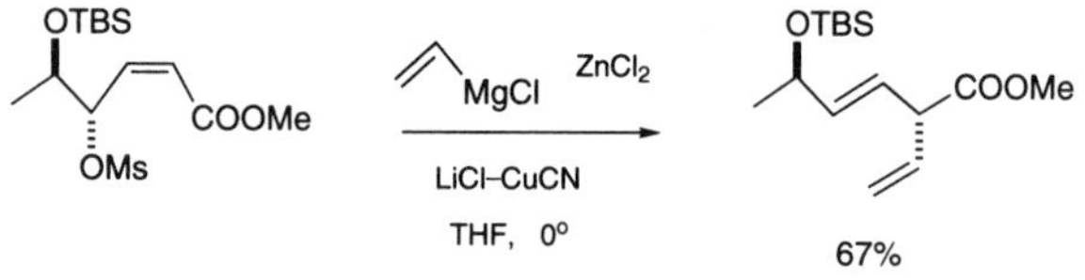

Cross-coupling. Reaction partners containing various functional groups have been used, which include pairings of acetate/nitro group[4] and oxazolidinone/ester.[5] Functionalized 2-haloalkenes are also readily made.[6]

Conjugate addition. Acceptors include nitroalkenes,[7] (tropone)iron tricarbonyl,[8] and *N*-acylpyridinium salts[9] in the conjugate addition. It may be followed by a cycloacylation process.[10]

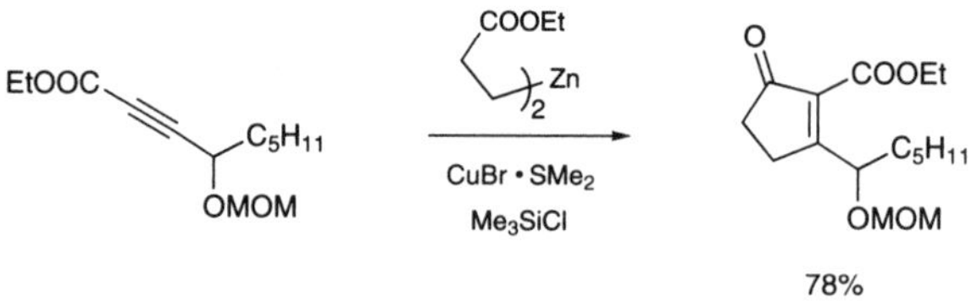

[1] P. Knochel, T.-S. Chou, C. Jubert, and D. Rajaopal, *JOC* **58**, 588 (1993).

[2] L.-H. Xu and E. P. Kündig, *HCA* **77**, 1480 (1994).

[3] T. Ibuka, K. Nakai, H. Habashita, K. Bessho, N. Fujii, Y. Chounan, and Y. Yamamoto, *T* **49**, 9479 (1993).

[4] C. E. Tucker and P. Knochel, *JOC* **58**, 4781 (1993).

[5] R. Duddu, M. Eckhardt, M. Furlong, H. P. Knoess, S. Berger, and P. Knochel, *T* **50**, 2415 (1994).

[6] L. Zhu, K. H. Shaughnessy, and R. D. Rieke, *SC* **23**, 525 (1993).

[7] S. E. Denmark and L. R. Marcin, *JOC* **58**, 3850 (1993).

[8] M.-C. P. Yeh, C.-C. Hwu, C.-H. Ueng, and H.-L. Lue, *OM* **13**, 1788 (1994).

[9] M.-J. Shiao, K.-H. Liu, and J.-S. Wang, *JCCS(T)* **40**, 175 (1993).

[10] M. T. Crimmins, P. G. Nantermet, B. W. Trotter, I. M. Vallin, P. S. Watson, L. A. McKerlie, T. L. Reinhold, A. W.-H. Cheung, K. A. Stetson, D. Dedopoulou, and J. L. Gray, *JOC* **58**, 1038 (1993).

Organocopper/zirconium reagents.

Conjugate addition.[1] Alkylzirconocenes in the presence of Cu(I) salt (3–10 mol %) add to enones, enals, and conjugated sulfones. Many functional groups are tolerated.

78%

[1] P. Wipf, W. Xu, J. H. Smitrovich, R. Lehmann, and L. M. Venenzi, *T* **50**, 1935 (1994).

Organogermanium reagents.

Allylation.[1] Aldehydes are converted to homoallylic alcohols by a combination of allyl bromide and germanium(II) iodide (also in the presence of ZnI_2). Ketones are less reactive, requiring an excess of the reagents and longer reaction times.

[1] Y. Hashimoto, H. Kagoshima, and K. Saigo, *TL* **35**, 4805 (1994).

Organoindium reagents.

Allylation. Aldehydes and aldimines are converted to homoallylic alcohols[1] and amines,[2] respectively, by the allylindium reagent generated from allyl bromide and the metal. Starting from glyceraldehyde acetonide and related aldehydes, the chain extension followed by ozonolysis and simple manipulations allows the elaboration of 2-deoxy sugars in a straightforward manner.[1]

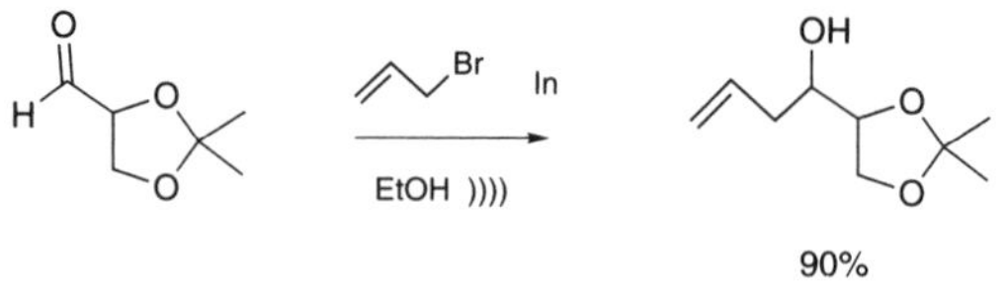

90%

[1] W. H. Binder, R. H. Prenner, and W. Schmid, *T* **50**, 749 (1994).
[2] S.-J. Jin, S. Araki, and Y. Butsugan, *BCSJ* **66**, 1528 (1993).

Organoiron reagents. 16, 241

Ketones from acyl chlorides.[1]   Either dialkylirons or lithium trialkylferrates can be employed to react with acyl chlorides to furnish ketones. These reagents are prepared in situ from $FeCl_3$ by reduction to $FeCl_2$ and reaction with RLi or RMgX. Actually RMgX + $FeCl_3$ (cat.) is more convenient to use.

[1] T. Kauffmann, K.-U. Vob, and G. Neiteler, *CB* **126**, 1453 (1993).

9-Organothio-9-borabicyclo[3.3.1]nonane.

Vinyl sulfides.[1]   2-Organothio-1-alkenes are produced from a Pd(0)-catalyzed reaction with terminal alkynes.

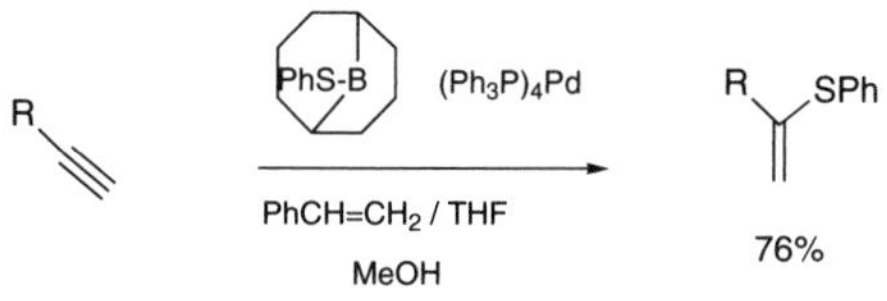

76%

[1] T. Ishiyama, K. Nishijima, N. Miyaura, and A. Suzuki, *JACS* **115**, 7219 (1993).

Organozinc reagents. 13, 220–222; 14, 233–235; 15, 238–240; 16, 246–248; 17, 228–234

Allenes.[1]   The reaction of organozinc/copper reagents with propargyl tosylate furnishes allene derivatives.

Carbocyclization.[2]   An organozinc derived from the appropriate iodoalkene is liable to cyclization. The new reactive center can be functionalized or quenched.

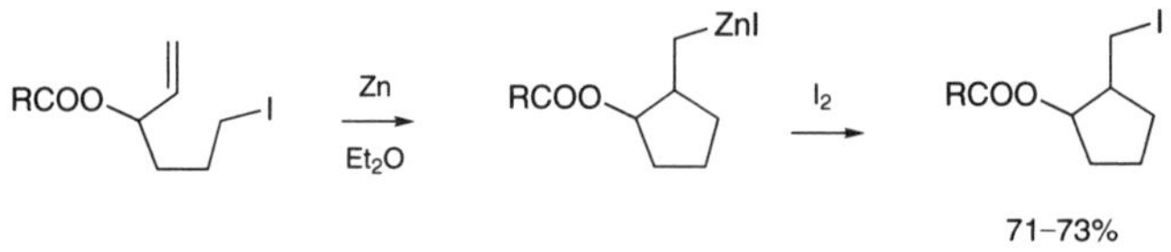

71–73%

gem-Functionalization of cyclopropanes.[3] Starting from *gem*-dibromocyclo-propanes, the systematic metallation to produce 1-halocyclopropylzincates enables stereoselective formation of C–C bonds.

[1] M. J. Dunn, R. F. W. Jackson, J. Pietruszka, N. Wishart, D. Ellis, and M. J. Wythes, *SL* 499 (1993).
[2] C. Meyer, I. Marek, G. Courtemanche, and J.-F. Normant, *SL* 266 (1993).
[3] T. Harada, T. Katsuhira, K. Hattori, and A. Oku, *JOC* **58**, 2958 (1993).

Osmium trichloride. 17, 235

α-Ketols.[1] Alkenes are oxidized to ketols by peracetic acid in the presence of $OsCl_3$.

[1] S.-i. Murahashi, T. Naota, and H. Hanaoka, *CL* 1767 (1993).

Osmium tetroxide. 13, 222–225; 14, 235–239; 15, 240–241; 16, 249–253; 17, 236–240

Asymmetric dihydroxylation. In the preparation of chiral diols, a typical mixture consisting of $(DHQD)_2$-PHAL, $K_2OsO_2(OH)_4$, $K_3Fe(CN)_6$, K_2CO_3, $MeSO_2NH_2$ in aqueous *t*-butanol can be applied successfully to many alkenes bearing different remote substituents. These include allyl halides,[1] tertiary allylic alcohols,[2] conjugated enynes,[3] vinyl- and allylsilanes,[4] enones, and α-substituted styrenes.[5] It should be noted that increasing the amounts of the ligand $(DHQD)_2$-PHAL (5 mol %) and $K_2OsO_2(OH)_4$ (1 mol %) leads to rapid turnover in the presence of methanesulfonamide.[6]

cis-Disubstituted and tetrasubstituted double bonds are most difficult classes of substrates for asymmetric dihydroxylation (AD), but several examples demonstrate that a level of 71–73% ee is reachable.[7] In a survey of 23 cyclic disubstituted olefins, the formation of diols with 5–99% ee has been observed.[8] Sulfur-containing alkenes (sulfides, disulfides, 1,3-dithianes) react on the double bonds only, but the enantioselectivity (61–98%) is highly dependent on the substitution pattern of the olefin.[9]

Allyl aryl ethers undergo AD with enantioselectivity affected by substituents in the aryl ring.[10] An *o*-substituent is deleterious (28–63% ee), whereas a *p*-substituent generally exerts favorable effects.

There are numerous applications of the method; perhaps mention should be made of the AD of a vinyldioxabenzosuberane[11] and a lactone enolcarbonate.[12] The dihydroxy aldehyde obtained from the former reaction has been converted to L-fructose by a chemoenzymatic procedure, and the latter transformation is related to synthesis of camptothecin.

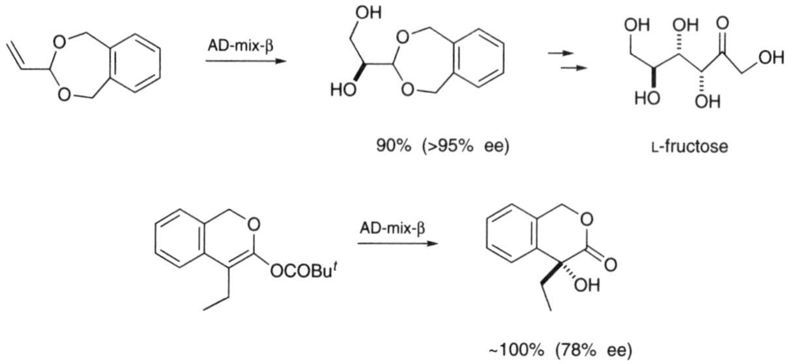

90% (>95% ee)   L-fructose

~100% (78% ee)

Allylic and homoallylic amines in which the nitrogen atom is masked with two Boc groups provide valuable building blocks for synthesis not only because of the multiple functionalities after AD, but also because the two hydroxyl groups emerge in differentiated forms in the products. The proximal oxygen atom becomes part of an oxazolidinone system.[13]

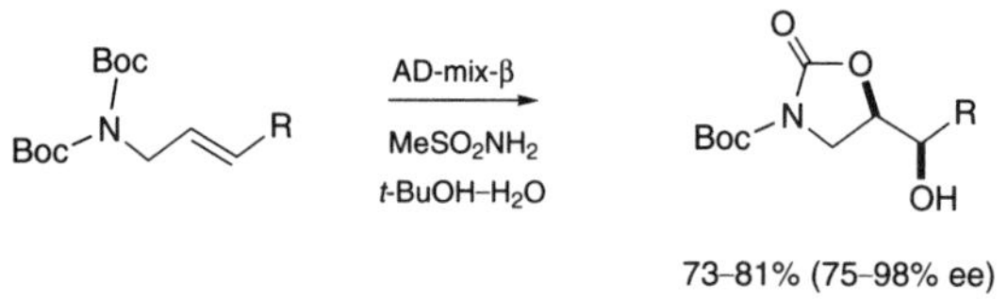

73–81% (75–98% ee)

It has been found that pyrimidine ligands show superior enantioselectivity in AD of terminal alkenes.[14] Covalent linkage of the quinine alkaloids to a polymer support generates heterogeneous catalysts.[15]

There is a report on the AD of 1,1-disubstituted allylic alcohol derivatives that give diols having an absolute configuration opposite to that predicted by Sharpless' steric model.[16]

[1] K. P. M. Vanhessche, Z.-M. Wang, and K. B. Sharpless, TL **35**, 3469 (1994).
[2] Z.-M. Wang and K. B. Sharpless, TL **34**, 8225 (1993).
[3] K. Tani, Y. Sato, S. Okamoto, and F. Sato, TL **34**, 4975 (1993).
[4] J. A. Soderquist, A. M. Rane, and C. J. Lopez, TL **34**, 1893 (1993).
[5] Z.-M. Wang and K. B. Sharpless, SL 603 (1993).

[6] Y. L. Bennani and K. B. Sharpless, *TL* **34**, 2079 (1993).

[7] M. S. Van Mieuwenhze and K. B. Sharpless, *TL* **35**, 843 (1994).

[8] Z.-M. Wang, K. Kakiuchi, and K. B. Sharpless, *JOC* **59**, 6895 (1994).

[9] P. J. Walsh, P. T. Ho, S. B. King, and K. B. Sharpless, *TL* **35**, 5129 (1994).

[10] Z.-M. Wang, X.-L. Zhang, and K. B. Sharpless, *TL* **34**, 2267 (1993).

[11] I. Henderson, K. B. Sharpless, and C.-H. Wong, *JACS* **116**, 558 (1994).

[12] D. P. Curran and S.-B. Ko, *JOC* **59**, 6139 (1994).

[13] P. J. Walsh, Y. L. Bennani, and K. B. Sharpless, *TL* **34**, 5545 (1993).

[14] G. A. Crispino, K.-S. Jeong, H. C. Kolb, Z.-M. Wang, D. Xu, and K. B. Sharpless, *JOC* **58**, 3785 (1993).

[15] D. Pini, A. Petri, and P. Salvadori, *T* **50**, 11321 (1994).

[16] K. J. Hale, S. Manaviazar, and S. A. Peak, *TL* **35**, 425 (1994).

Osmium tetroxide–Jones reagent.

Double bond cleavage.[1] The combination of reagents can be used to cleave mono- and 1,2-disubstituted alkenes, affording carboxylic acids directly.

[1] J. R. Henry and S. M. Weinreb, *JOC* **58**, 4745 (1993).

Oxabis(triphenylphosphonium) fluorosulfate.

Dehydration.[1] The salt (**1**) is an excellent dehydrating agent that effects esterification, amidation, and anhydride formation.

$$Ph_3\overset{+}{P}\diagdown_O\diagup\overset{+}{P}Ph_3 \qquad 2\,FSO_3^-$$

(**1**)

[1] D. G. Niyogi, S. Singh, and R. D. Varma, *JFC* **68**, 237 (1994).

Oxalyl chloride. 17, 241–242

Acid chlorides.[1] Oxalyl chloride converts silyl esters to acid chlorides under very mild conditions. The preparation of pyroglutamoyl chloride is an example.

2-Hydroxynaphthoquinones.[2] Under Friedel–Crafts reaction conditions aroylacetic esters react with oxalyl chloride to give the quinone skeleton. The ester group is removable through saponification.

[1] B. Rigo, S. El Ghammarti, P. Gautret, and D. Couturier, *SC* **24**, 2597 (1994).
[2] G. Sartori, F. Bigi, G. Canali, R. Maggi, G. Casnati, and X. Tao, *JOC* **58**, 840 (1993).

Oxygen.

Oxidation of primary alcohols.   The metal-catalyzed oxidation of primary alcohols (only) to carboxylic acids usually employs platinum, as shown in a procedure for the preparation of alkyl glucopyranuronates.[1] A palladium/carbon–lead(II) acetate system is also serviceable.[2]

Ring cleavage.[3]   Cyclic 1,3-diketones and β-keto esters undergo oxidative cleavage in a copper(II) perchlorate–mediated process.

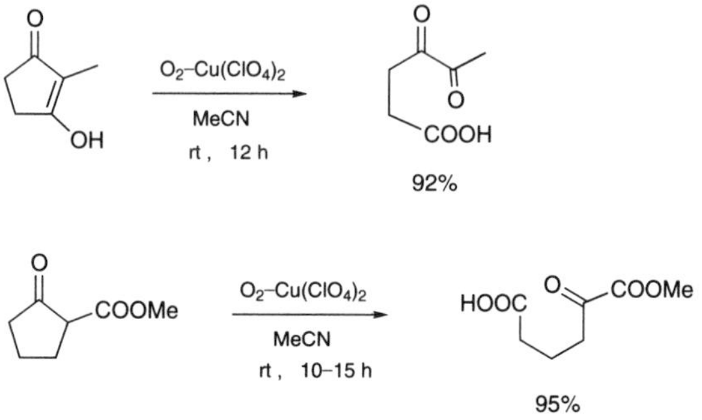

Baeyer–Villiger oxidation.   Cyclic ketones are converted to lactones by molecular oxygen in the presence of an aldehyde (benzaldehyde most frequently used), with[4] or without metal catalysts.[5] Hydrotalcite catalysts have also been evaluated.[6]

Epoxidation.   Cooxidation of an alkene and an aldehyde occurs in the presence of a metal catalyst.[7,8] Various other organic compounds are oxidized under selected conditions.[9]

Hydration of alkenes.   Iron–[10] and cobalt–porphyrin[11] systems mediate the delivery of oxygen to organic substrates. Reduction of the intermediates result in alcohols.

Additive oxidation of alkenylsilanes.[12]   Electrochemical oxidation in the presence of thiophenol gives α-benzenesulfenyl carbonyl products.

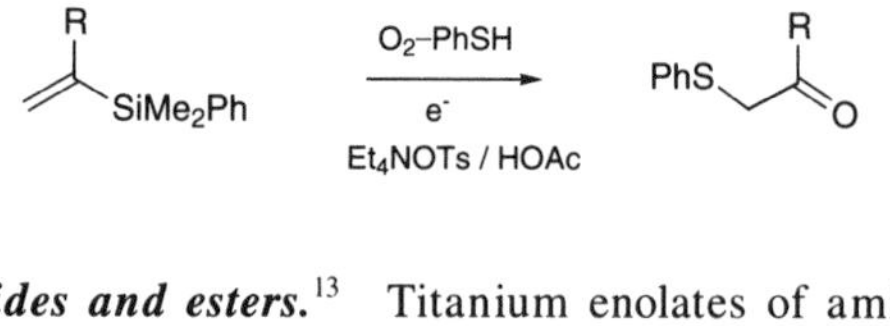

α-Hydroxy amides and esters.[13]   Titanium enolates of amides and esters undergo hydroxylation on treatment with oxygen at −30°C.

Alkyl hydroperoxides.[14]   Under basic conditions, the exposure of N-substituted sulfonhydrazides to oxygen results in deamination and formation of hydroperoxides.

[1] J. Fabre, D. Betbeder, F. Paul, and P. Monsan, SC 23, 1357 (1993).

[2] M. Akada, S. Nakano, T. Sugiyama, K. Ichitoh, H. Nakao, M. Akita, and Y. Moro-oka, BCSJ 66, 1511 (1993).

[3] J. Cossy, R. D. Belotti, V. Bellosta, and D. Brocca, TL 35, 6089 (1994).

[4] C. Bolm, G. Schlingloff, and K. Weickhardt, TL 34, 3405 (1993).

[5] K. Kaneda, S. Ueno, T. Imanaka, E. Shimotsuma, Y. Nishiyama, and Y. Ishii, JOC 59, 2915 (1994).

[6] K. Kaneda, S. Ueno, and T. Imanaka, CC 797 (1994).

[7] M. Hamamoto, K. Nakayama, Y. Nishiyama, and Y. Ishii, JOC 58, 6421 (1993).

[8] J. Iqbal, S. Bhatia, and M. M. Reddy, SC 23, 2285 (1993).

[9] P. Mastrorilli and C. F. Nobile, TL 35, 4193 (1994).

[10] T. Mori, T. Santa, T. Higuchi, T. Mashino, and M. Hirobe, CPB 41, 292 (1993).

[11] Y. Matsusita, K. Sugamoto, and T. Matsui, CL 925 (1993).

[12] S. Nakatani, J. Yoshida, and S. Isoe, T 49, 2011 (1993).

[13] W. Adam, M. Metz, F. Prechtl, and M. Renz, S 563 (1994).

[14] L. Collazo, F. S. Guziec, W.-X. Hu, A. Munoz, D. Wei, and M. Alvarado, JOC 58, 6169 (1993).

Oxygen, singlet. **13**, 228–229; **14**, 249; **15**, 243; **16**, 257–258; **17**, 251–253

Chloromethyl ketones.[1]   The CuCl$_2$-catalyzed photooxidation of 1-alkenes in pyridine/CH$_2$Cl$_2$ furnishes chloromethyl ketones in moderate to good yields.

Cleavage of alkenes.[2]   The reaction is sensitized by p-dimethoxybenzene.

Ene reactions.   Enones with an α-trialkylsilyl[3] or α-trialkylstannyl substituent[4] are accessible from alkenylsilanes and stannanes, respectively, by reaction with singlet oxygen and subsequent dehydration. Instead of dehydration, the addition of titanium(IV) isopropoxide to the allylic hydroperoxide intermediates induces epoxidation of the transposed alkene linkage.[5,6] The hydroperoxides can also be reduced and used for other synthetic purposes.[7]

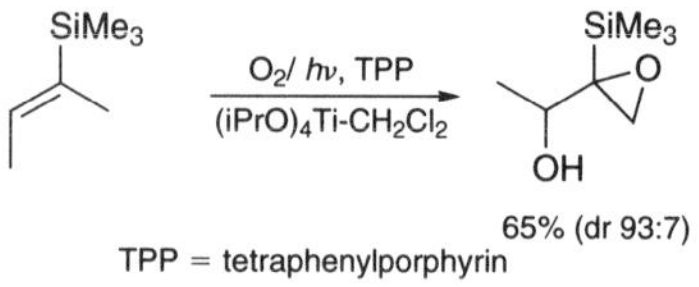

65% (dr 93:7)

TPP = tetraphenylporphyrin

Allylstannanes are converted to hydroperoxides with a vinylic stannyl group.[8] The reaction is stereoselective.

The reaction is regio- and diastereoselective for allylic alcohols.[9] The effect of a neighboring carboxyl group is also interesting.[10]

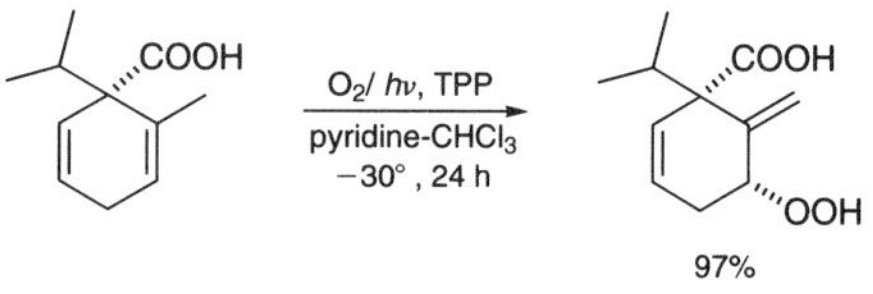

97%

[1] T. Sato and S. Yonemochi, T 50, 7375 (1994).
[2] U. T. Bhalerao and M. Sridhar, TL 34, 4341 (1993).
[3] W. Adam and M. J. Richter, S 176 (1994).
[4] W. Adam and P. Klug, S 557 (1994).
[5] W. Adam and M. J. Richter, JOC 59, 3335 (1994).
[6] W. Adam and P. Klug, CB 127, 1441 (1994).
[7] W. Adam and P. Klug, JOC 59, 2695 (1994).
[8] P. H. Dussault and R. J. Lee, JACS 116, 4485 (1994).
[9] W. Adam and B. Nestler, JACS 115, 5041 (1993).
[10] T. Linker and L. Fröhlich, ACIEE 33, 1971 (1994).

Ozone. 13, 229; 15, 243–244; 17, 253–254

Carbonyl compounds. Triethylamine is often superior to dimethyl sulfide as a quencher of ozonides, in terms of faster reactions and better yields of the products.[1]

Tertiary amines.[2] Using a secondary amine to decompose an ozonide derived from 1-alkene effects its alkylation. The amine initiates an eliminative fragmentation of the ozonide to generate an aldehyde and dialkylammonium formate. Schiff base formation from the aldehyde and another molecule of the amine is then followed by reduction by the formate ion.

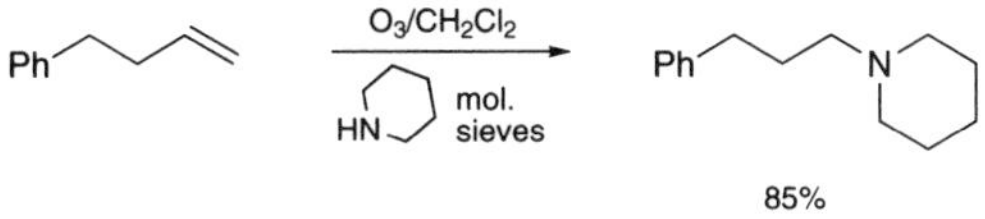

85%

Methyl esters by degradation of alkenes.[3]   Ozonolysis of mono-, di-, and trisubstituted alkenes in methanolic NaOH affords methyl esters.

Selective oxidation.   5-Hydroperoxy 1,3-dienes, after protection as peroxy-
acetals, undergo selective cleavage at the remote double bond.[4] The hydroperoxides
themselves are available from ene reaction of skipped dienes with singlet oxygen.

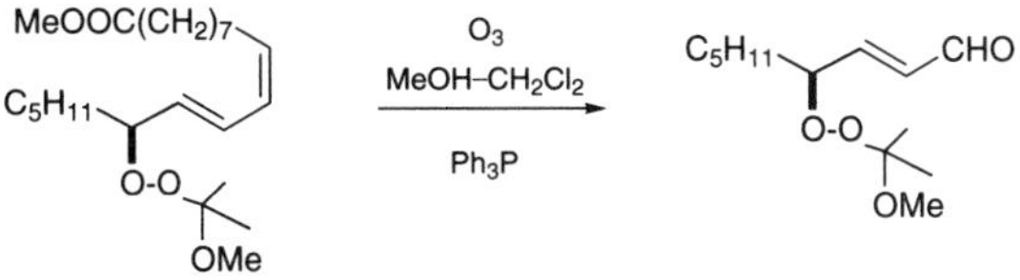

The phenylseleno group can be preserved in ozonolysis with the workup (Ph₃P as
quencher) at low temperatures to give selenenyl aldehydes.[5]

Cleavage of oximes.[6]   Carbonyl compounds are generated.

Cleavage of stabilized Wittig reagents.   Such compounds give carbonyl products; therefore, α-keto esters[7] and 2,3-diketo esters[8] are readily prepared.

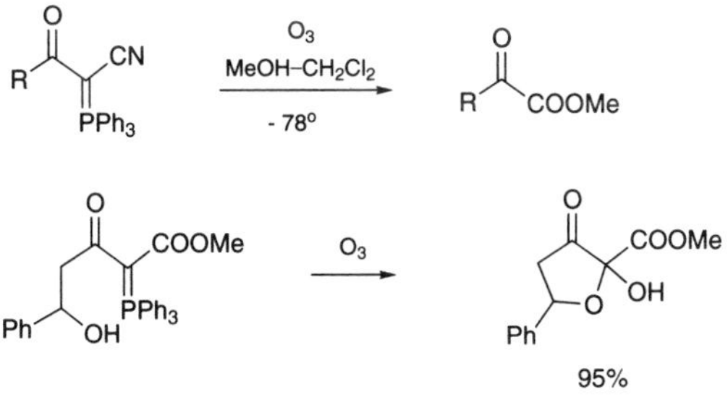

Esters from α-alkoxystannanes.[9]

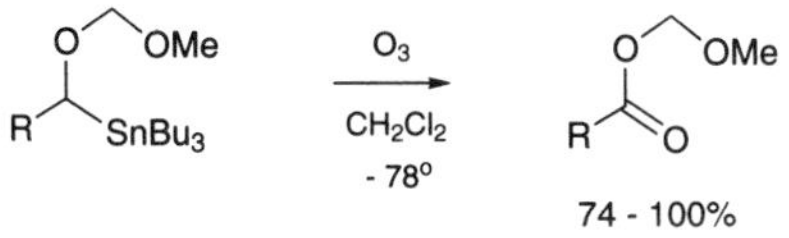

74 - 100%

[1] Y.-S. Hon, S.-W. Lin, and Y.-J. Chen, *SC* **23**, 1543 (1993).

[2] Y.-S. Hon and L. Lu, *TL* **34**, 5309 (1993).

[3] J. A. Marshall and A.W. Garofalo, *JOC* **58**, 3675 (1993).

[4] P. Dussault, A. Sahli, and T. Westermeyer, *JOC* **58**, 5469 (1993).

[5] D. L. J. Clive and M. H. D. Postema, *CC* 235 (1994).

[6] Y.-T. Yang, T.-S. Li, and Y.-L. Li, *SC* **23**, 1121 (1993).

[7] H. H. Wasserman and W.-B. Ho, *JOC* **59**, 4364 (1994).

[8] H. H. Wasserman and G. M. Lee, *TL* **35**, 9783 (1994).

[9] R. J. Linderman and M. Jaber, *TL* **35**, 378 (1994).

P

Palladium, colloidal.

Prepared from $(CF_3COCH_2COCF_3)_2Pd$ and SiH_4 or a Si–H-containing polymer in an organic solvent, it is an active and stable catalyst for hydrogenation and dehydrogenation of organic compounds.[1]

[1] L. A. Fowley, D. Michos, X.-L. Luo, and R. H. Crabtree, *TL* **34**, 3075 (1993).

Palladium/carbon. 13, 230–232; 15, 245

Heck and Suzuki couplings.[1,2]

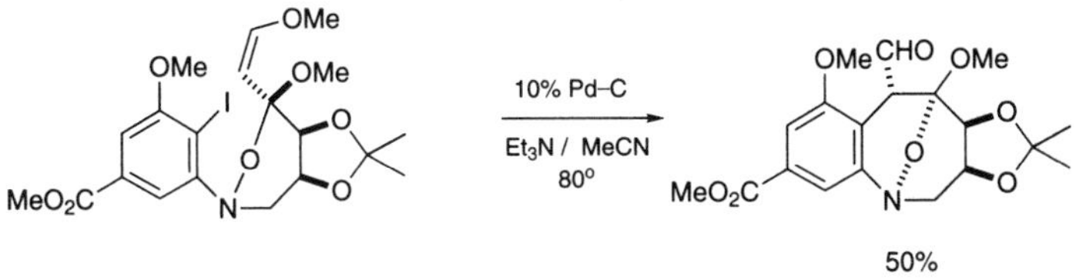

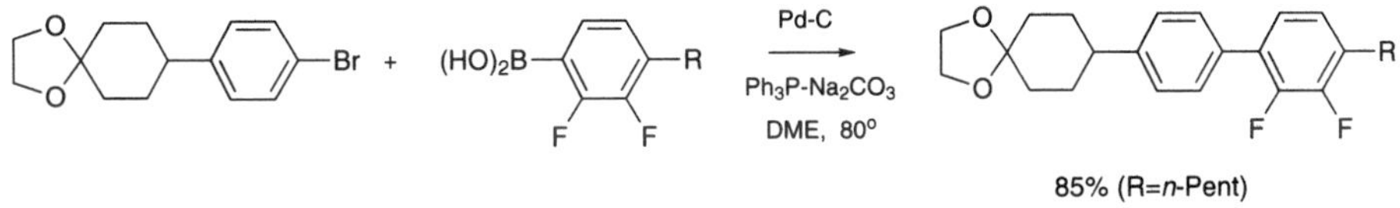

Dehydrogenation. A convenient route to 4-substituted benzoic esters[3] consists of heating mixtures of a coumalate ester with alkenes in the presence of Pd/C. A Diels–Alder reaction is followed by decarboxylation and dehydrogenation.

Hydrocarboxylation and chlorocarboxylation. Pd/C catalyzes the formation of carboxylic acids from conjugated dienes[4] and alkynes.[5] The reaction system also contains CO, HCOOH (or oxalic acid), Ph_3P, and 1,4-bis(diphenylphosphino)butane. For chlorocarboxylation, anhydrous HCl is used as a coreagent.[6]

[1] K. F. McClure, S. J. Danishefsky, and G. K. Schulte, *JOC* **59**, 355 (1994).
[2] G. Marck, A. Villiger, and R. Buchecker, *TL* **35**, 3277 (1994).
[3] Y.-i. Matsushita, K. Sakamoto, T. Murakami, and T. Matsui, *JCCS(T)* **41**, 3307 (1994).
[4] G. Vasapollo, A. Somasunderam, B. El Ali, and H. Alper, *TL* **35**, 6203 (1994).
[5] B. El Ali, G. Vasapollo, and H. Alper, *JOC* **58**, 4739 (1993).
[6] N. Benard, M. C. Bonnet, S. Lecolier, and I. Tkatchenko, *CC* 1448 (1993).

Palladium(II) acetate. **13**, 232–233, **14**, 248; **15**, 245–247; **16**, 259–263; **17**, 255–259

Heck reaction. For this extremely valuable method of coupling unsaturated fragments, there are many variants with respect to substrate structures, additives, and reaction conditions. For instance, a solid-phase synthesis of styrenecarboxylic acids has been reported.[1] The employment of diazonium salts[2] in the Heck reaction is successful; the presence of water does not seem to affect the efficiency (and in fact it was reported that water facilitates reactions of some other types of substrates).[3] It is also possible to conduct the coupling in tandem with diazotization,[4] which is carried out with BuONO–HOAc in CH_2Cl_2 at room temperature.

Since many other functional groups are tolerated in this reaction, a great variety of useful synthetic intermediates have been assembled: 4-styryl-2-azetidinones,[5] 4,5-diphenyl-3-styryloxazolin-2-ones,[6] β-alkoxystyrenes,[7,8] allylic[9] and homoallylic alcohols.[10] The coupling between vinylic iodides and tertiary allylic alcohols to construct isoprenoid segments has also been demonstrated.[11]

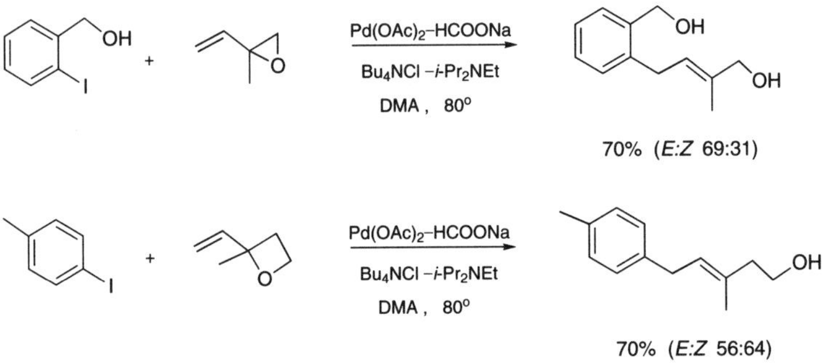

70% (*E:Z* 69:31)

70% (*E:Z* 56:64)

Secondary reactions sometimes also take place. Thus the arylation of 4-(ω-alkenyl)-β-lactams is accompanied by ring opening to give the 3-alkenamides.[12] Heterocycles[13,14] and carbocycles[15] are formed from functionalized substrates.

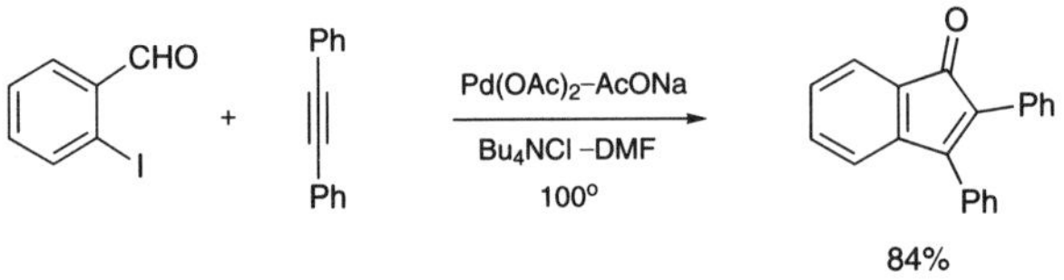

84%

Interestingly, 4-bromo-*N*-tosylindole reacts with ethyl acrylate in HOAc in the presence of $Pd(OAc)_2$ at C-3 only.[16]

Suzuki, Stille, and related couplings. The oxidative addition of the C–B bond in aryl and alkenyl boronic acids to Pd(0) initiates the coupling with aryl halides or alkenes. The synthesis of unsymmetrical biphenyls,[17] various styrenes,[18] and stilbenes[19] is expediently accomplished by this method. Ph$_4$BNa can be used as a phenylating agent when AgOAc is also present.

Biphenyls are also produced when diaryliodonium salts are coupled with aryltin compounds.[20]

Reductive acylation and transesterification.   Carbon monoxide acts as both reductant and acylating agent for *o*-substituted nitrobenzenes, delivering dihydrobenzoxazin-2-ones[21] from *o*-nitrobenzyl alcohols. Pd(OAc)$_2$ catalyzes formation of the vinyl esters of hydroxycarboxylic acids by reaction with vinyl acetate[22] without acetylating the hydroxyl group.

2-Benzenesulfenyl-1,3-dienes.   These compounds are now available from Pd(II)-catalyzed addition of PhSH to conjugated enynes.[23] They undergo oxidation to the sulfoxides, which are useful for the preparation of functionalized allylic alcohols.

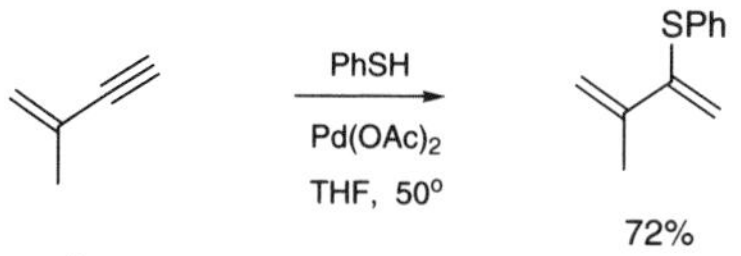

Allylic oxidation.   Allylic alcohols are oxidized to enones by Pd(OAc)$_2$ in wet DMF.[24] Allylic esters are obtained from alkene/carboxylic acid mixtures on treatment with Pd(OAc)$_2$–benzoquinone and an oxidant (MnO$_2$,[25] H$_2$O$_2$,[26] or *t*-BuOOH). Some unsaturated carboxylic acids undergo cyclization to give lactones.[27] Analogous oxidative cyclization leads to bicyclic azacycles.[28,29]

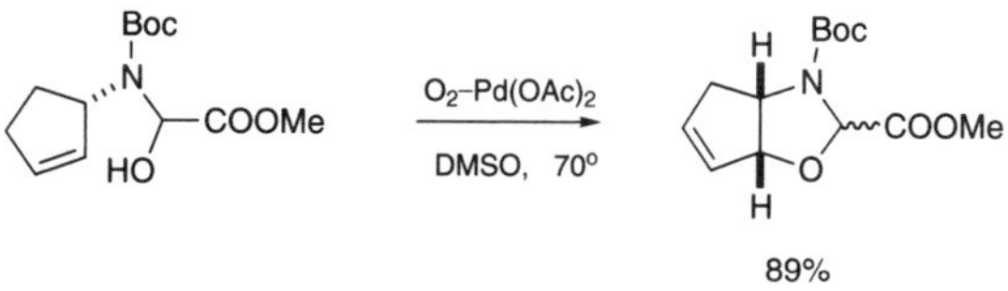

1,4-Difunctionalization of 1,3-dienes.[30]

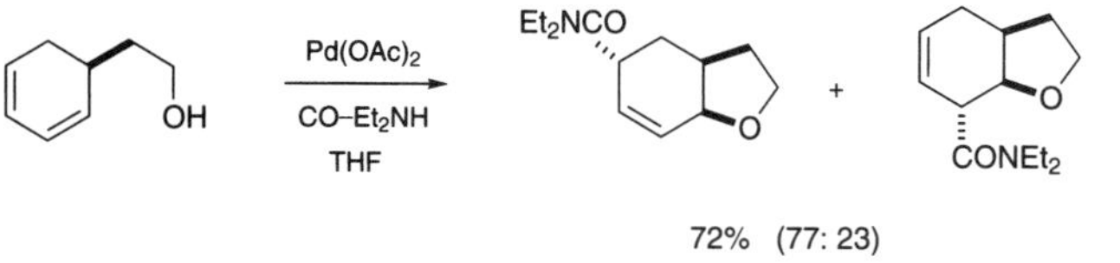

Cycloisomerization of enynes. Formation of 5-membered heterocycles[31,32] and carbocycles[33] from open-chain enynes may lead to 1,3- or 1,4-dienes by changing the substrates slightly. The product profiles are different from reactions catalyzed by the $(dba)_3Pd_2 \cdot CHCl_3-HOAc-(o\text{-}Tol)_3P$ system.[34]

[1] K.-L. Yu, M. S. Deshpande, and D. M. Vas, *TL* **35**, 8919 (1994).

[2] S. Sengupta and S. Bhattacharya, *JCS(PI)* 1943 (1993).

[3] H.-C. Zhang and G. D. Davies, *OM* **12**, 1499 (1993).

[4] M. Beller, H. Fischer, and K. Kühlein, *TL* **35**, 8773 (1994).

[5] A. Katake, K. Okano, I. Shimizu, and A. Yamamoto, *SL* 839 (1994).

[6] C. A. Busacca, R. E. Johnson, and J. Swestock, *JOC* **58**, 3299 (1993).

[7] M. Larhed, C.-M. Andersson, and A. Hallberg, *ACS* **47**, 212 (1993).

[8] D. Badone and U. Guzzi, *TL* **34**, 3603 (1993).

[9] R. C. Larock and S. Ding, *JOC* **58**, 804 (1993).

[10] R. C. Larock, S. Ding, and C. Tu, *SL* 145 (1993).

[11] H. Bienayme and C. Yezeguelian, *T* **50**, 3389 (1994).

[12] R. C. Larock and S. Ding, *TL* **34**, 979 (1993).

[13] R. C. Larock and H. Yang, *SL* 748 (1994).

[14] G. Dyker, *JOC* **58**, 6426 (1993).

[15] R. C. Larock, M. J. Doty, and S. Cacchi, *JOC* **58**, 4579 (1993).

[16] Y. Yokoyama, M. Takashima, C. Higaki, K. Shidori, S. Moriguchi, C. Ando, and Y. Murakami, *H* **36**, 1739 (1993).

[17] T. I. Wallow and B. M. Novak, *JOC* **59**, 5034 (1994).

[18] C. S. Cho, K. Itotani, and S. Uemura, *JOMC* **443**, 253 (1993).

[19] C. S. Cho and S. Uemura, *JOMC* **645**, 85 (1994).

[20] N. A. Bumagin, L. I. Sukhomlinova, S. O. Igushkina, A. N. Banchikov, T. P. Tolstaya, and I. P. Beletskaya, *BRAS* **41**, 2128 (1992).

[21] S. Cenini, S. Console, C. Crotti, and S. Tollari, *JOMC* **451**, 157 (1993).

[22] M. Lobell and M. P. Schneider, *S* 375 (1994).

[23] J.-E. Bäckvall and A. Ericsson, *JOC* **59**, 5850 (1994).

[24] V. Bellosta, R. Benhaddou, and S. Czemecki, *SL* 861 (1993).

[25] N. Ferret, L. Mussate-Mathieu, J.-P. Zahra, and B. Waegell, *CC* 2589 (1994).

[26] B. Akermark, E. M. Larsson, and J. D. Oslob, *JOC* **59**, 5729 (1994).

[27] R. C. Larock and T. R. Hightower, *JOC* **58**, 5298 (1993).

[28] R. A. T. M. van Benthem, H. Hiemstra, J. J. Michels, and W. N. Speckamp, *CC* 357 (1994).

[29] R. A. T. M. van Benthem, H. Hiemstra, G. R. Longarela, and W. N. Speckamp, *TL* **35**, 9281 (1994).

[30] P. G. Andersson and A. Aranyos, *TL* **35**, 4441 (1994).

[31] G. Zhu, S. Ma, and X. Lu, *JCR(S)* 366 (1993).

[32] S. Ma, G. Zhu, and X. Lu, *JOC* **58**, 3692 (1993).

[33] B. M. Trost, C. J. Tanoury, M. Lautens, C. Chan, and D. T. MacPherson, *JACS* **116**, 4255 (1994).

[34] B. M. Trost, D. L. Romero, and F. Rise, *JACS* **116**, 4268 (1994).

Palladium(II) acetate–tertiary phosphine. **13**, 91, 233–234; **14**, 249, 250–253; **15**, 247–248; **16**, 264–268; **17**, 259–269

Heck and related reactions. In certain couplings high pressure is required.[1] While Et$_3$N is used as base in many Heck reactions, it can be replaced by alkali metal acetate or bicarbonate together with a quaternary ammonium salt.[2] Base-free arylation of alkenes is subject to a remarkable chelate effect[3]; zinc is present in the system.

High-purity *o*- and *p*-methylstyrenes are obtained from the corresponding bromotoluenes by this method.[4] Some other useful preparations are based on the arylation of 1,3-dioxep-5-ene,[5] allyl carbamates,[6] and cyclization of vinylphosphonites.[7]

Benzodihydropyrans and tetrahydroquinolines are obtained[8] from reaction of 1,4-dienes with *o*-iodinated phenols and anilines, respectively. Formation of pyrido-[2,3-*b*]indoles from α-(*o*-bromoanilino)alkenenitriles[9] is unusual, as Et$_3$N contributes [CH$_3$CN] to form portion of the pyridine unit of the products.

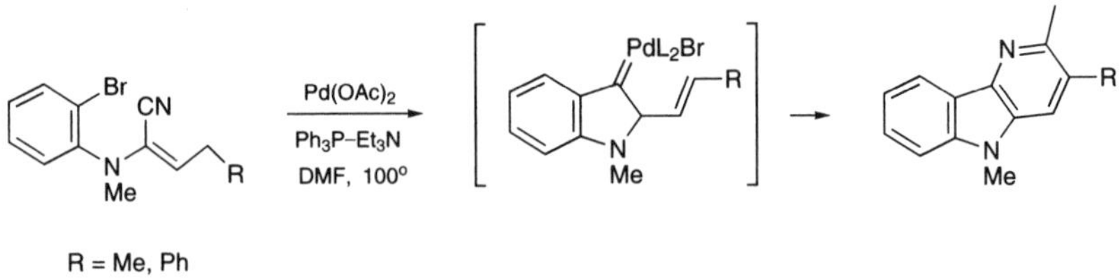

R = Me, Ph

An intramolecular reductive Heck reaction in the presence of HCOONa is useful for the access to exocyclic allylsilanes.[10]

78%

Heck reactions involving nonaromatic components are particularly versatile synthetic processes. For example, 1,2,4-trienes and 1,2-dien-4-ynes are readily prepared from propargyl carbonates.[11] Intramolecular Heck cyclization followed by Diels–Alder[12] or nucleophilic attack[13,14] leads to bicyclic products. Substituted pyrrolidines and piperidines are formed by coupling of alkenyl sulfonamides with vinylic halides.[15]

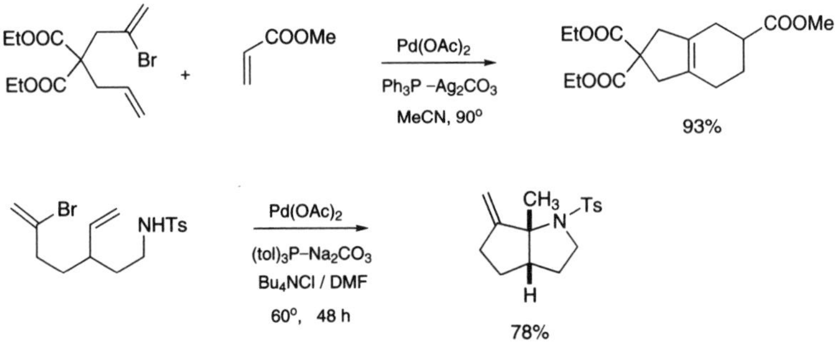

93%

78%

The Stille coupling is useful for the synthesis of electron-deficient dienes such as hexadienoic esters.[16]

1,2-Additions.   Hydrocarboxylation of alkenes[17] and alkynes[18] with HCOOH under CO is accomplished by Pd catalysis. A similar reaction undergone by a propargylic carbonate after an S_N2' process leads to an itaconic diester segment.[19] The intermediate is converted into a cyclopentenone[20] when the propargyl carbonate contains an additional double bond. It should be noted that simple elimination of popargylic carbonates to give enynes[21] occurs when the reaction is conducted in the absence of CO.

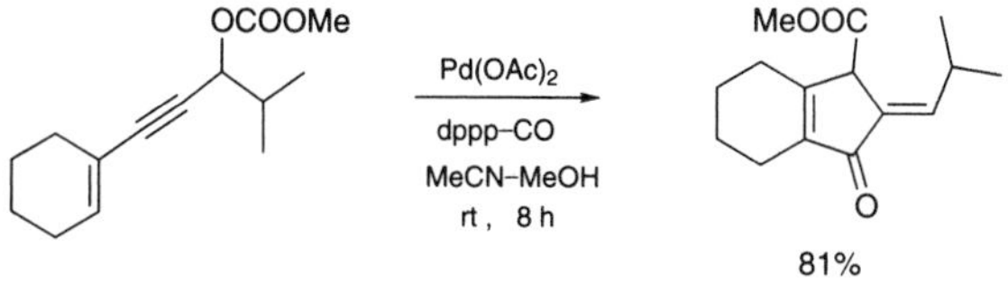

81%

In the acylcyanation of 1-alkynes[22] two different phosphines are used as ligands for Pd. For hydroarylation[23] of enals and enones the source of the aryl group is Ar_3Sb.

Deallylation. Allyl carbonates and carbamates are readily cleaved[24] by treatment with $Pd(OAc)_2$. There is a useful chemoselectivity for this reaction[25] in that dimethylallyl esters and cinnamyl esters are not affected. An expedient route to isoflavanones and isoflavones from allyl chroman-4-one-3-carboxylate involves arylation with $ArPb(OAc)_3$ and treatment with $Pd(OAc)_2$. The latter reaction removes the ester group to give isoflavanones[26] when it is carried out in the presence of Ph_3P, HCOOH, and Et_3N. Isoflavones are obtained when the keto esters are treated with $Pd(OAc)_2$ and $Ph_2PCH_2CH_2PPh_2$ in refluxing MeCN.

*Ph = sulfonated phenyl   96%

Allylic substitutions. The π-allylpalladium ions are electrophilic agents. Diallylation of (benzothiazol-2-ylthio)methyl ketones[27] is readily achieved using allyl methyl carbonate as reaction partner. The generation of a Pd-complexed trimethylenemethane from 2-(trimethylsilylmethyl)allyl esters is well known; the interception of this species by imines gives 3-methylenepyrrolidines.[28]

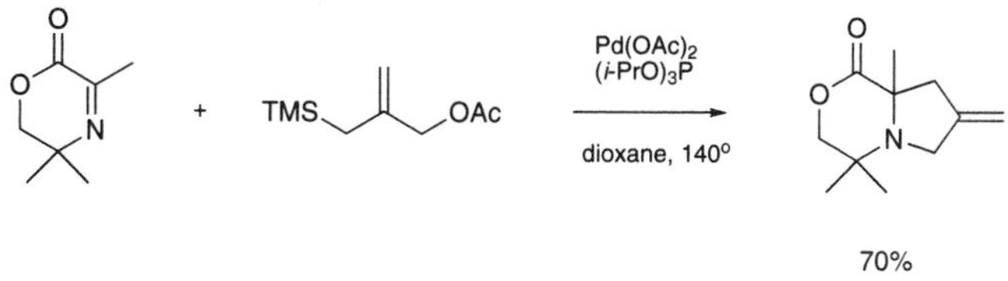

70%

The gradated reactivities of allylic acetates and carbonates have been exploited in synthesis. Thus chiral *cis*-4-acetoxycyclohex-2-enol is converted into either one of the enantiomers of 2,4-cyclohexadien-1-ylacetic acid[29] by direct displacement with sodiomalonate followed by acetylation, elimination, and saponification, or by displacement of the derived carbonate followed by elimination of the unreacted acetate.

Epoxide isomerization.[30] Epoxides undergo chemo- and regioselective isomerization to carbonyl compounds.

[1] K. Voight, U. Schick, F. E. Meyer, and A. de Meijere, *SL* 189 (1994).

[2] T. Jeffery and J.-C. Galland, *TL* **35**, 4103 (1994).

[3] M. Portnoy, Y. Ben-David, and D. Milstein, *OM* **12**, 4734 (1993).

[4] R. A. DeVries and A. Mendoza, *OM* **13**, 2405 (1994).

[5] T. Sakamoto, Y. Kondo, and H. Yamanaka, *H* **36**, 2437 (1993).

[6] K. Ono, K. Fugami, S. Tanaka, and Y. Tamaru, *TL* **35**, 4133 (1994).

[7] F. Hong, J. Xia, and Y. Xu, *JCS(P1)* 1665 (1994).

[8] R. C. Larock, N. G. Berrios-Pena, C. A. Fried, E. K. Yum, C. Tu, and W. Leong, *JOC* **58**, 4509 (1993).

[9] C.-C. Yang, P.-J. Sun, and J.-M. Fang, *CC* 2629 (1994).

[10] L. F. Tietze and R. Schimpf, *CB* **127**, 2235 (1994).

[11] T. Mandai and J. Tsuji, *SOC* **50**, 908 (1992).

[12] F. E. Meyer, K. H. Ang, A. G. Steinig, and A. de Meijere, *SL* 191 (1994).

[13] G. D. Harris, R. J. Herr, and S. M. Weinreb, *JOC* **58**, 5452 (1993).

[14] C. S. Nylund, J. M. Klopp, and S. M. Weinreb, *TL* **35**, 4287 (1994).

[15] R. C. Larock, H. Yang, S. M. Weinreb, and R. J. Herr, *JOC* **59**, 4172 (1994).

[16] I. N. Houpis, L. DiMichele, and A. Molina, *SL* 365 (1993).

[17] B. Al Ali and H. Alper, *JOC* **58**, 3595 (1993).

[18] D. Zagarian and H. Alper, *OM* **12**, 712 (1993).

[19] T. Mandai, Y. Tsujiguchi, S. Matsuoka, J. Tsuji, and S. Saito, *TL* **35**, 5697 (1994).

[20] T. Mandai, J. Tsuji, Y. Tsujiguchi, and S. Saito, *JACS* **115**, 5865 (1993).

[21] T. Mandai, Y. Tsujiguchi, S. Matsuoka, and J. Tsuji, *TL* **34**, 7615 (1993).

[22] K. Nozaki, N. Sato, and H. Takaya, *JOC* **59**, 2679 (1994).

[23] C. S. Cho, K. Tanabe, and S. Uemura, *TL* **35**, 1275 (1994).

[24] J. P. Genet, E. Blart, M. Savignac, S. Lemeune, and J.-M. Paris, *TL* **34**, 4189 (1993).

[25] S. Lemaire-Audoire, M. Savignac, E. Blart, G. Pourcelot, and J. P. Genet, *TL* **35**, 8783 (1994).

[26] D. M. X. Donnelly, J.-P. Finet, and B. A. Rattigan, *JCS(P1)* 1729 (1993).

[27] V. Calo, V. Fiandanese, A. Nacci, and A. Scilimati, *TL* **36**, 171 (1995).

[28] B. M. Trost and C. M. Marrs, *JACS* **115**, 6636 (1993).

[29] J.-E. Bäckvall, R. Gatti, and H. E. Schink, *S* 343 (1993).

[30] S. Kulasegaram and R. J. Kulawiec, *JOC* **59**, 7195 (1994).

Palladium(II) acetate–tin(II) acetate.

Side chain benzoyloxylation of aryl methyl ethers.[1]   Activation of the C–H bond is achieved in the presence of oxygen and Bz_2O, giving excellent yields of the benzoates.

$$PhOMe \ + \ Bz_2O \xrightarrow[\substack{Sn(OAc)_2-O_2 \\ 130°, \ 120 \ h}]{Pd(OAc)_2} PhOCH_2OBz$$

[1] T. Ohishi, J. Yamada, Y. Inui, T. Sakaguchi, and M. Yamashita, *JOC* **59**, 7521 (1994).

Palladium(II) acetate–tin(II) chloride.

Dimethylallylation of aldehydes.[1]   Allyltin reagents are formed in situ from isoprene, which then add to aldehydes to afford homoallylic alcohols.

17 - 85%

[1] Y. Masuyama, M. Tsunoda, and Y. Kurusu, *CC* 1451 (1993).

Palladium(II) acetylacetonate–tributylphosphine. 17, 269

Eliminations.   Allylic esters undergo reductive elimination. Myrtenyl formate gives 82% yield of a mixture of α-pinene and β-pinene[1] in a ratio of 2:98. However, propargylic formate is defunctionalized without isomerization.[2]

The displaced formate anion is the hydride source for the reduction. It should be noted that $HCOONH_4$ is added to the reaction of propargylic carbonates. In refluxing benzene the triple bond is also reduced.[3]

Elimination of unactivated sulfones to give alkenes[4] is mediated by $Pd(acac)_2$.

[1] T. Mandai, T. Matsumoto, and J. Tsuji, *SL* 113 (1993).
[2] T. Mandai, T. Matsumoto, Y. Tsujiguchi, S. Matsuoka, and J. Tsuji, *JOMC* **473**, 343 (1994).
[3] S.-K. Kang, D.-C. Park, D.-G. Cho, J.-U. Chung, and K.-Y. Jung, *JCS(P1)* 237 (1994).
[4] Y. Gai, L. Jin, M. Julia, and J.-N. Verpeaux, *CC* 1625 (1993).

Palladium(II) chloride. 13, 234–235; 15, 245–249; 16, 268–269

Redox processes. Hydrodehalogenation of organic halides is done using $PdCl_2$ anchored on poly(N-vinyl-2-pyrrolidone) as catalyst.[1] On the other hand, benzils are obtained[2] from diarylacetylenes in the presence of $PdCl_2$ and DMSO.

Biphenyls. Aryl halides can couple with aryltin[3] and silicon compounds[4] in aqueous media. The acylpalladium species obtained from carbonylation of 4-bromo-biphenyl can be intercepted in situ by 2,6-di-t-butylphenoxides.[5]

Heck-type reaction. Aryl and vinylic mercurials undergo coupling reactions with β-alkenyl-β-lactams[6] in the presence of $PdCl_2$, $CuCl_2$, and LiCl. Ring opening with double bond migration occurs during the reaction.

[1] Y. Zhang, S. Liao, and Y. Xu, *TL* **35**, 4599 (1994).
[2] K.-W. Chi, M. S. Yosubov, and V. D. Filimonov, *SC* **24**, 2119 (1994).
[3] A. I. Roshchin, N. A. Bumagin, and I. P. Beletskaya, *TL* **36**, 125 (1995).
[4] A. I. Roshchin, N. A. Bumagin, and I. P. Beletskaya, *DC* **334**, 602 (1994).
[5] Y. Kubota, T. Hanaoka, K. Takeuchi, and Y. Sugi, *CC* 1553 (1994).
[6] R. C. Larock and S. Ding, *JOC* **58**, 2081 (1993).

Palladium(II) chloride–copper(II) chloride–carbon monoxide. 13, 235–236

Oxidative carbonylation. This combination of reagents is commonly used for homologation of alkynes.[1,2] The products are usually obtained as the methyl esters. Heterocyclization[3–5] attends C–C bond formation when a proper functional group is present at a short distance.

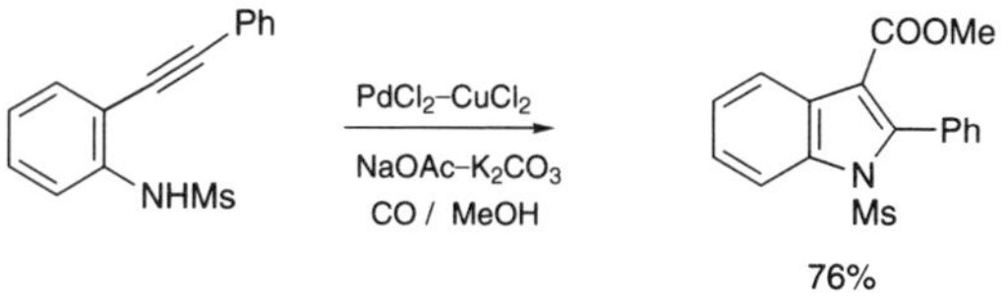

76%

[1] S. F. Vasilevsky, B. A. Trofimov, A. G. Mal'kina, and L. Brandsma, *SC* **24**, 85 (1994).

[2] T. T. Zung, L. G. Bruk, and O. N. Temkin, *MC* 2 (1994).

[3] Y. Kondo, F. Shiga, N. Murata, T. Sakamoto, and H. Yamanaka, *T* **50**, 11803 (1994).

[4] R. D. Walkup and M. D. Mosher, *T* **49**, 9285 (1993).

[5] M. Kimura, N. Saeki, S. Uchida, H. Harayama, S. Tanaka, K. Fugami, and Y. Tamaru, *TL* **34**, 7611 (1993).

Palladium(II) chloride–copper(I/II) chloride–oxygen.

Oxidation. The system is best known for oxidation of 1-alkenes to methyl ketones (Wacker oxidation). It has been employed to remove the allyl group of allyl glycosides directly[1] or after photolysis of the products.[2] In the presence of CO dihydroisobenzofuran is oxidized to phthalide in 86% yield.[3]

An electrochemical version[4] of the Wacker oxidation in an undivided cell uses tris(4-bromophenyl)amine as mediator instead of a Cu salt.

[1] H. B. Mereyala and S. Guntha, *TL* **34**, 6929 (1993).

[2] J. Lüning, U. Möller, N. Debski, and P. Welzel, *TL* **34**, 5871 (1993).

[3] M. Miyamoto, Y. Minami, Y. Ukaji, H. Kinoshita, and K. Inomata, *CL* 1149 (1994).

[4] T. Inokuchi, L. Ping, F. Hamaue, M. Izawa, and S. Torii, *CL* 121 (1994).

Palladium(II) iodide.

β-Lactones.[1] Vicinal *cis*-dicarbonylation of tertiary propargyl alcohols is observed in the presence of PdI_2 and KI under CO.

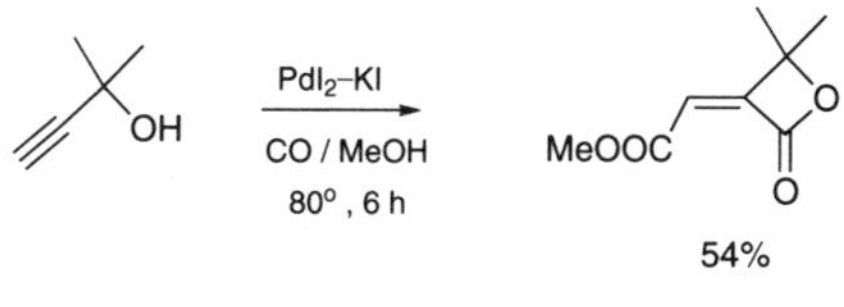

54%

[1] B. Gabriele, M. Costa, G. Salerno, and G. P. Chiusoli, *CC* 1429 (1994).

(R)-Pantolactone. 16, 269–270

α-Heterosubstituted esters. Chiral products are obtained from displacement reactions with racemic α-haloesters of (R)-pantolactone. Both inter- and intramolecular displacements give predominantly products with the (S)-configuration,[1] due to

1,4-asymmetric induction. It has been shown that the slow-reacting haloester prefers epimerization to direct reaction with the nucleophile.[2]

60% (*S:R* 10 : 1)

[1] K. Koh and T. Durst, *JOC* **59**, 4683 (1994).
[2] K. Koh, R. N. Ben, and T. Durst, *TL* **34**, 4473 (1993).

Paraformaldehyde.

Cleavage of oxime ethers.[1] The generation of carbonyl compounds in the presence of Amberlyst 15 in aqueous acetone proceeds at room temperature.

Synthesis of 5,6-dihydro-4 H-1,3-oxazines.[2] Acid-catalyzed depolymerization and condensation of the resulting formaldehyde with an amide gives the acyliminium ion, which is capable of undergoing a formal Diels–Alder reaction with alkenes.

Bromomethylation of arenes.[3] The introduction of up to three bromomethyl groups is achievable through varying the stoichiometry of paraformaldehyde and the reaction conditions.

Homologation of carbon chains. By a Grignard reaction with formaldehyde an organic halide is converted to the homologous alcohol. A preparation of the useful building block, 2-trimethylsilylacrolein,[4] is an example.

N-Halomethyl carboxamides. Chloromethylation of amides is accomplished in one step under anhydrous conditions by heating with $(HCHO)_n$ and Me_3SiCl in THF. These products are valuable synthetic intermediates; for example, they are converted to *N*-acyloxymethyl amides by reaction with sodium carboxylates.[5] A convenient preparation of *N*-bromomethyl imides is by warming the parent imides with $(HCHO)_n$, HBr, and HOAc at 70–80°C for several hours.[6]

[1] T. Sakamoto and Y. Kikugawa, *S* 563 (1993).
[2] A. R. Katritzky, I. V. Shcherbakova, R. D. Tack, and X.-Q. Dai, *T* **49**, 3907 (1993).
[3] A. W. van der Made and R. H. van der Made, *JOC* **58**, 1262 (1993).
[4] R. P. Hsung, *SC* **24**, 181 (1994).
[5] R. Moreira, E. Mendes, T. Calheiros, M. J. Bacelo, and J. Iley, *TL* **35**, 7107 (1994).
[6] R. C. Desai, R. P. Farrell, G.-H. Kuo, and D. J. Hlasta, *SL* 933 (1994).

Pentafluorobenzeneseleninic acid.

Allylic oxidation.[1] This novel oxidizing agent is prepared from C_6F_6 and NaSeH followed by exposing the product to O_2, O_3, and H_2O in sequence. It converts alkenes to enones, besides oxidizing alcohols to carbonyl compounds.

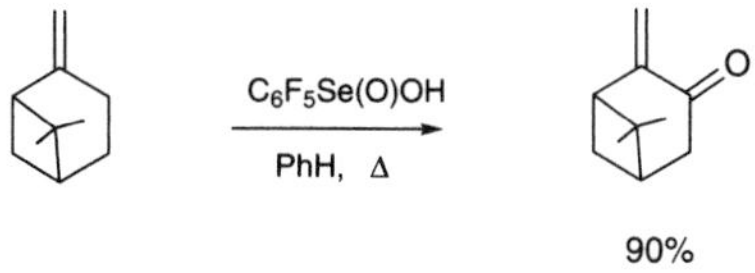

90%

[1] D. H. R. Barton and T.-L. Wang, *TL* **35**, 5149 (1994).

N,N,1,2,4-Pentamethyl-1,4-dihydronicotinamide.

Enantioselective reduction.[1] This compound mimics NADH and reduces α-keto esters and imino esters in the presence of $Mg(ClO_4)_2$. The hydroxyl products have an (*R*) configuration.

[1] J. P. Versleijen, M. S. Sanders-Hoven, S. A. Vanhommerig, J. A. Vekemans, and E. M. Meijer, *T* **49**, 7793 (1993).

Perfluorodialkyloxaziridines.

The reagents are available in two steps from commercial $(R_f)_3N$ on treatment with SbF_5 [to obtain $R_fN=C(F)R_f'$] and then with $ArCO_3H$.[1]

Oxyfunctionalization of hydrocarbons.[1] The insertion of an oxygen atom in tertiary C–H bonds by these neutral and stable reagents at room temperature proceeds smoothly.

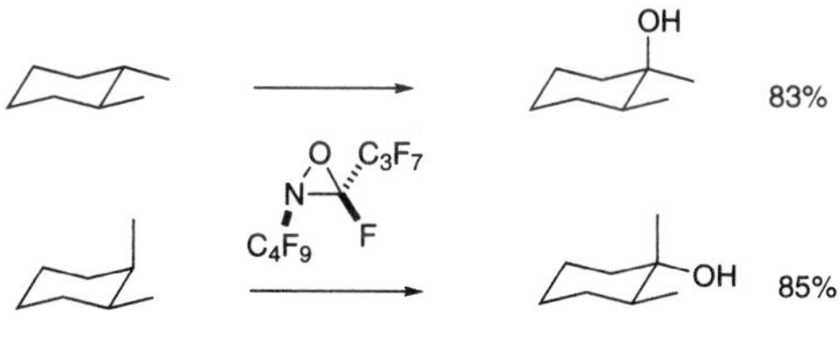

83%

85%

Oxidation of sulfides and sulfoxides.[2] The oxidation can be controlled by temperature. For example, benzenesulfenylmethyl azide is converted to the sulfoxide (96% yield) at −40°C, but further oxidation at −20°C gives the sulfone (93% yield).

[1] D. D. DesMarteau, A. Donadelli, B. Montanari, V. A. Petrov, and G. Resnati, *JACS* **115**, 4897 (1993).
[2] D. D. DesMarteau, V. A. Petrov, B. Montanari, M. Pregnolato, and G. Resnati, *JOC* **59**, 2762 (1994).

Periodic acid. **13**, 238–239; **16**, 292

Acetonide hydrolysis—cleavage. Terminal acetonides are selectively hydrolyzed and cleaved with H_5IO_6. Sometimes, a one-pot reaction can be accomplished.[1]

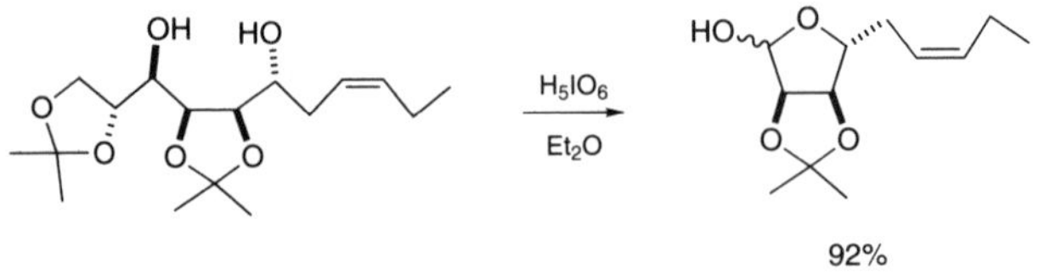

92%

Iodohydrins.[2] In the presence of $NaHSO_3$, hypoiodous acid is formed, which adds to alkenes. Hypobromous acid is similarly generated from $NaBrO_3$.

[1] W.-L. Wu and Y.-L. Wu, *JOC* **58**, 3586 (1993).
[2] H. Masuda, K. Takase, M. Nishio, A. Hasegawa, Y. Nishiyama, and Y. Ishii, *JOC* **59**, 5550 (1994).

Periodinane of Dess and Martin. 14, 254; 15, 252; 16, 271–272; 17, 271

Alcohol oxidation. Water has a rate-enhancing effect on the oxidation.[1] The oxidation of 2-alkenylcyclopropylmethanol gives dihydrooxepins directly by way of a Claisen rearrangement.[2] The cyclobutane analogs also undergo ring-expanding rearrangement.

[1] S. D. Meyer and S. L. Schreiber, *JOC* **59**, 7549 (1994).
[2] R. K. Boeckman, Jr., M. D. Shair, J. R. Vargas, and L. A. Stolz, *JOC* **58**, 1295 (1993).

Phase-transfer catalysts. 13, 239–240; 15, 252–253

Cyclophosphazenic polypodands.[1] Obtained as a mixture from the commercial Brij 30, these cheap and efficient catalysts are comparable to most of the common phase-transfer catalysts.

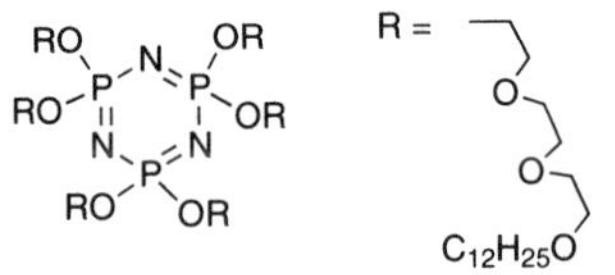

(1)

Perfluorocarbon fluids.[2] A commercially available liquid (b.p. 97°C) is successfully used in 9 different reactions.

Alkylation of heterofunctionalities. Etherification of alcohols can be conducted in the absence of solvent using a polyether.[3] The *O*-alkylation of 1-perfluoroalkyl-2-fluoroethanols is accompanied by dehydrofluorination; thus the products are enol ethers.[4] Selective *O*-alkylation of *o*-aminophenols is observed,[5] and an efficient method for the synthesis of triaryl phosphates from sodium phenolates and

POCl₃ also involves the phase-transfer technique,[6] employing both BnNEt₃Cl and dibenzo[18]crown-6 as catalyst.

$$\text{F}\diagdown\text{CH(C}_8\text{H}_{17}\text{)OH} \xrightarrow[\text{Bu}_4\text{NHSO}_4]{\text{50\% NaOH / CH}_2\text{Cl}_2} \text{product}$$

51%

A remarkable *t*-butyl ester formation of *N*-protected amino acids is based on solid–liquid phase transfer (*t*-BuBr, K₂CO₃, BnNEt₃Cl, MeCONMe₂).[7] Mixed dialkyl peroxides are also readily synthesized in good yields and under mild conditions.[8] Although the preparation of alkyl nitrates from tosylates under phase-transfer conditions requires somewhat higher temperatures and long reaction times, the yields are still useful.[9] Thus 3-methyl-3-nitratomethyloxetane has been obtained in 71% yield.

N-Alkylation of azoles with propargyl bromide is interesting. Pyrazole forms the normal product at room temperature, but it gives *N*-allenylpyrazole at 40°C using a stronger base (KOH vs. K₂CO₃).[10] *p*-Nitrophenylazoles may be obtained from an ultrasonic reaction of the azoles with *p*-fluoronitrobenzene in the presence of KOH and Bu₄NBr without solvent.[11]

Sulfides are readily formed from thiols on reaction with halides (including glycosyl halides)[12] or epoxides[13] in the presence of a phase-transfer catalyst. In aqueous micelles of cetyltrimethylammonium bromide the reaction of Na₂S with haloalkanes generates symmetrical sulfides.[14] Alkynyl phenyliodonium salts are suitable electrophiles for *S*-alkylation of *O,O′*-dialkyl phosphorodithiolates.[15]

C-Alkylations. The alkylation of ethyl phenylthioacetate is aided by simultaneous application of microwave radiation.[16] Sometimes unexpected products appear from alkylation because secondary reactions such as dehydrochlorination[17] and hydrolysis[18] take place.

Cyclopropanation. A list of carbenes generated under phase-transfer conditions includes :CF₂ (from CHBr₃–CBr₂F₂),[19] :C=C=CHCl (from HCCCHCl₂),[20] TolS-C-SO₂Me (from TolSCHCl[SO₂Me]),[21] and :CHSePh (from ClCHSePh).[22] The last case also employs ultrasound.

1,1-Dihalo-2-aryltellurocyclopropanes are the cycloadducts of dihalocarbenes and alkenyl tellurides.[23] Tetramethylammonium salts are highly selective catalysts for *gem*-dichlorocyclopropane synthesis from electrophilic alkenes.[24]

Condensations. The transformation of acid chlorides to anhydrides[25] by (Ph₃P)₂CoCl₂ and the direct synthesis of unsymmetrical carbonic esters[26] from alcohols, CO₂, and alkyl halides by solid–liquid phase-transfer reactions are most expedient.

Additions and eliminations. Some arylpropenes are found to give adducts with arylacetonitriles, due to isomerization and addition.[27] Asymmetric induction by a chiral micellar system (prepared from a quaternary ammonium salt of ephedrine) has been observed in the hydratomercuration of alkenes.[28] Cyclization of 2-alkynylanilines in a two-phase system to give indoles[29] is catalyzed by $PdCl_2$ and takes place in the presence of Bu_4NCl.

67-98%

In a convenient route to cyclic ketene acetals[30] dehydrobromination is conducted in the presence of Bu_4NBr while being irradiated with ultrasound.

Oxidations. Allylic and benzylic alcohols are converted to the aldehydes with great selectivity by a Cr-mediated oxidation with sodium percarbonate.[31] The phase-transfer oxidation of anilines with $KMnO_4$ results in azoarenes.[32]

Amide degradation.[33] In a liquid triphasic system made up of NaOCl, NaBr, a phase-tranfer agent, Na_3PO_4, benzene, and water, primary amides are degraded to nitriles of one fewer carbon atom. After Hofmann rearrangement the amine products undergo bromination and eventually elimination.

48 - 68%

[1] A. Gobbi, D. Landini, A. Maia, G. Delogu, and G. Podda, *JOC* **59**, 5059 (1994).
[2] D.-W. Zhu, *S* 953 (1993).
[3] B. Abribat, Y. Le Bigot, and A. Gaset, *SC* **24**, 2091 (1994).
[4] C. Driss, M. M. Chaabouni, and A. Baklouti, *JFC* **67**, 137 (1994).
[5] R. Carrillo and E. Diez-Barra, *SC* **24**, 945 (1994).
[6] A. R. Kore, A. D. Sagar, and M. M. Salunkhe, *BSCB* **103**, 85 (1994).
[7] P. Chavallet, P. Garrouste, B. Malawska, and J. Martinez, *TL* **34**, 7409 (1993).
[8] C. Navarro, M. Degueil-Castaing, D. Colombani, and B. Maillard, *SC* **23**, 1025 (1993).
[9] J. R. Hwu, K. A. Vyas, H. M. Patel, C.-H. Lin, and J. C. Yang, *S* 471 (1994).
[10] E. Diez-Barra, A. de la Hoz, A. Loupy, and A. Sanchez-Migallon, *H* **38**, 1367 (1994).
[11] M. L. Cerrada, J. Elguero, J. de la Fuente, C. Pardo, and M. Ramos, *SC* **23**, 1947 (1993).
[12] S. Cao, S. J. Meunier, F. O. Andersson, M. Letellier, and R. Roy, *TA* **5**, 2303 (1994).
[13] A. K. Maiti and P. Bhattacharyya, *T* **50**, 10483 (1994).
[14] C. D. Muduliar and S. H. Mashraqui, *JCR(S)* 174 (1994).
[15] Z.-D. Liu and Z.-C. Chen, *JOC* **58**, 1924 (1993).
[16] R.-H. Deng and Y.-Z. Jiang, *SC* **24**, 1917 (1994).
[17] A. Jonczyk and T. Kulinski, *SC* **23**, 1801 (1993).
[18] C. Sarangi and Y. R. Rao, *JCR(S)* 392 (1994).
[19] P. Balcerzak and A. Jonczyk, *JCR(S)* 200 (1994).
[20] K. N. Shavrin, I. B. Shvedova, and O. M. Nefedov, *MC* 50 (1993).

[21] K. Schank, A.-M. A. Abdel Wahab, S. Bulger, P. Eigen, J. Jager, and K. Jost, *T* **50**, 3721 (1994).
[22] C. C. Silveira, A. L. Braga, and G. L. Fiorin, *SC* **24**, 2074 (1994).
[23] X. Huang and S.-H. Jiang, *SC* **23**, 431 (1993).
[24] M. Fedorynski, W. Ziolkowska, and A. Jonczyk, *JOC* **58**, 6120 (1993).
[25] J.-X. Wang, Y. Hu, and W. Cui, *SC* **24**, 3261 (1994).
[26] S. Oi, Y. Kuroda, S. Matsuno, and Y. Inoue, *NKK* 985 (1993).
[27] W. Lasek and M. Makosza, *S* 780 (1993).
[28] Y. Zhang, W. Bao, and H. Dong, *SC* **23**, 3029 (1993).
[29] S. Cacchi, V. Carnicelli, and F. Marinelli, *JOMC* **475**, 289 (1994).
[30] A. Diaz-Ortiz, E. Diez-Barra, A. de la Hoz, and P. Prieto, *SC* **23**, 1935 (1993).
[31] J. Muzart, A. A. Ajiou, and S. Ait-Mohand, *TL* **35**, 1989 (1994).
[32] M. Hedayatullah and A. Roger, *BSCB* **102**, 59 (1993).
[33] J. Correia, *S* 1127 (1994).

Phenyl(cyano)iodine(III) tosylate.

Alkenyl(phenyl)iodonium salts.[1] These salts are prepared from alkenylstannanes by reaction with PhI(CN)OTs in CH_2Cl_2 at $-23°C$.

[1] R. J. Hinkle and P. J. Stang, *S* 313 (1994).

Phenyliodine(III) bis(trifluoroacetate). 13, 241–242; 14, 257; 15, 257–258; 16, 274–275

Cleavage of thioglycosides.[1] Oxidation with the hypervalent iodine reagent facilitates replacement of the thio group. *O*-Glycosylation occurs via the sulfonium ions but not the oxonium ions.

Dearomatization of p-substituted phenols and derivatives. 2,5-Cyclohexadienones are the major products in the reaction. 4-Halo derivatives or *p*-quinols are formed in the presence of pyridinium poly(hydrogen fluoride),[2] aqueous $NaCl$,[3] or water[4] alone as additive besides solvents.

R = Me, H

α-Keto esters.[5] Phenylhydrazones of α-keto esters undergo oxidative cleavage at 0°C in aqueous MeCN. The yields are good (7 examples, 74–98%).

Radical alkylation of heteroaromatic bases.[6] Carboxylic acids undergo decarboxylation to generate free radicals, which can be captured by heterocycles. Thus 2-(1-adamantyl)-4-cyanopyridine is obtained in 88% yield from 1-adamantylcarboxylic acid and 4-cyanopyridine.

[1] L. Sun, P. Li, and K. Zhao, *TL* **35**, 7147 (1994).
[2] O. Karam, J.-C. Jacquesy, and M.-P. Jouannetaud, *TL* **35**, 2541 (1994).
[3] O. Karam, M.-P. Jouannetaud, and J.-C. Jacquesy, *NJC* **18**, 1151 (1994).
[4] A. McKillop, L. McLaren, and R. J. K. Taylor, *JCS(P1)* 2047 (1994).
[5] D. H. R. Barton, J. C. Jaszberenyi, and T. Shinada, *TL* **34**, 7191 (1993).
[6] H. Togo, M. Aoki, T. Kuramochi, and M. Yokoyama, *JCS(P1)* 2417 (1993).

Phenyliodine(III) diacetate. **13**, 241–243; **14**, 258–259; **15**, 258; **16**, 275–276; **17**, 280–281

Diaryliodonium triflates.[1] An improved preparative method for aryl phenyliodonium triflates (10 examples, 74–98% yield) involves reaction of arenes with this reagent in the presence of triflic acid.

Oxidation of phenols.[2] *p*-Substituted phenols give *p*-quinols. Catechols and hydroquinones are oxidized to the corresponding quinones.

Cleavage of azones. The regeneration of ketones from semicarbazones[3] and tosylhydrazones[4] is rapid and high yielding. It is performed at room temperature in aqueous MeCN. The acylhydrazones of *o*-hydroxyaryl ketones are transformed into 1,2-diacylbenzenes.[5]

Methyl carbamates.[6] Primary amides undergo a Hofmann rearrangement at 0°C to room temperature. The isocyanates react with MeOH in situ.

2-Aryl-4-quinolones and 3-hydroxyflavones. The quinolones are obtained from 2,3-dihydro compounds and flavones are hydroxylated at C-3.[8] KOH is added to the reaction medium.

Phenyl alkynyl selenides.[9] Phenylselenylation of 1-alkynes is easily accomplished with the PhI(OAc)₂–PhSeSePh combination.

Bridgehead functionalization.[10] A method for introducing an iodine atom at a bridgehead adjacent to a ketone is by oxidation of the derived acetal such as (**1**) with PhI(OAc)₂–I₂.

Diaryltelluronium diacetates.[11] This carboxylate group transfer reaction is common to various phenyliodine(III) dicarboxylates. Accordingly, not only the diacetates of diaryltellurides can be prepared in this manner.

[1] T. Kitamura, J.-i. Matsuyuki, and H. Taniguchi, *S* 147 (1994).
[2] A. Pelter and S. M. A. Elgendy, *JCS(P1)* 1891 (1993).
[3] D. W. Chen and Z. C. Chen, *S* 773 (1994).
[4] H. Zeng and Z.-C. Chen, *SC* **23**, 2497 (1993).
[5] R. M. Moriarty, B. A. Berglund, and M. S. C. Rao, *S* 318 (1993).
[6] R. M. Moriarty, C. J. Chany, R. K. Vaid, O. Prakash, and S. M. Tuladhar, *JOC* **58**, 2478 (1993).
[7] O. Prakash, D. Kumar, R. K. Saini, and S. P. Singh, *SC* **24**, 2167 (1994).
[8] O. Prakash, S. Pahuja, and M. P. Tanwar, *IJC(B)* **33B**, 272 (1994).
[9] M. Tingoli, M. Tiecco, L. Testaferri, and R. Balducci, *SL* 211 (1993).
[10] U. P. Spitz and P. E. Eaton, *ACIEE* **33**, 2220 (1994).
[11] Z.-D. Liu and Z.-C. Chen, *HC* **3**, 559 (1992).

Phenyliodine(III) dichloride–lead(II) thiocyanate.

p-Thiocyanation of phenols.[1] The substitution occurs in CH_2Cl_2 at 0°C. Usually the reaction is complete within 1 h.

[1] Y. Kita, T. Okuno, M. Egi, K. Iio, Y. Takeda, and S. Akai, *SL* 1039 (1994).

Phenyliodine(III) dimethoxide.

(Arenesulfonyliminoiodo)benzenes.[1] Arenesulfonamides condense with PhI-$(OMe)_2$ with elimination of methanol. The reaction is conducted in MeOH in the presence of 3A molecular sieves, and then in CH_2Cl_2.

[1] G. Besenyei, S. Nemeth, and L. I. Simandi, *TL* **34**, 6105 (1993).

Phenylmanganese *N*-methylanilide.

Kinetic enolization.[1] Ketones form kinetic enolates with this manganese amide, which is prepared from $MnCl_2 \cdot 2LiCl$, Ph(Me)NLi, and PhLi in ether. Manganese dialkylamides are much less efficient.

[1] G. Cahiez, B. Figadere, and P. Clery, *TL* **35**, 6295 (1994).

9-Phenylphosphabicyclo[4.2.1]nonane.

(E)-Selective Wittig reactions. Ylides prepared from the bridged tetrahydrophosphole (**1**) react with aldehydes in an (*E*)-selective fashion.

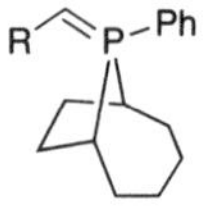

$$R \diagup\!\!\!\!\diagdown P\!\!-\!Ph$$

(1)

[1] E. Vedejs and M. J. Peterson, *JOC* **58**, 1985 (1993).

Phenylthiomethyl isocyanide.

Vicarious nucleophilic substitutions.[1]   The reagent is useful for introducing an isocyanomethyl group to an *o*-position of nitroarenes in the presence of *t*-BuOK in DMF. The products are readily converted into formamides or amines.

[1] M. Makosza, A. Kinowski, and S. Ostrowski, *S* 1215 (1993).

Phenyl tosyl selenide.

Radical cyclizations.[1]   Homolysis of PhSeTs (initiated by AIBN) in refluxing benzene in the presence of a diene or enyne leads to difunctional cyclic products.

[1] J. E. Brumwell, N. S. Simpkins, and N. K. Terrett, *T* **50**, 13533 (1994).

Phenyl *p*-toluenesulfinylmethyl sulfone.

Alkenyl sulfones.[1]   The sulfonyl reagent condenses with aldehydes (catalyzed by piperidine at $-20°C$).

[1] J. C. Carretero and E. Dominguez, *JOC* **58**, 1596 (1993).

Phenyl(2,2,2-trifluoroethyl)iodonium triflate.

N-Trifluoroethylation of amines.[1]   The usefulness of the reagent is shown by the selective mono-*N*-alkylation of amino alcohols at room temperature.

[1] V. Montanari and G. Resnati, *TL* **35**, 8015 (1994).

N-Phenyltrifluoromethanesulfonimide.

Alkynes.[1]   A method for synthesizing alkynes involves enoltriflylation of α-sulfinyl ketones and elimination. Both steps are done in one operation using LDA, PhNTf$_2$ in the presence of HMPA.

[1] T. Satoh, N. Itoh, S. Watanabe, H. Matsuno, and K. Yamakawa, *CL* 567 (1994).

Phenyl vinyl sulfoxide.

Ene reactions.[1]   With $(CF_3CO)_2O$ as an activating agent, the sulfoxide undergoes electrophilic ene-type reaction with alkenes.

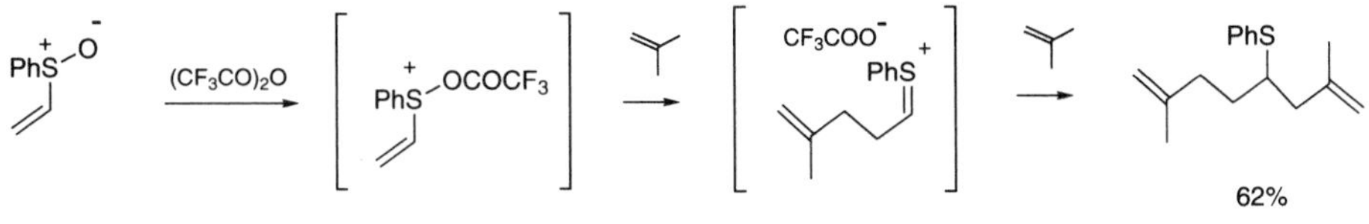

[1] J. Harvey, M.-H. Brichard, and H. G. Viehe, *JCS(PI)* 2275 (1993).

Phosphine.

Addition to alkenes. PH_3 is generated from red phosphorus and KOH in water at 50°C. In alkaline solution it adds to electron-deficient alkenes such as acrylonitrile to give tertiary phosphines.[1] It also forms 1:2 adducts with styrenes.[2]

[1] D. Semenzen, G. Etemad-Moghadem, D. Albouy, and M. Koenig, *TL* **35**, 3297 (1994).
[2] B. A. Trofimov, L. Brandsma, S. N. Arbutzova, S. F. Malysheva, and N. K. Gusarov, *TL* **35**, 7647 (1994).

Phosphorus(V) oxide–acetonitrile.

Mukaiyama–Michael addition.[1]   The reaction of silyl ketene acetals with enones is catalyzed with a species formed from P_4O_{10} and MeCN.

[1] V. Berl, G. Helmchen, and S. Preston, *TL* **35**, 233 (1994).

Phosphorus oxychloride. 13, 249; 15, 267; 17, 288

Isocyanates.[1]   Primary amines form unstable carbamic acids in the presence of CO_2 and a base. On treatment with $POCl_3$ the acids afford isocyanates.

5-Chloro-3-furaldehydes.[2]   γ-Keto acids undergo cyclization, chlorodehydration, and formylation with $POCl_3$–DMF (Vilsmeier reagent).

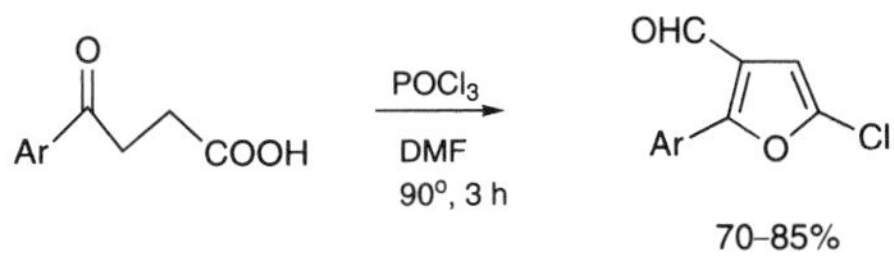

[1] T. E. Waldman and W. D. McGhee, *CC* 957 (1994).
[2] M. Venugopal, B. Balasundaran, and P. T. Perumal, *SC* **23**, 2593 (1993).

Pivaloyl chloride.

THF ring opening.[1] Pivaloyl chloride and acetyl chloride induce ring opening of 2-substituted tetrahydrofurans in different fashions. The S_N2 pathway is favored in the reaction with *t*-BuCOCl.

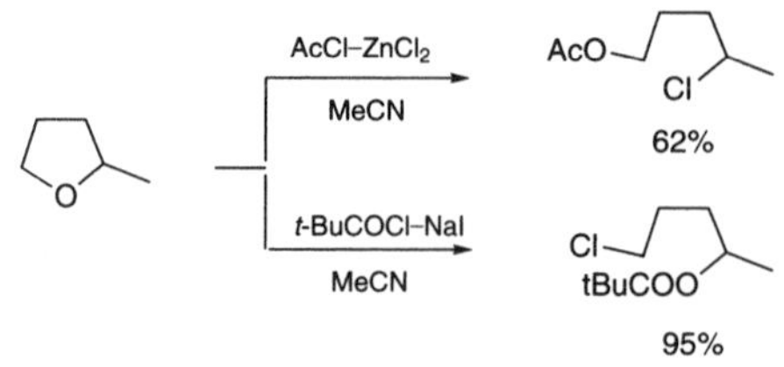

[1] P. Mimero, C. Saluzzo, and R. Armouroux, *TL* **35**, 1553 (1994).

(1S,3R)-2-Pivaloyl-3-methylbenzoisothiazoline 1-oxide.

Asymmetric sulfinylation.[1] The reaction of ketone enolates with this reagent (**1**) gives chiral sulfoxides. With the chelation assistance of the sulfinyl group the reduction of the ketones is stereoselective, and a subsequent thermolysis leads to allylic alcohols with an (*R*) configuration.

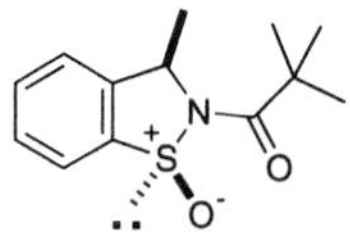

(1)

[1] I. D. Linney, H. Tye, M. Wills, and R. J. Butlin, *TL* **35**, 1785 (1994).

Polymethylhydrosiloxane.

Reduction of esters. Alcohols are produced from Zr-[1] and Ti-[2] catalyzed reduction of esters at temperatures ranging from 0°C to ambient.

[1] K. J. Barr, S. C. Berk, and S. L. Buchwald, *JOC* **59**, 4323 (1994).
[2] S. W. Breeden and N. J. Lawrence, *SL* 833 (1994).

Polyphosphoric acid.

N-Acylanilines.[1] When electron-rich arenes are treated with a carboxylic acid, hydroxylamine hydrochloride in PPA, several reactions occur in sequence. These are the Friedel–Crafts acylation, oxime formation, and Beckmann rearrangement. Anilides are obtained in 45–95% overall yields (11 examples).

4-Pyridones.[2]   The condensation of dianions of acyclic β-enamino ketones with esters followed by acid treatment may lead to 4-pyranones (with HCl) or 4-pyridones (with PPA, 70°C).

[1] T. Cablewski, P. A. Gurr, K. D. Raner, and C. R. Strauss, *JOC* **59**, 5814 (1994).
[2] G. Bartoli, M. Bosco, C. Cimarelli, R. Dalpozzo, G. Guercio, and G. Palmieri, *JCS(P1)* 2081 (1993).

Polyphosphoric acid trimethylsilyl ester. 15, 269

Cyclic conjugated imines.[1]   Cyclization of *N*-alkenyl amides in which the double bonds are properly distanced to the carbonyl group is effected by this reagent. It is prepared by heating P_4O_{10} (0.1 mol) and $(Me_3Si)_2O$ (30 mL) in CCl_4 (70 mL) for 1.5 h.

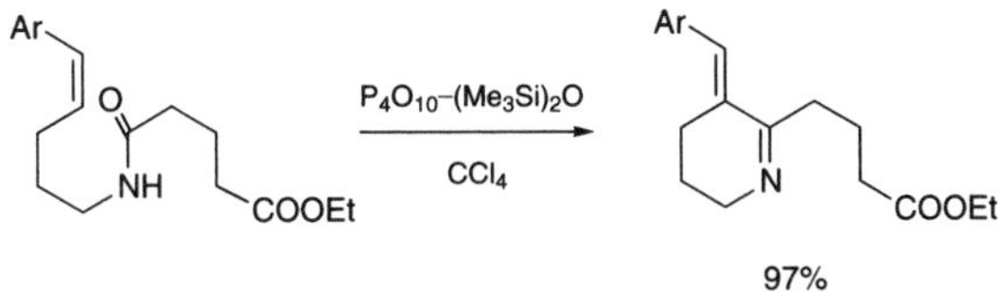

[1] A. L. Marquart, B. L. Podlogar, E. W. Huber, D. A. Demeter, N. P. Neet, H. J. R. Weintraub, and M. R. Angelastro, *JOC* **59**, 2092 (1994).

Poly(4-vinylpyridinium) poly(hydrogen fluoride).

Solid equivalent of HF.[1]   The extensive reactions mediated by $pyH(HF)_n$ can be conducted with the polymer-bound reagent.

[1] G. A. Olah, X.-Y. Li, Q. Wang, and G. K. S. Prakash, *S* 693 (1993).

Potassium. 15, 269

Cleavage of N-substituted 2-aryl-1,3-oxazolidines.[1]   This reduction to afford *N*-(2-hydroxycthyl)-*N*-benzylamines occurs at room temperature via α-amino carbanions.

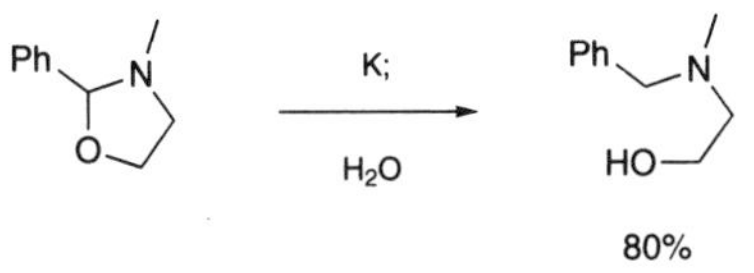

[1] U. Azzena, G. Melloni, and C. Nigra, *JOC* **58**, 6707 (1993).

Potassium *t*-butoxide. **13**, 252–254; **15**, 271–272; **17**, 289–290

Eliminations. Treatment of *o*-alkyl *N'*-(*t*-butylthio)azoarenes with *t*-BuOK accomplishes cyclization and delivers 1*H*-indazoles.[1] Unsaturated α-amino nitriles are obtained[2] on alkylation of 2-(*N*-methylanilino)-2-benzenesulfenylacetonitrile, due to elimination of benzenethiol in situ.

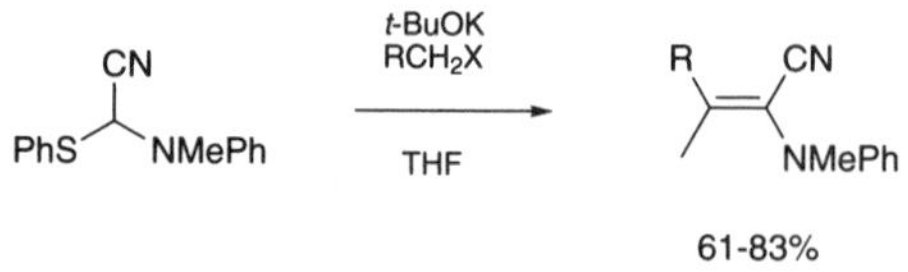

61-83%

Allylic phosphonates. A deconjugative isomerization of vinylic phosphonates[3] is catalyzed by *t*-BuOK in DMSO.

Heterocyclizations. Hemiacetals derived from 5-hydroxy-2-alkenoic esters under basic conditions can cyclize to form 1,3-dioxanes.[4] This method is suitable for the synthesis of *syn*-1,3-diol units. The base-catalyzed reaction of 2-hydroxymethyl 1,3-enynes and the thiol analogs provide furans[5] and thiophenes.[6] 2,3-Disubstituted furans are accessible from isomeric enynes,[7] by way of an S_N2' reaction.

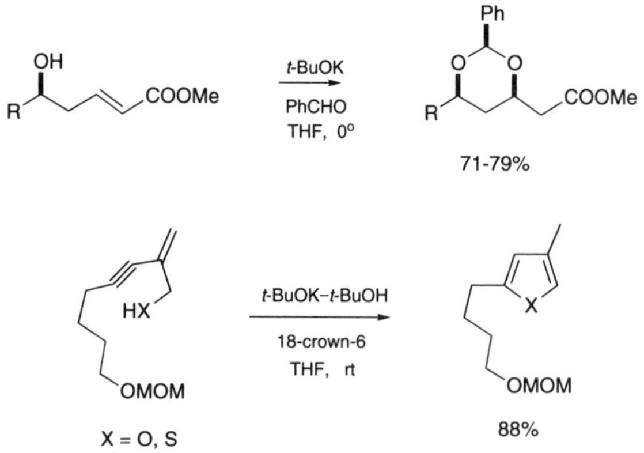

71-79%

X = O, S

88%

Hydrodebromination of 1-bromo-1-chlorocyclopropanes.[8] This chemoselective reduction is actually promoted by dimsyl anion, which is present when *t*-BuOK is dissolved in DMSO.

[1] C. Dell'Erba, M. Novi, G. Petrillo, and C. Tavani, *T* **50**, 3529 (1994).
[2] C.-C. Chen, S.-T. Chen, T.-H. Chuang, and J.-M. Fang, *JCS(P1)* 2217 (1994).
[3] J. J. Kiddle and J. H. Babler, *JOC* **58**, 3572 (1993).
[4] D. A. Evans and J. A. Gauchet-Prunet, *JOC* **58**, 2446 (1993).
[5] J. A. Marshall and W. J. DuBay, *JOC* **58**, 3435 (1993).

[6] J. A. Marshall and W. J. DuBay, *SL* 209 (1993).
[7] J. A. Marshall and C. E. Bennett, *JOC* **59**, 6110 (1994).
[8] G. W. Wijsman, W. H. de Wolf, and F. Bickelhaupt, *RTC* **113**, 53 (1994).

Potassium dicarbonyl(cyclopentadienyl)ferrate.

Tishchenko reaction.[1]   Aromatic aldehydes undergo redox coupling to give benzyl aroates in THF. In some cases the yield can be as high as 99%.

Pinacol silyl ethers.[2]   In the presence of Me_3SiCl the reaction (vide supra) takes another path, as the adducts are trapped as α-siloxybenzyliron complexes. At a higher temperature (60°C in benzene) demetallative coupling occurs.

[1] T. Ohishi, Y. Shiotani, and M. Yamashita, *OM* **13**, 4641 (1994).
[2] R. M. Vargas and M. M. Hossain, *TL* **34**, 2727 (1993).

Potassium diphenylphosphide.

Functionalized arylphosphines.[1] Ph_2PK is highly nucleophilic, and its reaction with some aryl fluorides gives substituted triphenylphosphines.

[1] S. J. Coote, G. J. Dawson, C. G. Frost, and J. M. J. Williams, *SL* 509 (1993).

Potassium fluoride. 13, 256–257; 15, 272

As base.   In the absence of solvent KF promotes the condensation of trimethylsilylacetonitrile with aldehydes to give the β-cyanohydrin silyl ethers.[1] With microwave as energy source the reaction is complete within minutes. KF in the presence of 18-crown-6 successfully effects the Michael reaction[2] between α-acetamidomalonic esters and propynoic esters also.

An interesting olefination of electron-deficient ketones[3] by alkanesulfonyl halides using KF probably involves sulfene formation. The removal of the fluorenylmethoxycarbonyl (Fmoc) group[4–6] from amino acids and peptides with KF/18-crown-6 does not affect methyl, ethyl, *t*-butyl, benzyl, and *p*-methoxybenzyl groups. Actually transcarbamoylation (into Boc and Cbz groups) can be accomplished in one flask, by presenting the proper electrophiles in the reaction mixture.[5]

As nucleophile.   An enantioselective synthesis of α-fluoro esters[7] from the corresponding sulfonates is based on an essentially pure S_N2 reaction using KF in $HCONH_2$. The replacement of activated aromatic nitro groups requires more vigorous conditions (DMSO, 130°C).[8]

The silaphilicity of the fluoride ion has been used to advantage in the β-lactam synthesis from silyl ketene acetals and imines,[9] and in the rearrangement of chloromethylsilanes.[10]

[1] R. Latouche, F. Texier-Boullet, and J. Hamelin, *BSCF* 535 (1993).
[2] V. Tolman and P. Sedmera, *CCCC* **58**, 1430 (1993).
[3] B. S. Nader, J. A. Cordova, K. E. Reese, and C. L. Powell, *JOC* **59**, 2898 (1994).
[4] J. Jiang, W.-R. Li, and M. M. Joullie, *SC* **24**, 187 (1994).
[5] W.-R. Li, J. Jiang, and M. M. Joullie, *TL* **34**, 1413 (1993).
[6] W.-R. Li, J. Jiang, and M. M. Joullie, *SL* 362 (1993).
[7] E. Fritz-Langhals, *TA* **5**, 981 (1994).
[8] A. J. Beaumont, J. H. Clark, and N. A. Boechat, *JFC* **63**, 25 (1993).
[9] F. Texier-Boullet, R. Latouche, and J. Hamelin, *TL* **34**, 2123 (1993).
[10] J. J. Eisch and C. S. Chiu, *HC* **5**, 265 (1994).

Potassium fluoride–alumina. 16, 282

As base. Numerous reactions that are initiated by deprotonation have been conducted with KF–Al$_2$O$_3$. These include the synthesis of diaryl ethers, amines, and sulfides by nucleophilic aromatic substitutions,[1] N-alkylation of 2,4-dinitrophenyl-hydrazones,[2] condensation of 3-phenylisoxazol-5-one with aldehydes,[3] and ring closure of N-(ω-chloroalkyl) carboxamides to afford 1,3-oxazolines and 1,3-oxazines.[4]

A sequence of Michael addition, iodination, and intramolecular alkylation is involved in the formation of 2-substituted cyclopropane-1,1-diphosphonate esters[5] from tetraethyl methylenediphosphonate. The reagents are KF–Al$_2$O$_3$ and I$_2$.

55% (R=CN)

[1] E. A. Schmittling and J. S. Sawyer, *JOC* **58**, 3229 (1993).
[2] K. Thangaraj and L. R. Morgan, *SC* **24**, 2063 (1994).
[3] D. Villemin, B. Martin, and B. Garrigues, *SC* **23**, 2251 (1993).
[4] M. A. Mitchell and B. C. Benicewicz, *S* 675 (1994).
[5] D. Villemin, F. Thibault-Starzyk, and M. Hachemi, *SC* **24**, 1425 (1994).

Potassium hexamethyldisilazide. 13, 257; 16, 282–283

Emmons–Wadsworth condensation. Using chiral phosphonoacetates, the reaction with *meso* aldehydes[1] and ketones[2] leads mainly to one type of product.

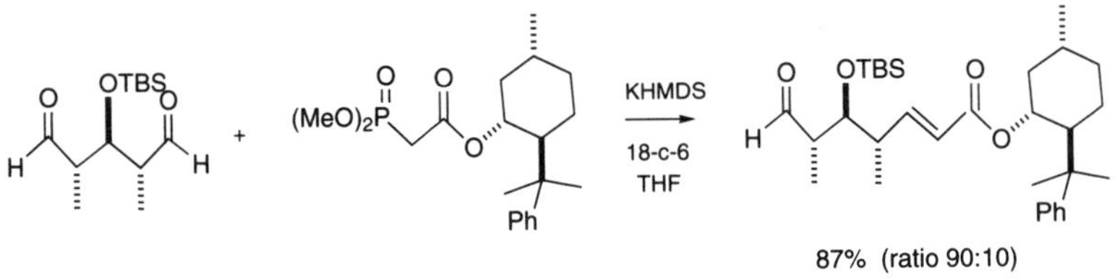

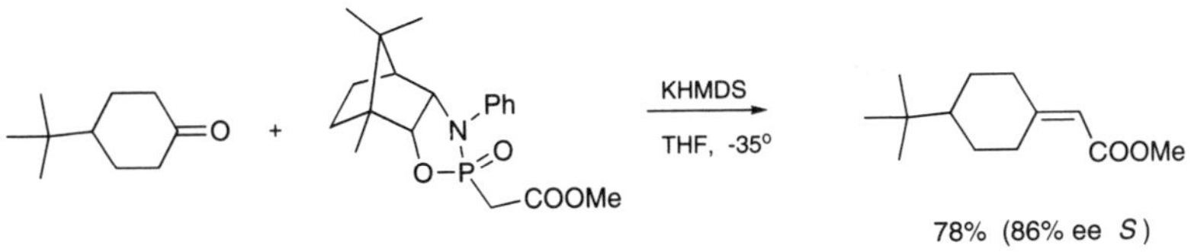

87% (ratio 90:10)

78% (86% ee *S*)

***Cyclopropylation.*[3]**   Introduction of a cyclopropyl group to an α-position of a ketone in one operation is not easy. Surprisingly, a direct alkylation with π-allylpalladium complexes to give such results has been observed. Nucleophilic attack at the central carbon of the alkylating agents is involved.

[1] N. Kann and T. Rein, *JOC* **58**, 3802 (1993).
[2] S. E. Denmark and I. Rivera, *JOC* **59**, 6887 (1994).
[3] A. Wilde, A. R. Otte, and H. M. R. Hoffmann, *CC* 615 (1993).

Potassium hydride. **13**, 257–258; **14**, 265; **17**, 290

***Alkylation of sulfoximines.*[1]**   Deprotonation of sulfoximines with KH in DME and addition of primary alkyl bromides together with Bu$_4$NBr complete the alkylation. Most of the solid–liquid phase-transfer reactions proceed in the 90% yield range.

***1,3-Shift of nonenolizable β-ketoesters.*[2]**   The very efficient ester group shift induced by KH–crown ether at room temperature proceeds via cyclobutanone intermediates. β-Cyano ketones are decomposed under these conditions, while β-keto sulfones do not react.

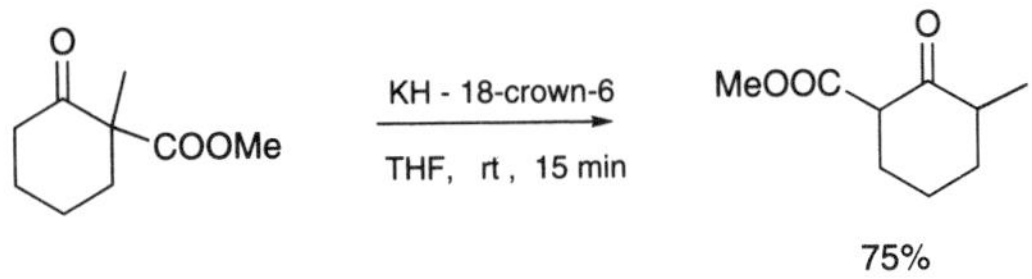

75%

***β-Nitro ethers.*[3]**   The Michael addition of alkali metal alkoxides to nitroalkenes is strongly dependent on the metal ion. Lithium alkoxides give only moderate yields due to competing reactions. Potassium and sodium alkoxides are far superior (yields in the 70–96% range) as dimerization and trimerization of nitroalkenes (even nitroethene) are not detected.

[1] C. R. Johnson and O. M. Lavergne, *JOC* **58**, 1922 (1993).
[2] A. Habi and D. Gravel, *TL* **35**, 4315 (1994).
[3] J. L. Duffy, J. A. Kurth, and M. J. Kurth, *TL* **34**, 1259 (1993).

Potassium 9-(*O*-[1,2-isopropylidene-5-deoxy-D-xylofuranosyl])-9-boratabicyclo[3.3.1]nonane.

Reduction of α-keto acetals. Enantioselective reduction by this borohydride (**1**) in THF at −78°C gives alcohols of (*S*)-configuration with high enantiomer excess.

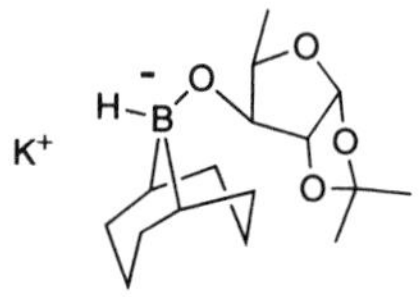

(1)

[1] B. T. Cho and Y. S. Chun, *TA* **5**, 1147 (1994).

Potassium monoperoxysulfate. 13, 259; 14, 267; 15, 274–275; 16, 285

Oxidation of sulfides and sulfoxides.[1] The oxidation with Oxone® is controllable by reaction temperature and time. For example, *p*-methylthiobenzoic acid is oxidized to the sulfoxide at temperature below 5°C in 5 min, and to the sulfone at room temperature while extending the reaction time to 1 h. Both reactions are carried out in aqueous acetone in the presence of NaHCO$_3$ and NaOH.

Epoxidation. Oxone® is used to generate dioxirane from a ketone added to the reaction medium. Such dioxiranes epoxidize alkenes stereoselectively.[2] 2-Cyclohexenol gives two epoxy alcohols in a ratio of 77:23 (*trans:cis*).

[1] K. S. Webb, *TL* **35**, 3457 (1994).
[2] M. Kurihara, S. Ito, N. Tsutsumi, and N. Miyata, *TL* **35**, 1577 (1994).

Potassium nitrate.

Alkyl nitrates.[1] A convenient preparation of RONO$_2$ free of RONO uses KNO$_3$ and boron trifluoride hydrate as nitrating agent in CH$_2$Cl$_2$.

[1] G. A. Olah, Q. Wang, X.-y. Li, and G. K. S. Prakash, *S* 207 (1993).

Potassium permanganate. **13**, 258–259; **14**, 267; **15**, 273–274

Cleavage of C=N and C=C bonds. Cleavage of ketoximes[1] occurs with $KMnO_4$ in aqueous MeCN at room temperature. Calixarenes facilitate C=C cleavage by $KMnO_4$ in aqueous CH_2Cl_2.[2]

[1] A. Wali, P. A. Ganeshpure, and S. Satish, *BCSJ* **66**, 1847 (1993).
[2] E. Nomura, H. Taniguchi, and Y. Otsuji, *BCSJ* **67**, 309 (1994).

Potassium permanganate–alumina.

Cleavage of C=C bonds.[1] 1-Alkenes and 1,2-disubstituted alkenes give aldehydes without overoxidation.

Selective oxidation of arenes.[2] Substituted arenes are selectively oxidized at a benzylic position.

[1] D. G. Lee, T. Chen, and Z. Wang, *JOC* **58**, 2918 (1993).
[2] D. Zhao and D. G. Lee, *S* 915 (1994).

Potassium permanganate–copper(II) sulfate. **16**, 283–284

Degradation of sugars.[1] The new C–C bond cleavage of 5-oxygenated furanosides leads to γ-lactones. The presence of $CuSO_4$ is essential.

[1] S. B. Mandal, B. Achari, and P. P. G. Dastidar, *TL* **34**, 1979 (1993).

Potassium permanganate–montmorillonite.

Aromatization.[1] Hantzsch 1,4-dihydropyridines undergo dehydrogenation with the supported $KMnO_4$, with ultrasonic irradiation.

[1] J.-J. Vanden Eynde, R. D'Orazio, and Y. V. Haverbeke, *T* **50**, 2479 (1994).

Potassium phenyltrifluoroborate.

Stereoselective alkylation of amino acids.[1] *N*-Alkylidene (including amidine) derivatives of α-amino acids form zwitterionic cyclic adducts with $K(PhBF_3)$ in which the phenyl group attaching to the boron atom is *cis* to the α-H in order to minimize bonding interactions. Alkylation with inversion of configuration occurs owing to the same steric effects; entry of the alkylating agent from the less hindered face of the heterocycle is favored.

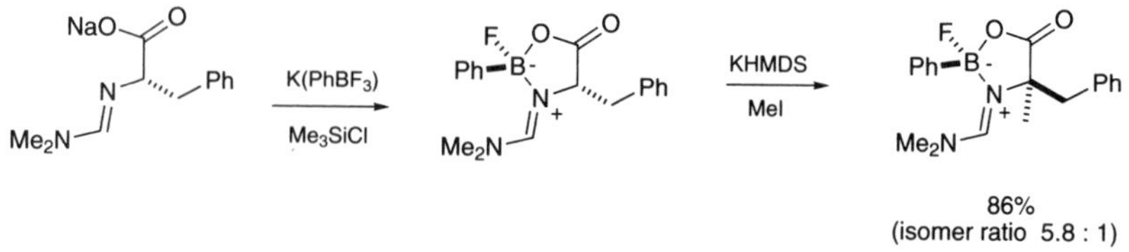

86%
(isomer ratio 5.8 : 1)

[1] E. Vedejs, S. C. Fields, and M. R. Schrimpf, *JACS* **115**, 11612 (1993).

Potassium tetracarbonylhydridoferrate.

Selective reduction.[1] Selective reduction of electron-deficient carbonyl compounds with this reagent has been reported. Trifluoroacetophenone is reduced, while acetophenone itself remains untouched. The more reactive carbonyl group of benzils, α-ketoesters, and *N*-methylisatin is reduced to the alcohol.

[1] J.-J. Brunet, R. Chauvin, F. Kindela, and D. Neibecker, *TL* **35**, 8801 (1994).

Propargyl chloride.

N-Allenylation.[1] Pyrrole, 2-substituted and 2,3-disubstituted pyrroles are allenylated on reaction with propargyl chloride in DMSO, using KOH as base. Actually 2,3-dichloropropene and 1,2,3-trichloropropane are also effective allenylating agents at slightly elevated temperatures (40°–50°C).

[1] O. A. Tarasova, L. Brandsma, and B. A. Trofimov, *S* 571 (1993).

Pyridinium chlorochromate. 14, 269; 15, 276

Oxidation of allylic and homoallyic alcohols. A convenient route to 3-thioalkyl-2-cycloalkenones[1] is based on the established pattern of transpositional oxidation of tertiary allylic alcohols. The substrates are readily prepared by reaction of the enones with phenylthiomethyllithium or 1,3-dithian-2-yllithium reagents.

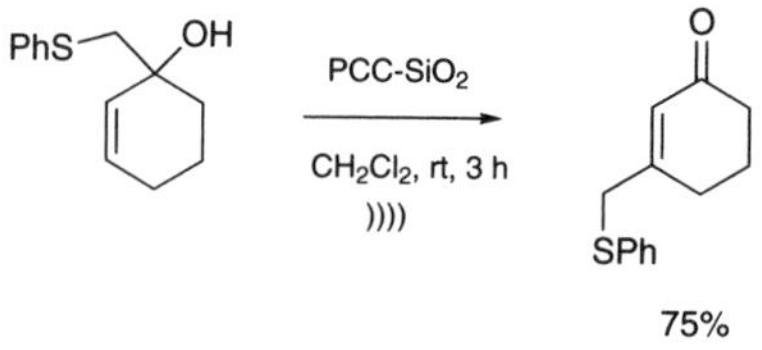

75%

Homoallylic steroidal alcohols[2] and their THP ethers[3] give enediones.

Elimination of silanes from α-silyl alcohols.[4]   By this oxidation acylsilanes can be prepared in two steps from esters and amides: reaction with silyllithiums and PCC oxidation. Secondary silyl alcohols are oxidized to aldehydes.

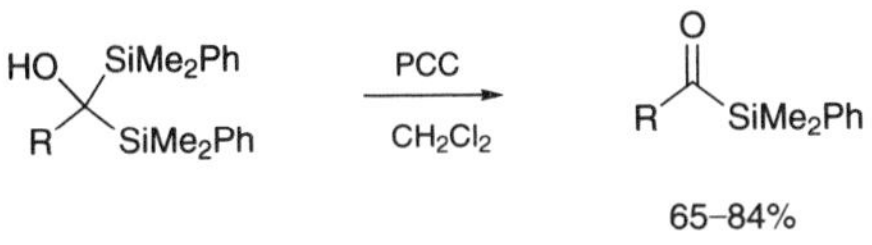

65–84%

Oxidative cyclization of δ,ε-unsaturated tertiary alcohols.[5]   Tetrahydrofuran units are formed in moderate yields by this oxidation.

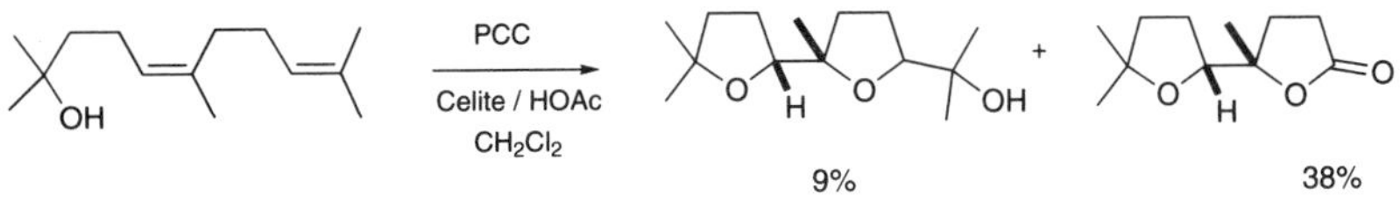

9%   38%

[1] F. A. Luzzio and W. J. Moore, *JOC* **58**, 2966 (1993).
[2] K. Blaszczyk and Z. Paryzek, *SC* **24**, 3255 (1994).
[3] E. J. Parish, S. A. Kizits, and R. W. Heidepriem, *SC* **23**, 223 (1993).
[4] I. Fleming and U. Ghosh, *JCS(P1)* 257 (1994).
[5] F. E. McDonald and T. B. Towne, *JACS* **116**, 7921 (1994).

Pyridinium tosylate. 15, 276; 16, 287–288

Dehydration of aldols.[1]   Dehydration of β-silyl aldols with pyridinium tosylate (PPTS) and MsCl–Et$_3$N give very different results. The PPTS reaction is regioselective and stereoselective, affording mainly the (Z)-isomers.

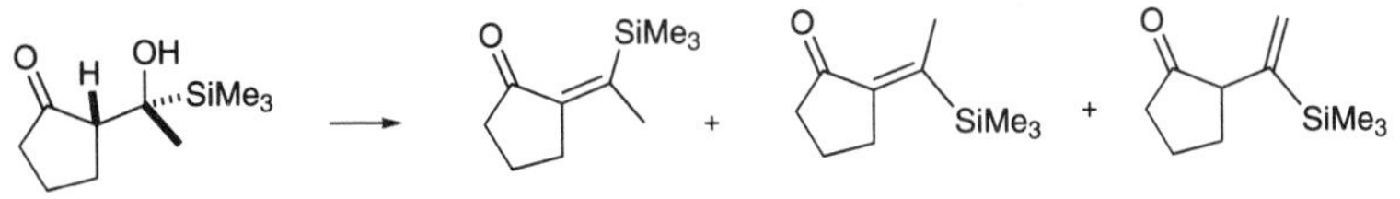

pyHOTs	82	18	1
MsCl–Et$_3$N	7	81	12

Selective O-desilylation.[2]   PPTS in ethanol removes silyl groups (e.g., TBS) from aliphatic silyl ethers in preference to the aryl silyl ethers; thus the chemoselectivity is opposite to the reaction with K$_2$CO$_3$–Kriptofix 222 (in MeCN).

[1] K. Nakatani, T. Izawa, Y. Odagaki, and S. Isoe, *CC* 1365 (1993).
[2] C. Prakash, S. Saleh, and I. A. Blair, *TL* **35**, 7565 (1994).

Pyrylium tetrafluoroborate.

Extensively conjugated aldehydes.[1]   The pyrylium salt reacts with $CH_2=PPh_3$ at C-2. The resulting 2H-pyran-2-methylphosphonium salt is a precursor of the masked ω-formylpentadienyl Wittig reagent.

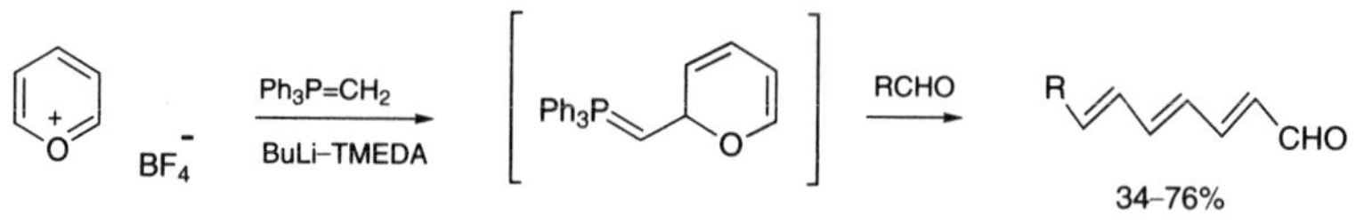

[1] K. Hemming and R. J. K. Taylor, *CC* 1409 (1993).

R

Rhenium(VII) oxide. 17, 296–297

Spirocyclization.[1] 2-(ω-Hydroxyalkyl)-1-oxa-2-cycloalkenes are converted into hydroxyspiroketals.

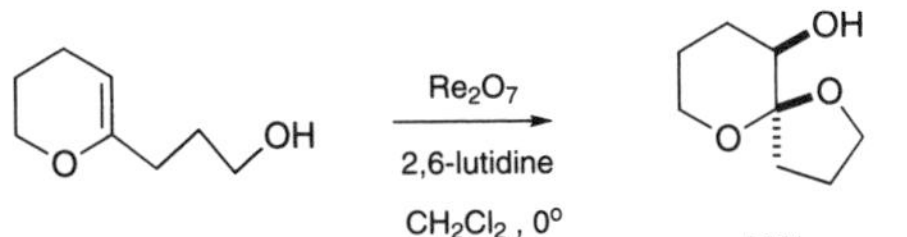

56%

Bisbenzocyclooctadienes.[2] The nonphenolic coupling used in the synthesis of lignan lactones is accomplished with Re_2O_7 in TFA–TFAA media.

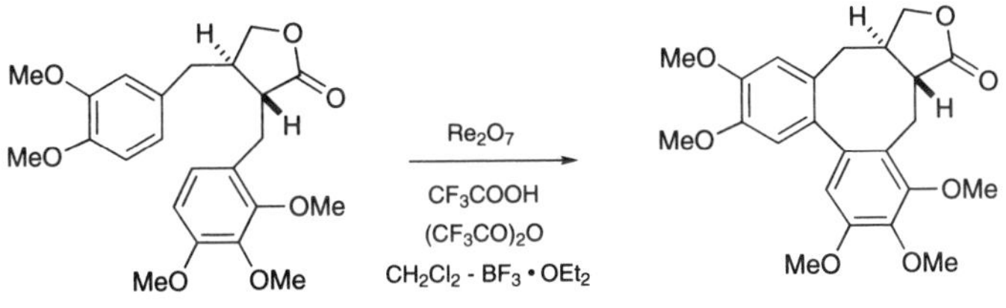

98%

[1] R. S. Boyce and R. M. Kennedy, *TL* **35**, 5133 (1994).
[2] D. Planchenault, R. Dhal, and J.-P. Robin, *T* **49**, 5823 (1993).

Rhodium carbonyl clusters. 13, 288; **15**, 334

Hydrogenation.[1] $Rh_6(CO)_{16}$ catalyzes the saturation of conjugated carbonyl compounds with CO and H_2O at atmospheric pressure and near ambient temperatures.

Homologations. Hydroformylation[2] of *S*- and *O*-containing alkenes is mediated by $Rh_4(CO)_{12}$. Homologous carboxylic acids and derivatives (esters, amides) are obtained from allyl phosphates.[3]

Bicyclo[3.3.0]octenones.[4] Silylative cyclocarbonylation of alkynes provides an easy access to these substances.

[1] T. Joh, K. Fujiwara, and S. Takahashi, *BCSJ* **66**, 978 (1993).
[2] E. M. Campi, W. R. Jackson, P. Perlmutter, and E. E. Tasdelen, *AJC* **46**, 995 (1993).
[3] Y. Imada, O. Shibata, and S.-i. Murahashi, *JOMC* **451**, 183 (1993).
[4] I. Matsuda, H. Ishibashi, and N. Ii, *TL* **36**, 241 (1995).

Rhodium carboxylates. 13, 266–269; **15**, 278–280; **16**, 289–292; **17**, 298–302

α-Alkoxy esters.[1,2] Rhodium carbenoids derived from α-diazo esters undergo O–H bond insertion in the reaction with alcohols or phenols. Low to moderate asymmetric induction from chiral esters is observed.[3]

Insertion into Si–H bonds.[4] α-Silyl esters are similarly obtained. From $R_2Si(Cl)H$ the products are readily converted into α-(alkoxysilyl)alkanoic esters.[5]

Heterocyclization. Formation of α-alkylidene lactams[6] by carbonylation of alkynyl amines and cyclization of 2-allyloxybenzylamines with allylic rearrangement[7] are reported.

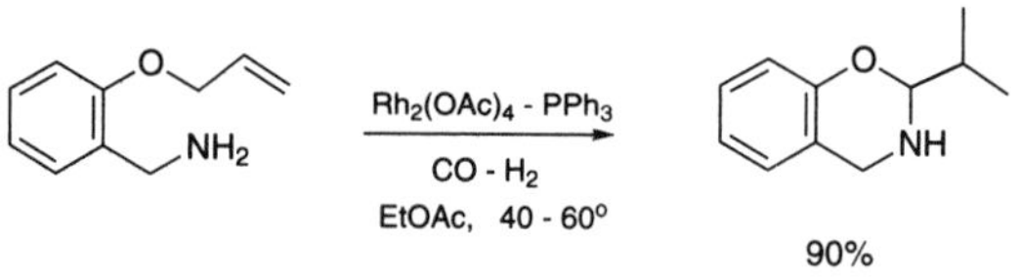

Other interesting cyclizations include the synthesis of oxazoles[8] and 3-keto-pyrrolidines.[9]

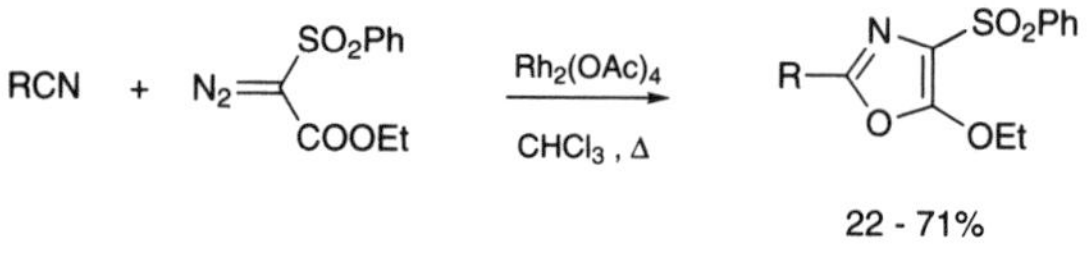

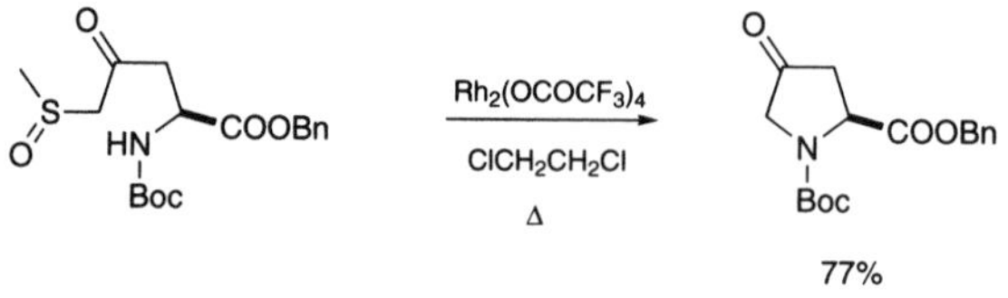

Silylformylation. (*Z*)-Selective addition of R_3Si/CHO fragments to alkynes[10] from a mixture of R_3SiH and CO is promoted by $(C_4F_9COO)_4Rh_2$ under mild conditions.

[1] G. G. Cox, D. J. Miller, C. J. Moody, E.-R. H. B. Sie, and J. J. Kulagowski, *T* **50**, 3195 (1994).
[2] D. Haigh, *T* **50**, 3177 (1994).
[3] E. Aller, G. G. Cox, D. J. Miller, and C. J. Moody, *TL* **35**, 5949 (1994).
[4] Y. Landais, D. Planchenault, and V. Weber, *TL* **35**, 9549 (1994).
[5] O. Andrey, Y. Landais, and D. Planchenault, *TL* **34**, 2927 (1993).
[6] E. M. Campi, J. M. Chong, W. R. Jackson, and M. Van Der Schoot, *T* **50**, 2533 (1994).
[7] E. M. Campi, W. R. Jackson, Q. J. McCubbin, and A. E. Trnacek, *CC* 2763 (1994).
[8] K. J. Doyle and C. J. Moody, *T* **50**, 3761 (1994).
[9] J. E. Baldwin, R. M. Adlington, C. R. A. Godfrey, D. W. Gollins, and J. G. Vaughan, *CC* 1434 (1993).
[10] M. P. Doyle and M. S. Shanklin, *OM* **13**, 1081 (1994).

Rhodium(I) chloride, COD-complexed.

Carbonylation. Insertion of CO into a C–S bond,[1] and formation of lactams from unsaturated amines[2] as well as indanones from 1-aryl-2-(trimethylsilyl)ethynes[3] are realized in the presence of $[Rh(cod)Cl]_2$.

α-Siloxy aldehydes.[4] The concurrent reductive formylation and *O*-silylation of aldehydes by R_3SiH and CO proceeds at room temperature. Interestingly, the cationic complex $Rh(cod)_2BF_4$ catalyzes hydrosilylation of propargylic alcohols to give only (*E*)-γ-silyl allylic alcohols.[5]

[1] K. Khumtaveeporn and H. Alper, *JOC* **59**, 1414 (1994); *JACS* **116**, 5662 (1994).
[2] Z. Zhang and I. Ojima, *JOMC* **454**, 281 (1993).
[3] R. Takeuchi and H. Yasue, *JOC* **58**, 5386 (1993).
[4] M. E. Wright and B. B. Cochran, *JACS* **115**, 2059 (1993).
[5] R. Takeuchi, S. Nitta, and D. Watanabe, *CC* 1777 (1994).

Rhodium–phosphine complexes. 13, 144; 14, 123–124; 15, 69–70, 90–92; 16, 86–88

Hydrogenation. The cationic catalyst (**1**)[1] is useful for OH-directed saturation of proximal alkenylsilanes and stannanes.

(1)

Hydroformylation.   A number of Rh complexes[2–4] have been studied. In situ generation of such complexes from others[5–8] are also effective for hydroformylation.

[1] M. Lautens, C. H. Zhang, B. J. Goh, C. M. Crudden, and M. J. A. Johnson, *JOC* **59**, 6208 (1994).
[2] L. Kollar and P. Sandor, *JOMC* **445**, 257 (1993).
[3] C. Botteghi and S. Paganelli, *JOMC* **451**, C18 (1993).
[4] T. J. Kwok and D. J. Wink, *OM* **12**, 1954 (1993).
[5] G. D. Cuny and S. L. Buchwald, *JACS* **115**, 2066 (1993).
[6] N. Sakai, S. Mano, K. Nozaki, and H. Takaya, *JACS* **115**, 7033 (1993).
[7] K. Totland and H. Alper, *JOC* **58**, 3326 (1993).
[8] H. Alper and J.-Q. Zhou, *CC* 316 (1993).

Ruthenium–carbene complexes.

Metathetic ring closure.   Five-, six-, and seven-membered carbocycles as well as N and O heterocycles are constructed from dienes[1] in the presence of a Ru–carbene complex. For unhindered amines, prior protonation is necessary. Fused bicyclic structures are accessible from dienynes.[2]

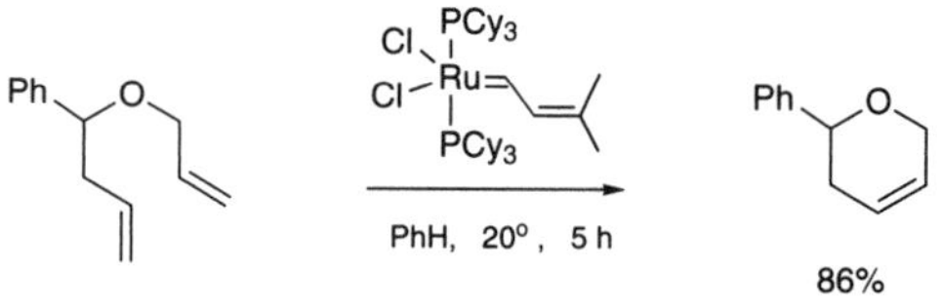

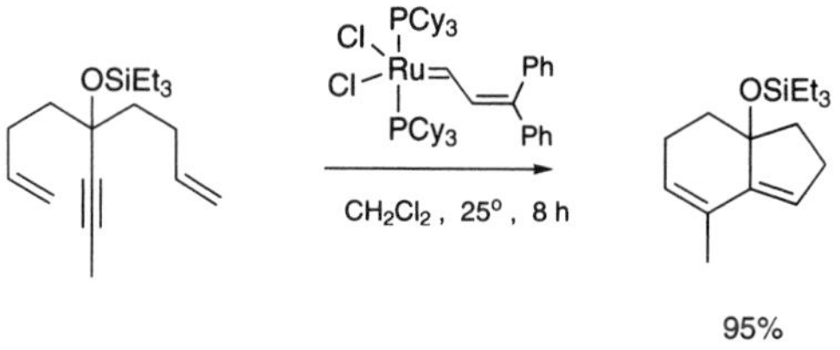

[1] G. C. Fu, S. T. Nguyen, and R. H. Grubbs, *JACS* **115**, 9856 (1993).
[2] S.-H. Kim, N. Bowden, and R. H. Grubbs, *JACS* **116**, 10801 (1994).

Ruthenium–carbonyl clusters.

Reduction. $Ru_3(CO)_{12}$ catalyzes the reduction of nitroarenes with CO. 4(3H)-Quinazolinones are obtained from N-(2-nitrobenzoyl)amides.[1] In the presence of montmorillonite-bipy · Pd(II) the reaction leads to methyl carbamates.[2]

Condensation of alkynes with carboxylic acids.   Addition of RCOOH to alkynes forms enol carboxylates.[3] On the other hand, heterocyclization occurs with acetylenedicarboxylic esters.[4]

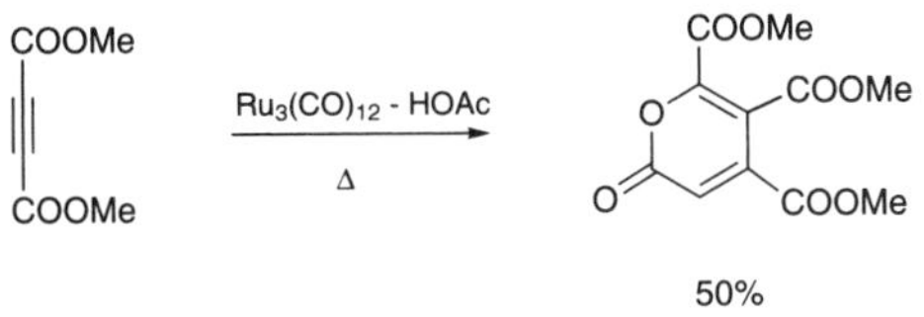

COOMe
COOMe
$Ru_3(CO)_{12}$ - HOAc
Δ
COOMe
COOMe
COOMe
50%

[1] M. Akazome, T. Kondo, and Y. Watanabe, *JOC* **58**, 310 (1993).
[2] V. L. K. Valli and H. Alper, *JACS* **115**, 3778 (1993).
[3] M. Rotem and Y. Shvo, *JOMC* **448**, 189 (1993).
[4] N. Menashe and Y. Shvo, *H* **35**, 611 (1993).

Ruthenium(II) chloride, tris(triphosphine) complex. 13, 107; **14**, 130–131; **15**, 128; **16**, 126–128

Alkene isomerization. Allylic alcohols are converted to saturated ketones.[1] Isomerization of the allyl group of *N,O*-acetals paves way to a synthesis of 2-(α-amidoalkyl)propanals.[2]

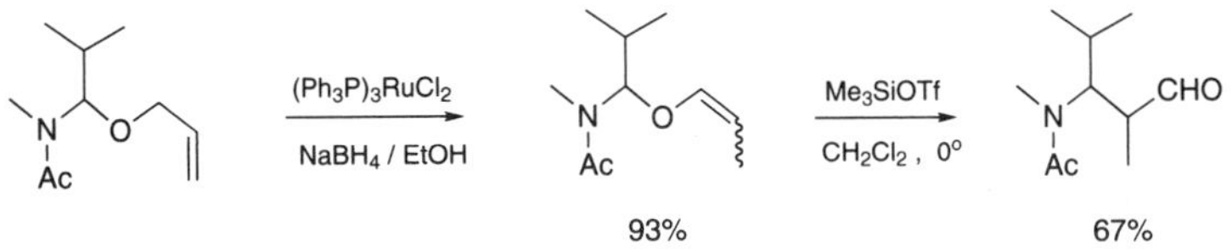

$(Ph_3P)_3RuCl_2$
$NaBH_4$ / EtOH
Ac
93%
Me_3SiOTf
CH_2Cl_2, 0°
CHO
Ac
67%

Perfluoroalkylation. Arenes and heteroarenes give perfluoroalkyl derivatives[3] upon reaction with R_fSO_2Cl under ruthenium complex catalysis. Pyrroles also undergo this reaction.[4]

Acylcyanides.[5] The oxidation of cyanohydrins with *t*-BuOOH to furnish acylcyanides is catalyzed by $(PPh_3)_3RuCl_2$. The products can be used for selective *N*-acylation of amino alcohols and polyamines.

Carbonylation. A new route to 2(5*H*)-furanones is by oxidative cyclocarbonylation of allylic alcohols.[6]

Alkenylsilanes.[7] By metathesis the introduction of a carbon chain to vinylsilanes is realized.

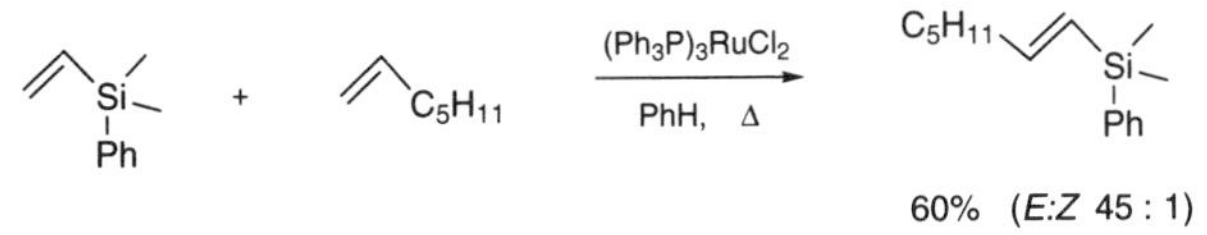

Si
Ph
+
C_5H_{11}
$(Ph_3P)_3RuCl_2$
PhH, Δ
C_5H_{11}
Si
Ph
60% (E:Z 45 : 1)

[1] J.-E. Bäckvall and U. Andreasson, *TL* **34**, 5459 (1993).
[2] T. Arenz, H. Frauenrath, G. Raabe, and M. Zorn, *LA* 931 (1994).
[3] N. Kamigata, T. Ohtsuka, T. Fukushima, M. Yoshida, and T. Shimizu, *JCS(P1)* 1339 (1994).
[4] N. Kamigata, T. Ohtsuka, M. Yoshida, and T. Shimizu, *SC* **24**, 2049 (1994).
[5] S.-I. Murahashi and T. Naota, *S* 433 (1993).

[6] T. Kondo, K. Kodoi, T. Mitsudo, and Y. Watanabe, *CC* 755 (1994).
[7] B. Marciniec and C. Pietraszuk, *JOMC* **447**, 163 (1993).

Ruthenium(III) chloride–peracetic acid.

α-Ketols.[1]   The Ru-catalyzed oxidation of alkenes to ketols is useful for a synthesis of cortisone acetate.

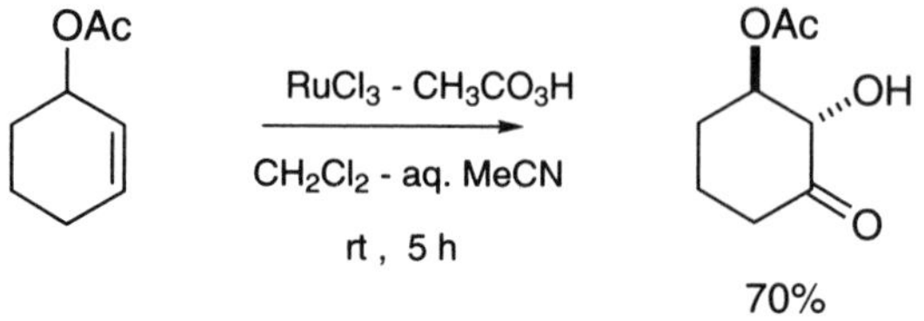

70%

[1] S.-I. Murahashi, T. Saito, H. Hanaoka, Y. Murakami, T. Naota, H. Kumobayashi, and S. Akutagawa, *JOC* **58**, 2929 (1993).

Ruthenium(III) chloride–sodium periodate.

$R_2S \rightarrow R_2SO_2$.[1]   (84–100%)

Dihydroxylation of alkenes.[2]   (36–76%)

[1] W. Su, *TL* **35**, 4955 (1994).
[2] T. K. M. Shing, V. W.-F. Tai, and E. K. W. Tam, *ACIEE* **33**, 2312 (1994).

Ruthenium(IV) oxide. 13, 268–269; 14, 272–273; 15, 281; 16, 292–293; 17, 304

Nonphenolic coupling.[1]   In TFA–TFAA media diarylbutanes afford mixtures of 1-aryltetralins and dibenzocyclooctadienes.

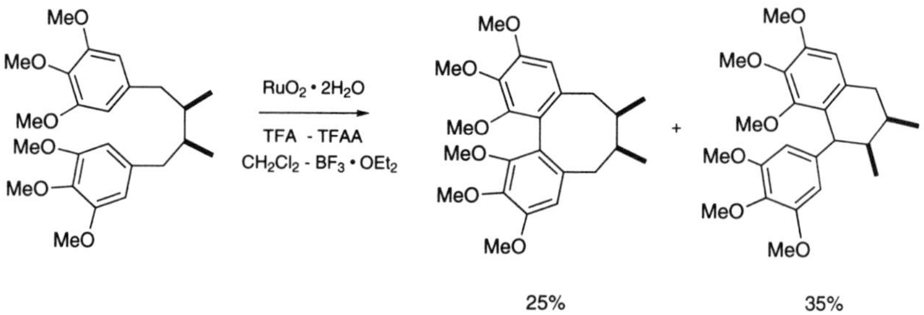

25%   35%

Baeyer–Villger oxidation.[2]   The oxidation system includes RuO_2, O_2, and PhCHO.

[1] R. Dhal, Y. Landais, A. Lebrun, V. Lenain, and J.-P. Robin, *T* **50**, 1153 (1994).
[2] T. Inokuchi, M. Kanazaki, T. Sugimoto, and S. Torii, *SL* 1037 (1994).

S

Samarium. 14, 275; **17**, 305–307

Reduction.[1] Carboxylic acids, esters, amides, and nitriles are reduced very rapidly by Sm and hydrochloric acid. Ytterbium may be used instead of samarium. The reduction of α-halocarbonyl compounds and the pinacol coupling with Sm in an aprotic system (Me$_3$SiCl–NaI/MeCN)[2] are as efficient as those mediated by SmI$_2$.

Catalytic aldol reactions.[3]

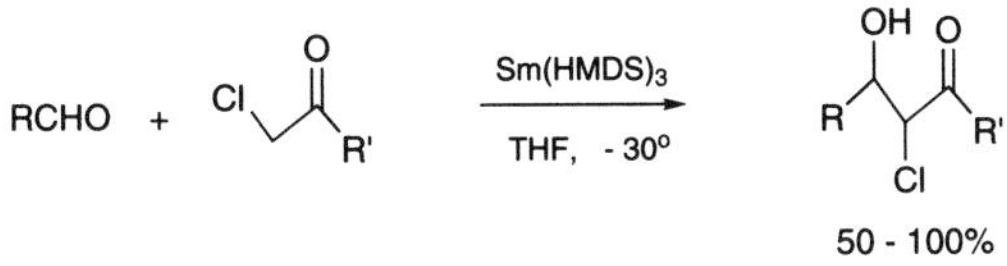

$$\text{RCHO} + \text{Cl}\overset{O}{\underset{R'}{\bigwedge}} \xrightarrow[\text{THF, } -30°]{\text{Sm(HMDS)}_3} \overset{OH}{\underset{Cl}{R}}\overset{O}{\underset{}{R'}}$$

50 - 100%

Alkylation of alkynes.[4] Ethynylsilanes are alkylated at the unsubstituted terminus in the presence of Sm and SmI$_2$.

Allylation. Amalgamated Sm promotes reaction of allyl halides (bromide, iodide) with carbonyl compounds.[5] Diallylation of esters occurs.[6]

Cyclopropanation.[7] The Sm–HgCl$_2$ couple can be used in place of Zn–Cu in the Simmons–Smith reaction on allylic alcohols.

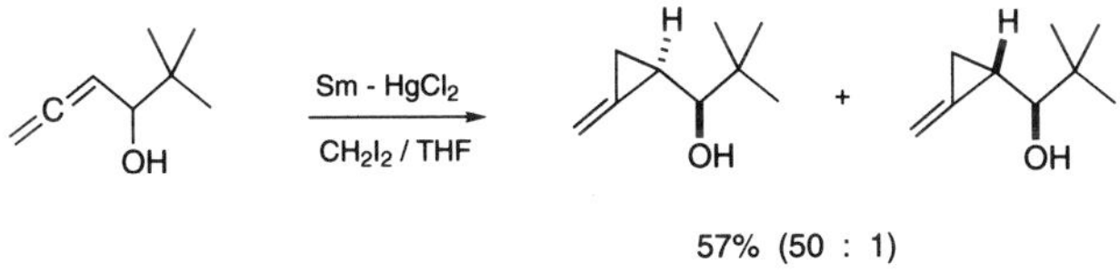

$$\xrightarrow[\text{CH}_2\text{I}_2 \text{ / THF}]{\text{Sm - HgCl}_2}$$

57% (50 : 1)

[1] Y. Kamochi and T. Kudo, *CPB* **42**, 402 (1994).
[2] N. Akane, T. Hatano, H. Kusui, Y. Nishiyama, and Y. Ishii, *JOC* **59**, 7902 (1994).
[3] H. Sasai, S. Arai, and M. Shibasaki, *JOC* **59**, 2661 (1994).
[4] M. Murakami, M. Hayashi, and Y. Ito, *SL* 179 (1994).
[5] X. Gao, X. Wang, R.-F. Cai, J.-D. Wei, and S.-H. Wu, *HX* **51**, 1139 (1993).
[6] X. Gao, J. Zeng, J.-Y. Zhou, and S.-H. Wu, *HX* **51**, 1191 (1993).
[7] M. Lautens and P. H. M. Delanghe, *JACS* **116**, 8526 (1994).

Samarium(II) bromide.

Pinacol coupling.[1] SmBr$_2$ prepared from Sm$_2$O$_3$ proves to be an excellent reagent for cross-coupling of carbonyl compounds.

45%

[1] A. Lebrun, J.-L. Namy, and H. B. Kagan, *TL* **34**, 2311 (1993).

Samarium(III) chloride. 14, 275–276; 15, 282

Deacetalization.[1]   Cleavage of acetals under anhydrous conditions (in the presence of AcCl) is catalyzed by $SmCl_3$.

Barbier-type reactions.   The in situ generation of organometallic reagents from organohalides and their reaction with carbonyl substrates are achievable electrochemically with a Mg anode and $SmCl_3$ as catalyst in DMF.[2,3]

25 - 76%

[1] S.-H. Wu and Z.-B. Ding, *SC* **24**, 2173 (1994).
[2] H. Hebri, E. Dunach, and J. Perichon, *CC* 499 (1993).
[3] H. Hebri, E. Dunach, and J. Perichon, *TL* **34**, 1475 (1993).

Samarium(II) iodide. 13, 270–272; 14, 276–281; 15, 282–284; 16, 294–300; 17, 307–311

Preparation.[1]   From Sm and diiodoethane in tetrahydropyran, which is superior to THF.

Reductions.[2]   Aromatic carboxylic acids, esters, and amides are reduced to $ArCH_2OH$ and nitriles to $ArCH_2NH_2$ by the SmI_2–H_2O system. Pyridine derivatives are completely reduced to afford piperidines[3] in aqueous THF. With MeOH as a proton source the labile 2-aminonitroalkanes are converted to the diamines.[4]

Reaction of organochalcogenides.   Elemental selenium is reduced by SmI_2, and a subsequent reaction with RCOCl or arenediazonium salts gives diacyl diselenide[5] and diaryl diselenides,[6] respectively. On the other hand, diorganyl disulfides,[7] diselenides, and ditellurides[8] undergo reductive cleavage, furnishing acyl and alkyl chalcogenides. The samarium chalcogenides are active Michael donors.[9]

Thiolsulfonic acids[10] and thiocyanates[11] suffer S–S and C–S bond severance, respectively, by SmI_2 to furnish RS^- and thence disulfides. Similarly, the cleavage of sulfonamides[12] under these mild conditions makes such compounds more valuable as synthetic intermediates.

Vinylsamarium species are generated from vinyl sulfides[13] and sulfones.[14] Protonation or alkylation may then be performed. Formation of *C*-disaccharides from a *Si*-tethered glycosyl sulfone has been effected with SmI_2.[15] Intervention of a glycosyl radical intermediate is implicated.

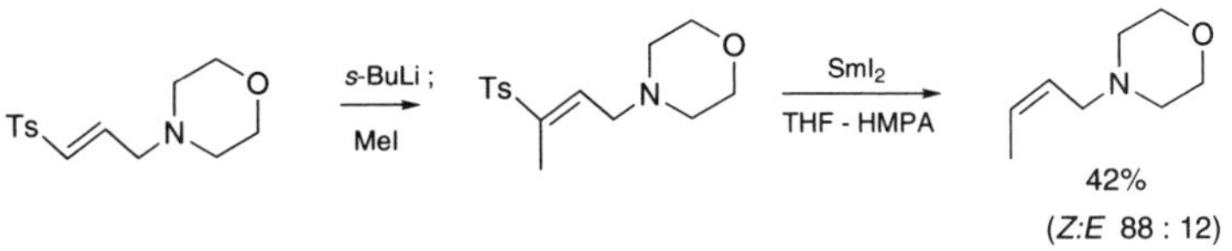

42%
(*Z:E* 88 : 12)

Treatment of bisbenzenesulfonylmethane and a ketone with SmI_2 leads to a β-hydroxy sulfone.[16] Allylic sulfones are a source of allyl anions because they are cleaved by SmI_2 in HMPA; thus a new preparative method for homoallylic alcohols[17] is available. However, the desulfurative cyclization[18] of allylic sulfides toward a carbonyl group may belong to a different mechanistic category.

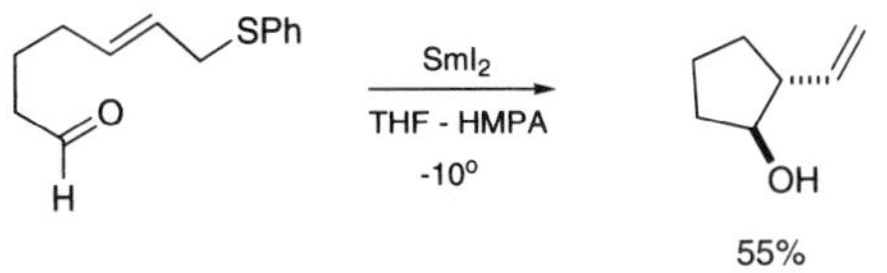

55%

Organosamarium reactions with carbonyl compounds. The facile formation of organosamarium reagent from various halides makes it possible to substitute more conventional metals by Sm. Actually SmI_2 can be used instead of the metal to form the active Sm(III) reagents for such processes as Reformatsky-type reactions,[19-21] allylation of ketenes[22] (with enantioselective protonation of the resulting Sm enolates), coupling of glycosyl halides with ketones,[23] and preparation of β-hydroxy sulfides.[24]

Intramolecular Barbier reactions have been effected to give cyclic products.[25,26] The conversion of iodoalkyl esters to hydroxy ketones (isolated as the acetates)[27,28] represents a reversal of the Baeyer–Villiger reaction with transposition.

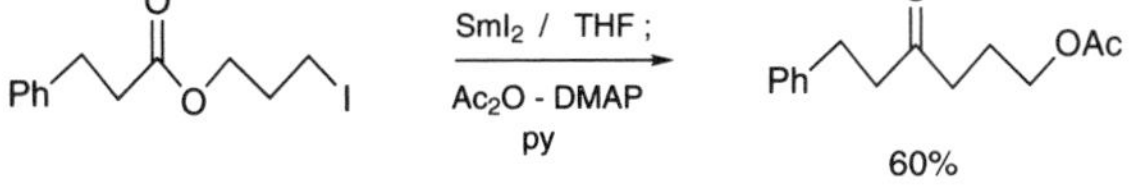

60%

Organosamariums also add to enones in the Michael fashion[29] when a copper salt is present. α-Amino radicals generated from cleavage of *N*-[1-(*N'*,*N'*-dialkylamino)-alkenyl]benzotriazoles on exposure to SmI_2 may add to the electron-deficient double bond.[29a] The $S_{RN}1$ reaction of haloarenes with phenone enolates in DMSO is catalyzed by SmI$_2$.[30]

Three-component coupling. A very efficient α-hydroxy imine synthesis[31] is the SmI$_2$-mediated coupling of an organic halide, an isocyanide, and a carbonyl compound. Such products readily undergo autoxidation, and unsymmetrical α-diketones are accessible on hydrolysis.[32] On the other hand, the α-hydroxy imines can be reduced to provide 2-amino alcohols.[33]

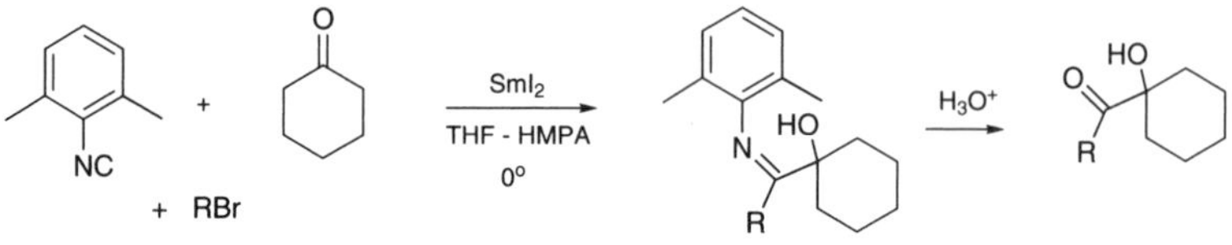

Reductive enolization. Like α-halocarbonyl compounds, α-ketols[34] and their esters[35] are also subject to reductive enolization, and some of these enolates have been used in aldol condensations.[36] Deconjugative deoxygenation of cephalosporins has been observed.[37] Cyclic derivatives of γ,δ-dihydroxy α,β-unsaturated esters are also deoxygenated with double bond migration to afford the δ-hydroxy β,γ-unsaturated esters.[38] If Mg is used instead of SmI$_2$, the double bond also suffers reduction.

Carbonyl couplings. Many variations of cross couplings are possible; these include aldehydes with α-dicarbonyl compounds.[39,40] Ketyl radicals derived from carbonyl compounds also add to alkenes such as acrylonitrile[41] or N-allyl moieties.[42] Intramolecular cyclizations on terminal alkenes[43] or allene species[44] have also been exploited for synthetic purposes.

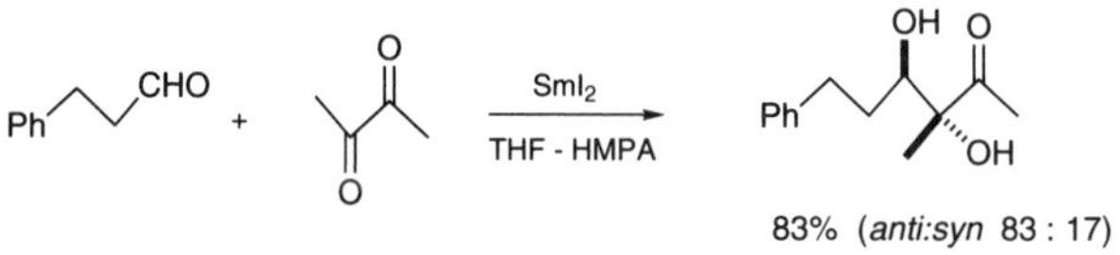

83% (*anti:syn* 83 : 17)

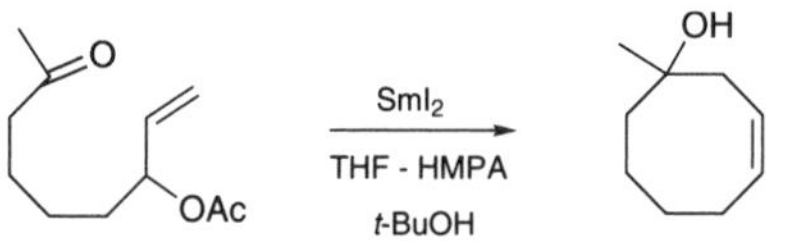

81%

Eliminations. Alkenes are formed rapidly from *vic*-dibromides,[45] β-chloroethers,[46] β-iodoethyl esters,[47] and β-acetoxy sulfones[48] on treatment with SmI$_2$. The formation of furan derivatives[49] from epoxypropargyl esters by reaction with

SmI_2–$Pd(PPh_3)_4$ may involve elimination to give hydroxy enynes, which undergo cyclization.

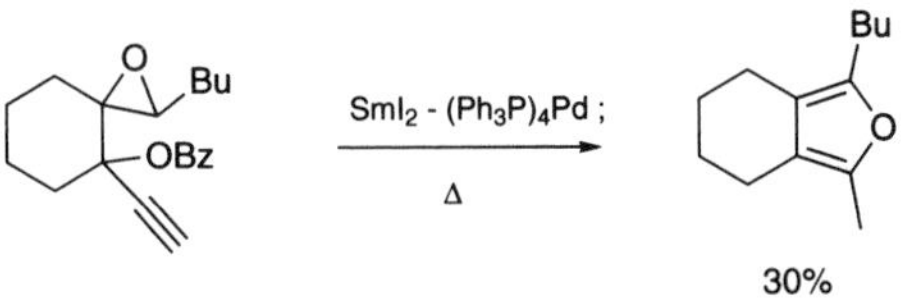

30%

Carbon radical generation. Radicals that are formed during dehalogenation can be intercepted intramolecularly by a C=C bond[50] or C=N bond.[51] A cyclopentene synthesis[52] from 1,1-dihaloalkenes apparently involves two-stage reduction, which is intervened by hydrogen abstraction.

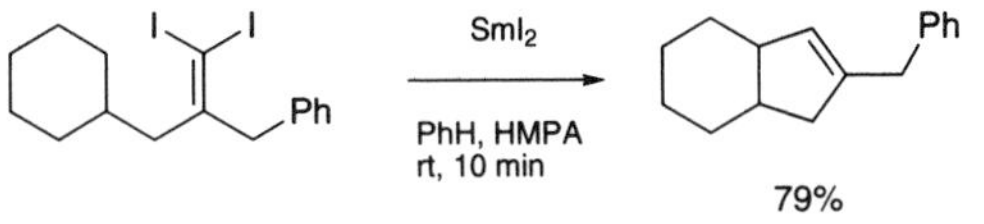

79%

Cyclopropane fission.[53] Cyclopropane-1,1-diesters suffer reductive opening, and this process can be used to generate ester bishomoenolates.[54]

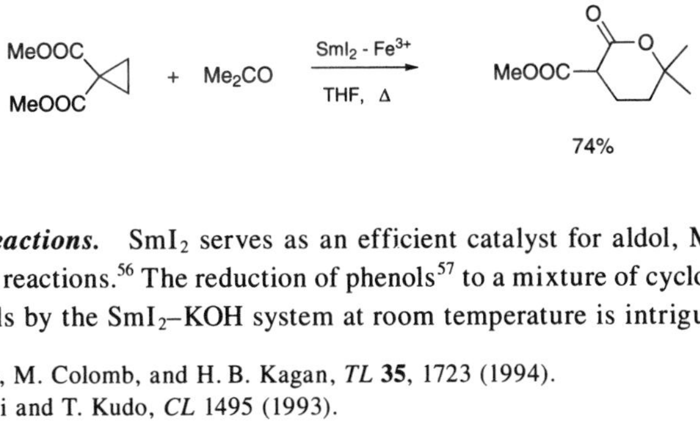

74%

Other reactions. SmI_2 serves as an efficient catalyst for aldol, Michael,[55] and Diels–Alder reactions.[56] The reduction of phenols[57] to a mixture of cyclohexanols and cyclohexenols by the SmI_2–KOH system at room temperature is intriguing.

[1] J.-L. Namy, M. Colomb, and H. B. Kagan, *TL* **35**, 1723 (1994).
[2] Y. Kamochi and T. Kudo, *CL* 1495 (1993).
[3] Y. Kamochi and T. Kudo, *H* **36**, 2383 (1993).
[4] M. A. Sturgess and D. J. Yarberry, *TL* **34**, 4743 (1993).
[5] X. Jia, Y. Zhang, and X. Zhou, *SC* **23**, 1403 (1993).
[6] Y. Zhang, X. Jia, and X. Zhou, *SC* **24**, 1247 (1993).
[7] X. Jia, Y. Zhang, and X. Zhou, *SC* **23**, 387 (1993).
[8] Y. Zhang, Y. Yu, and R. Lin, *SC* **23**, 189 (1993).
[9] H.-J. Jiang and Y.-M. Zhang, *HX* **14**, 307 (1994).
[10] X. Jia, Y. Zhang, and X. Zhou, *SC* **24**, 2893 (1994).

[11]　X. Jia, Y. Zhang, and X. Zhou, *TL* **35**, 8833 (1994).

[12]　E. Vedejs and S. Lin, *JOC* **59**, 1602 (1994).

[13]　M. Hojo, H. Harada, J. Yoshizawa, and A. Hosomi, *JOC* **58**, 6541 (1993).

[14]　P. L. Tbanez and C. Najera, *TL* **34**, 2003 (1993).

[15]　A. Chenede, E. Perrin, E. D. Rekai, and P. Sinay, *SL* 420 (1994).

[16]　S. Chandrasekar, J. Yu, J. R. Falck, and C. Mioskowski, *TL* **35**, 5441 (1994).

[17]　J. Clayden and M. Julia, *CC* 2261 (1994).

[18]　T. Kan, S. Nara, S. Ito, F. Matsuda, and H. Shirahama, *JOC* **59**, 5111 (1994).

[19]　S. Hanessian and C. Girard, *SL* 865 (1994).

[20]　T. Arime, N. Kato, F. Komadate, H. Saegusa, and N. Mori, *JCCC(T)* **41**, 3315 (1994).

[21]　S. Fukuzawa and S. Sakai, *NKK* 513 (1993).

[22]　S. Takeuchi, A. Ohira, N. Miyoshi, H. Mashio, and Y. Ohgo, *TA* **5**, 1763 (1994).

[23]　P. de Pouilly, A. Chenede, J.-M. Mallet, and P. Sinay, *BSCF* 256 (1993).

[24]　M. Yamashita, K. Kitagawa, T. Ohhara, Y. Iida, A. Masumi, I. Kawasaki, and S. Ohta, *CL* 653 (1993).

[25]　A. Fadel, *TA* **5**, 531 (1994).

[26]　G. A. Molander and J. A. McKie, *JOC* **58**, 7216 (1993).

[27]　G. A. Molander and J. A. McKie, *JACS* **115**, 5821 (1993).

[28]　G. A. Molander and S. R. Shakya, *JOC* **59**, 3445 (1994).

[29]　P. Wipf and S. Venkatraman, *JOC* **58**, 3455 (1993).

[29a]　J. M. Aurrecoechea and A. Fernandez-Acebes, *TL* **34**, 549 (1993).

[30]　M. A. Nazareno and R. A. Rossi, *TL* **35**, 5185 (1994).

[31]　M. Murakami, T. Kawano, H. Ito, and Y. Ito, *JOC* **58**, 1458 (1993).

[32]　M. Murakami, I. Komoto, H. Ito, and Y. Ito, *SL* 511 (1993).

[33]　M. Murakami, H. Ito, and Y. Ito, *JOC* **58**, 6766 (1993).

[34]　S. Hanessian and C. Girard, *SL* 861 (1996).

[35]　E. J. Enholm and S. Jiang, *H* **32**, 2247 (1992).

[36]　E. J. Enholm, S. Jiang, and K. Abboud, *JOC* **58**, 4061 (1993).

[37]　H.-Y. Kang, Y. S. Cho, H. Y. Koh, and M. H. Chang, *SC* **23**, 2977 (1993).

[38]　S.-K. Kang, S.-G. Kim, D.-C. Park, J.-S. Lee, W.-J. Yoo, and C. S. Park, *JCS(PI)* 9 (1993).

[39]　N. Miyoshi, S. Takeuchi, and Y. Ohgo, *CL* 2129 (1993).

[40]　N. Miyoshi, S. Takeuchi, and Y. Ohgo, *CL* 959 (1993).

[41]　M. Kawatsura, F. Matsuda, and H. Shirahama, *JOC* **59**, 6900 (1994).

[42]　J. E. Baldwin, S. C. M. Turner, and M. G. Moloney, *T* **50**, 9411 (1994).

[43]　G. A. Molander and J. A. McKie, *JOC* **59**, 3186 (1994).

[44]　J. M. Aurrecoechea and R. F.-S. Anton, *JOC* **59**, 702 (1994).

[45]　R. Yanada, K. Bessho, and K. Yanada, *CL* 1279 (1994).

[46]　L. Crombie and L. J. Rainbow, *JCS(PI)* 673 (1994).

[47]　A. J. Pearson and K. Lee, *JOC* **59**, 2257 (1994).

[48]　M. Ihara, S. Suzuki, T. Taniguchi, Y. Tokunaga, and K. Fukumoto, *SL* 859 (1994).

[49]　J. M. Aurrecoechea and M. Solay-Ispizua, *H* **37**, 223 (1994).

[50]　S. Fukuzawa and T. Tsuchimoto, *SL* 803 (1993).

[51]　C. F. Sturino and A. G. Fallis, *JACS* **116**, 7447 (1994).

[52]　M. Kunishima, K. Hioki, S. Tani, and A. Kato, *TL* **35**, 7253 (1994).

[53]　Y. H. Kim and I. S. Lee, *HC* **3**, 509 (1992).

[54]　T. Imamoto, T. Hatajima, and T. Yoshizawa, *TL* **35**, 7805 (1994).

[55]　P. Van de Weghe and J. Collin, *TL* **34**, 3881 (1993).

[56]　P. Van de Weghe and J. Collin, *TL* **35**, 2545 (1994).

[57]　Y. Kamochi and T. Kudo, *TL* **35**, 4169 (1994).

Samarium(III) iodide.

Enone synthesis.[1] The SmI₃-catalyzed condensation of α-haloketones and aldehydes (Reformatsky-type reaction) leads to enones directly.

Cyclic ether cleavage.[2] SmI₃ is a typical Lewis acid, which induces reaction with acid chlorides. It also contributes an iodide ion, giving ω-iodoalkyl carboxylates.

Boronate esters.[3] Hydroboration of alkenes with catecholborane is catalyzed by SmI₃.

[1] Y. Yu, R. Lin, and Y. Zhang, *TL* **34**, 4547 (1993).
[2] Y. Yu, Y. Zhang, and R. Lin, *SC* **23**, 1973 (1993).
[3] D. A. Evans, A. R. Muci, and R. Sturmer, *JOC* **58**, 5307 (1993).

Samarium(II) menthoxide.

Aldol condensation.[1] Alkoxides of Sm(II) give better results than Sm(III) in promoting aldolization, e.g., higher diastereoselectivity (*anti*-selective for both (*E*)- and (*Z*)-keteneacetals). Moderate enantioselectivities are observed when Sm(II) L-menthoxide is used as catalyst.

[1] Y. Makioka, I. Nagkagawa, Y. Taniguchi, K. Takaki, and Y. Fujiwara, *JOC* **58**, 4771 (1993).

Samarium(II) triflate.

Preparation.[1] The reagent is obtained by treatment of Sm(OTf)₃ with s-BuLi or EtMgBr at room temperature and used in situ.

Pinacol coupling.[2] Compared with SmI₂-mediated couplings, a higher efficiency has been observed.

[1] S. Fukuzawa, T. Tsuchimoto, and T. Kanai, *CL* 1981 (1994).
[2] T. Hanamoto, Y. Sugimoto, A. Sugino, and J. Inanaga, *SL* 377 (1994).

Scandium(III) perchlorate.

Glycosylation.[1] α-Ribofuranosides undergo selective α-glycosylation with C- and N-nucleophiles at room temperature in the presence of Sc(ClO₄)₃.

[1] I. Hachiya and S. Kobayashi, *TL* **35**, 3319 (1994).

Scandium(III) triflate.

Aldol and Michael reactions.[1] This Lewis acid catalyzes aldol and Michael reactions smoothly (20 examples, 66–97% yield), using silyl enol ethers as the donors. The catalyst is recoverable and reusable.

Allylation.[2] The highly efficient reaction (16 examples, 77–98% yield) between tetraallyltin and carbonyl compounds in MeCN at room temperature to produce homoallylic alcohols is promoted by Sc(OTf)₃.

Friedel–Crafts acylation.[3] The reaction is carried out in nitromethane at 50°C. Yields are in the 79–99% range (6 examples).

Diels–Alder reactions. The reaction can be performed in aqueous THF at 0°C.[4] A chiral catalyst prepared with (+)-1,1'-binaphthol, and 1,2,6-trimethylpiperidine is useful for promoting enantioselective reactions.[5]

[1] S. Kobayashi, I. Hachiya, H. Ishitani, and M. Araki, *SL* 472 (1993).
[2] I. Hachiya and S. Kobayashi, *JOC* **58**, 6958 (1993).
[3] A. Kawada, S. Mitamura, and S. Kobayashi, *SL* 545 (1994).
[4] S. Kobayashi, I. Hachiya, M. Araki, and H. Ishitani, *TL* **34**, 3755 (1993).
[5] S. Kobayashi, M. Araki, and I. Hachiya, *JOC* **59**, 3758 (1994).

Selenium.

Dialkyl diselenides. The synthesis[1] involves treatment of an alkyl halide with Se and Zn dust in aqueous NaOH under nitrogen at 80°C. Access to dibenzyl diselenides[2] by a phase-transfer process (Se, NaOH, polyethyleneglycol-400/benzene, 65°C) is also convenient (9 examples, 73–96%).

Isoselenocyanates and selenoamides. Formamides are converted to isoselenocyanates[3] by reaction with Se, Et$_3$N, and phosgene in toluene. Selenoamides are obtained[4] from secondary amines, Se, and 1,1-dihaloalkanes in the presence of NaH in HMPA. An oxidation is involved.

S-Alkylthiocarbamates.[5] Selenium is a catalyst for the condensation of amines with S and carbon monoxide. Alkylation of the adducts with alkyl halides gives the thiocarbamates (12 examples, 41–100% yield).

[1] P. Lue and X. Zhou, *SC* **23**, 1721 (1993).
[2] J. X. Wang, C.-H. Wang, W. Cui, and Y. Hua, *JCS(PI)* 2341 (1994).
[3] D. H. R. Barton, S. I. Parekh, M. Tajbakhsh, E. A. Theodorakis, and C.-L. Tse, *T* **50**, 639 (1994).
[4] Y. Takikawa, M. Yamaguchi, T. Sasaki, K. Ohnishi, and K. Shimada, *CL* 2105 (1994).
[5] T. Mizuno, I. Nishiguchi, and N. Sonoda, *T* **50**, 5669 (1994).

Selenium dioxide. 13, 272–273; 17, 312–313

Dethioacetalization.[1] Excellent yields have been obtained for recovering carbonyl compounds from dithioacetals by oxidative cleavage with SeO$_2$ in HOAc at room temperature (8 examples, 90–98%).

Allylic hydroxylation.[2]   The SeO_2–HCOOH combination in dioxane has found use in converting alkenes to allylic alcohols.

[1] S. A. Haroutounian, *S* 39 (1995).
[2] K. Shibuya, *SC* **24**, 2923 (1994).

Silica gel. 15, 285

Hydrolysis.   The anomeric acetate is the only hydrolyzable group in polyacetyl amino sugars on treatment with SiO_2–MeOH.[1] Oxazolidinones such as those derived from proline are similarly cleaved,[2] allowing for a very simple purification of the end products in a synthetic approach to α-alkylprolines.

> 90%

Isomerization of epoxides. α,β-Epoxy ketones[3] and β,γ-epoxy nitroalkanes[4] undergo ring opening to give α-diketones and γ-hydroxy nitroalkenes, respectively.

Desilylation.[5]   Trimethylsilylethynyl sulfones lose the silyl group on flash silica gel chromatography. A desilylation step is thus saved if the free alkynes are the desired products; on the other hand, one must be careful about the lability of such silyl derivatives toward SiO_2.

Diels–Alder reactions of inverse electron demands.[6]   The catalytic activity of SiO_2 to promote such reactions (e.g., enol ethers with α-pyrones) has been noted.

[1] M. Avalos, R. Babiano, P. Cintas, J. L. Jimenez, J. C. Palacios, and C. Valencia, *TL* **34**, 1359 (1993).
[2] M. J. Genin, P. W. Baures, and R. L. Johnson, *TL* **35**, 4967 (1994).
[3] T. B. Rao and J. M. Rao, *SC* **23**, 1527 (1993).
[4] J. Boelle, R. Schneider, P. Gerardin, and B. Loubinoux, *SC* **23**, 2563 (1993).
[5] Z. Chen and M. L. Trudell, *SC* **24**, 3149 (1994).
[6] G. H. Posner, J.-C. Carry, J. K. Lee, D. S. Bull, and H. Dai, *TL* **35**, 1321 (1994).

Silver bromate.

Oxidation.[1]   Trimethylsilyl ethers are oxidized to aldehydes or acids in good yields by varying the reaction temperature. In refluxing CH_2Cl_2 an aldehyde is obtained, while in refluxing MeCN the carboxylic acid is produced.

[1] H. Firouzabadi and I. Mohammadpoor-Baltork, *SC* **24**, 1065 (1994).

Silver carbonate.

α,α-Dialkoxy ketones and imines.[1] Heating α-bromo-α-chloro ketones or imines with silver carbonate in an alcohol effects the replacement of the halogen atoms.

Carbonyl regeneration from oximes.[2] Silver carbonate on bentonite is a useful reagent to convert oximes in refluxing benzene to carbonyl compounds.

[1] N. De Kimpe, E. Stanoeva, and M. Boeykens, *S* 427 (1994).
[2] R. Sanabria, R. Miranda, V. Lara, and F. Delgado, *SC* **24**, 2805 (1994).

Silver fluoride.

Cyclic azomethine ylides.[1] Saturated α,α'-bistrimethylsilyl azacycles lose both silyl groups to form the 1,3-dipolar species on exposure to AgF. The silver salt apparently acts both as an oxidant and a desilylating agent. In the presence of dipolarophiles 7-azabicyclo[2.2.1]heptane derivatives are readily prepared from *N*-benzyl-2,5-bistrimethylsilylpyrrolidine. A short route to epibatidine based on this chemistry has been developed.

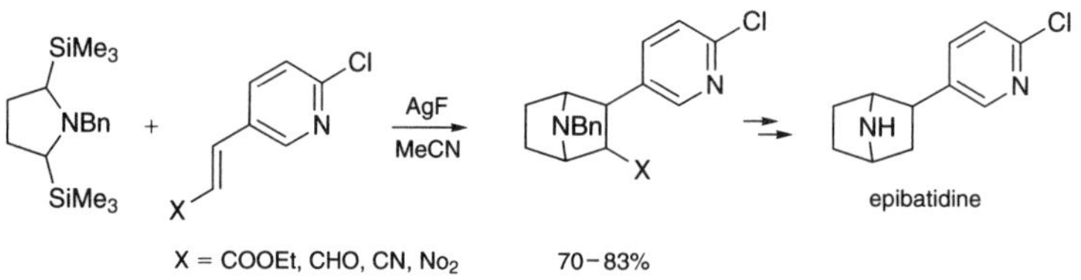

X = COOEt, CHO, CN, NO_2 70–83%

[1] G. Pandey, T. D. Bagul, and G. Lakshmaiah, *TL* **35**, 7439 (1994).

Silver nitrate.

Allenones to furans.[1] The isomerization typified by the following equation can be used to prepare 3-deuteriofurans.

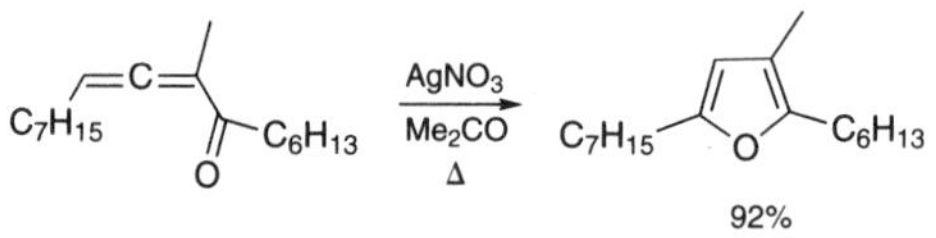

92%

Butenolides from β-(1-bromoalkyl)-β-lactones.[2] The dehydrobromination with ring expansion works in general. The bromo-β-lactones are the kinetic products of bromolactonization of β,γ-unsaturated acids.

[1] J. A. Marshall and G. S. Bartley, *JOC* **59**, 7169 (1994).
[2] T. H. Black and J. Huang, *TL* **34**, 1411 (1993).

Silver nitrite–iodine.

Dethioacetalization.[1]   The mild conditions (aqueous THF at room temperature) are favorable for regeneration of the carbonyl group from a monothioacetal or dithioacetal. The $AgClO_4$–I_2 system is even better, but it suffers from potential explosion hazard on a large-scale reaction.

[1] K. Nishide, K. Yokota, D. Nakamura, T. Sumiya, M. Node, M. Ueda, and K. Fuji *TL* **34**, 3425 (1993).

Silver(I) oxide.

RCONH₂ → RCN.[1]   The dehydration of primary amides under nonacidic conditions employs Ag_2O, molecular sieves and EtI.

Condensation with thiocarbonyl compounds.[2]   Activation of the C=S bond with Ag_2O renders such compounds reactive toward nucleophiles.

p-Quinone methides.[3]   4-Alkylphenols are oxidized to these reactive substances. It is a useful way to induce cyclization by a weak nucleophile such as an aromatic ring.

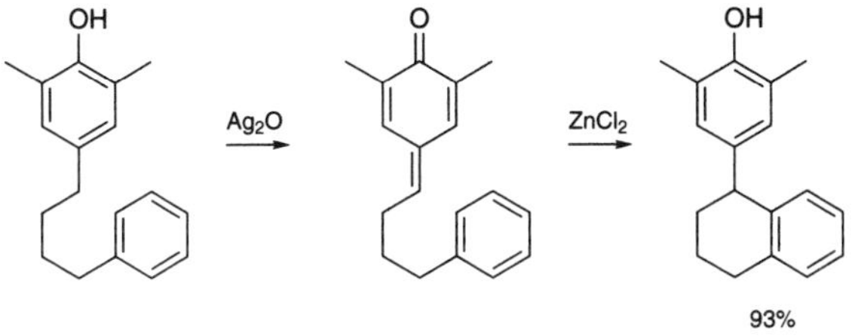

93%

[1] M. L. Sznaidman, C. Crasto, and S. M. Hecht, *TL* **34**, 1581 (1993).
[2] I. Shibuya, Y. Taguchi, T. Tsuchiya, A. Oishi, and E. Katoh, *BCSJ* **67**, 3048 (1994).
[3] S. R. Angle, D. O. Arnaiz, J. P. Boyce, R. P. Frutos, M. S. Louie, H. L. Mattson-Arnaiz, J. D. Rainier, K. D. Turnbull, and W. Yang, *JOC* **59**, 6322 (1994).

Silver(II) oxide.

Etherification.[1]   Methyl and isopropyl benzyl ethers are formed when the benzyl bromides are heated with AgO in the corresponding alcohols.

[1] B. Ortiz, F. Walls, F. Yuste, H. Barrios, R. Sanchez-Obregon, and L. Pinelo, *SC* **23**, 749 (1993).

Silver perchlorate.   16, 300–301

Replacement of anomeric α-haloacetoxy groups.   A combination of $AgClO_4$ with $LiClO_4$ or PhSnS provokes the heterolysis of such esters.[1] A silyl ether can be used as nucleophile for the glycosylation.[2]

Glycosylation by trityl sugars.[3] Disaccharide formation catalyzed by $AgClO_4$–$SnCl_4$ makes it possible to use the more stable trityl sugars (in situ detritylation) as nucleophiles.

Friedel–Crafts acylation.[4] Using the $AgClO_4$–$SiCl_4$ combination as catalyst, the acylation of arenes with free carboxylic acids and an aroic anhydride proceeds via the mixed anhydrides.

[1] T. Mukaiyama and N. Shimomura, *CL* 781 (1993).
[2] N. Shimomura and T. Mukaiyama, *BCSJ* **67**, 2532 (1994).
[3] S. Houdier and P. J. A. Vottero, *TL* **34**, 3283 (1993).
[4] K. Suzuki, H. Kitagawa, and T. Mukaiyama, *BCSJ* **66**, 3729 (1993).

Silver tetrafluoroborate. 13, 273–274

Aliphatic Friedel–Crafts acylations.[1] The Ag(I)-catalyzed reaction of an acyl chloride with functionalized alkenes proceeds under mild conditions. Functional groups such as halogens, ketone, and ester can be tolerated.

S-Glycosides.[2] Thioglycosides can be prepared from glycosyl bromides by reaction with β-trimethylsilylethyl sulfides in the presence of $AgBF_4$. The silver salt provides Ag^+ to assist ionization of the halogen, and also a F^- to initiate removal of the silylethyl group of the intermediately formed glycosylsulfonium salts.

[1] D. Barbry, C. Faven, and A. Ajana, *SC* **23**, 2647 (1993).
[2] A. Mahadevan, C. Li, and P. L. Fuchs, *SC* **24**, 3099 (1994).

Silver triflate. 13, 274–275; 14, 282–283; 16, 302; 17, 314

Glycosylation. Anomeric trichloroacetimidates,[1] phenylselenides,[2] and carbonates[3] are substituted by oxygen and nitrogen nucleophiles in the presence of AgOTf. For the displacement of glycosyl bromides,[4] *t*-butyl ethers instead of the hydroxyl compounds have also been used successfully.

Intramolecular *C*-glycosylation by a silyl enol ether leads to bicyclic acyloxetanes.[5]

Etherification.[6]   Various alcohols (primary, secondary, tertiary) can be etherified with primary alkyl halides in a reaction mediated by AgOTf and the base 2,6-di-*t*-butylpyridine.

Areneselenenyl triflates.   Conversion of selenenyl bromides to the triflates by AgOTf makes it possible to avoid introduction of the halogen atom to the alkene substrates on their selenylation. For example, stereoselective selenomethoxylation is readily achieved.[7]

Functionalized alkynes from alkynyl(phenyl)iodine(III) esters.[8]   The hypervalent iodinated acetylenes are formed by reaction of tributylstannylalkynes with PhI(CN)X (X = OTs, OBz, SCN). Elimination of PhI from these derivatives occurs on treatment with AgOTf. The reaction is applicable to diyne substrates.

Free radicals from α-haloketones.[9]   Photodechlorination in the presence of alkenes results in coupling products.

[1] S. P. Douglas, D. M. Whitfield, and J. J. Krepinsky, *JCC* **12**, 131 (1993).
[2] S. Mehta and B. M. Pinto, *JOC* **58**, 3269 (1993).
[3] N. Shimomura, T. Matsutani, and T. Mukaiyama, *BCSJ* **67**, 3100 (1994).
[4] A. Vargas-Berenguel, M. Meldal, H. Paulsen, K. J. Jensen, and K. Bock, *JCS(P1)* 3287 (1994).
[5] D. Craig and V. R. N. Munsinghe, *CC* 901 (1993).
[6] R. M. Burk, T. S. Gac, and M. B. Roof, *TL* **35**, 8111 (1994).
[7] R. Deizel, S. Goulet, L. Grenier, J. Bordeleau, and J. Bernier, *JOC* **58**, 3619 (1993).
[8] R. R. Tykwinski and P. J. Stang, *T* **49**, 3043 (1993).
[9] S.-H. Oh and T. Sato, *JOC* **59**, 3744 (1994).

Silver trifluoroacetate.   **16**, 301–302; **17**, 313–314

Photochemical trifluoromethylation.[1]   Arenes undergo trifluoromethylation in trifluoroacetic acid with $AgOCOCF_3$–TiO_2.

[1] C. Lai and T. E. Mallouk, *CC* 1359 (1993).

Sodium.   **13**, 277

Carbonyl reductions.   In an alcohol solvent sodium reduces ketones, esters, and related compounds. 3-Anisyl-methylenecamphor gives 3-*endo*-(*p*-methoxybenzyl)-isoborneol,[1] which is a potential chiral auxiliary. 1,3-Amino alcohols are obtained[2] from β-enamino ketones, but there is lack of stereoselectivity in the reduction.

Carboxamides are usually reduced to primary amines. However, α,α-disubstituted amino acid amides afford the 2-aminoethanols[3] on reduction with Na in propanol.

Sodium deposited on alumina is also a convenient reducing agent.[4] A nonprotonic solvent such as THF is used in the reduction.

[1] B. I. Seo, L. K. Wall, H. Lee, J. W. Buttrum, and D. E. Lewis, *SC* **23**, 15 (1993).

[2] G. Bartoli, C. Cimarelli, and G. Palmieri, *JCS(P1)* 537 (1994).

[3] H. M. Moody, B. Kaptein, Q. B. Broxterman, W. J. H. Boesten, and J. Kamphuis, *TL* **35**, 1777 (1994).

[4] S. Singh and S. Dev, *T* **49**, 10959 (1993).

Sodium amalgam.

Reductive elimination.[1]   A method for the synthesis of conjugated dienes and trienes involves reaction of 1,4-dibenzoyloxy-2-alkenes and 1,6-dibenzoyloxy-2,4-alkadienes with Na–Hg.

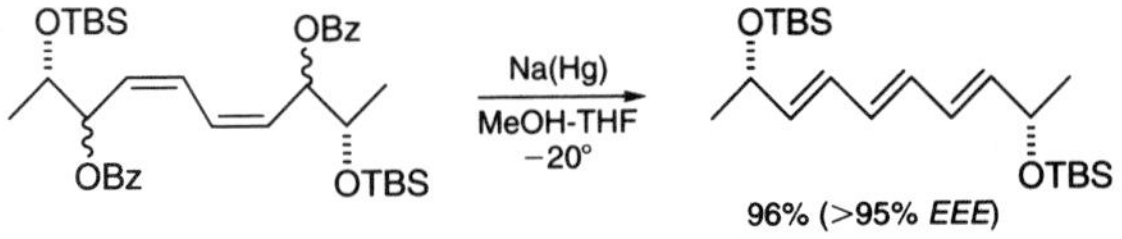

[1] G. Solladie, G. B. Stone, J.-M. Andres, and A. Urbano, *TL* **34**, 2835 (1993).

Sodium–ammonia. 16, 303–304

3-Arylbutyraldehydes.[1]   These aldehydes are produced when 5-aryl-4,5-dihydro-1,3-dioxepins are subjected to reduction with Na–NH$_3$. The transformation may involve elimination to furnish dienolate ions, which undergo reduction at the styrenic double bond.

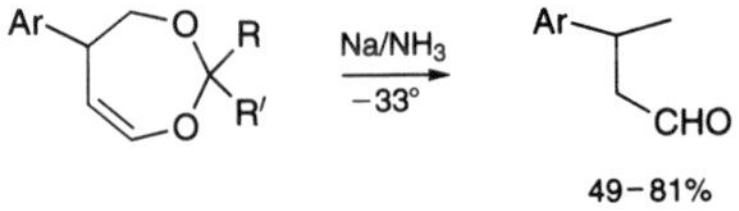

49–81%

[1] K. Samizu and K. Ogasawara, *TL* **35**, 7989 (1994).

Sodium arenesulfinates.

Heteroarylmethyl p-tolyl sulfones.[1]   These compounds can be prepared from the corresponding benzylic alcohols by reaction with 4-MeC$_6$H$_4$SO$_2$Na and HCOOH. The method is useful for substrates possessing σ^+ values between −1.90 to −0.95 in the aryl ring.

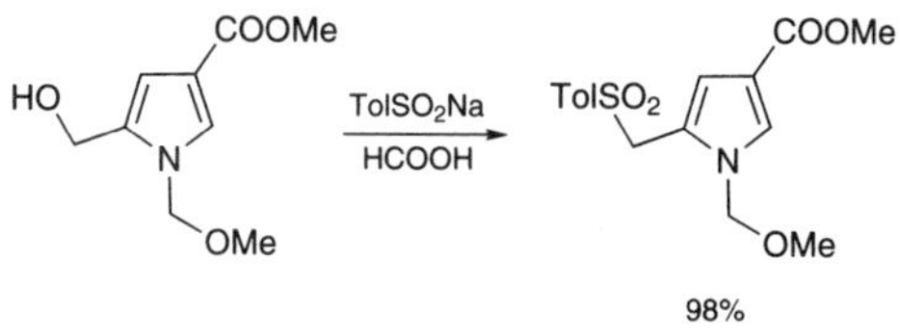

98%

Replacement of an unsaturated phenyliodo group. Formation of unsaturated sulfones by a tandem Michael addition–elimination is a highly efficient process that allows the synthesis of (Z)-1,2-bis(benzenesulfonyl)alkenes from (Z)-β-(benzenesulfonyl)alkenyliodonium salts.[2] In β-ketoethynyl(phenyl)iodonium salts the electron-withdrawing power of the ketone group is weaker, the Michael addition is followed by carbene formation. Cyclopentenones are formed.[3] A seemingly direct substitution of alkynyl(phenyl)iodonium salts gives alkynyl sulfones efficiently.[4]

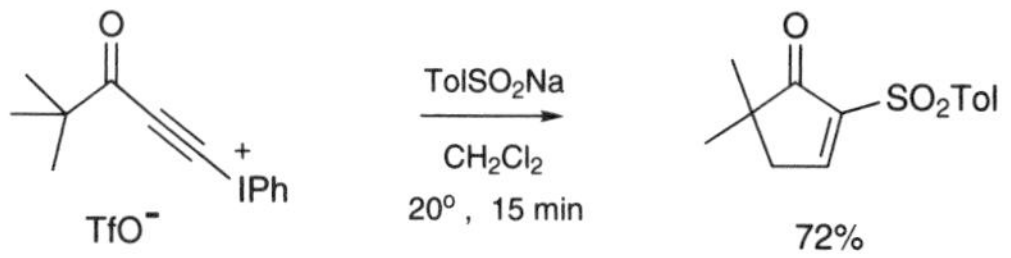

Free radical addition to alkenes. Generation of the toluenesulfonyl radical in the presence of proper alkenes leads to sulfones,[5] and cyclization occurs[6] when a free radical can be created in a remote position.

[1] L. C. Castedo, J. Delamano, C. Lopez, M. B. Lopez, and G. Tojo, *H* **38**, 495 (1994).
[2] M. Ochiai, K. Oshima, Y. Masaki, M. Kunishima, and S. Tani, *TL* **34**, 4829 (1993).
[3] B. L. Williamson, R. R. Tykwinski, and P. J. Stang, *JACS* **116**, 93 (1994).
[4] R. R. Tykwinski, B. L. Williamson, D. R. Fischer, P. J. Stang, and A. M. Arif, *JOC* **58**, 5235 (1993).
[5] C.-P. Chuang, *SC* **23**, 2371 (1993).
[6] I. W. Harvey and G. H. Whitham, *JCS(P1)* 185 (1993).

Sodium azide.

α-Azido amides. An unusual preparation[1] is the reaction of *N*-mesyloxy amides with NaN$_3$ in the presence of 15-crown-5. α-Lactams may be the intermediates. On the other hand, the displacement of α-(4-nitrobenzenesulfonyloxy) ketones to generate α-azido ketones[2] is a simple S$_N$2 reaction.

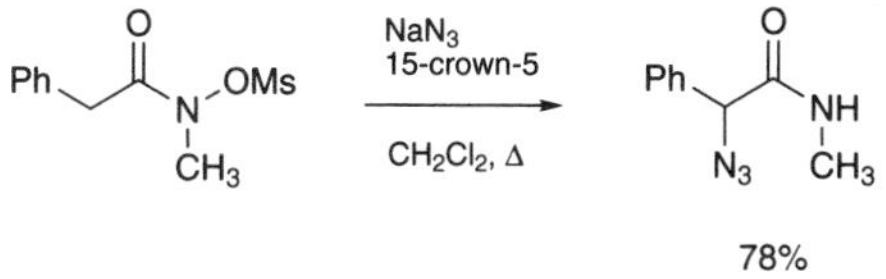

78%

Acyl azides.[3] A one-pot synthesis of $RCON_3$ from carboxylic acids is through reaction with Ph_3P, NCS, and NaN_3 in acetone.

Tetrazoles. The transformation of secondary amides to tetrazoles[4] requires triflic anhydride and NaN_3, probably to form imino triflates, which then undergo 1,3-dipolar cycloaddition and elimination. Ketones are also converted to 1,5-disubstituted 1H-tetrazoles directly when they are heated with NaN_3 and $TiCl_4$ in MeCN.[5]

[1] R.V. Hoffman, N.K. Nayyar, and W. Chen, *JOC* **58**, 2355 (1993).
[2] T. Patonay and R.V. Hoffman, *JOC* **59**, 2902 (1994).
[3] P. Froyen, *PSS* **89**, 57 (1994).
[4] E.W. Thomas, *S* 767 (1993).
[5] H. Suzuki, Y.S. Hwang, C. Nakaya, and Y. Matano, *S* 1218 (1993).

Sodium borohydride. 13, 278–279; 15, 290; 16, 304

Ketones from nitroalkenes.[1] The reduction also enables the assembly of spiroacetals[1] using the Nef condensation.

Reduction of C=N bonds. A simple method for the synthesis of α-stannyl-amines[2] containing a secondary amino group involves borohydride reduction of the corresponding imidoylstannanes. 3-Unsubstituted 4-isoxazolines are available by a reductive elimination[3] of 3-thioisoxazolium salts with $NaBH_4$.

A polyamine synthesis depends on the reductive cleavage of 2,3-dihydropyrimidines. The products have a syn-1,3 configuration.[4] 2-Aminocycloalkanone oximes give aliphatic amino nitriles,[5] which are the reduction products from Beckmann fragmentation.

Anthrones → anthracenes.[6] This reduction is carried out in two stages, first in diglyme and then in methanol. The addition of MeOH requires cooling to moderate the exothermic reaction.

o-Acyloxytoluenes.[7] Complete deoxygenation of the formyl group of several acylated salicylaldehydes has been observed in the borohydride reduction in aqueous DMSO.

Debromination.[8] α-Bromoethers undergo debromination, whereas ordinary alkyl halides remain untouched.

95%

[1] R. Ballini, G. Bosica, and R. Schaafstra, *LA* 1235 (1994).
[2] H. Ahlbrecht and V. Baumann, *S* 770 (1994).
[3] H. B. Jeon and K. Kim, *TL* **34**, 1939 (1993).
[4] J. Barluenga, V. Kouznetsov, E. Rubio, and M. Tomas, *TL* **34**, 1981 (1993).
[5] P. A. Petukhov and A. V. Tkachev, *SL* 580 (1993).
[6] D. J. Marquardt and F. A. McCormick, *TL* **35**, 1131 (1994).
[7] H.-B. Zhou, R. D. Xie, F. Chao, X.-D. Feng, and Q.-F. Zhou, *HX* **50**, 1182 (1992).
[8] N. De Kimpe, M. Boelens, and J. Baele, *JOC* **59**, 5485 (1994).

Sodium borohydride–aqueous base.

Enone reduction.[1] 2,6-Di-*t*-butyl-*p*-quinols give the 4-hydroxycyclohexenones.

Formation of aryl radicals.[2] Reduction of aryl bromides proceeds via radical intermediates. Addition to a side-chain double bond in such reactions is observed.

Organotelluride anions.[3] Ditellurides are reduced to generate the anions that are useful for phenylseleno removal from the α-position of a carbonyl group.

[1] A. Nishinaga, S. Kojima, T. Mashino, and K. Maruyama, *CL* 961 (1994).
[2] R. Rai and D. B. Collum, *TL* **35**, 6221 (1994).
[3] C. C. Silveira, E. J. Lenardao, and J. V. Comasseto, *SC* **24**, 575 (1994).

Sodium borohydride–bismuth(III) chloride.

Reduction of nitrogenous compounds.[1] This reducing system transforms nitroarenes and imines to amines.

[1] H. N. Borah, D. Prajapati, and J. S. Sandhu, *JCR(S)* 228 (1994).

Sodium borohydride–calcium chloride.

Stereoselective reduction.[1] Epoxy ketones are reduced to give predominantly the *erythro* alcohols. Lanthanum chloride can be used instead of $CaCl_2$.

93% (9 : 91)

[1] M. Taniguchi, H. Fujii, K. Oshima, and K. Utimoto, *T* **51**, 679 (1995).

Sodium borohydride–cerium(III) chloride. 15, 291; **17**, 314–316

Cyclic imide reduction.[1]   The reagent reduces imides to give hydroxy lactams in a regioselective manner, which is different from LiBHEt$_3$. The complementary is very useful in synthesis.

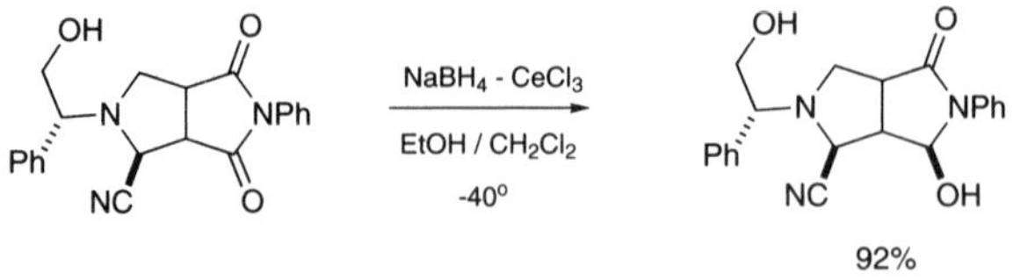

92%

[1] P. Deprez, J. Royer, and H.-P. Husson, *T* **49**, 3781 (1993).

Sodium borohydride–copper(II) salts. 13, 279

RCON$_3$ → RCONH$_2$.[1]   The presence of CuSO$_4$ facilitates the conversion of acyl azides to amides (80–95% yield) by NaBH$_4$.

Reductions in the presence of Cu(II) exchanged resins.[2]   Activation of ketones toward reduction under such conditions has been claimed.

[1] H. S. P. Rao and P. Siva, *SC* **24**, 549 (1994).
[2] A. Sarkar, B. R. Rao, and B. Ram, *SC* **23**, 291 (1993).

Sodium borohydride–iodine. 17, 316

2-Amino alcohols.[1]   α-Amino acids can be reduced by this system, which is actually a borane generator.

[1] M. tJ. McKennon, A. I. Meyers, K. Drauz, and M. Schwarm, *JOC* **58**, 3568 (1993).

Sodium borohydride–manganese(II) chloride.

β-Hydroxy esters and -amides.[1]   β-Keto esters and amides undergo reduction with good stereoselectivity. Thus the 2-methyl derivatives give predominantly the *erythro* products, which is opposite to the results of a Bu$_4$NBH$_4$ reduction.

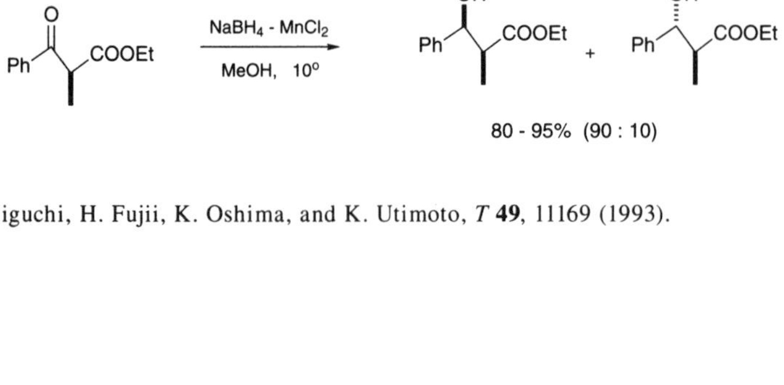

80 - 95% (90 : 10)

[1] M. Taniguchi, H. Fujii, K. Oshima, and K. Utimoto, *T* **49**, 11169 (1993).

Sodium borohydride–pentafluorophenol.

Allylic alcohols. The tendency of $NaBH_4$ to effect conjugate reduction of α,β-unsaturated compounds is greatly reduced by C_6F_5OH. With borane scavenged by TMEDA or 1-hexene the selectivity is further improved.[1] Only highly reactive ketones and acid chlorides are reduced by this reagent.[2]

[1] J. C. Fuller, S. M. Williamson, and B. Singaram, *JFC* **68**, 265 (1994).
[2] J. C. Fuller, M. L. Karpinski, S. M. Williamson, and B. Singaram, *JFC* **66**, 123 (1994).

Sodium borohydride–propanedithiol–triethylamine.

$RN_3 \rightarrow RNH_2$.[1] The isoxazole ring system to which an azidomethyl group is attached is not affected by the reducing milieu.

[1] Y. Pei and B. O. S. Wickham, *TL* **34**, 7509 (1993).

Sodium borohydride–sulfuric acid or trifluoroacetic acid.

Reductive alkylation of arylamines. The $NaBH_4$–H_2SO_4 system may be used[1] instead of the more expensive $NaBH_3CN$. Monoalkylation or peralkylation may be controlled.[2]

Reductive demetallation of Cr–aminocarbene complexes.[3] The C=Cr bond of such complexes is cleaved and replaced by the CH_2 group when they are reduced with $NaBH_4$–CF_3COOH at 0°C.

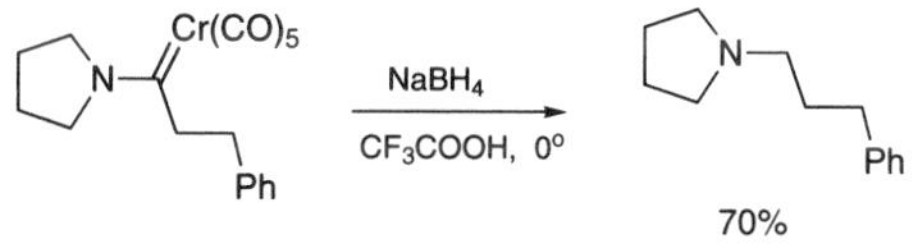

[1] G. Verardo, A. G. Giumanini, P. Strazzolini, and M. Poiana, *S* 121 (1993).
[2] G. Verardo, A. G. Giumanini, P. Strazzolini, *SC* **24**, 609 (1994).
[3] C. Baldoli, P. del Buttero, E. Licandro, S. Maiorana, A. Papagni, and A. Zanotti-Gerosa, *SL* 677 (1994).

Sodium borohydride–tellurium.

$ArSO_2Cl \rightarrow ArSSAr$.[1] The tellurated borohydride $NaBH_2Te_3$ is generated in situ. The reductive dimerization is carried out in DMF with moderate heating.

[1] H. Suzuki, T. Nakamura, and M. Yoshikawa, *JCR(S)* 70 (1994).

Sodium borohydride–trimethylsilyl chloride.

Acetic acid α-radical.[1] The radical is generated through reduction of iodoacetic acid with this combination of reagents in the presence of AIBN. The radical is intercepted by acceptors such as acrylic esters.

[1] F. Foubelo, F. Lioret, and M. Yus, *T* **49**, 8465 (1993).

Sodium bromate.

N-Bromo amides and imides.[1] The very valuable brominating and oxidizing agents can be regenerated from the spent amides and imides by *N*-bromination. Using $NaBrO_3$ and HBr (or NaBr) in the presence of sulfuric acid is an economical and convenient method.

[1] S. Fujisaki, S. Hamura, H. Eguchi, and A. Nishida, *BCSJ* **66**, 2426 (1993).

Sodium bromite. 14, 287; 15, 292

Oxidation of alcohols. The oxidation of primary alcohols to aldehydes with aqueous $NaBrO_2$ is catalyzed either by Cu or Bu_3SnCl.[1] $α,ω$-Diols give lactones (54–95%)[2] when the oxidation is carried out with $NaBrO_2 \cdot 3H_2O-Al_2O_3$ in an organic solvent.

Vicinal tricarbonyl compounds have been obtained from 2,3-dihydroxy-alkanoic esters with $NaBrO_2$ in combination with 4-benzoyloxy-2,2,6,6-tetramethylpiperidin-1-oxyl.[3] In an alternative method bromide ion is used to mediate electrooxidation.

Epoxidation of alkenes.[4] This reaction proceeds at room temperature in the presence of $CuSO_4$.

[1] J. Yatabe, T. Sugisaki, O. Moriya, and T. Kageyama, *NKK* 359 (1993).
[2] T. Morimoto, M. Hirano, K. Iwasaki, and T. Ishikawa, *CL* 53 (1994).
[3] T. Inokuchi, P. Liu, and S. Torii, *CL* 1411 (1994).
[4] J. Yatabe, T. Sugisaki, O. Moriya, and T. Kageyama, *NKK* 1446 (1992).

Sodium N-(t-butoxycarbonyl)-N-diethylphosphoramide.

Protected allylic amines. The reagent, which has been used in lieu of potassium phthalimide in the synthesis of primary amines,[1] is also active in Pd(0)-catalyzed substitution of allylic acetates.[2] The Boc group of the products can be cleaved by CF_3COOH at 0°C, whereas both protecting groups are removed by HCl in benzene.

[1] A. Zwierzak and S. Pilichowska, *S* 922 (1982).
[2] R. O. Hutchins, J. Wei, and S. J. Rao, *JOC* **59**, 4007 (1994).

Sodium t-butylthiolate.

t-Butylsulfenamides.[1] Reductive sulfenylation of azoxybenzenes is achieved by this thiolate, but not by other alkanethiolates. Thus the reaction with *i*-PrSNa results in the anilines (due to the instability of the products).

[1] M. T. Dario, S. Montanari, C. Paradisi, and G. Scorrano, *TL* **35**, 301 (1994).

Sodium cyanoborohydride. 14, 287–288; 16, 305–306

Reduction of cyclic imines.   More examples of stereoselective formation of cyclic amines[1] by reduction have appeared in the literature. Another useful feature of the reducing agent is the possible preservation of the N–O bond, which permits the synthesis of perhydro-1,2-oxazines[2] and *N*-alkylation of hydroxylamines.[3]

Macrolactone formation.[4]   ω-Iodoalkyl acrylates are transformed into macrolactones (10- to 16-membered) by a photoinduced reaction in the presence of NaBH₃CN. The yields (73–90%) are remarkably high for the intramolecular process.

[1] K. Pal, M. L. Behenke, and L. Tong, *TL* **34**, 6205 (1993).
[2] R. Zimmer, T. Arnold, K. Homann, and H.-U. Reissig, *S* 1050 (1994).
[3] W. Oppolzer, P. Cintas-Moreno, O. Tamura, and F. Cardinaux, *HCA* **76**, 187 (1993).
[4] M. Abe, T. Hayashikoshi, and T. Kurata, *CL* 1789 (1994).

Sodium cyanoborohydride–boron trifluoride etherate.

Deoxygenation of aryl ketones.[1]   The convenient reduction is particularly distinguished by its mildness, in comparison with Clemmensen and Wolff–Kishner reductions.

[1] A. Srikrishna, J. A. Sattigeri, R. Viswajanani, and C. V. Yelamaggad, *SL* 93 (1995).

Sodium dithionite. 13, 281

α-Hydroxy sulfinates.[1]   The adducts formed by admixture of RCHO, $Na_2S_2O_4$, and NaOH are similar to bisulfite complexes of aldehydes.

Reduction of perfluoroalkyl iodides.[2]   In the presence of a mild base deiodination takes place. If propargyl alcohol is present, adducts are formed. The allylic alcohol derived from iodoperfluoroethane and propargyl alcohol is a useful precursor of 3-trifluoromethylpyrazole-2-methanol.

ArNO₂ → ArNH₂.[3]   Excellent yields (91–97%) of arylamines are produced in the reduction using viologen, which acts as phase-transfer catalyst and electron-transfer agent.

[1] M. Mulliez and C. Naudy, *T* **49**, 2469 (1993).
[2] X.-Q. Tang and C.-M. Hu, *CC* 631 (1994).
[3] K. K. Park, C. H. Oh, and W. K. Joung, *TL* **34**, 7445 (1993).

Sodium ethoxide.

Cationic aza-Cope rearrangement.[1] Doubly allylic ammonium salts may be induced to undergo sigmatropic rearrangement after equilibration to the allyl vinyl ammonium isomers. Hydrolysis of the resulting iminium ions leads to γ,δ-unsaturated carbonyl compounds.

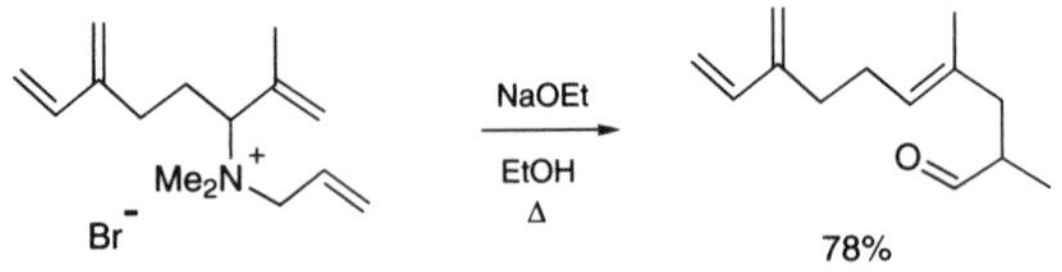

[1] K. Honda and S. Inoue, *SL* 739 (1994).

Sodium formate.

Selective dehalogenation.[1] Aromatic halogen atoms can be removed catalytically while using sodium formate as hydrogen source. For example, *N,N*-diacetyl-2,4,6-trichloroaniline gives the 2,6-dichloro compound on refluxing with Pd–C and HCOONa in MeCN.

Allylsilanes from trialkylsilylallyl esters.[2] The Pd(0)-catalyzed cleavage of allylic functionalities requires a nucleophile to complete a cycle. For reduction, HCOONa is adequate, and the presence of 15-crown-5 is beneficial.

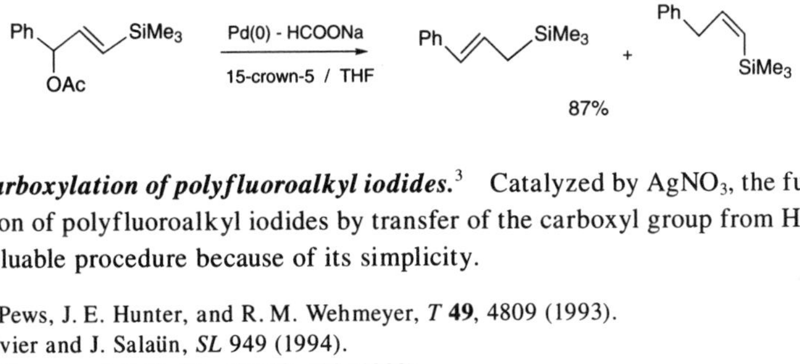

Carboxylation of polyfluoroalkyl iodides.[3] Catalyzed by AgNO₃, the functionalization of polyfluoroalkyl iodides by transfer of the carboxyl group from HCOONa is a valuable procedure because of its simplicity.

[1] R. G. Pews, J. E. Hunter, and R. M. Wehmeyer, *T* **49**, 4809 (1993).
[2] J. Ollivier and J. Salaün, *SL* 949 (1994).
[3] B.-N. Huang and F.-H. Wu, *YH* **13**, 403 (1993).

Sodium hexamethyldisilazide.

1-Bromoalkynes.[1] Dehydrobromination of 1,1-dibromoalkenes with the base is quite efficient. An epoxide survives such treatment.

α-Iodovinyl epoxides.[2] Iodination of α-allenyl alcohols and treatment of the diiodides with the strong base lead to the epoxides. Under mild conditions the vinylic iodide does not undergo dehydroiodination.

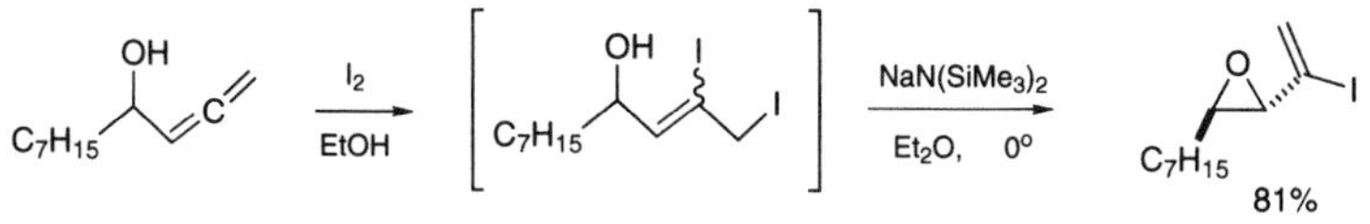

[1] D. Grandjean, P. Pale, and J. Chuche, *TL* **35**, 3529 (1994).
[2] W. Friesen and M. Blouin, *JOC* **58**, 1653 (1993).

Sodium hydride. 14, 288; 16, 307–308

Enol carbonates.[1]   *O*-Acylation is accomplished by rapid addition of ketone enolates, which are generated with NaH–TMEDA in refluxing THF, to chloroformate esters at 0°C. Aryl and α,β-enone enolates are better prepared with NaN(SiMe$_3$)$_2$–TMEDA at −78°.

Allenes from α-hydroxyalkyl vinylsilanes.[2]   Successful extension of the β-elimination process to an allene synthesis has been demonstrated.

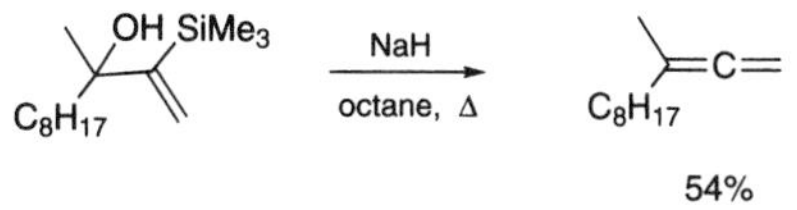

Complex reagent for transesterification.[3]   The aggregate prepared from NaH, *t*-BuONa, and Ni(OAc)$_2$ is an effective catalyst for transesterification.

[1] L. M. Harwood, Y. Houminer, A. Manage, and J. I. Seeman, *TL* **35**, 8027 (1994).
[2] P. F. Hudrlik, A. M. Kassim, E. L. O. Agwaramgbo, K. A. Doonquah, R. R. Roberts, and A. M. Hudrlik, *OM* **12**, 2367 (1993).
[3] Y. Fort, M. Remy, and P. Caubere, *JCR(S)* 418 (1993).

Sodium hydrogen selenide.

Aromatic selenoamides.[1]   NaSeH prepared from NaBH$_4$ and Se can be used in situ in refluxing ethanol to achieve the transformation of nitriles to selenoamides.

Reduction of nitrogenous aromatic compounds.[2]   Nitro-, nitroso- and dinitro-soarenes are reduced to the anilines in 70–94% yield.

[1] X.-R. Zhao, M.-D. Ruan, W.-Q. Fan, and X.-J. Zhou, *SC* **24**, 1761 (1994).
[2] D. K. Dutta, D. Konwar, and J. S. Sandhu, *JCR(S)* 388 (1994).

Sodium hydrogen telluride. 13, 282

Conjugate reduction.[1]   NaTeH in ethanol at room temperature can reduce α,β-unsaturated carbonyl compounds in the 1,4 sense. Conjugated triple bonds are saturated stepwise.

Desulfonylation.[2] (2E)-α-Amido-α,β-unsaturated sulfones are converted to (2Z)-cinnamanilides in 67–81% yield (9 examples) in a mixture of EtOH and THF.

Transesterification.[3] Various esters can be transformed into ethyl esters in hot ethanol and acetic acid in the presence of NaTeH.

[1] M. Yamashita, Y. Tanaka, A. Arita, and M. Nishida, *JOC* **59**, 3500 (1994).
[2] X. Huang, J.-H. Pi, and Z.-Z. Huang, *HC* **3**, 535 (1992).
[3] J. R. Suresh, P. S. Mohan, and P. Shanmugam, *IJC(B)* **33B**, 290 (1994).

Sodium hydroxide.

Azetidines and pyrrolidines.[1] Base-promoted cyclization of epoxy tosylamides is sensitive to the structures of the substrates. Thus products having either a primary or a secondary hydroxyl group may be obtained.

95%

Elimination/cyclization.[2] Propargyl alcohols in which the other propargyl position contains a benzotriazolyl group is subject to elimination by base to afford furans.

81%

[1] J. Moulines, J.-P. Bats, P. Hautefaye, A. Nuhrich, and A.-M. Lamidey, *TL* **34**, 2315 (1993).
[2] A. R. Katritzky, J. Li, and M. F. Gordeev, *JOC* **58**, 3038 (1993).

Sodium hypochlorite. 15, 293; 16, 308; 17, 316

Epoxidation.[1] NaOCl provides the active oxygen for delivery to alkenes through interaction with metal–porphyrin complexes. Typically, 5,10,15,20-tetrakisaryl-porphyrins and their 2,3,7,8,12,13,17,18-octahalo derivatives form active catalysts with metal ions such as Mn. Another heterocyclic ligand (e.g., 4-methylpyridine) and a phase-transfer agent are also added to the reaction medium.

Carbonyl regeneration from oximes.[2] This transformation is usually carried out in MeCN, but for aldoximes the basicity of the solution must be regulated to pH5–7 in order to avoid oxidation of the released aldehydes to carboxylic acids.

ArI → ArCl.[3]   The halogen exchange occurs under phase-transfer conditions at pH 8.2–9.2.

[1] A. M. d'A. R. Gonsalves, M. M. Pereira, A. C. Serra, R. A. W. Johnstone, and M. L. P. G. Nunes, *JCS(P1)* 2053 (1994).

[2] J. M. Khurana, A. Ray, and P. K. Sahoo, *BCSJ* **67**, 1091 (1994).

[3] T. O. Bayraktaroglu, M. A. Gooding, S. F. Khatib, H. Lee, M. Kourouma, and R. G. Landolt, *JOC* **58**, 1264 (1993).

Sodium hypochlorite–2,2,6,6-tetramethylpiperidin-1-oxyl.

Oxidation of primary alcohols.[1]   This selective oxidation system allows the preparation of uronic acids from monosaccharide derivatives.

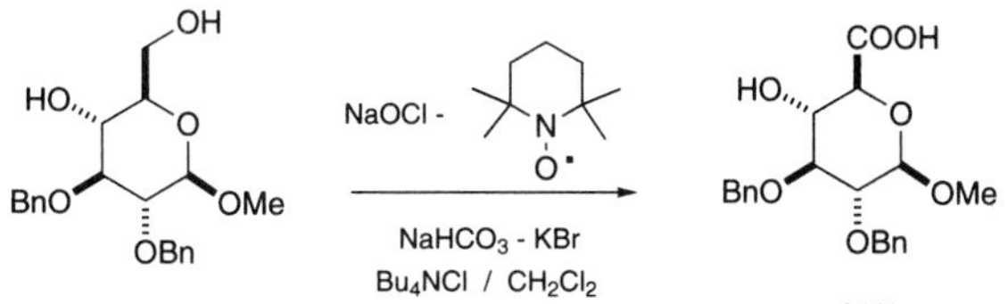

61%

Sulfoxides.[2]   In the phase-transfer oxidation of sulfides, saturated $NaHCO_3$ is added. The stoichiometry of NaOCl controls the oxidation at one or both sulfur atoms of bis(phenylthio)alkanes.

[1] N. J. Davis and S. L. Flitsch, *TL* **34**, 1181 (1993).

[2] R. Siedlecka and J. Skarzewski, *S* 401 (1994).

Sodium hypophosphite–Raney nickel.

Ketones from nitroalkenes.[1] NaH_2PO_2 is a good hydrogen donor in the presence of a catalyst. Its combination with Raney nickel to convert nitroalkenes to ketones has been put to use in the general synthesis of 1,3-, 1,4-, and 1,5-diketones from functionalized nitroalkanes.

[1] R. Ballini and G. Bartoli, *S* 965 (1993).

Sodium methoxide.

Cleavage of cyclic α-chloro-α-sulfonyl ketones.[1]   The base causes ring opening to give methyl ω-chloro-ω-sulfonylalkanoates. On the other hand, Favorskii rearrangement is the favored pathway when piperidine is used instead of NaOMe.

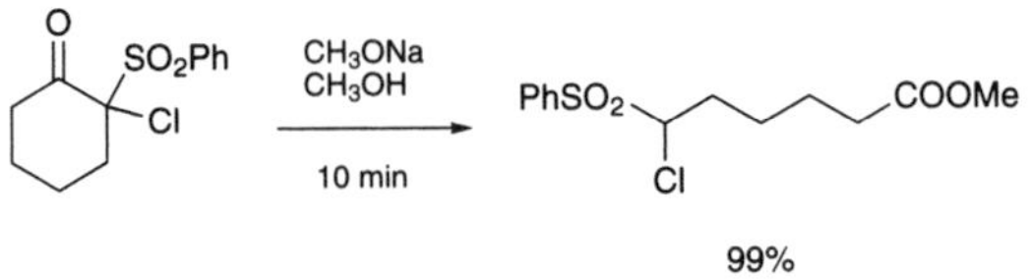

99%

Selective alcoholysis of acylureas.[2]   Different modes of reactions are witnessed when acylureas are treated with either NaOMe in MeOH or BnOLi in ether. Methyl esters and amides are the respective products.

1-(Benzylideneamino)cyclopropanecarboxylic esters.[3]   γ-Chloro α-imino esters in which the β-position is fully substituted undergo base-promoted cyclization.

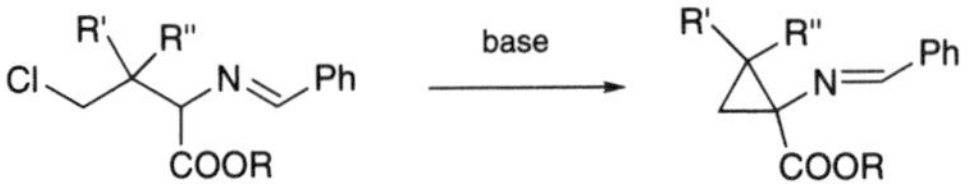

2-Methoxy-4-haloquinolines.[4]   Selective displacement of the 2-halogen from 2,4-dihaloquinolines may be achieved with solid NaOMe (or NaOEt) in refluxing toluene.

[1] T. Satoh, K. Oguro, J. Shishikura, N. Kanetaka, R. Okada, and K. Yamakawa, *BCSJ* **66**, 2339 (1993).
[2] K. Kishikawa, H. Eida, S. Kohmoto, M. Yamamoto, and K. Yamada, *S* 173 (1994).
[3] M. Boeykens, N. De Kimpe, and K. A. Tehrani, *JOC* **59**, 6973 (1994).
[4] A. G. Osborne and L. A. D. Miller, *JCS(P1)* 181 (1993).

Sodium nitrate–trimethylsilyl chloride–aluminum chloride.

Nitration of arenes.[1]   The active reagent may be nitryl chloride (NO_2Cl). It can nitrate PhCOOMe and $PhCF_3$ in good yields at room temperature.

[1] G. A. Olah, P. Ramaiah, G. Sandford, A. Orlinkov, and G. K. S. Prakash, *S* 468 (1994).

Sodium nitrite–potassium ferricyanide.

Nitration.[1,2]   The synthesis of *gem*-dinitroalkanes by reaction of nitronate anions with the combined salts is efficient.

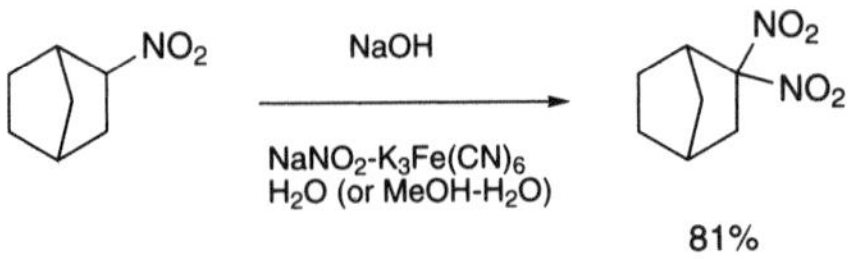

81%

[1] A. P. Marchand, R. Sharma, U. R. Zope, W. H. Watson, and R. P. Kashyap, *JOC* **58**, 759 (1993).
[2] G. A. Olah, P. Ramaiah, G. K. S. Prakash, and R. Gilardi, *JOC* **58**, 763 (1993).

Sodium nitrosoferricyanide.

Diazotization.[1]   2-Amino alcohols are diazotized by this reagent and converted to epoxides. The method has been applied to the preparation of enantiomeric propylene oxides from threonine.

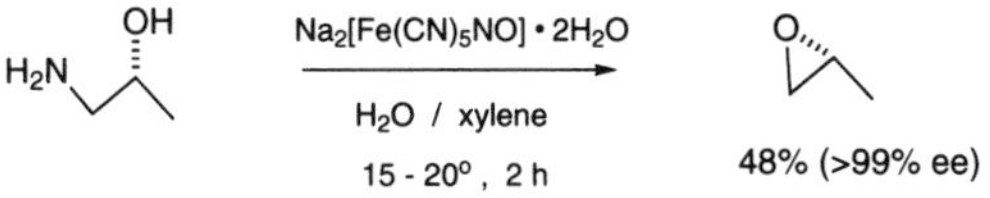

[1] K. Rossen, P. M. Simpson, and K. M. Wells, *SC* **23**, 1071 (1993).

Sodium perborate. 14, 290–291; 16, 310

Oxidative deselenylation.[1]   In the presence of Ac_2O sodium perborate converts α-selenyl carbonyl compounds to the conjugated systems. The possible oxidant is (1).

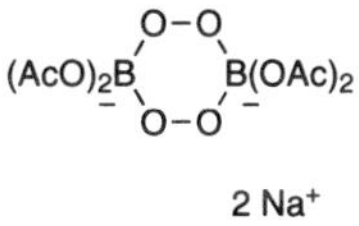

(1)

Hydration of nitriles.[2]   Formation of the quinazolinone system from *o*-amido benzonitriles in one step is shown.

Oxidative rearrangements.   In the presence of an acid, *C,N*-diarylaldimines give *N,N*-diarylformamides.[3] Benzylic alcohols form the corresponding hydroperoxides, which split off the alkyl residue to furnish phenols[4] by the well-known rearrangement pathway.

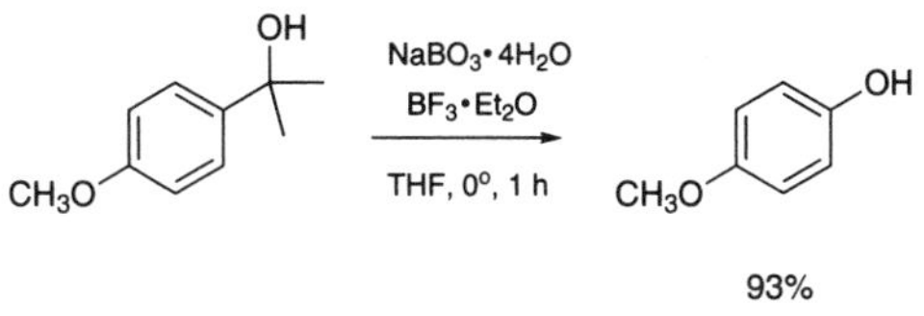

[1] G.W. Kabalka, N. K. Reddy, and C. Narayana, *SC* **23**, 543 (1993).
[2] B. Baudoin, Y. Ribeill, and N. Vicker, *SC* **23**, 2833 (1993).
[3] P. Nongkunsarn and C. A. Ramsden, *TL* **34**, 6773 (1993).
[4] G.W. Kabalka, N. K. Reddy, and C. Narayana, *TL* **34**, 7667 (1993).

Sodium periodate. 15, 294

Chalcogenide oxidation. The oxidation and selenoxide elimination sequence is applicable to the generation of functionalized alkenes which are prone to Claisen rearrangement. Medium-sized rings including tetrahydroazepin-2-ones are accessible in this manner.[1]

Allylic sulfoxides are more stable then sulfenate esters; therefore, the former species do not show tendency to undergo [2,3]sigmatropic rearrangement. However, when the double bond belongs to an enol ether, the sulfenate ester becomes geminal to the alkoxy group and expellable. Accordingly, α,β-unsaturated aldehydes are isolated from oxidation of the alkoxyallyl sulfides.[2]

Cleavage of enamines.[3] The reagent is particularly suitable for the preparation of *o*-nitroaraldehydes from *o*-nitrotoluenes via ω-aminostyrenes.

Boronic acids from pinanediol boronates.[4] The liberation of a boronic acid from its esters is sometimes hampered by the favorable equilibrium toward the latter. The problem is solved by destroying the diol, and one method is to use $NaIO_4$.

Chiral amines.[5] Synthesis of chiral amines from ketones through reductive amination with (+)- or (-)-norephedrine requires selective C–N bond cleavage. Sacrificial dissection of the chiron by periodate oxidation completes the operation.

Selenophenes.[6] Ring contraction of dihydroselenopyrans is effected by $NaIO_4$.

2'-Hydroxychalcones → flavones.[7] The oxidative cyclization proceeds in DMSO at 100–120°C.

[1] P. A. Evans, A. B. Holmes, and K. Russell, *JCS(P1)* 3397 (1994).
[2] T. Sato and J. Otera, *TL* **35**, 6701 (1994).

[3] M. G. Vetelino and J. W. Coe, *TL* **35**, 219 (1994).
[4] S. J. Coutts, J. Adams, D. Krowlikowski, and R. J. Snow, *TL* **35**, 5109 (1994).
[5] R. Sreekumar and C. N. Pillai, *TA* **4**, 2095 (1993).
[6] T. Kataoka, Y. Ohe, A. Umeda, T. Iwamura, M. Yoshimatsu, and H. Shimizu, *CC* 577 (1993).
[7] N. Hands and S. K. Grover, *SC* **23**, 1021 (1993).

Sodium periodate–tetraphenylporphyrin–MnCl.

Epoxidation of alkenes.[1]   The oxidation is carried out under phase-transfer conditions.

[1] D. Mohajer and S. Tangestaninejad, *CC* 240 (1993).

Sodium persulfate.

Sulfonylation.[1]   The iodine atom of fluoroalkyl iodides suffers oxidation and is then replaced by a sulfonic acid group on reaction with $Na_2S_2O_8$. On further treatment with chlorine and then with KF the sulfonyl fluorides are readily synthesized.

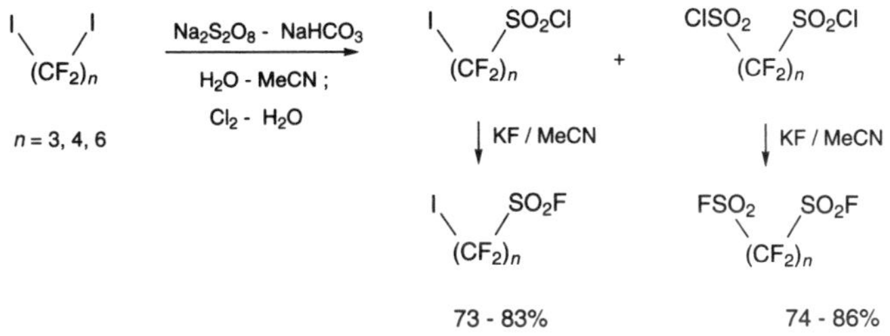

Iodolactones.[2]   Iodolactonization of unsaturated acids may employ KI as the source of iodine when $Na_2S_2O_8$ is added to the reaction medium.

Quinones.[3]   For the oxidation of dihydroxyarenes the persulfate–Cu(II) system is efficient and mild. Good yields are obtained (9 examples, 91–100%).

[1] W. Qiu and D. J. Burton, *JFC* **60**, 93 (1993).
[2] A. C. Royer, R. C. Mebane, and A. M. Swafford, *SL* 898 (1993).
[3] C. Costantini, M. d'Ischia, and G. Prota, *S* 1399 (1994).

Sodium piperidino(diethyl)aluminum hydride.

Aldehydes.   The reduction of carboxylic acid derivatives such as esters and nitriles is stopped at the aldehyde stage with the modified sodium aluminum hydride. The reagent is prepared from $NaAlH_2Et_2$ and piperidine in THF–toluene at 0°C.[1] Interestingly, acid chlorides can be converted to aldehydes by treatment with piperidine and then with $NaAlH_2Et_2$.[2]

[1] N. M. Yoon, J. H. Ahn, D. K. An, and Y. S. Shon, *JOC* **58**, 1941 (1993).
[2] N. M. Yoon, K. I. Choi, Y. S. Gyoung, and W. S. Jun, *SC* **23**, 1775 (1993).

Sodium telluride.

Deoxygenation of epoxides.[1] Na_2Te is generated from $NaBH_4$ and Te. The epoxide deoxygenation with this reagent giving alkenes is accompanied by ester hydrolysis when such a group is present.

Reduction of ArCHO.[2] *N*-Methyl-2-pyrrolidone is used as solvent. When an ArCN is treated similarly, a pyrrolo[2,3-*d*]pyrimidine is formed.

[1] D. C. Dittmer, Y. Zhang, and R. P. Discordia, *JOC* **59**, 1004 (1994).
[2] H. Suzuki and T. Nakamura, *JOC* **58**, 241 (1993).

Sodium triacetoxyborohydride. 13, 283; 16, 309–310

β-Amino esters.[1] Diastereoselective and enantioselective reduction of enamino esters with a chiral *N*-substituent by this reagent manifests 1,3-asymmetric induction. The de values range from 39% to 85% for 7 examples.

Reductive amination–lactamization.[2] Reaction of carbonyl compounds with the proper ω-amino esters in the presence of $NaBH(OAc)_3$ leads to lactams directly.

Dialkyl 1-aminoalkylphosphonates.[3] Reduction of α-oximinophosphonate esters with $NaBH(OAc)_3$ and $TiCl_3$ in a buffered MeOH solution constitutes a convenient method for the synthesis of the aminophosphonates.

[1] C. Cimarelli, G. Palmieri, and G. Bartoli, *TA* **5**, 1455 (1994).
[2] A. F. Abdel-Magid, B. D. Harris, and C. A. Maryanoff, *SL* 81 (1994).
[3] A. Ryglowski and P. Kafarski, *SC* **24**, 2725 (1994).

Sodium trialkylborohydrides.

Preparation.[1] From $NaAlH_2Et_2$ and R_3B in the presence of 1,4-diazabicyclo-[2.2.2]octane. Such reagents with bulky alkyl groups exhibit high stereoselectivities in reduction reactions.

[1] J. L. Hubbard, J. C. Fuller, T. C. Jackson, and B. Singaram, *T* **49**, 8311 (1993).

Sodium tri-*t*-butoxyaluminum hydride.

Aldehydes.[1] Reduction of acid chlorides in diglyme proceeds at $-78°C$. A great variety of aldehydes can be acquired in very good yields.

[1] J. S. Cha and H. C. Brown, *JOC* **58**, 4732 (1993).

Sodium trimethylsilanethiolate.

Demethylation of methoxyarenes.[1] Me_3SiSNa is prepared from $(Me_3Si)_2S$ and NaOMe, and the demethylation occurs in the presence of 1,3-dimethyl-2-imidazoli-dinone at 180°C. Certain phenolates thus generated can undergo cyclization to give dibenzofuran derivatives.[2]

In situ method for reduction and sulfurization.[3] The combination of Me_3SiCl and Na_2S successfully reduces arylamines and sulfoxides, converts nitriles to thioamides, alkyl halides to sulfides, as well as demethylates aryl methyl ethers.

[1] M.-J. Shiao, W.-S. Ku, and J. R. Hwu, *H* **36**, 328 (1993).
[2] L.-L. Lai, P.-Y. Lin, W.-H. Huang, M.-J. Shiao, and J. R. Hwu, *TL* **35**, 3545 (1994).
[3] M.-J. Shiao, L.-L. Lai, W.-S. Ku, P.-Y. Lin, and J. R. Hwu, *JOC* **58**, 4742 (1993).

Succinimidyl diazoacetate.

This new diazoacylating agent (**1**) is prepared in 65% yield[1] by condensation of *N*-hydroxysuccinimide with *p*-tosylhydrazonoacetic acid (DCC method).

(**1**)

[1] A. Ouihia, L. Rene, J. Guilhem, C. Pascard, and B. Badet, *JOC* **58**, 1641 (1993).

Sulfur. 15, 297

α-Keto dithio esters.[1] Starting from methyl ketones, the functionalization involves iodination, sulfurization (with S_8, Et_3N in DMSO–DMF), and *S*-methylation.

Trifluoromethyl aryl sulfides.[2] The preparation of these substances from aryl halides requires a Cu(I)-catalyzed reaction with sulfur and methyl fluorosulfonyldi-fluoroacetate in HMPA or *N*-methylpyrrolidinone.

RCOOH → RSH.[3] The degradative thiolation is performed by conversion into the thiohydroxamates, further treatment with sulfur under photochemical conditions, and reduction with $NaBH_4$. Good overall yields (88–98%) of thiols are obtained.

[1] D. Villemin and F. Thibault-Starzyk, *SL* 148 (1993).
[2] Q.-Y. Chen and J.-X. Duan, *CC* 918 (1993).
[3] D. H. R. Barton, E. Castagnino, and J. C. Jaszberenyi, *TL* **35**, 6057 (1994).

Sulfuric acid.

Carbamates from thiocarbamates.[1] *O*-Alkyl thiocarbamates undergo rearrangement to give the *S*-alkylthiol carbamates in refluxing $CHCl_3$ containing conc. H_2SO_4 as catalyst. Replacement of the alkylthio group is accomplished by heating the rearrangement products in an alcohol (+Na).

H_2SO_4 on silica.[2] Dehydration, acetalization, etherification, and many common acid-promoted processes can take advantage of the superb properties of the catalyst.

[1] S. K. Tandel, S. Rajappa, and S. V. Pansare, *T* **49**, 7479 (1993).
[2] F. Chavez, S. Suarez, and M. A. Diaz, *SC* **24**, 2325 (1994).

Sulfur tetrafluoride.

$RCOOH \rightarrow RCF_3$.[1] This transformation is suitable for the synthesis of trifluoromethylcyclopropanes.

[1] Y. M. Pustovit, P. I. Ogojko, V. P. Nazaretian, L. B. Faryat'eva, a. B. Rozhenko, and A. N. Alexeeko, *JFC* **69**, 225 (1994).

Sulfuryl chloride. 13, 284; 14, 291–292; 16, 311

Chlorination of sulfur compounds. Conversion of sulfides to carbonyl compounds by way of α-chlorination with SO_2Cl_2 is quite useful. Thus acetylenic aldehydes can be prepared by a two-step process.[1] In the presence of $AgNO_3$ or KNO_3 sulfuryl chloride transforms sulfides directly into α-chloro sulfoxides.[2]

Sulfinyl chlorides are produced[3] when disulfides are treated with SO_2Cl_2 and trimethylsilyl acetate at 0°C.

[1] C. C. Fortes and C. F. D. Garrote, *SC* **23**, 2869 (1993).
[2] Y. H. Kim, H. H. Shin, and Y. J. Park, *S* 209 (1993).
[3] J. Drabowicz, B. Bujnicki, and B. Dudzinski, *SC* **24**, 1207 (1994).

T

Tantalum(V) chloride–zinc. 16, 312; 17, 321

Reductive coupling of alkynes with alkenes and imines. Alkynes form $TaCl_3$ complexes, which add to unsaturated alcohols regioselectively.[1] The reaction of such complexes with metalloimines give allylic amines.[2]

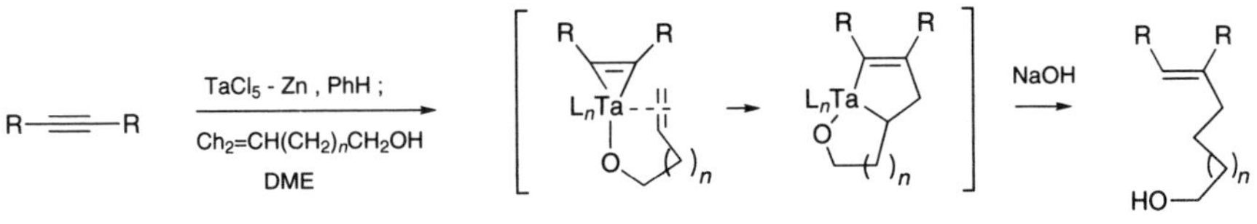

Reformatsky reaction.[3] Low-valent Ta instead of Zn can be used in this reaction.

[1] K. Takai, M. Yamada, H. Odaka, and K. Utimoto, *JOC* **59**, 5852 (1994).
[2] K. Takai, H. Odaka, Y. Kataoka, and K. Utimoto, *TL* **35**, 1893 (1994).
[3] Y. Aoyagi, W. Tanaka, and A. Ohta, *CC* 1225 (1994).

Tellurium. 15, 298

Allyl- and benzyllithiums.[1] Wurtz coupling of the halides is avoided during formation of the organolithium reagents, when they react with BuTeLi (from Te + BuLi) and then BuLi. The one-pot process involves RTeBu intermediates.

[1] T. Kanda, S. Kato, T. Sugino, N. Kambe, and N. Sonoda, *JOMC* **473**, 71 (1994).

Tellurium(IV) chloride. 16, 316

Dialkyl chlorophosphates.[1] $TeCl_4$ is an efficient chlorinating agent that converts dialkyl and trialkyl phosphites into chlorophosphates at room temperature.

[1] Y. J. Koh and D. Y. Oh, *SC* **23**, 1771 (1993).

Tellurium(IV) ethoxide.

Nitriles.[1] Thioamides and amides eliminate H_2X on exposure to $Te(OEt)_4$ at room temperature and high temperatures, respectively. Below 80°C amides give esters.

[1] K. Omote, Y. Aso, T. Otsubo, and F. Ogura, *BCSJ* **67**, 1759 (1994).

Tetraallyltin.

Homoallyl alcohols.[1] The tin reagent effects chemoselective allylation of carbonyl compounds in aqueous acidic media.

[1] A. Yanagisawa, H. Inoue, M. Morodome, and H. Yamamoto, *JACS* **115**, 10356 (1993).

Tetrabutylammonium borohydride.

Reduction of conjugated iminum salts.[1] Amines are obtained in this reduction in acidic methanol. Thus, by reduction and demetalation, the alkylation product of an osmium complex of *N,N*-dimethylaniline is transformed into the 4-substituted cyclohex-2-enylamine.

[1] J. Gonzalez, M. Sabat, and W. D. Harman, *JACS* **115**, 8857 (1993).

Tetrabutylammonium fluoride (TBAF). 13, 286–287; 14, 293–294; 15, 298, 304; 17, 324–326

As base. Because of its high basicity in aprotic media, TBAF is capable of cleaving 2-cyanoethyl esters[1] in THF–DMF and acts as a catalyst in epoxide opening.[2]

Eliminations. When functionalized silanes in which a potential leaving group is attached to a β-atom or to a vinylogously related atom are treated with TBAF, fragmentation ensues. New uses of this process are preparations of 2,3-dimethylene-2,3-dihydrothiophene,[3] substituted 1,2,3-butatrienes,[4] chiral allylic alcohols,[5] and α-fluoroketones.[6] The precursors for the allylic alcohols are the alkylation products (with aldehydes) of 2-(trimethylsilyl)ethyl sulfoxides, and those for the fluoroketones are 1-silyl-1-hydroxymethyloxiranes.

Generation of carbanion equivalents. The desilylation method enables the access to blocked aldols from α-trimethylsilyl α,β-unsaturated ketones.[7] The conversion of 5-trimethylsilyl-2,3-dihydroisoxazoles to β-lactams[8] is also initiated by desilylation, which is followed by fragmentation to ketene intermediates and reclosure of the ring.

31%

Of some significance is the extension of aromatic aldehydes into conjugated dienals by reaction of a disilyl imine,[9] and the generation of the (2-pyridylthio)methyl carbanion.[10]

46%

Miscellaneous reactions. TBAF liberates a stannylthiolate anion from $(Bu_3Sn)_2S$ to react with α,α-dichloroesters. α-(Alkoxycarbonyl)thioaldehydes (ROOC–CHS) emerge.[11]

Both the conversion of 2,2,2-trifluoroethanesulfonates into mesylates,[12] and the reductive allylation[13] of dithio esters and trithio carbonates with allyl fluoride are mediated by TBAF. It is also used to promote silylation of alcohols[14] by silazines or hydrosilanes or disilanes.

[1] Y. Kita, H. Maeda, F. Takahashi, S. Fukui, T. Ogawa, and K. Hatayama, *CPB* **42**, 147 (1994).
[2] D. Albanese, D. Landini, and M. Penso, *S* 34 (1994).
[3] K. J. van den Berg and A. M. van Leusen, *RTC* **112**, 7 (1993).
[4] H.-F. Chow, X.-P. Cao, and M.-K. Leung, *CC* 2121 (1994).
[5] S. Kusuda, Y. Ueno, and T. Toru, *T* **50**, 1045 (1994).
[6] M. M. Kabat, *TA* **4**, 1417 (1993).
[7] K. Matsumoto, K. Oshima, and K. Utimoto, *CL* 1211 (1994).
[8] A. Ahn, J. W. Kennington, and P. DeShong, *JOC* **59**, 6282 (1994).
[9] M. Bellassoued and A. Majidi, *JOMC* **444**, C7 (1993).
[10] S. Kohra, H. Ueda, and Y. Tominaga, *H* **36**, 1497 (1993).
[11] T. Takahashi, N. Kurose, T. Koizumi, and M. Shiro, *H* **36**, 1601 (1993).
[12] Y. S. Choe and J. A. Katzenellenbogen, *TL* **34**, 1579 (1993).
[13] A. Capperucci, A. deg'Innocenti, M. C. Ferrara, B. F. Bonini, G. Mazzanti, and A. Ricci, *TL* **35**, 161 (1994).
[14] Y. Tanabe, M. Murakami, K. Kitaichi, Y. Yoshida, H. Okumura, and A. Maeda, *TL* **35**, 8409 (1994).

Tetrabutylammonium peroxydisulfate.

Tetrahydrofuranyl ethers.[1] Formation of such ethers (81–97%) from alcohols in refluxing THF is the result of radical coupling.

Cleavage of N,N-dimethylhydrazones.[2] Ketones are formed in very good yields (7 examples, 89–97% yield).

[1] J. C. Jung, H. C. Choi, and Y. H. Kim, *TL* **34**, 3581 (1993).
[2] H. C. Choi and Y. H. Kim, *SC* **24**, 2307 (1994).

Tetrabutylammonium perrhenate(VII)–triflic acid.

Beckmann rearrangement.[1]   The combined reagents catalyze Beckmann rearrangement of oximes in refluxing $MeNO_2$. Addition of 20–50 mol % of $NH_2OH \cdot HCl$ prevents hydrolysis of the oximes.

[1] K. Narasaka, H. Kusama, Y. Yamashita, and H. Sato, *CC* 489 (1993).

Tetrabutylammonium triphenyldifluorostannate.

1-Halo-1-fluoroalkanes.[1]   The conversion of alkane 1,1-bistriflates to the *gem*-fluorohaloalkanes is achieved by sequential treatment with $[Bu_4N][Ph_3SnF_2]$ in CH_2Cl_2 at 0°C and then with $[Bu_4N]X$ (X = Cl, Br, I) at room temperature.

[1] A. G. Martinez, J. O. Barcina, A. Z. Rys, and L. R. Subramanian, *SL* 587 (1993).

Tetraethylammonium cyanoborohydride.

Ketone reduction.[1]   The diastereoselective reduction by this borohydride can take advantage of oxygen substituents at a β-position. Addition of $TiCl_4$ to form chelates before reduction occurs leads to the *syn* isomers as major products.

[1] C. R. Sarko, I. C. Guch, and M. DiMare, *JOC* **59**, 705 (1994).

Tetrachlorosilane. 16, 317

Esters.[1]   The formation of esters RCOOR′ from $RCOOSiMe_3$ and $R'OSiMe_3$ is catalyzed by $SiCl_4$ and AgOTf (7 examples, 83–96% yield).

[1] I. Shiina and T. Mukaiyama, *CL* 2319 (1992).

Tetrakis(alkoxycarbonyl)palladacyclopentadiene.

Enyne metathesis.[1]   A profound skeletal transformation results in the formation of unusual ring systems.

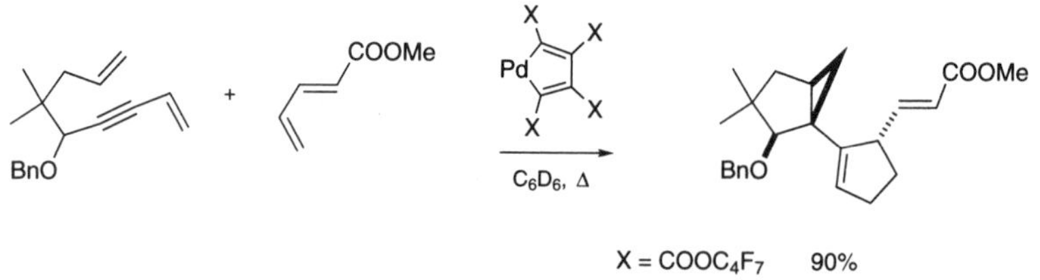

X = $COOC_4F_7$ 90%

[1] B. M. Trost and A. S. K. Hashmi, *JACS* **116**, 2183 (1994).

Tetrakis(triphenylphosphine)palladium (0). 13, 289–294; 14, 295–299; 15, 300–304; 16, 317–323; 17, 327–331

Allylic substitutions. The catalyzed substitution reaction is well suited for the preparation of N-protected allylic amines[1] and the conversion of allylic carbonates to cyclic carbamates.[2] Asymmetric α-allylation of cyclic ketones[3] is conveniently carried out by treating the enamines derived from an allyl ester of proline with $(Ph_3P)_4Pd$ and subsequent hydrolysis. The rearrangement of allyl 2-pyrimidinyl carbonates to 1-allyl-2(1H)-pyrimidinones[4] and the introduction of allylic side chains to peptides[5] through ester enolate Claisen rearrangement belong to the same reaction set. 1-Arylprop-2-enols are suitable for C-cinnamylation of 1,3-diketones.[6] For aryl allyl ether cleavage[6a] treatment with a combination of the Pd(0) catalyst with $NaBH_4$ suffices.

73%

Arylboronic acids and derivatives surrender their aryl moieties on displacing allenylmethyl carbonates and propargylic carbonates, furnishing 2-aryl-1,3-dienes[7] and arylallenes,[8] respectively.

88%

Allylation of carbonyl compounds and imines. Allylzinc reagents are used.[9] However, other allylic components such as benzoates[10] and sulfones[11] can be used together with Et_2Zn.

Coupling of organozincs with aryl halides. The method is very useful for the synthesis of biaryls,[12] including heterobiaryls.[13] Vinylic,[14] aliphatic zinc reagents,[15] and even $Zn(CN)_2$[16] are reactive toward ArX in the presence of $(Ph_3P)_4Pd$. Perfluoroalkenylzinc bromides[17] and cyclopropenylzinc chlorides[18] also couple efficiently with iodoarenes.

Coupling of organozincs with alkenyl halides. The reaction tolerates many functional groups. Thus (dialkoxyboryl)methyl zinc reagents[19] can be used, and iodoallyl alcohols need only to be converted to the zinc alkoxides before coupling.[20] The alternative mode involving alkenylzinc chlorides and 2-(β-bromoalkenyl)-1,3,2-dioxaborinanes can also be used successfully to afford 1,3-dienes.[21] 2,3-Dibromoacrylic esters undergo regioselective (at the β-carbon) and stereoselective coupling with ArZnX.[22]

Coupling involving organoborons.[23,24] Saturated and unsaturated organoboranes as well as arylboronic acids are equally effective in coupling with various halides and enol triflates. Biaryl synthesis on a solid support[25] is successful. In some other cases the reaction is assisted by CsF.[26]

Coupling of organotins with unsaturated halides. Functionalized reagents are widely applicable for the coupling such that α-fluorostyrenes,[27] α-arylacrylic esters,[28] and cinnamyl amines[29,30] are readily obtained.

Enol triflates have about the same reactivity as vinylic halides in the coupling reaction.[31] For phenylations, $(Bu_4N)^+(Ph_3SnF_2)^-$ is a useful starting material.[32]

Destannylative acylation. Unsaturated tin compounds react with acid chlorides to give unsaturated ketones. When the tin atom is attached to a polymer,[33] the ketone products released into solution are very easy to isolate.

The reaction of β-stannyl enones with acid chlorides provides 1,4-diketones.[34] This is a reductive coupling.

Coupling of organosilanes with aryl halides. Disilanes are split and coupled to aryl halides under rather vigorous conditions.[35,36] When alkyltrifluorosilanes are used, the coupling[37] also requires Bu_4NF as a promoter.

Cyclopropanation of norbornene.[38] An unusual and high-yielding cycloaddition occurs when α-ketol carbonates and norbornene are heated with $(Ph_3P)_4Pd$ in DMF.

R = Me 99%
R = Ph 82%

Intramolecular zinc-ene reaction.[39] Metal exchange of a π-allylpalladium complex to form a σ-bonded organozinc species permits ene reaction with a proximal alkene unit. The cyclic product can be functionalized.

E = H 79%

[1] R. O. Hutchins, J. Wei, and S. J. Rao, *JOC* **59**, 4007 (1994).
[2] T. Bando, H. Harayama, Y. Fuzukawa, M. Shiro, K. Fugami, S. Tanaka, and Y. Tamaru, *JOC* **59**, 1465 (1994).

[3] K. Hiroi, J. Abe, K. Suya, S. Sato, and T. Koyama, *JOC* **59**, 203 (1994).

[4] M. L. Falck-Pedersen, T. Benneche, and K. Undheim, *ACS* **47**, 63 (1993).

[5] U. Kazmaier, *JOC* **59**, 6667 (1994).

[6] M. Sakakibara and A. Ogawa, *TL* **35**, 8013 (1994).

[6a] R. Beugelmans, S. Bourdet, A. Bigot, and J. Zhu, *TL* **35**, 4349 (1994).

[7] T. Moriya, T. Furuuchi, N. Miyaura, and A. Suzuki, *T* **50**, 7961 (1994).

[8] T. Moriya, N. Miyaura, and A. Suzuki, *SL* 149 (1994).

[9] T. A. J. van der Heide, J. L. van der Baan, V. de Kimpe, F. Bickelhaupt, and G. W. Klumpp, *TL* **34**, 3309 (1993).

[10] K. Yasui, Y. Goto, T. Yajima, Y. Taniseki, K. Fugami, A. Tanaka, and Y. Tamaru, *TL* **34**, 7619 (1993).

[11] J. Clayden and M. Julia, *CC* 1905 (1994).

[12] K. Takagi, *CL* 469 (1993).

[13] T. Sakamoto, Y. Kondo, N. Murata, and H. Yamanaka, *T* **49**, 9713 (1993).

[14] L.-L. Gundersen, A. K. Bakkestuen, A. J. Aasen, H. Overas, and F. Rise, *T* **50**, 9743 (1994).

[15] D. P. G. Hamon, R. A. Massy-Westropp, and J. L. Newton, *TL* **34**, 5333 (1993).

[16] D. M. Tschaen, R. Desmond, A. O. King, M. C. Fortin, B. Piprik, S. King, and T. R. Verhoeven, *SC* **24**, 887 (1994).

[17] P. A. Morken and D. J. Burton, *JOC* **58**, 1167 (1993).

[18] S. Untiedt and A. de Meijere, *CB* **127**, 1511 (1994).

[19] T. Watanabe, N. Miyaura, and A. Suzuki, *JOMC* **444**, C1 (1993).

[20] E. Negishi, M. A. Y. V. Gulevich, and Y. Noda, *TL* **34**, 1437 (1993).

[21] C. Mazal and M. Vaultier, *TL* **35**, 3089 (1994).

[22] F. Bellina, A. Carpita, M. Desantis, and R. Rossi, *TL* **35**, 6913 (1994).

[23] T. Oh-e, N. Miyaura, and A. Suzuki, *JOC* **58**, 2201 (1993).

[24] Q. Zheng, Y. Yang, and A. R. Martin, *TL* **34**, 2235 (1993).

[25] R. Frenette and T. W. Friesen, *TL* **35**, 9177 (1994).

[26] S. W. Wright, D. L. Hageman, and L. D. McClure, *JOC* **59**, 6095 (1994).

[27] D. P. Matthews, P. P. Waid, J. S. Sabol, and J. R. McCarthy, *TL* **35**, 5177 (1994).

[28] J. I. Levin, *TL* **34**, 6211 (1993).

[29] R. J. P. Corriu, G. Bolin, and J. J. E. Moreau, *BSCF* 273 (1993).

[30] R. J. P. Corriu, B. Geng, and J. J. E. Moreau, *JOC* **58**, 1443 (1993).

[31] Y. Al-Abed, T. H. Al-Tel, C. Schröder, and W. Volter, *ACIEE* **33**, 1499 (1994).

[32] A. G. Martinez, J. O. Barcina, A. de Fresno Cerezo, and L. R. Subramanian, *SL* 1047 (1994).

[33] H. Kuhn and W. P. Neumann, *SL* 123 (1994).

[34] A. M. Echavarren, M. Perez, A. M. Castano, and J. M. Cuerva, *JOC* **59**, 4179 (1994).

[35] P. Babin, B. Bennetau, M. Theurig, and J. Dunogues, *JOMC* **446**, 135 (1993).

[36] S. Cros, B. Bennetau, J. Dunogues, and P. Babin, *JOMC* **468**, 69 (1994).

[37] H. Matsuhashi, M. Kuroboshi, Y. Hatanaka, and T. Hiyama, *TL* **35**, 6507 (1994).

[38] S. Ogoshi, T. Morimoto, K.-i. Nishio, K. Ohe, and S. Murai, *JOC* **58**, 9 (1993).

[39] W. Oppolzer and F. Schröder, *TL* **35**, 7939 (1994).

Tetrakis(triphenylphosphine)palladium (0)−copper(I) iodide.

Coupling of alkynes with unsaturated halides. This coupling represents the simplest stereoselective assembly of alkynylarenes[1] and enynes.[2,3] An amine is also added to neutralize the liberated acid.

Allenes. Alkynyl allenes are available from S_N2' displacement of propargylic derivatives.[4,5]

91%

[1] R. W. Bates, C. J. Gabel, and J. Ji, *TL* **35**, 6993 (1994).
[2] M. Alami, B. Crousse, and G. Linstrumelle, *TL* **35**, 3543 (1994).
[3] R. S. Paley, J. A. Lafontaine, and M. P. Ventura, *TL* **34**, 3663 (1993).
[4] S. Gueugnot and G. Linstrumelle, *TL* **34**, 3853 (1993).
[5] C. Darcel, C. Bruneau, and P. H. Dixneuf, *CC* 1845 (1994).

Tetramethylammonium fluoride.

Aryl fluorides.[1]   The fluorodenitration of activated nitroarenes with Me_4NF in DMSO leads to ArF. The reagent is azeotropically dried in situ.

[1] N. Boechat and J. H. Clark, *CC* 921 (1993).

Tetramethylammonium triacetoxyborohydride. 14, 299–300; 16, 324; 17, 331–332

syn-1,2-Amino alcohols.[1]   The reduction of α-hydroxy oximino ethers by this borohydride is highly diastereoselective, and complements the reduction with $LiAlH_4$ or i-Bu_2AlH, which affords predominantly the *anti* products.

[1] D. R. Williams, M. H. Osterhout, and J. P. Reddy, *TL* **34**, 3271 (1993).

Tetramethylguanidinium azide.

Glycosyl azides.[1]   Invertive displacement to give the azides in 90–99% yields.

[1] C. Li, A. Arasappan, and P. L. Fuchs, *TL* **34**, 3535 (1993).

Tetranitromethane.

Sulfoxides.[1]   Sulfides undergo photooxidation in the presence of $C(NO_2)_4$.
α-Nitro ketones.[2]   Enol silyl ethers undergo *C*-nitration at room temperature.

[1] D. Ramkumar and S. Sankararaman, *S* 1057 (1993).
[2] R. Rathore, Z. Lin, and J. K. Kochi, *TL* **34**, 1859 (1993).

Tetraphenylantimony bromide.

Stereoselective aldol condensation.[1]   The Ph_4SbBr-catalyzed condensation of tin enolates with α-chlorocycloalkanones furnishes predominantly aldol products with a *cis*-chlorohydrin unit.

[1] M. Yasuda, T. Oh-hata, I. Shibata, A. Baba, M. Matsuda, and N. Sonoda, *TL* **35**, 8627 (1994).

Tetraphenylstibonium methoxide.

Alkylation of cyclic 1,3-diketones.[1]   The *C*-allylation and propargylation of 2-substituted 1,3-cycloalkanediones is promoted by Ph$_4$SbOMe. *O*-Butylation occurs with BuI.

[1] M. Fujiwara, K. Hitomi, A. Baba, and H. Matsuda, *TL* 875 (1994).

Tetrapropylammonium perruthenate–*N*-methylmorpholine *N*-oxide. 14, 302; **16**, 325

Oxidation of sulfides.[1]   The mixture of reagents (TPAP as catalyst, NMO as co-oxidant) oxidizes sulfides to sulfones in MeCN in the presence of 4Å-molecular sieves. Isolated double bonds are generally not affected.

Oxidation of amines and hydroxylamines.[2]   Secondary amines undergo dehydrogenation to afford imines, whereas *N,N*-dialkylhydroxylamines are converted to nitrones at room temperature.

[1] K. R. Guertin and A. S. Kende, *TL* **34**, 5369 (1993).
[2] A. Goti, M. Romani, and F. De Sarlo, *TL* **35**, 6567, 6571 (1994).

Thallium(III) nitrate. 16, 326

Ring contraction of chroman-4-ones.[1]   In the same fashion as the formation of arylacetic esters from open-chain phenones, dihydrobenzofuran-3-carboxylic esters are obtained when chroman-4-ones are oxidized with TTN in the presence of trimethyl orthoformate.

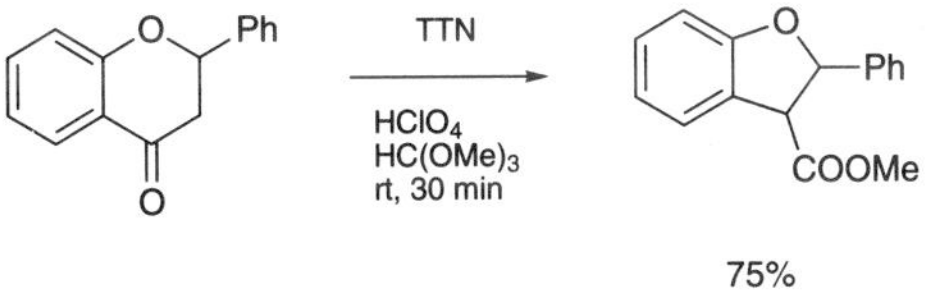

75%

[1] M. S. Khanna, O. V. Singh, C. P. Garg, and R. P. Kapoor, *SC* **23**, 585 (1993).

Thallium(III) tosylate.

Dehydrogenation of 2,3-dihydroquinol-4-ones.[1]   The dehydrogenation proceeds in excellent yields (6 examples, 90–96%) in refluxing DME.

[1] O. V. Singh and R. S. Kapil, *SC* **23**, 277 (1993).

Thianthrenium tetrafluoroborate.

α-Hydroxy aldehyde acetals.[1]   The salt (1) is a radical cation that effects α-oxidation of aldehydes. On reaction of the products with NaOMe–MeOH acetals are isolated.

(1)

[1] M. Schulz, R. Kluge, and J. Michaelis, *SL* 669 (1994).

Thionyl chloride. 16, 297

Reaction with alcohols.[1] Alcohols are converted into alkyl iodides by $SOCl_2$–KI in DMF at 50°C, but at 0°C only the formate esters are obtained (LiI is used instead of KI in these cases).

Nitrile oxide preparation.[2] Primary nitroalkanes undergo dehydration on treatment with $SOCl_2$–Et_3N in CH_2Cl_2. In the presence of alkenes, isoxazolines are formed.

[1] I. Fernandez, B. Garcia, S. Munoz, J. R. Pedro, and R. de la Salud, *SL* 489 (1993).
[2] Y.-J. Chen and C.-N. Li, *JCCS(T)* **40**, 203 (1993).

Tin. 13, 298; 17, 333–334

Homoallylic alcohols. Allylic bromides are transformed into allyltin reagents that react with aromatic aldehydes with very high *erythro*-selectivity.[1] However, the stereoselectivity is only moderate in the reaction with aliphatic aldehydes. The intramolecular version is useful for the synthesis of cyclic alcohols.[2]

[1] J.-Y. Zhou, Z.-G. Chen, and S.-H. Wu, *SC* **24**, 2661 (1994).
[2] J.-Y. Zhou, Z.-G. Chen, and S.-H. Wu, *CC* 2783 (1994).

Tin(II) bromide. 14, 303–304

Transetherification and transesterification. Tin(II) bromide is a catalyst for converting p-methoxybenzyl ethers[1] into methoxymethyl ethers by $CH_2(OMe)_2$ and $MeOCH_2Br$. Benzyl ethers[2] and trimethylsilyl ethers[3] are cleaved and acetylated in one operation by the action of $SnBr_2$–AcBr in CH_2Cl_2 at room temperature.

[1] T. Oriyama, M. Kimura, and G. Koga, *BCSJ* **67**, 885 (1994).
[2] T. Oriyama, M. Kimura, M. Oda, and G. Koga, *SL* 437 (1993).
[3] T. Oriyama, M. Oda, J. Gono, and G. Koga, *TL* **35**, 2027 (1994).

Tin(IV) bromide.

Allylation. Long-range asymmetric induction by an oxygen functionality is observed in the reaction of allylstannanes with aldehydes using tin(IV) bromide as catalyst.[1,2] Irrespective of the original configuration of the allylic double bond, the homoallylic alcohol products always display the (Z) configuration. SnCl$_4$ has a similar efficiency.[3,4]

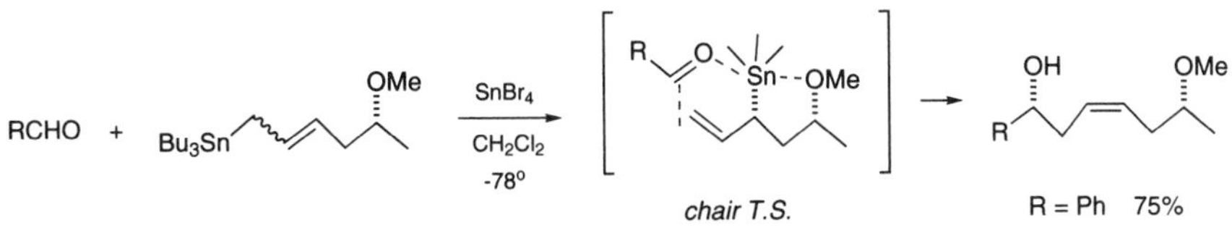

chair T.S.   R = Ph   75%

[1] J. S. Carey, T. S. Coulter, and E. J. Thomas, *TL* **34**, 3933, 3935 (1993).
[2] J. S. Carey, E. J. Thomas, and S. J. Stanway, *CC* 283, 285 (1994).
[3] A. Terrawutgulrag and E. J. Thomas, *JCS(PI)* 2863 (1993).
[4] A. H. McNeill and E. J. Thomas, *S* 322 (1994).

Tin(II) chloride. 13, 298–299; 15, 309–310; 16, 329

Carbonyl masking and regeneration.   Tin(II) chloride is effective as catalyst for dithioacetalization[1] as well as hydrolysis of acetals.[2]

Reductions.   Disulfides are cleaved,[3] and nitroaryl azides are reduced to the amines[4] without affecting the nitro group or other substituents. However, β,γ-unsaturated nitroalkenes are reduced to enones,[5] and ω-nitrostyrenes give α-allyloxy arylacetaldoximes in the presence of allyl alcohol.[6]

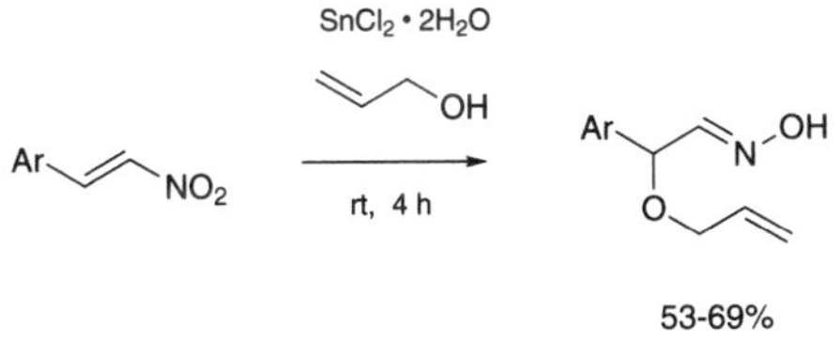

53-69%

Reformatsky-type reactions.   In the condensation of α-halo ketones with aldehydes to form enones, SnCl$_2$ and Na$_2$SO$_3$ can be used as promoters,[7] whereas SnCl$_2$ alone is used to effect synthesis of β,β-dichloro β-nitroalcohols from trichloronitromethane and aldehydes.[8]

Allylation of aldehydes.   In this Barbier-type condensation different catalysts are also added to promote the activity of tin(II) chloride. These include NaI,[9] CuX,[10] and Mg.[11]

β-Keto esters.[12]   The homologation of aldehydes with diazoalkanes is catalyzed by tin(II) chloride. The reaction with α-diazoacetic esters leading to β-keto esters is useful for the synthesis of α-acyl-β-tetronic acids.

78%

72%

Ribofuranosides and ribonucleosides. The anomeric substituents of ribo-furanyl esters[13] and carbonates[14] are subject to replacement by various nucleophiles in a reaction catalyzed by $SnCl_2$. β-Isomers are formed selectively.

3-Arylpyridines.[15] The reduction of 6-aryl-5-nitrobicyclo[2.2.1]hept-2-enes with $SnCl_2$ in refluxing THF or dioxane generates 3-arylpyridines in moderate yields. A 7-step tandem process is proposed.

[1] N. B. Das, A. Nayak, and R. P. Sharma, *JCR(S)* 242 (1993).
[2] K. L. Ford and E. J. Roskamp, *JOC* **58**, 4142 (1993).
[3] Y.-H. Chang, J. D. Peak, S. W. Wierschke, and W. A. Feld, *SC* **23**, 663 (1993).
[4] K. R. Gee and J. F. W. Keana, *SC* **23**, 357 (1993).
[5] N. B. Das, J. C. Sarma, R. P. Sharma, and M. Bordolos, *TL* **34**, 869 (1993).
[6] A. Hassner, K. M. L. Rai, and W. Dehaen, *SC* **24**, 1669 (1994).
[7] R. Lin, Y. Yu, and Y. Zhang, *SC* **23**, 271 (1993).
[8] A. S. Demir, C. Tanyeli, A. S. Mahasneh, and H. Aksoy, *S* 155 (1994).
[9] T. Imai and S. Nishida, *S* 395 (1993).
[10] T. Imai and S. Nishida, *CC* 277 (1994).
[11] C. Sarangi, N. B. Das, and R. P. Sharma, *JCR(S)* 398 (1994).
[12] K. Nomura, T. Iida, K. Hori, and E. Yoshii, *JOC* **59**, 488 (1994).
[13] N. Shimomura, M. Saitoh, and T. Mukaiyama, *CL* 433 (1994).
[14] T. Mukaiyama, T. Matsutani, and N. Shimomura, *CL* 2089 (1994).
[15] T.-L. Ho and P.-Y. Liao, *TL* **35**, 2211 (1994).

Tin(IV) chloride. 13, 300–301; 14, 304–306; 15, 311–313; 17, 335–340

Reaction with epoxy alcohols. The Lewis acidity of $SnCl_4$ induces opening of the epoxide ring. The subsequent transformation depends on the nearby structure. Thus rearrangement and cyclization products can be obtained.[1,2]

Heterocyclizations. Tetrahydropyran derivatives are formed[3] stereoselectively via α-oxycarbenium ion intermediates through π-cyclization. The cations are generated with the assistance of $SnCl_4$. In refluxing mesitylene, lactams condense with 2-aminoacetaldehyde diethyl acetal[4] in the presence of $SnCl_4$, forming bicyclic imidazoles.

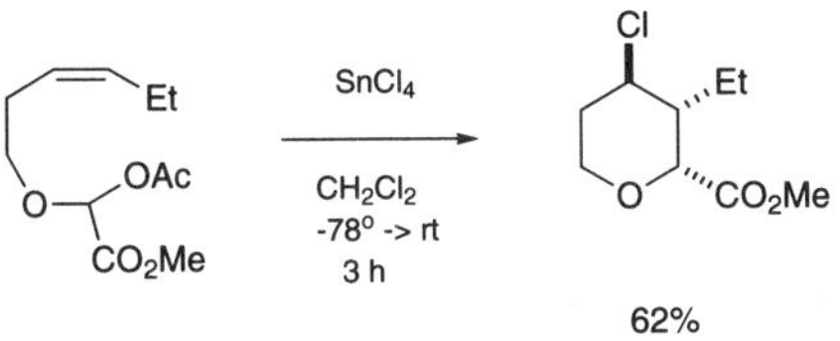

62%

A synthesis of β-lactams[5] from 2-pyridylthio esters with imines apparently involves reaction of tin enolates.

Annulations involving allylsilanes. Addition of allylsilanes to carbonyl compounds under the influence of Lewis acids results in tetrahydrofurans.[6,7] During the ring formation the silyl group is migrated to the central carbon atom of the original allyl fragment. Because of its highly stereoselective nature, the process is useful for the synthesis of complex structures.

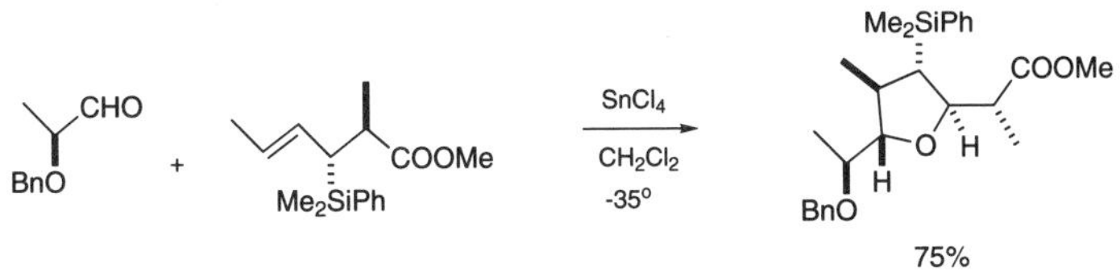

75%

Silyltetralins are obtained from a formal [3+3]cycloaddition[8] initiated by attack of allylsilanes on benzylic cations.

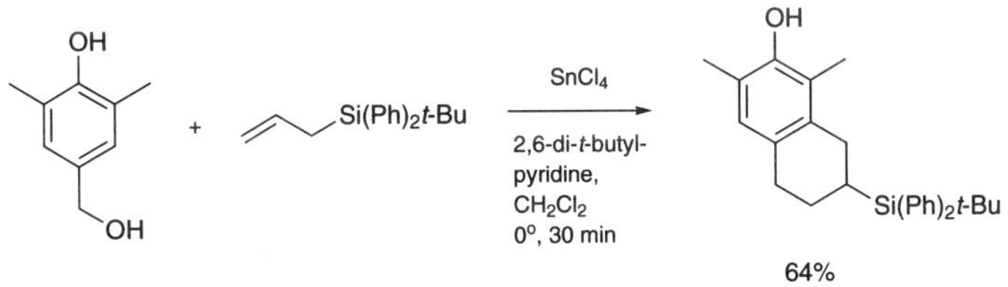

64%

Diels–Alder reactions. Diels–Alder reactions are catalyzed by Lewis acids, including SnCl₄. Selective complexation of SnCl₄ to the less hindered ester group of an unsymmetrical chiral maleic ester directs asymmetric Diels–Alder reactions.[9]

Hetero-Diels–Alder reactions of unsaturated α-keto esters also benefit from the presence of SnCl₄, as unactivated alkenes can be used as dienophiles.[10]

Ene reaction.[11] Formation of 8-benzylaminoindolizidine and 1-benzylamino-quinolizidine derivatives from the imines of 2-pyrrolidinecarbaldehyde and 2-piperidinecarbaldehyde is mediated by SnCl₄. On the other hand, TiCl₄ converts the same imines to bicyclic products that possess benzylideneimino and isopropyl groups in a *cis* arrangement.

Glycosylation. Glycosidation with tin phenoxides[12] and silyl ethers[13] may use SnCl₄ as catalyst. It is interesting that either α- or β-glucosides are accessible from the same glycosyl donor by slight variation of conditions.

[1] C. M. Marson, A. J. Walker, J. Pickering, A. D. Hobson, R. Wrigglesworth, and S. J. Edge, *JOC* **58**, 5944 (1993).

[2] C. M. Marson, S. Harper, A. J. Walker, J. Pickering, J. Campbell, R. Wrigglesworth, and S. J. Edge, *T* **49**, 10339 (1993).

[3] L. D. M. Lolkema, H. Hiemstra, C. Semeyn, and W. N. Speckamp, *T* **50**, 7115 (1994).

[4] D. H. Hua, F. Zhang, J. Chen, and P. D. Robinson, *JOC* **59**, 5084 (1994).

[5] R. Annunziata, M. Benaglia, M. Cinquini, F. Cozzi, and L. Raimondi, *T* **50**, 5821 (1994).

[6] J. S. Panek and R. Beresis, *JOC* **58**, 809 (1993).

[7] T. Akiyama, K. Ishikawa, and S. Ozaki, *CL* 627 (1994).

[8] S. R. Angle and J. P. Boyce, *TL* **35**, 6461 (1994).

[9] K. Maruoka, M. Akakura, S. Saito, T. Ooi, and H. Yamamoto, *JACS* **116**, 6153 (1994).

[10] A. Sera, M. Ohara, H. Yamada, E. Egashira, N. Ueda, and J.-i. Setsune, *BCSJ* **67**, 1912 (1994).

[11] S. Laschat and M. Grehl, *CB* **127**, 2023 (1994).

[12] F. Clerici, M. L. Gelmi, and S. Mottadelli, *JCS(PI)* 985 (1994).

[13] K. Matsubara, T. Sasaki, and T. Mukaiyama, *CL* 1373 (1993).

Tin(IV) chloride–tributylamine.

Alkynyltrichlorotin reactions.[1] The reagents are generated from 1-alkynes by treatment with SnCl₄–Bu₃N in CH₂Cl₂. Thus alkynylation of aldehydes, acetals, and enones (1,4-mode) is achieved under mild conditions using a one-pot operation.

Condensation of silyl enol ethers and ketones with 1-alkynes. Both alkylidenenation[2] and cyclopentenone formation[3] are observed. The annulation requires the treatment of the initial condensation products with a strong base (e.g., DBU).

Even with essentially the same reaction components, the "in situ" version produces β,γ-unsaturated ketones.[4]

[1] M. Yamaguchi, A. Hayashi, and M. Hirama, *CL* 2479 (1992).

[2] M. Yamaguchi, A. Hayashi, and M. Hirama, *JACS* **115**, 3362 (1993).

[3] M. Yamaguchi, M. Sehata, A. Hayashi, and M. Hirama, *CC* 1708 (1993).
[4] A. Hayashi, M. Yamaguchi, and M. Hirama, *SL* 51 (1995).

Tin(II) hexamethyldisilazide.

N,N-Dialkyl enamines. Upon transsilylation, amines are activated such that their reaction with ketones forms enamines.[1]

Amides and lactams. Conversion of acids[2] and esters[3] to amides is quite easy in the presence of $[(Me_3Si)_2N]_2Sn$. This method is also applicable to ring closure of β-amino esters to give β-lactams.[4]

[1] C. Burnell-Curty and E. J. Roskamp, *SL* 131 (1993).
[2] C. Burnell-Curty and E. J. Roskamp, *TL* 34, 5193 (1993).
[3] L. A. Smith, W.-B. Wang, C. Burnell-Curty, and E. J. Roskamp, *SL* 850 (1993).
[4] W.-B. Wang and E. J. Roskamp, *JACS* 115, 9417 (1993).

Tin(IV) oxide–isopropanol.

Deoxygenation of carbonyl compounds.[1] Hydrous SnO_2 is prepared from $SnCl_4$ by precipitation with aqueous ammonia then drying and calcination at 300°C for 5 h. The reduction of aliphatic carboxylic acids gives the corresponding alcohols, whereas aromatic acids are further reduced to the hydrocarbons. Aromatic ketones also give hydrocarbons.

[1] K. Takahashi, M. Shibagaki, H. Kuno, and H. Matsushita, *BCSJ* 67, 1107 (1994).

Tin(II) triflate. 13, 301–302; 14, 306–307; 15, 313–314; 17, 341–344

Propargyl amines.[1] α-Alkoxy amines undergo ionization in the presence of $Sn(OTf)_2$, enabling reaction with alkynylstannanes to give propargyl amines.

Glycosylation.[2] Lactols (sugars) react with various nucleophiles when activated with $Sn(OTf)_2$ and $(Me_3Si)_2O$. Other effective catalysts are $SnCl_2$, $Yb(OTf)_3$, and $La(OTf)_3$.

Aldol reactions. Asymmetric aldol reactions of ketene silyl acetals with catalytic systems consisting of $Sn(OTf)_2$, a chiral diamine, and SnO or $Bu_2Sn(OAc)_2$ are very efficient.[3,4] It is interesting that slight changes in the structure of the chiral diamine ligand (derived from proline) can have opposite effects on the enantioselectivity.[5]

[1] T. Nagasaka, S. Nishida, K. Adachi, T. Kawahara, S. Sugihara, and F. Hamaguchi, *H* **36**, 2657 (1993).

[2] T. Mukaiyama, K. Matsubara, and M. Hora, *S (Spec. Issue)* 1368 (1994).

[3] T. Mukaiyama, I. Shina, H. Uchiro, and S. Kobayashi, *BCSJ* **67**, 1708 (1994).

[4] I. Shina, H. Uchiro, S. Kobayashi, and T. Mukaiyama, *T* **49**, 1761 (1993).

[5] S. Kobayashi and M. Horibe, *ACS* **48**, 9805 (1994).

Titanium, low-valent. 13, 303–304, 310–311; 14, 307–309; 15, 315, 316, 317; 16, 330–332

Preparation.[1]   Reduction of $TiCl_3$ by high-surface Na (supported on Al_2O_3, TiO_2, NaCl, . . .) is a convenient procedure. In the McMurry reaction it shows a template effect for the cyclization of dicarbonyl compounds.

Cleavage of propargyl ethers.[2]   With the low-valent Ti species (from $TiCl_3$, Mg in THF) the reductive cleavage of propargyl ethers is selective, since allyl, benzyl, and methyl ethers remain intact under these conditions.

Indoles from o-acylanilides.[3]   The reductive coupling is brought about by $TiCl_3$ and Zn in DME.

Reductive elimination.[4]   All (*E*) trienes are formed by treatment of 1,6-dibenzoyloxy-2,4-dienes with the low-valent titanium species prepared from $TiCl_3$ and $LiAlH_4$. This method is applicable to the synthesis of a leukotriene.

[1] A. Fürstner and G. Seidel, *S* 63 (1995).

[2] S. K. Nayak, S. M. Kadam, and A. Banerji, *SL* 581 (1993).

[3] A. Fürstner, A. Hupperts, A. Ptock, and E. Janssen, *JOC* **59**, 5215 (1994).

[4] G. Solladie, G. B. Stone, and A. Rubio, *TL* **34**, 1803, 1807 (1993).

Titanium(III) chloride. 13, 302; 16, 330

Cleavage of N–O bonds.   *O*-Methyl hydroxamates are reduced to primary amides[1] in ethanol. Hydroxylamines are transformed into secondary amines or imines.[2]

Benzoins.   The cross-coupling of aroyl cyanides with aldehydes by aqueous $TiCl_3$ gives convenient access to benzoins,[3] as well as 4-hydroxy-3-phenyl-coumarins.[4]

In the synthesis of α-hydroxy β-keto esters[5] from acid chlorides and methyl phenylglyoxylate the sensitivity of acid chlorides requires maintenance of an anhydrous condition and the presence of pyridine to capture HCl.

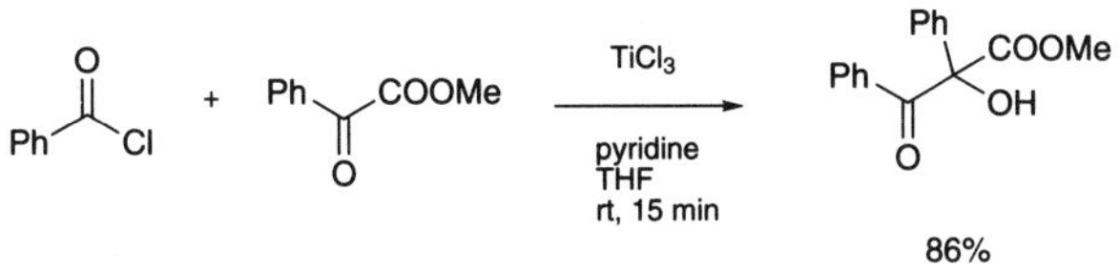

86%

[1] L. E. Fisher, J. M. Caroon, J. S. R. Stabler, S. Lundberg, and J. M. Muchowski, *JOC* **58**, 3643 (1993).
[2] Y. Kodera, S. Watanabe, Y. Imada, and S.-i. Murahashi, *BCSJ* **67**, 2542 (1994).
[3] A. Clerici and O. Porta, *JOC* **58**, 2889 (1993).
[4] A. Clerici and O. Porta, *S* 99 (1993).
[5] S. Araneo, A. Clerici, and O. Porta, *TL* **35**, 2213 (1994).

Titanium(IV) chloride. **13**, 304–309; **14**, 309–311; **15**, 317–320; **16**, 332–337; **17**, 344–347

Nucleophilic reactions. Chiral epoxides are converted into 2,2-dimethyl-1,3-dioxolanes[1] with inversion of configuration by reaction with acetone. An efficient procedure for imine formation[2] from ketones and amines specifies $TiCl_4$ as promoter. Hydrolysis (or alcoholysis) of $RCONH_2$[3] is achieved in the presence of $TiCl_4$ in acidic media.

Allylations. In reactions using allylsilane as nucleophiles $TiCl_4$ is one of the most popular catalysts. It is now employed in allylating squaric acid derivatives,[4] 8-phenylmenthyl pyruvate,[5] and *N*-acyliminium ions.[6] A remarkable observation is that the four stereocenters of the products from 1,8-bis(trimethylsilyl)-2,6-octadiene and two aldehyde molecules are generated with good control.[7] The stereoselective intramolecular reaction is most suitable for the synthesis of bicyclic α-methylene-γ-lactones.[8]

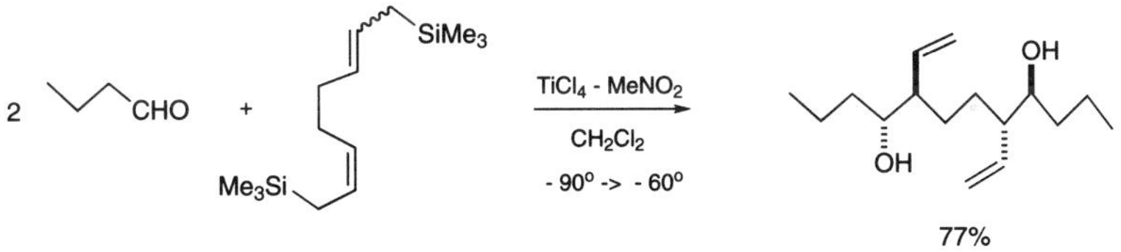

77%

For reactions involving epoxy allylsilanes, $TiCl_4$ is definitely the preferred catalyst[9] (other catalysts such as $BF_3 \cdot OEt_2$ also induce transformations of the epoxide ring). Tetrahydropyridines[10] result from an intramolecular attack of an allylsilane on an iminium ion.

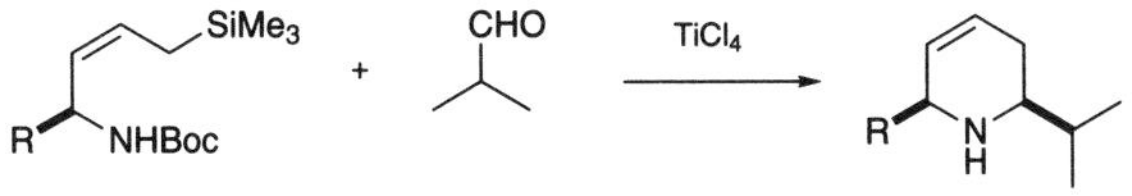

Cycloadditions. The ability of TiCl₄ to catalyze [2+2]-, [3+2]-, and [4+2]-cycloadditions is well known. The [2+2]-[11] and [3+2]-cycloadditions[12,13] involving allylsilanes are useful synthetic methods for assembling polyfunctional cyclic compounds. In Diels–Alder reactions,[14] the catalyst not only promotes cycloaddition of specified addends, but cyclohexadienes can be obtained directly through β-elimination of the cycloadducts.[15]

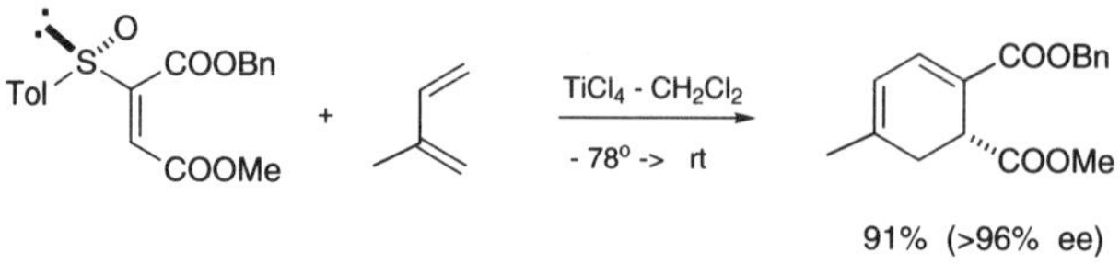

91% (>96% ee)

Aldol reactions. Highly functionalized carbonyl compounds are produced using diketene[16] and 2,2-dialkoxycyclopropanecarboxylic esters[17] in these condensations. An intramolecular reaction provides access to 2-alkoxycarbonyl-2-cycloalkenones,[18] and an aromatic version completes the cyclopentannulation of a podocarpic acid derivative,[19] making available analogs of C-aromatic steroids.

Aldols derived from propargylic aldehydes are also versatile synthetic intermediates. The *anti*-selective condensation is changed to a *syn*-selective when the hexacarbonyldicobalt-complexed acceptors are used.[20]

syn-Selective aldolization due to chelation control and the *anti*-selective variant due to nonchelation control are realized by changing catalysts as with a camphor derivative.[21]

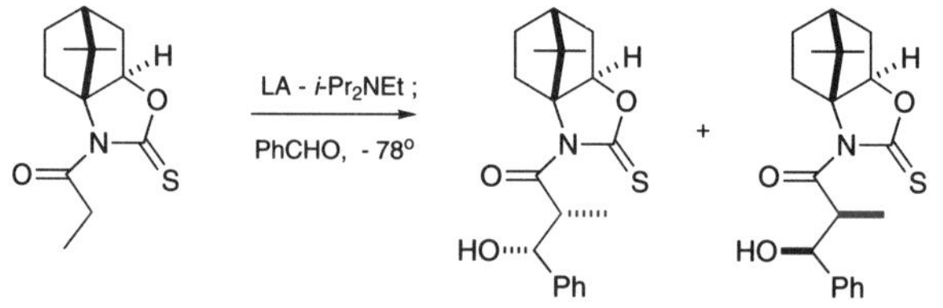

LA = TiCl₄	84%	97	:	3
LA = Bu₂BOTf	86%	>99	:	1

Oximoyl chlorides. Through simultaneous dichlorination and reduction, nitroalkenes are converted to α-chloro-oximoyl chlorides as precursors of nitrile oxides.[22] The presence of the α-chlorine may be useful in some synthetic situations. Oximoyl chlorides are obtained when the reaction is conducted in the presence of triethylsilane.[23]

64-82%

[1] T. Nagata, T. Takai, T. Yamada, K. Imagawa, and T. Mukaiyama, *BCSJ* **67**, 2614 (1994).
[2] R. Carlson, U. Larsson, and L. Hansson, *ACS* **46**, 1211 (1992).
[3] L. E. Fisher, J. M. Caroon, S. R. Stabler, S. Lundberg, S. Zaidi, C. M. Sorensen, M. L. Sparacino, and J. M. Muchowski, *CJC* **72**, 142 (1994).
[4] M. Ohno, Y. Yamamoto, Y. Shibasaki, and S. Eguchi, *JCS(P1)* 263 (1993).
[5] M.-Y. Chen and J.-M. Fang, *JCS(P1)* 1737 (1993).
[6] Y. Ukaji, K. Tsukamoto, Y. Nasada, M. Shimizu, and T. Fujisawa, *CL* 221 (1993).
[7] H. Pellissier, L. Toupet, and M. Santelli, *JOC* **59**, 1709 (1994).
[8] K. Nishitani, Y. Nakamura, R. Orii, C. Arai, and K. Yamakawa, *CPB* **41**, 822 (1993).
[9] M. Hojo, N. Ishibashi, K. Ohsumi, K. Miura, and A. Hosomi, *JOMC* **473**, C1 (1994).
[10] M. Franciotti, A. Mann, A. Mordini, and M. Taddei, *TL* **34**, 1355 (1993).
[11] H. Monti, G. Audran, J.-P. Monti, and G. Leandri, *SL* 403 (1994).
[12] H.-J. Knölker, G. Baum, and R. Graf, *ACIEE* **33**, 1613 (1994).
[13] H.-J. Knölker, N. Foitzik, H. Goesmann, and R. Graf, *ACIEE* **32**, 1081 (1993).
[14] Y. Hashimoto, T. Nagashima, K. Kobayashi, M. Hasegawa, and K. Saigo, *T* **49**, 6349 (1993).
[15] I. Alonso, J. C. Carretero, and J. L. G. Ruano, *JOC* **58**, 3231 (1993).
[16] B. Loubinoux, A. C. O'Sullivan, J.-L. Sinnes, and T. Winkler, *T* **50**, 2047 (1994).
[17] S. Shimada, Y. Hashimoto, T. Nagashima, M. Hasegawa, and K. Saigo, *T* **49**, 1589 (1993).
[18] R. L. Funk, J. F. Fitzgerald, T. A. Olmstead, K. S. Para, and J. A. Wos, *JACS* **115**, 8849 (1993).
[19] R. C. Cambie, P. S. Rutledge, R. J. Stevenson, and P. D. Woodgate, *JOMC* **471**, 133 (1994).
[20] C. Mukai, O. Kataoka, and M. Hanaoka, *JCS(P1)* 563 (1993).
[21] T.-H. Yan, C.-W. Tan, H.-C. Lee, H.-C. Lo, and T.-Y. Huang, *JACS* **115**, 2613 (1993).
[22] G. Kumaran and G. H. Kulkarni, *TL* **35**, 5517 (1994).
[23] G. Kumaran and G. H. Kulkarni, *TL* **35**, 9099 (1994).

Titanium(IV) chloride–diisopropylethylamine.

Claisen condensation.   2,3-Dihydro-4*H*-pyran-4-ones can be made[1] by an intramolecular condensation without affecting existing stereocenters.

81%

β-Amino esters and β-lactams.   Chiral *N*-acyloxazolidinones undergo enolization on treatment with TiCl$_4$–*i*-Pr$_2$NEt, and the resulting titanium enolates add to imines stereoselectively.[2] The adducts are converted to β-lactams in two steps. On the other hand, reaction of the enolates derived from thioesters gives β-lactams directly.[3]

[1] W. Oppolzer and I. Rodriguez, *HCA* **76**, 1282 (1993).
[2] I. Abrahams, M. Motevalli, A. J. Robinson, and P. B. Wyatt, *T* **50**, 12755 (1994).
[3] R. Annunziata, M. Bengalia, M. Cinquini, F. Cozzi, and L. Raimondi, *T* **50**, 9471 (1994).

Titanium(IV) chloride–lithium aluminum hydride. 13, 310; 14, 307–308; 15, 320–321

Sulfides from sulfones.[1]   The reduction is rapid (30 min) and gives the products in high yields (62–93%).

α-Chloro enones.[2]   The $TiCl_4$–$LiAlH_4$ reagent system is capable of generating dichlorocarbene from CCl_4, which reacts with silyl enol ethers to form α-chloro enones.

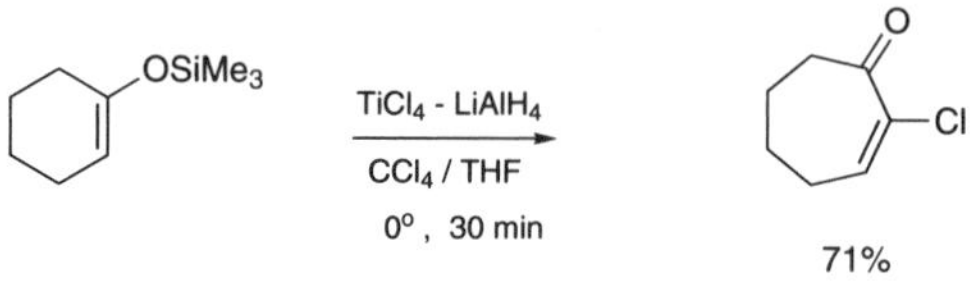

[1] E. Akgun, K. Mahmood, and C. A. Mathis, *CC* 761 (1994).
[2] M. Mitani and Y. Kobayashi, *BCSJ* **67**, 284 (1994).

Titanium(IV) chloride–p-trifluoromethylbenzoic anhydride–silver(I) triflate.

Transacylation.   Silyl esters behave as acyl donors to amines[1] in the presence of the dual-catalyst system and the aromatic anhydride. Silyl ethers instead of the alcohols can be converted to esters under these conditions,[2] and for lactonization,[3] a similar system ($AgClO_4$ instead of AgOTf) is effective.

[1] M. Miyashita, I. Shiina, and T. Mukaiyama, *CL* 1053 (1993).
[2] M. Miyashita, I. Shiina, S. Miyoshi, and T. Mukaiyama, *BCSJ* **66**, 1516 (1993).
[3] T. Mukaiyama, J. Izumi, M. Miyashita, and I. Shiina, *CL* 907 (1993).

Titanium(IV) chloride–titanium tetraisopropoxide. 14, 87–88; 15, 321–322, 335–336; 16, 337–338, 353–354

Admixture of these reagents generates chlorotitanium isopropoxides in situ.

1,3-Diarylallenes.[1]

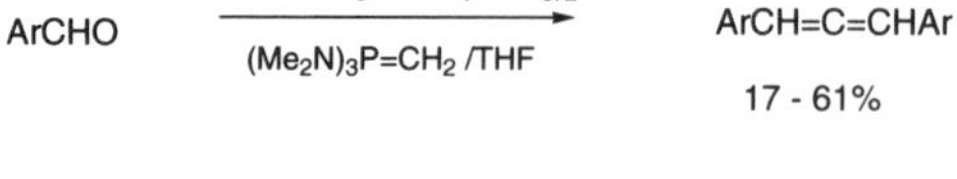

cis-4,5-Disubstituted δ-lactones.[2]

85% (*cis: trans* 92 : 8)

2-Aryl-2,3-dihydrobenzofurans.[3] The [3+2]cycloaddition of benzoquinones with styrenes is high yielding, and it has been applied to the synthesis of pterocarpans and neolignans.

Diels–Alder reactions. C_2-Symmetrical 2-alkenyl-1,3-dioxanes are excellent dienophiles for asymmetric Diels–Alder reactions.[4] In the presence of appropriate Lewis acid catalyst(s) the dienophiles ionize and are thereby activated.

Aldol condensation. The reaction of *N*-protected α-amino aldehydes with ketene silyl acetals is *syn*-selective[5] in the presence of $(i\text{-PrO})_2TiCl_2$. *N,N*-Disubstituted amides and thioamides behave similarly, although the less acidic Lewis acid $(i\text{-PrO})_3TiCl$ is used as catalyst.[6]

[1] K. A. Reynolds, P. G. Dopico, M. J. Sundermann, K. A. Hughes, and M. G. Finn, *JOC* **58**, 1298 (1993).
[2] S. Shimada, I. Tohno, Y. Hashimoto, and K. Saigo, *CL* 1117 (1993).
[3] T. A. Engler, K. D. Combrink, M. A. Letavic, K. O. Lynch, and J. E. Ray, *JOC* **59**, 6567 (1994).
[4] T. Sammakia and M. A. Berliner, *JOC* **59**, 6890 (1994).
[5] S. Kiyooka, K. Suzuki, M. Shirouchi, Y. Kaneko, and S. Tanimori, *TL* **34**, 5729 (1993).
[6] L. Z. Viteva, T. S. Gospodova, and Y. N. Stefanovsky, *T* **50**, 7193 (1994).

Titanium tetraisopropoxide. 13, 311–313; 14, 311–312; 15, 322; 16, 339; 17, 347–348

Alcoholysis of thioesters.[1] When thioesters are reacted with $(i\text{-PrO})_4Ti$ as a catalyst in an alcohol under reflux, ordinary esters are formed.

Methylimines.[2] An excellent catalyst for the condensation of amines with paraformaldehyde prior to reduction with $NaBH_4$ is $(i\text{-PrO})_4Ti$. Weakly basic amines (e.g., nitroanilines) do not react.

β,γ-Epoxy alcohols. Treatment of an allylic hydroperoxide with $(i\text{-PrO})_4Ti$ and molecular sieves results in the formation of the epoxy alcohol.[3] The epoxy oxygen atom comes from the OOH group; therefore, it has a defined stereochemistry. Starting from an allylic alcohol, an epoxy diol having at least three consecutive stereocenters can be prepared through photooxygenation and the catalyzed oxygen atom transfer.

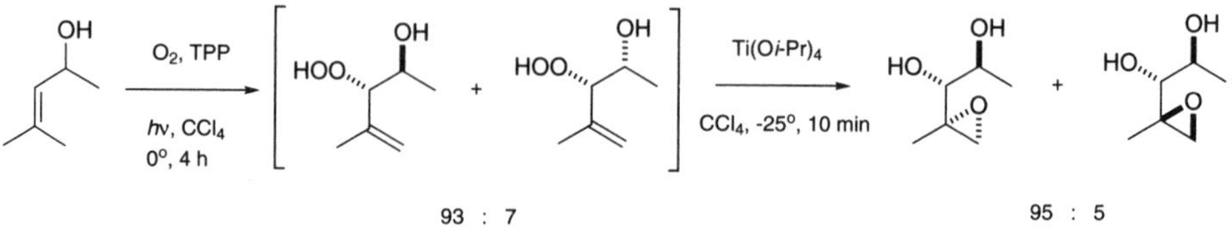

93 : 7   95 : 5

[1] M. Muzard and C. Portella, *JOC* **58**, 29 (1993).
[2] S. Bhattacharyya, *TL* **35**, 2401 (1994).
[3] W. Adam and B. Nestler, *JACS* **115**, 7226 (1993).

Titanium(IV) chloride–zinc.

Reductive cross-coupling of carbonyl compounds.[1]   The low-valent titanium species couples ketones to nitriles and carboxylic acid derivatives such as esters and acid chlorides. The products represent reductive acylation products of ketones.[2]

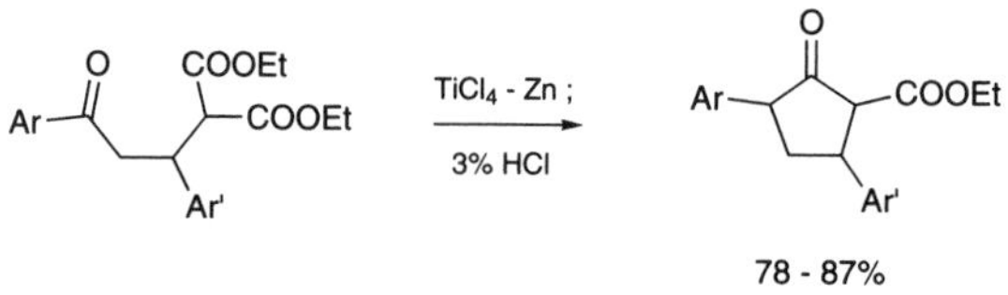

78 - 87%

Amidines.[3]   Coreduction of nitroarenes and nitriles leads to amidines. For aliphatic γ-nitronitriles an intramolecular coupling furnishes 2-amino-1-pyrrolines.

[1] J. Gao, M.-Y. Hu, J.-X. Chen, S. Yuan, and W.-X. Chen, *TL* **34**, 1617 (1993).
[2] D.-Q. Shi, J.-X. Chen, W.-Y. Chai, W.-X. Chen, and T.-Y. Kao, *TL* **34**, 2963 (1993).
[3] J.-X. Chen, W.-Y. Chai, J.-L. Zhu, J. Gao, W.-X. Chen, and T.-Y. Kao, *S* 87 (1993).

Titanocene derivatives. 13, 102; 14, 120–121; 15, 32–33, 81–82, 120; 16, 72–77

7-Hydroxynorbornenes.[1]   Titanocene dichloride provides a diene unit in Diels–Alder reactions. Hydrolysis of the adducts introduces a hydroxyl group at C-7.

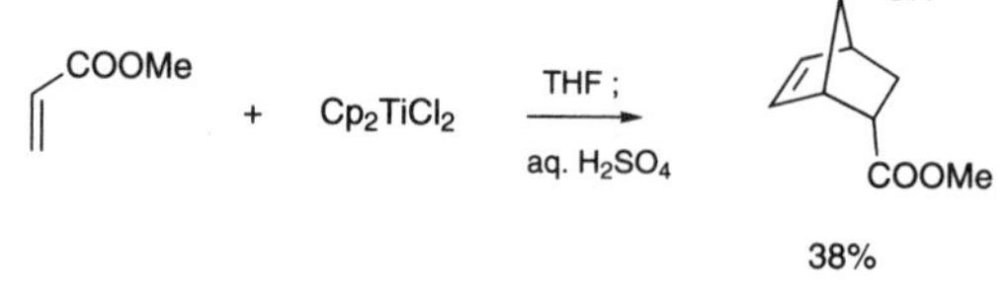

38%

Allylation.   The reaction of allylsilanes with carbonyl compounds, acetals, and ortho esters is catalyzed by $Cp_2Ti(OTf)_2$. Rapid reaction rates and good conversion are observed for these reactions.[2] Allylzinc reagents are obtained from dienes in THF by hydrozincation with Cp_2TiCl_2–ZnI_2–LiH in a ratio of $(0.1:1.2:2.0)$ with respect to the carbonyl compound to be alkylated.[3]

Pauson–Khand-type reaction.[4] Using t-$BuMe_2SiCN$ as a surrogate for carbon monoxide, $Cp_2Ti(PMe_3)_2$ scaffolds alkene and alkyne units to form cyclopentenones.

[1] C. A. Merlic and H. D. Bendorf, *OM* **12**, 559 (1993).
[2] T. K. Hollis, N. P. Robinson, J. Whelan, and B. Bosnich, *TL* **34**, 4309 (1993).
[3] Y. Gao, H. Urabe, and F. Sato, *JOC* **59**, 5521 (1994).
[4] S. C. Berk, R. B. Grossman, and S. L. Buchwald, *JACS* **115**, 4912 (1993).

1-(p-Toluenesulfinyl)propyne.

1,3-Dipolar cycloadditions.[1] This dipolarophile reacts with nitrones to give 4-toluenesulfinyl-5-methylisoxazolines, which on hydrogenation are transformed into β-amino ketones. Some simple piperidine alkaloids can be synthesized by this method.

[1] C. Louis, S. Mill, V. Mancuso, and C. Hootele, *CJC* **72**, 1347 (1994).

2-(p-Toluenesulfonyl)ethylamine.

Amides.[1] Reductive alkylation (with an aldehyde and $NaBH_4$) followed by N-acylation and treatment with t-BuOK in THF constitutes a route to $RCONHCH_2R'$.

[1] D. DiPietro, R. M. Borzilleri, and S. M. Weinreb, *JOC* **59**, 5856 (1994).

p-Toluenesulfonylimino iodobenzene. 17, 348

Aziridines.[1] Alkenes are transformed into aziridines by copper-mediated transfer of the TsN group from TsN=IPh at room temperature. Both $Cu(OTf)_2$ and $CuClO_4$ are highly efficient catalysts.

Allyl tosyl amides.[2] Reaction of allyl phenyl tellurides which are obtained from the corresponding halides (in situ) with TsN=IPh (or chloramine T) furnishes the amides.

[1] D. A. Evans, M. M. Faul, and M. T. Bilodeau, *JACS* **116**, 2742 (1994).
[2] Y. Nishibayashi, S. K. Srivastava, K. Ohe, and S. Uemura, *TL* **36**, 6725 (1995).

Trialkylaluminums. 15, 341–342; 16, 374; 17, 372–375

Addition to carbonyl compounds. A useful group transformation is reductive *gem*-dimethylation of a ketone. It can be performed in one step using the Me_3Al–Me_3SiOTf combination.[1] In the presence of CuBr, trialkylaluminums add to conjugated ketones and aldehydes in the 1,4-manner.[2,3]

Chiral β-keto sulfoxides react with Me_3Al under chelation control when $ZnCl_2$ is present.[4] α-Sulfonyl-α,β-unsaturated esters undergo conjugate addition with Me_3Al even without a Cu catalyst.[5]

Functional group exchange. A synthesis of cyanoguanidines[6] from O-substituted N-cyanoisoureas is promoted by Me$_3$Al.

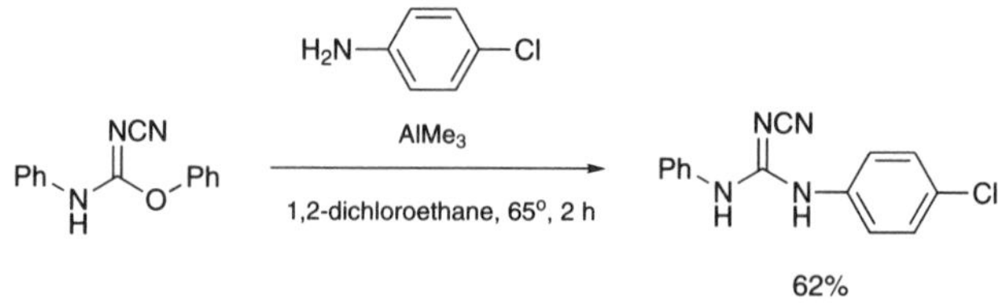

62%

Removal of chiral auxiliaries is very successful with lithium benzylthio-(trimethyl)aluminate, which is prepared in situ from BnSLi and Me$_3$Al.[7]

Carboalumination.[8] Alkylation of 1-alkynes using a mixture of Me$_3$Al and Cp$_2$ZrCl$_2$ to afford 2-methyl-1-alkenes can be accomplished in the presence of water. The regioselectivity is excellent.

β-Lactams.[9] The cyclization of β-amino esters is induced by i-Bu$_3$Al. Although the yields are not high, many functional groups are tolerated.

Ring cleavage. Diastereoselective cleavage of C_2-symmetrical 1,3-dioxanes has been effected by reagents prepared from organoaluminums. For example, reduction with a species obtained from Me$_3$Al and C$_6$F$_5$OH is useful for synthesis of chiral secondary alcohols.[10]

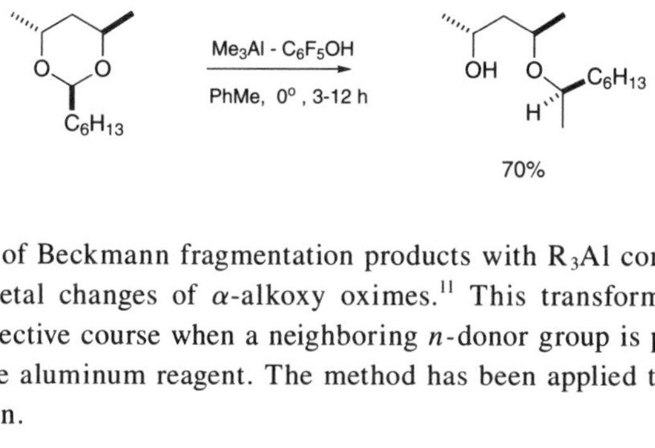

70%

Interception of Beckmann fragmentation products with R$_3$Al completes homologation and skeletal changes of α-alkoxy oximes.[11] This transformation follows a highly stereoselective course when a neighboring n-donor group is present to direct the attack of the aluminum reagent. The method has been applied to a synthesis of *endo*-brevicomin.

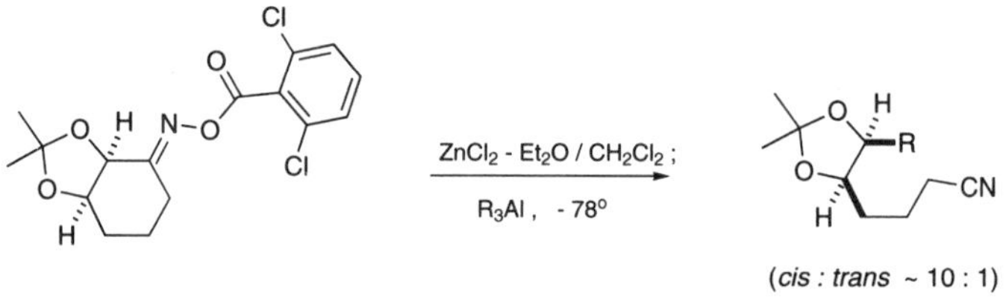

(*cis : trans* ~ 10 : 1)

[1] C. U. Kim, P. F. Misco, B. Y. Luh, and M. M. Mansuri, *TL* **35**, 3017 (1994).

[2] J. Kabbara, S. Fleming, K. Nickisch, H. Neh, and J. Westermann, *T* **50**, 743 (1994).

[3] J. Kabbara, S. Fleming, K. Nickisch, H. Neh, and J. Westermann, *SL* 679 (1994).

[4] M. C. Carreno, J. L. G. Ruano, M. C. Maestro, M. P. Gonzalez, A. B. Bueno, and J. Fischer, *T* **49**, 11009 (1993).

[5] J. Rojo, M. Garcia, and J. C. Carretero, *T* **49**, 9787 (1993).

[6] K. S. Atwal, F. N. Ferrara, and S. Z. Ahmed, *TL* **35**, 8085 (1994).

[7] O. Miyata, Y. Fujiwara, A. Nishiguchi, H. Honda, T. Shinada, I. Ninomiya, and T. Naito, *SL* 637 (1994).

[8] P. Wipf and S. Lim, *ACIEE* **33**, 1068 (1994).

[9] H. Vorbrüggen and R. B. Woodward, *T* **49**, 1625 (1993).

[10] K. Ishihara, N. Hanaki, and H. Yamamoto, *JACS* **115**, 10695 (1993).

[11] H. Fujioka, M. Miyazaki, H. Kitagawa, T. Yamanaka, H. Yamamoto, K. Takuma, and Y. Kita, *CC* 1634 (1993).

Trialkylsilyldiazomethane. 13, 327–328; 15, 344; 16, 361

Alkynes.[1] Deoxygenative homologation of ketones is accomplished simply by treatment with lithiated trimethylsilyldiazomethane.

$$Ph-\overset{O}{\underset{}{C}}< \quad \xrightarrow[\text{THF, } -78° \to \Delta]{Me_3SiC(Li)=N_2} \quad Ph-\!\!\equiv\!\!- \quad 58\%$$

Esterification.[2] Trisopropylsilyldiazomethane is very easy to prepare and stable. Contrary to Me_3SiCHN_2, its reaction with acids yields the pure $i\text{-}Pr_3SiCH_2$-esters.

[1] K. Miwa, T. Aoyama, and T. Shioiri, *SL* 107 (1994).

[2] J. A. Soderquist and E. I. Miranda, *TL* **34**, 4905 (1993).

Triarylbismuthines.

Esters and amides.[1] Condensation of carboxylic acids bearing an α-H with alcohols and amines proceeds via triacyloxybismuthines and ketenes under neutral conditions.

Oxidation.[2] Bismuthines accept oxygen from iodosylbenzene under ultrasonic irradiation. These Ar_3BiO are mild oxidants, which convert benzylic, allylic, and secondary alcohols to carbonyl compounds, hydrazobenzenes to azobenzenes. By contrast, oxides of lower pnictogen elements are devoid of oxidizing power for organic substances.

[1] T. Ogawa, T. Hikasa, T. Ikegami, N. Ono, and H. Suzuki, *CL* 815 (1993).

[2] H. Suzuki, T. Ikegami, and Y. Matano, *TL* **35**, 8197 (1994).

Tribromomethyl(trimethyl)silane.

Methyl ketones from aldehydes.[1]   The homologation is mediated by Cr(II).

[1] D. M. Hodgson and P. J. Comina, *SL* 663 (1994).

Tributylstannyloxyalkenes.

γ-Imino ketones and diketones.[1]   Tin enolates are very reactive toward α-halo imines, forming iminoketones that are readily hydrolyzed.

[1] M. Yasuda, Y. Katoh, I. Shibata, A. Baba, H. Matsuda, and N. Sonoda, *JOC* **59**, 4386 (1994).

Tributyltin halides. 13, 315

Vinyltins.   Bu$_3$SnCl is commonly used as an electrophile to trap α-lithioimines,[1] enolates of trimethylsilylacetic esters,[2] and alkynyltrialkylborate salts.[3]

β,γ-Epoxy ketones.[4]   The reaction of tin enolates with α-halo carbonyl compounds generally affords 1,4-dicarbonyl products. However, in the presence of Bu$_3$SnBr and Bu$_4$NBr the major pathway is switched to attack on the carbonyl group instead of on the α-carbon, resulting in β,γ-epoxy ketones.

[1] B. Jousseaume, M. Pereyre, N. Petit, J.-B. Verlhac, and A. Ricci, *JOMC* **443**, C1 (1993).
[2] A. J. Zapata, C. Fortoul, and R. C. Acuna, *JOMC* **448**, 69 (1993).
[3] M.-Z. Deng, N.-S. Li, and Y.-Z. Huang, *JOC* **58**, 1949 (1993).
[4] M. Yasuda, T. Oh-hata, I. Shibata, A. Baba, and H. Matsuda, *JCS(P1)* 859 (1993).

Tributyltin hydride. 13, 316–319; **14**, 312–318; **15**, 325–333; **16**, 343–350; **17**, 351–361

Reductions.   Bu$_3$SnH is capable of reducing epoxides[1] in the presence of MgI$_2$, giving alcohols. On the other hand, epoxy ketones suffer reduction at the carbonyl group.[2] Bu$_3$SnH–SiO$_2$ is useful for reduction of aldehydes.[3]

Hydrodehalogenation.   A synthesis of chiral epoxides[4] from trichloromethyl ketones is through enantioselective reduction to the trichloromethyl carbinols and tinhydride-mediated conversion to the chlorohydrins before base treatment to close the three-membered ring. Dehalogenation of halocarboxylic acids[5] in water is apparently a method that deserves wide application.

Other defunctionalizations.   Deoxygenation of alcohols via thioxocarbamate derivatives[6] or benzothiazol-2-yl sulfides[7] (formed by Mitsunobu reaction) is readily achieved.

Isonitriles have been prepared from carboxylic acids via selenocarbamates.[8] When a leaving group is present at the β-position of the selenium atom, the homolysis of the C–Se bond is followed by expulsion of the β-substituent. Glycal formation[9] from azidosugar selenides on treatment with Bu$_3$SnH occurs in excellent yields.

Chain extension.   α-Alkoxy radicals and β-acetoxy radicals generated from the selenides[10] and iodides,[11] respectively, add to alkenes. Since the precursors are quite readily available, assembly of difunctional compounds by free radical reactions is now a recognized avenue. Furthermore, alkenyl radicals can be used to initiate the chain extension process.[12,13]

Free radical cyclization of 1,6-dienes[14] with functionalization at the termini of the double bonds is much valued because the two different functional groups can be elaborated further. Stannylformylation is one possibility.

A very important tactic for α-amino radical generation involves translocation via hydrogen abstraction by an aromatic free radical generated from a halogen five bonds away. Thus chain extension at the α-position of an amide[15] or amine[16] can be accomplished with great success.

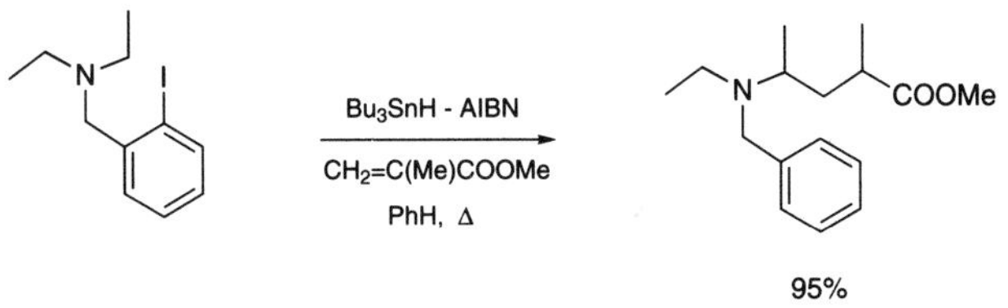

Construction of carbocycles.   Formation of cyclopentane rings can be initiated by treatment of unsaturated thiocarbamates,[17] selenides,[18] and halo compounds with Bu$_3$SnH.[19–21] When the radical adds to a triple bond, an alkylidenecyclopentane unit is formed.

Although the goal of constructing the lysergic acid skeleton was thwarted, the tricyclization by a tandem free radical cyclization of an aryl bromide[22] still attests to the remarkable power of such processes.

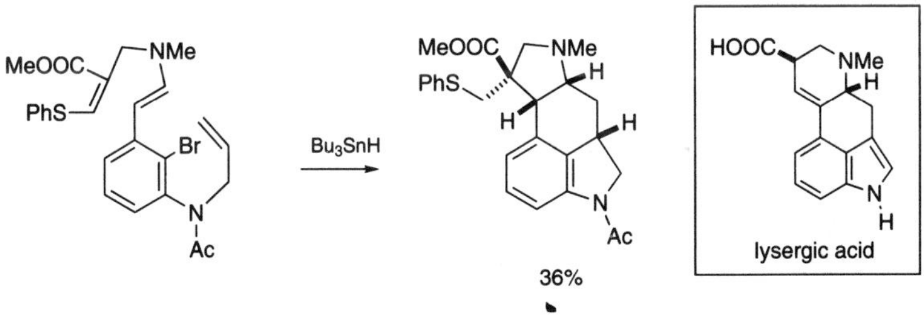

A three-component free radical reaction[23] leading to a cyclohexene-dicarboxylic ester is also quite useful.

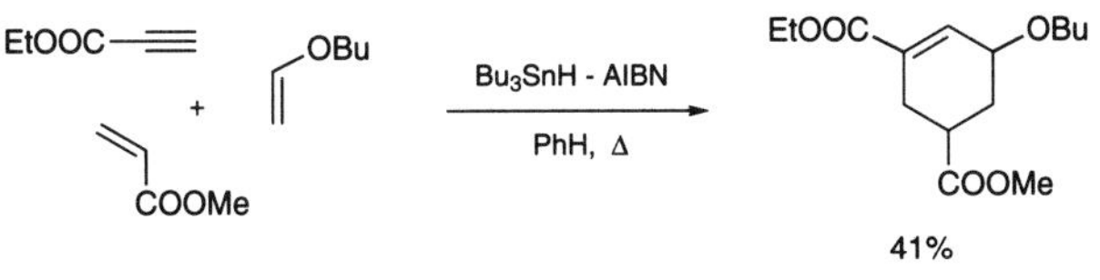

Heterocyclizations. Tetrahydrofuran rings,[24-26] tetrahydropyran ethers,[27] indoles,[28] and even β-lactams[29] are formed in high yields by intramolecular free radical cyclizations.

Much work has been devoted to the synthesis of the pyrrolidine and pyrrolidone systems by free radical methods. For pyrrolidine derivatives it is preferable to initiate the cyclization from an aminyl radical generated from N-chloro-[30] or N-sulfenylamine.[31,32]

A formal intramolecular displacement appears to be the most favorable pathway available to an α-amide radical.[33] Such an α-amide radical also adds to a distal unsaturated ester.[34] Higher yields of cyclized products are obtained when the α-amide radical is stabilized, for example, by a benzenesulfenyl group.[35]

The cyclization involving carbon radical and a remote azide group opens a route to pyrrolidines.[36] Macrolactams are formed[37] by the addition of an aminyl radical to a β-keto ester.

Aldehydes form O-stannyl ketyls with the Bu₃Sn radical. A β-alkoxyacrylate unit five bonds away from the ketyl is well positioned to be involved in a radical transfer process; therefore, cyclization ensues.[38] Fused oxacycles of well-defined stereochemistry are assembled by this reaction. Heterocyclic 1,2-amino alcohols are similarly procured from dialdehyde monooxime ethers.[39]

Rearrangements. Some rearrangement processes start from intramolecular addition reactions. These include the formation of ketones from enol ethers,[40] ω-arylalkanols from ω-bromoalkyl aryl ethers,[41] and cycloalkyl silyl ethers from ω-bromoalkyl silyl ketones.[42]

Free radical ring expansion of α-halo-α-bromoalkylcycloalkanones[43] is synthetically useful. For example, in three steps an α,α-dichlorocyclobutanone is transformed to a cycloheptanone.

74%

[1] G. Guanti, L. Banfi, V. Merlo, and E. Narisano, *T* **50**, 2219 (1994).

[2] T. Kawakami, I. Shibata, A. Baba, H. Matsuda, and N. Sonoda, *T* **50**, 8625 (1994).

[3] B. Figadere, C. Chaboche, X. Franck, J.-F. Peyrat, and A. Cave, *JOC* **59**, 7138 (1994).

[4] E. J. Corey and C. J. Helal, *TL* **34**, 5227 (1993).

[5] U. Maitra and K. D. Sarma, *TL* **35**, 7861 (1994).

[6] M. Oba and K. Nishiyama, *S* 624 (1994).

[7] I. Dancy, L. Laupichler, P. Rollin, and J. Thiem, *LA* 343 (1993).

[8] A. G. M. Barrett, H. Kwon, and E. M. Wallace, *CC* 1760 (1993).

[9] F. Santoyo-Gonzalez, F. G. Calvo-Flores, F. Hernandez-Mateo, P. Garcia-Mendoza, J. Isac-Garcia, and M. D. Perez-Alvarez, *SL* 454 (1994).

[10] Y. Nishiyama, H. Yamamoto, S. Nakata, and Y. Ishii, *CL* 841 (1993).

[11] F. Foubelo, F. Lloret, and M. Yus, *T* **50**, 5131 (1994).

[12] K. Miura, D. Itoh, T. Hondo, and A. Hosomi, *TL* **35**, 9605 (1994).

[13] F. Foubelo, F. Lloret, and M. Yus, *T* **50**, 6715 (1994).

[14] I. Ryu, A. Kurihara, H. Muraoka, S. Tsunoi, N. Kambe, and N. Sonoda, *JOC* **59**, 7570 (1994).

[15] L. Williams, S. E. Booth, and K. Undheim, *T* **50**, 13697 (1994).

[16] K. Undheim and L. Williams, *CC* 883 (1994).

[17] T. Morikawa, M. Uejima, Y. Kobayashi, and T. Taguchi, *JFC* **65**, 79 (1993).

[18] D. L. J. Clive, H. W. Manning, T. L. B. Boivin, and M. H. D. Postema, *JOC* **58**, 6857 (1993).

[19] Y. Tanabe, Y. Nishi, and K.-i. Wakimura, *CL* 1757 (1994).

[20] L. A. Buttle and W. B. Motherwell, *TL* **35**, 3995 (1994).

[21] C.-K. Sha, C.-Y. Shen, T.-S. Jean, R.-T. Chiu, and W.-H. Tseng, *TL* **34**, 7641 (1993).

[22] Y. Özlü, D. E. Cladingboel, and P. J. Parsons, *T* **50**, 2183 (1994).

[23] E. Lee, C. U. Hur, Y. H. Rhee, Y. C. Park, and S. Y. Kim, *CC* 1466 (1993).

[24] D. Schinzer, P. G. Jones, and K. Obierey, *TL* **35**, 5853 (1994).

[25] V. H. Rawal, S. P. Singh, C. Dufour, and C. Michoud, *JOC* **58**, 7718 (1993).

[26] S. D. Burke and K. W. Jung, *TL* **35**, 5837 (1994).

[27] J. C. Lopez, A. M. Gomez, and B. Fraser-Reid, *CC* 1533 (1994).

[28] T. Fukuyama, X. Chen, and G. Peng, *JACS* **116**, 3127 (1994).

[29] H. Ishibashi, C. Kameoka, A. Yoshikawa, R. Ueda, K. Kodama, T. Sato, and M. Ikeda, *SL* 649 (1993).

[30] M. Tokuda, H. Fujita, and H. Suginome, *JCS(P1)* 777 (1994).

[31] W. R. Bowman, D. N. Clark, and R. J. Marmon, *T* **50**, 1275 (1994).

[32] J. Boivin, E. Froquet, and S. Z. Zard, *T* **50**, 1745 (1994).

[33] K. Goodall and A. F. Parsons, *JCS(P1)* 3257 (1994).

[34] A. F. Parsons and R. J. K. Taylor, *JCS(P1)* 1945 (1994).

[35] T. Sato, K. Tsujimoto, K.-i. Matsubayashi, H. Ishibashi, and M. Ikeda, *CPB* **40**, 2308 (1992).

[36] S. Kim, G. H. Joe, and J. Y. Do, *JACS* **116**, 5521 (1994).

[37] S. Kim, G. H. Joe, and J. Y. Do, *JACS* **115**, 3328 (1993).

[38] E. Lee, J. S. Tae, Y. H. Chong, Y. C. Park, M. Yun, and S. Kim, *TL* **35**, 129 (1994).

[39] T. Naito, K. Tajiri, T. Harimoto, I. Ninomiya, and T. Kiguchi, *TL* **35**, 2205 (1994).

[40] D. Crich and Q. Yao, *CC* 1265 (1993).

[41] E. Lee, C. Lee, J. S. Tae, H. S. Whang, and K. S. Li, *TL* **34**, 2343 (1993).
[42] Y.-M. Tsai, K.-H. Tang, and W.-T. Jiaang, *TL* **34**, 1303 (1993).
[43] W. Zhang, Y. Hua, S. J. Geib, G. Hoge, and P. Dowd, *T* **50**, 12579 (1994).

Tributyltin hydride–tetrakis(triphenylphosphine)palladium.

Reductions.[1]   Thioesters undergo chemoselective conversion to aldehydes. A conjugated vinylic thio group is not affected.

[1] H. Kuniyasu, A. Ogawa, and N. Sonoda, *TL* **34**, 2491 (1993).

Tributyltin hydride–triethylborane. 15, 333; 16, 350; 17, 363–364

Desulfurization.[1]   Episulfides are converted into alkenes under very mild conditions ($-78°C$).

Reductive cyclization of ω-ethynyl α,β-unsaturated esters.[2]   The Lewis acid–catalyzed radical cyclization is subject to 1,3-asymmetric induction by the chiral moiety of the ester. Thus (-)-8-phenylmenthyl esters give mainly the 2-methylenecycloalkylacetic esters of (R) configuration.

[1] J. Uenishi and Y. Kubo, *TL* **35**, 6697 (1994).
[2] M. Nishida, E. Ueyama, H. Hayashi, Y. Ohtake, Y. Yamaura, E. Yanaginuma, O. Yonemitsu, A. Nishida, and N. Kawahara, *JACS* **116**, 6455 (1994).

Tributyl(trimethylsilyl)stannane.

α-Hydroxystannanes.[1]   Aldehydes but not ketones are converted into α-trimethylsiloxystannanes on treatment with $Bu_3SnSiMe_3$ in the presence of Bu_4NCN. The silyl ether groups of the adducts are readily hydrolyzed to afford the hydroxylstannanes.

o-Quinodimethanes.[2]   The tributylstannyl anion triggers 1,4-elimination of α,α'-dibromo-o-xylenes. A fluoride source (CsF or $[Et_2N]_3S^+SiMe_3F_2^-$) is required to generate the anion from the silylstannane.

Barbier-type reactions.[3]   Cyclization of iodoketones in which the iodine atom is allylic or vinylic can be accomplished with the same combination of $Bu_3SnSiMe_3$ and CsF.[4]

Cyclization by tandem Michael–Dieckmann reaction.[5]   Ring construction from compounds with juxtaposed ketone (or ester) and conjugated ester groups is made easy. It is rendered irreversible by trapping the alkoxide ion as a trimethylsilyl ether.

[1] R. K. Bhatt, J. Ye, and J. R. Falck, *TL* **35**, 4081 (1994).
[2] H. Sato, N. Isono, K. Okamura, T. Date, and M. Mori, *TL* **35**, 2035 (1994).
[3] A. Kinoshita and M. Mori, *CL* 1475 (1994).
[4] M. Mori, N. Isono, N. Kaneta, and M. Shibasaki, *JOC* **58**, 2972 (1993).
[5] T. Honda and M. Mori, *CL* 1013 (1994).

Tricarbonylruthenium(II) chloride.

Cycloisomerization of enynes.[1] Cyclization of 1,6- and 1,7-enynes under CO to afford the cyclic isomers proceeds in excellent yields in the presence of $[RuCl_2(CO)_3]_2$.

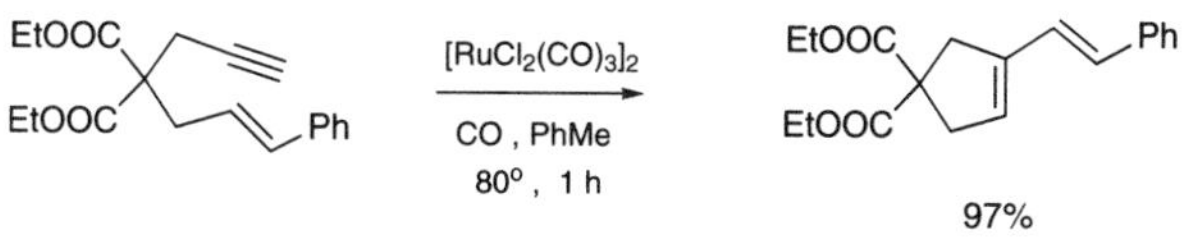

[1] N. Chatani, T. Morimoto, T. Muto, and S. Murai, *JACS* **116**, 6049 (1994).

Trichlorosilane.

Homoallylic alcohols.[1,2]

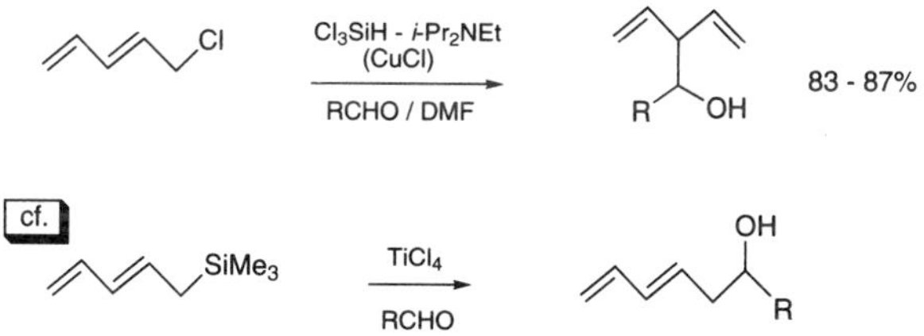

Semihydrogenation of alkynes.[3] The surface of silica gel is covered with hydrosilane groups on reaction with trichlorosilane. The solid-supported silane effects hydrogenation of alkynes to (Z)-alkenes selectively.

[1] S. Kobayashi and K. Nishio, *S* 457 (1994).
[2] S. Kobayashi and K. Nishio, *CL* 1773 (1994).
[3] A. D. Kini, D. V. Nadkarni, and J. L. Fry, *TL* **35**, 1507 (1994).

Triethylsilane. 14, 322; 15, 338; 16, 356

Reductions. Ionic reduction by Et_3SiH usually requires a protic acid or Lewis acid to help ionize the substrate. Thus reduction of vinylstannanes to alkylstannanes uses a Et_3SiH–TMSOTf combination.[1] A preparation of 4-hydroxy α-amino acid derivatives from the corresponding ketones by such reduction is *cis*-selective.[2] Reductive etherification of ketones is accomplished in the presence of silyl ethers, which supply the alkyl moiety. Isobornyl ethers are obtained in good yields from camphor.[3]

Deoxygenation of allylic alcohols[4] without affecting isolated tertiary alcohols, double bonds, and acetals is accomplished by reaction with Et_3SiH in the presence of $LiClO_4$. A route to β-*C*-glycosyl aldehydes involves reaction of sugar lactones with 2-lithio-1,3-dithiane, reduction with Et_3SiH, and hydrolysis of the dithiane.[5]

Deoxygenation of alcohols by the radical pathway requires prior conversion into derivatives such as thionocarbamates.[6] The reduction is carried out in the presence of $(t\text{-}BuO)_2$. Alkenes are obtained from 1,2-dixanthates[7] under such conditions.

Oxazolidin-5-ones are readily cleaved to give amino acids.[8] The lactam carbonyl group can be removed from the *N*-Boc derivatives in a two-stage reduction, the second stage featuring silane as reducing agent.[9] Ester groups are tolerated in the transformation.

Diaryl ketones are directly reduced to hydrocarbons[10] with $Et_3SiH–BF_3 \cdot OEt_2$. In the reduction of phosphine oxides[11] (i-PrO)$_4$Ti is used as the Lewis acid catalyst.

C–X Bond cleavage. *S*-Benzylthio esters of amino acids undergo reductive cleavage to provide the aldehydes[12] in the presence of Pd-C. The 2-arylthiazolidines obtained from cysteine are cleaved at the C–N bond.[13] Thus this process completes a two-step *S*-protection.

82% (R=Me)

Silylation. Hydrosilylation of 1-alkynes in the presence of a Rh(I) catalyst is regioselective. A reversal of the stereoselectivity is possible by adding Ph_3P and changing the solvent.[14] Silylformylation is conducted at 1 atm. CO at 0°C.[15]

[Rh(cod)Cl]₂ / EtOH	85%	94	:	4	:	2
[Rh(cod)Cl]₂ -Ph₃P / MeCN	93%	2	:	97	:	1

Fischer carbene complexes undergo demetallative hydrosilylation. It can be adapted to a synthesis of allylsilanes.[16]

[1] Y. Zhao, P. Quayle, and E. A. Kuo, *TL* **35**, 4179 (1994).

[2] R. F.W. Jackson, A. B. Rettie, A. Wood, and M. J. Wythes, *JCS(P1)* 1719 (1994).

[3] S. Hatakeyama, H. Mori, K. Kitano, H. Yamada, and M. Nishizawa, *TL* **35**, 4367 (1994).

[4] D. J. Wustrow, W. J. Smith, and L. D. Wise, *TL* **35**, 61 (1994).

[5] M. E. L. Sanchez, V. Michelet, I. Besnier, and J. P. Genet, *SL* 705 (1994).

[6] K. Nishiyama and M. Oba, *TL* **34**, 3745 (1993).

[7] D. H. R. Barton, D. O. Jang, and J. C. Jaszberenyi, *T* **49**, 2793 (1993).

[8] J. A. Robl, M. P. Cimarusti, L. M. Simpkins, H. N. Weller, Y. Y. Pan, M. Malley, and J. D. DiMarco, *JACS* **116**, 2348 (1994).

[9] C. Pedregal, J. Ezquerra, A. Escribano, M. C. Carreno, and J. L. G. Ruano, *TL* **35**, 2053 (1994).

[10] I. Simonou, *SC* **24**, 1999 (1994).

[11] T. Coumbe, N. J. Lawrence, and F. Muhammad, *TL* **35**, 625 (1994).

[12] P. T. Ho and K.-Y. Ngu, *JOC* **58**, 2313 (1993).

[13] L. S. Richter, J. C. Marsters, Jr., and T. R. Gadek, *TL* **35**, 1631 (1994).

[14] R. Takeuchi and N. Tanouchi, *JCS(P1)* 2909 (1994).

[15] M. P. Doyle and M. S. Shanklin, *OM* **12**, 11 (1993).

[16] C. C. Mak, M. K. Tse, and K. S. Chan, *JOC* **59**, 3585 (1994).

N-(Triethylsilylethylidene) *t*-butylamine, *t*-BuN=CHCH$_2$SiEt$_3$.

α,β-Unsaturated aldehydes.[1]   The LDA-promoted Peterson reaction of the reagent with carbonyl compounds followed by mild acid (citric acid) hydrolysis completes a two-carbon homologation.

[1] D. P. Provencal and J. W. Leahy, *JOC* **59**, 5496 (1994).

Triethylsilyl iodide.

A convenient preparation method[1] involves PdCl$_2$-catalyzed reaction of triethylsilane and MeI at room temperature. Other silyl iodides are obtainable in the same manner (6 examples, 77–97% yield).

[1] A. Kunai, T. Sakurai, E. Toyoda, M. Ishikawa, and Y. Yamamoto, *OM* **13**, 3233 (1994).

Trifluoroacetic acid. 14, 322–323; 15, 338–339

Trifluorocetates.[1]   Alkenes undergo Markovnikov addition with CF$_3$COOH in the presence of V$_2$O$_5$ at reflux temperature.

De-O-tritylation.[2]   Trityl ether cleavage can be performed by treatment with CF$_3$COOH–(CF$_3$CO)$_2$O without affecting other reducible and acid-hydrolyzable functionalities.

Diarylamines.[3]   A method for phenylamination of arenes is by warming with PhN$_3$–TfOH.

Desulfurization of 3-indolyl sulfides.[4]   Nonreductive removal of thio groups uses CF$_3$COOH to protonate the 3-indolyl sulfides, thus allowing thiophilic attack by a thiol reagent (e.g., *o*-mercaptobenzoic acid). The corresponding 3-unsubstituted indoles are produced.

[1] B. M. Choudhary and P. N. Reddy, *CC* 405 (1993).
[2] E. Krainer, F. Naider, and J. Becker, *TL* **34**, 1713 (1993).
[3] G. A. Olah, P. Ramaiah, Q. Wang, and G. K. S. Prakash, *JOC* **58**, 6900 (1993).
[4] P. Hamel, N. Zajac, J. G. Atkinson, and Y. Girard, *TL* **34**, 2059 (1993).

Trifluoroacetic anhydride.

Trifluorobutynamides.[1] *C*-Acylation of triphenylphosphoranylideneacetamides with $(CF_3CO)_2O$ gives adducts that on pyrolysis at 200–240°C provide the alkynoic amides. Other perfluoroalkanoic anhydrides can be used.

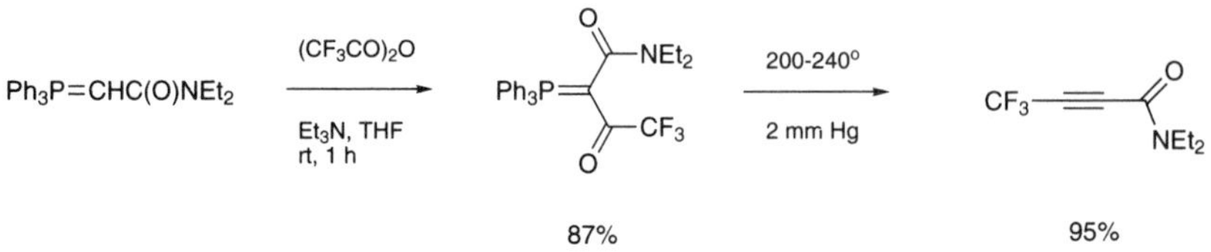

α-Trifluoromethyl acyloins.[2,3] These compounds are obtained from α-hydroxy acids on treatment with $(CF_3CO)_2O$ and pyridine in benzene.

Trifluoropyruvamides.[4] A simple synthesis of these compounds is from the 1:1 adducts of isocyanides and $(CF_3CO)_2O$ at low temperature.

α-Hydroxy aldehydes.[5] The Pummerer reaction of alkenyl sulfoxides is abnormal, as it furnishes α,β-bistrifluoroacetoxy sulfides, which are readily hydrolyzed.

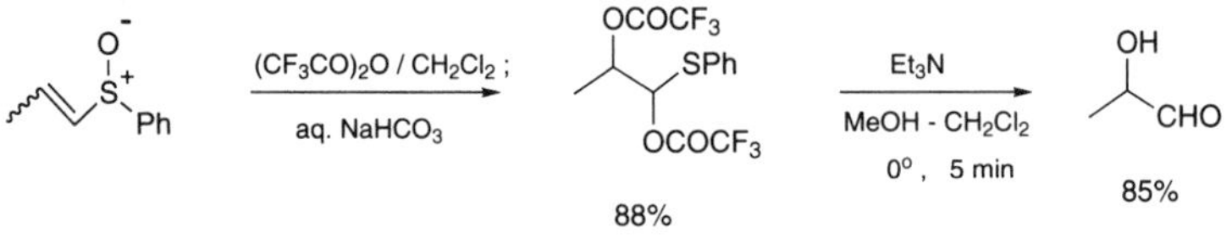

Sulfoxide reduction.[6] In the presence of Me_2S sulfoxides are reduced in very high yields to sulfides. This process as a followup reaction to sulfinylethylation of cyclic ketones is synthetically useful.

Trifluoroethylidenation of activated CH_2.[7] The Polonovski reaction of dimethyl(trifluoroethyl)amine oxide with $(CF_3CO)_2O$ in the presence of a compound containing an active methylene group accomplishes trifluoroethylidenation at that site.

Nitration of silyl enol ethers.[8] Bu_4NNO_3 is activated by $(CF_3CO)_2O$ toward reaction with triisopropylsilyl enol ethers.

[1] Y. Shen and S. Gao, *JFC* **61**, 105 (1993).
[2] M. Kawase and T. Kurihara, *TL* **35**, 8209 (1994).

[3] M. Kawase, *TL* **35**, 149 (1994).
[4] L. El Kaim, *TL* **35**, 6669 (1994).
[5] D. Craig, K. Daniels, and A. R. MacKenzie, *T* **49**, 11263 (1993).
[6] J. Montgomery and L. E. Overman, *JOC* **58**, 6476 (1993).
[7] C. Ates, Z. Janousek, and H. G. Viehe, *TL* **34**, 5711 (1993).
[8] P. A. Evans and J. M. Longmire, *TL* **35**, 8345 (1994).

Trifluoromethanesulfonic acid (triflic acid). 14, 323–324; 15, 339

Trimethylsilyl triflate.[1]   Admixture of FSO_3H and $MeC(OSiMe_3)=NSiMe_3$ or $CO(NHSiMe_3)_2$ generates TMSOTf in situ, which is useful in various catalytic reactions.

Vinyl triflates.[2]   1-Alkynes add TfOH at room temperature in the Markovnikov fashion.

Allyl sulfones → vinyl sulfones.[3] α-Trimethylsilylation (BuLi/TMSCl) followed by protodesilylation with TfOH or TsOH achieves the transformation. The major products have an (E) configuration.

Intramolecular Schmidt reactions.[4]   Bicyclic tertiary amines are obtained from unsaturated azides.

82%

[1] M. El Gihani and H. Heaney, *SL* 583 (1993).
[2] G. T. Crisp and A. G. Meyer, *S* 667 (1994).
[3] R. L. Funk, J. Umstead-Daggett, and K. M. Brummond, *TL* **34**, 2867 (1993).
[4] W. H. Pearson, R. Walavalkar, J. M. Schkeryantz, W.-K. Fang, and J. D. Blickensdorf, *JACS* **115**, 10183 (1993).

Trifluoromethanesulfonic anhydride (triflic anhydride). 13, 324–325; 14, 324–326; 15, 339–340; 16, 357–358

Perfluoroalkynylphosphonate esters.[1]   Phosphonylmethyl perfluoroalkyl ketones are dehydrated to give the alkynes on reaction with Tf_2O and $i\text{-}Pr_2NEt$ via triflyl ethers.

Isoquinolines.[2]   A one-step synthesis of these heterocyclic compounds requires phenylacetic esters, nitriles, and Tf_2O.

Conjugated triflones.[3]   Reaction of Me_3SiCH_2Li (from Me_3SiCH_2I and $t\text{-}BuLi$) with Tf_2O and then with aldehydes gives the triflones in 56–84% yield.

[1] Y. Shen and M. Qi, *JCS(P1)* 2153 (1993).
[2] A. G. Martinez, A. H. Fernandez, D. M. Vilchez, L. L. Gutiererez, and L. R. Subramanian, *SL* 229 (1993).
[3] A. Mahadevan and P. L. Fuchs, *TL* **35**, 6025 (1994).

S-(Trifluoromethyl)dibenzothiophenium triflate.

Trifluoromethylation of enolates. Bulky enol boronates undergo trifluoromethylation with (**1**) in a stereoselective manner in cases such as the dienolates of 8a-methyl-Δ^4-octal-3-one.

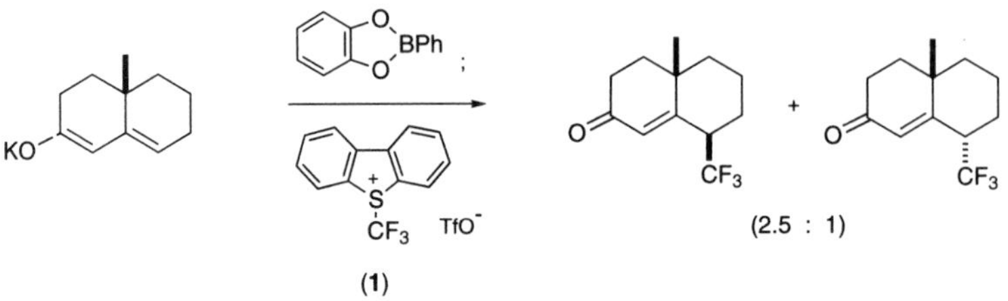

(2.5 : 1)

(**1**)

[1] T. Umemoto and K. Adachi, *JOC* **59**, 5692 (1994).

(Trifluoromethyl)trimethylsilane. **15**, 341
Preparation.[1]

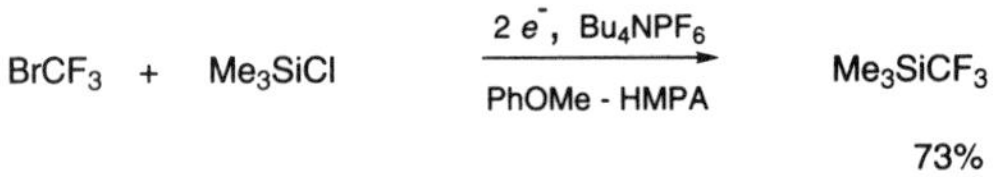

$$BrCF_3 \ + \ Me_3SiCl \ \xrightarrow[\text{PhOMe - HMPA}]{2\,e^-,\ Bu_4NPF_6} \ Me_3SiCF_3$$

73%

Trifluoromethylation.[2] Carbonyl compounds are converted into the trifluoromethyl carbinols in excellent yields by this reagent in the presence of a quaternary ammonium fluoride. Moderate asymmetric induction is observed when a chiral catalyst is used.

Addition to azirines[3] giving 2-trifluoromethylaziridines is also smooth.

2,2-Difluoroenoxysilanes.[4] The synthesis from acylsilanes is promoted by $Bu_4N^+(Ph_3SnF_2)^-$. With Bu_4NF the primary products undergo further aldol reactions.

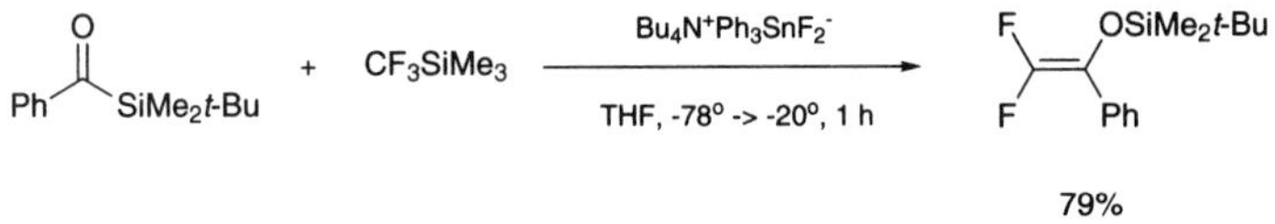

79%

[1] G. K. S. Prakash, D. Deffieux, A. K. Yudin, and G. A. Olah, *SL* 1057 (1994).
[2] K. Iseki, T. Nagai, and Y. Kobayashi, *TL* **35**, 3137 (1994).
[3] C. P. Felix, N. Khatimi, and A. J. Laurent, *TL* **35**, 3303 (1994).
[4] T. Brigaud, P. Doussot, and C. Portella, *CC* 2117 (1994).

Triisopropoxytitanium acetate.

Regioselective cleavage of epoxides.[1] 2,3-Epoxy alcohols react with (*i*-PrO)₃TiOAc regioselectively to give 3-acetoxy 1,2-diols. This result is useful for devising a route to 2-deoxy-D-ribose.

[1] Y. E. Raifeld, A. Nikitenko, and B. M. Arshava, *T* **49**, 2510 (1993).

2,4,6-Triisopropylphenylborane (tripylborane, TripBH₂).

Hydroborations.[1] This borane is a stable solid that can react regioselectively with alkenes to form either TripBHR or TripBR₂; TripBHR can be converted into mixed boranes TripBRR'. Oxidation of the adducts furnishes alcohols. Treatment of TripBR₂ with cyanide leads to R₂C=O.

[1] K. Smith, A. Pelter, and Z. Jin, *JCS(P1)* 395 (1993).

β-(Trimethylsilyl)alkyl phenyl sulfones.

Vinyl anion equivalents.[1] After the derived sulfonyl anions react with electrophiles, elimination of the sulfonyl/silyl groups may be induced. With an epoxide as electrophile the treatment with NaH in refluxing hexane actually triggers a C/O-transsilylation to prelude expulsion of the sulfinate anion.

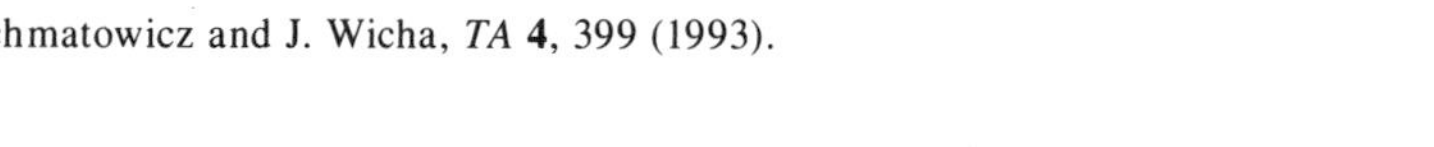

[1] B. Achmatowicz and J. Wicha, *TA* **4**, 399 (1993).

Trimethylsilyl azide. 13, 24–25; 14, 25; 15, 342–343; 16, 17

Tetrazoles. Formation of tetrazoles from nitriles is catalyzed by Bu₂SnO[1] or Me₃Al.[2]

β-Glycosyl azides.[3] A high-yielding preparation from peracylated sugars involves Me₃SiN₃, SnCl₄, and AgClO₄ as reagents.

Cleavage of allyl esters.[4] The Pd(0)-catalyzed ionization generates π-allylpalladium species, which must be processed to return the metal in an active state. The Me_3SiN_3–Bu_4NF combination fulfills this purpose.

β-Azido alcohols.[5] The epoxide opening by Me_3SiN_3 is catalyzed by (t-BuN)$_2$CrCl$_2$.

[1] S. J. Wittenberger and B. G. Donner, *JOC* **58**, 4139 (1993).
[2] B. E. Huff and M. A. Staszak, *TL* **34**, 8011 (1993).
[3] K, Matsubara and T. Mukaiyama, *CL* 247 (1994).
[4] G. Shapiro and D. Buechler, *TL* **35**, 5421 (1994).
[5] W.-H. Leung, E. K. F. Chow, M.-C. Wu, P.W.Y. Kum, and L.-L. Yeung, *TL* **36**, 107 (1995).

Trimethylsilyl azide–iodosylbenzene. 17, 378

Azidonation. The α-carbon atom of *N,N*-dialkylanilines is subject to rapid and efficient introduction of an azide group.[1] Similarly, carbamates also undergo this transformation.[2]

p-Alkyl anisoles form benzylic azides[3] at room temperature in a reaction with a similar reagent pair (PhI[OCOCF$_3$]$_2$ instead of PhIO).

vic-Diazides.[4] The rapid addition of two azido groups to the α- and β-positions of triisopropylsilyl enol ethers proceeds in a *trans* manner. The 1-azido group of these compounds is readily replaced by various carbon nucleophiles.

α-Azidonation of amides.[5] Introduction of an azide group to the carbon atom attached to the nitrogen atom of amides, carbamates, and ureas can use the PhIO–Me_3SiN_3 combination. *N*-Acyliminium ions are the intermediates.

[1] P. Magnus, J. Lacour, and W. Weber, *JACS* **115**, 9347 (1993).
[2] P. Magnus and C. Hulme, *TL* **35**, 8097 (1994).
[3] Y. Kita, H. Tohma, T. Takada, S. Mitoh, S. Fujita, and M. Gyoten, *SL* 427 (1994).
[4] P. Magnus, M. B. Roe, and C. Hulme, *CC* 263 (1995).
[5] P. Magnus, C. Hulme, and W. Weber, *JACS* **116**, 4501 (1994).

N-Trimethylsilyl(bisfluorosulfonyl)imide. 15, 51; 16, 50

Mukaiyama aldolization.[1] This substance catalyzes reactions between silyl enol ethers and acetals.

[1] A. Trehan, A. Vij, M. Walia, G. Kaur, R. D. Verma, and S. Trehan, *TL* **34**, 7335 (1993).

Trimethylsilyl bromide.

vic-Dibromides and bromolactones. The combination of Me_3SiBr with an oxidant represents a bromine source. Thus with a $R_4N^+MnO_4^-$ it converts carvone to the 7,8-dibromide.[1] Bromolactonization of unsaturated carboxylic acids is accomplished in DMSO with Et_3N as base.[2]

[1] B. G. Hazra, M. D. Chordia, B. B. Bahule, V. S. Pore, and S. Basu, *JCS(PI)* 1667 (1994).
[2] K. Miyashita, A. Tanaka, H. Mizuno, M. Tanaka, and C. Iwata, *JCS(PI)* 847 (1994).

Trimethylsilyl chloride. 15, 89; **16**, 85–86

Iminium salts.[1] The preparation from aldehydes and silyldialkylamines is promoted by Me$_3$SiCl.

Cleavage of oximes and semicarbazones.[2] Using Me$_3$SiCl in combination with DMSO in refluxing MeCN, the cleavage method is relatively simple.

Deconjugation of acyclic β-halo enones.[3] Me$_3$SiCl catalyzes the isomerization of β-halo α,β-unsaturated ketones to the corresponding β,γ-unsaturated ketones.

Silyl nitronates.[4] Intramolecular cycloaddition of β-propargyloxy nitroalkanes is induced by Me$_3$SiCl–Et$_3$N. It is a convenient way to dihydrofuraldehydes. Dihydropyranaldehydes are obtainable from the homologous nitro ethers.

Nitration of arenes.[5] The mixture of Me$_3$SiCl, NaNO$_3$, and AlCl$_3$ form an effective nitrating agent in CCl$_4$, which is probably NO$_2{}^+$AlCl$_4{}^-$ (8 examples, 62–97% yield).

[1] W. Jahn and W. Schroth, *TL* **34**, 5863 (1993).
[2] F. Ghelfi, R. Grandi, and U. M. Pagnoni, *SC* **23**, 2279 (1993).
[3] F.-T. Luo and L.-C. Hsieh, *TL* **35**, 9585 (1994).
[4] J. L. Duffy and M. J. Kurth, *JOC* **59**, 3783 (1994).
[5] G. A. Olah, P. Ramaiah, G. Sandford, A. Orlinkov, and G. K. S. Prakash, *S* 468 (1994).

Trimethylsilyl cyanide. 13, 87–88; **14**, 107; **15**, 102–104; **17**, 89

Cyanohydrins and ethers. Tetracyanoethylene has been introduced as a new catalyst for the carbonyl derivatization.[1] KCN–ZnI$_2$ is used to achieve *syn*-selective derivatization of β-hydroxy ketones.[2] Exchange of one alkoxy group of an acetal in a TMSOTf-catalyzed process is subject to remote asymmetric induction.[3] The monoadduct of benzoquinones can be reduced by SmI$_2$ to give *p*-hydroxybenzonitriles.[4]

96%   (5 : 1)

Nitriles. Allylic cyanides are formed from the acetates or carbonates in (Ph$_3$P)$_4$Pd-catalyzed transformations.[5] Intramolecular cyanosilylation of alkynes provides a way of controlled functionalization of such compounds using a tether tactic.[6] α-Cyano selenides are products from SnCl$_4$-catalyzed reaction of diselenoacetals with Me$_3$SiCN. Seleno-ortho esters also exchange one seleno group for the cyanide.[7]

57%

Isonitriles. In anodic reactions of α-heterosubstituted organotin compounds the CN group is delivered to the position vacated by the tin atom in the form of an isonitrile.[8] Interestingly, nitriles are obtained when the supporting electrolyte is changed from Bu_4NBF_4 (in THF) to Bu_4NClO_4 (in CH_2Cl_2).

t-Butylation of amines.[9] The condensation of acetone, amine, and Me_3SiCN affords an α-aminonitrile, which on further reaction with MeLi results in the formation of a *t*-butylamine.

Pauson–Khand-type reaction.[10] In a cyclopentenone synthesis catalyzed by the titanocene $Cp_2Ti(PMe_3)_2$, trimethylsilyl cyanide supplies the carbonyl unit.

[1] T. Miura and Y. Masaki, *JCS(PI)* 1659 (1994).
[2] M. S. Batra, F. J. Aguilar, and E. Bruner, *T* **50**, 8169 (1994).
[3] G. A. Molander and J. P. Haar, *JACS* **115**, 40 (1993).
[4] S. H. Olsen and S. J. Danishefsky, *TL* **35**, 7901 (1994).
[5] J. Tsuji, N. Yamada, and S. Tanaka, *JOC* **58**, 16 (1993).
[6] M. Suginome, H. Kinugusa, and Y. Ito, *TL* **35**, 8635 (1994).
[7] M. Yoshimatsu, T. Yoshiuchi, H. Shimizu, M. Hori, and T. Kataoka, *SL* 121 (1993).
[8] J. Yoshida, M. Itoh, Y. Morita, and S. Isoe, *CC* 549 (1994).
[9] M. J. Genin, C. Biles, and D. L. Romero, *TL* **34**, 4301 (1993).
[10] S. C. Berk, R. B. Grossman, and S. L. Buchwald, *JACS* **116**, 8593 (1994).

Trimethylsilyl(diethyl)amine.

α-Alkylation of conjugated carbonyl compounds.[1] Upon Mukaiyama aldolization and silica gel chromatography. The 1,4-adducts from reaction of enones with Me_3SiNEt_2 regenerate the original chromophore. A net α-alkylation on nonenolizable enals and enones is realized.

[1] M. Hojo, M. Nagayoshi, A. Fujii, T. Yanagi, N. Ishibashi, K. Miura, and A. Hosomi, *CL* 719 (1994).

Trimethylsilyl fluorosulfonate.

Silyl donor and catalyst.[1] This compound FSO_3SiMe_3, readily obtained by reaction of allyltrimethylsilane with fluorosulfonic acid at $-78°$, is comparable and sometimes superior to trimethylsilyl triflate in catalytic activity.

[1] B. H. Lipshutz, J. Burgess-Henry, and G. P. Roth, *TL* **34**, 995 (1993).

Trimethylsilyl iodide. 16, 188–189

Vinyl iodides.[1]   A convenient synthesis from ketones features enolphosphorylation and treatment of the products with Me_3SiI (or Me_3SiCl–NaI in MeCN) for a few minutes at room temperature.

Conjugate addition.[2] π-Allyliron complexes donate the allyl group to conjugated carbonyl compounds in the presence of Me_3SiI.

Iodination of β-lactams.   The *N*-tosylates give 3-iodo-β-lactams[3] as a result of S_N2' displacement on the enol form.

[1] K. Lee and D. F. Wiemer, *TL* **34**, 2433 (1993).
[2] K. Itoh, S. Nakanishi, and Y. Otsuji, *JOMC* **473**, 215 (1994).
[3] M. Teng and M. J. Miller, *JACS* **115**, 548 (1993).

Trimethylsilyl trifluoromethanesulfonate. 13, 329–331; 14, 333–335; 15, 346–350; 16, 363–364; 17, 379–386

Dealkoxylative condensations.   Numerous glycosylation reactions are catalyzed with TMSOTf. Glycosyl donors include isopropenyl,[1] trichloroacetimidinyl,[2] and phosphinoxy glycosides.[3]

Under the influence of TMSOTf acyclic acetals react with allylsilanes,[4] while cyclic counterparts remain unaffected. Formation of 5,6-dihydropyrans[5] is interesting because new O–C and C–C bonds are created. The transformation of allylic acetals to mixed acetals and oxathioacetals is mediated by Me_2S via 1,3-transposed sulfonium ions, which then undergo S_N2' reactions with alkoxides or thiolates.[6] On the other hand, oxathioacetals are cleaved when *p*-nitrobenzaldehyde is present.[7]

The Mukaiyama aldolization of α-sulfenyl acetals is *anti*-selective.[8]

A related reaction is the opening of β-methoxy-β-lactams[9] with TMSOTf–MeCN involving C–N bond cleavage, trapping with MeCN and transfer of a water molecule.

Carbonyl condensations.   Two-carbon homologation of carbonyl compounds by condensation with β-silyl enol ethers gives α,β-unsaturated aldehydes.[10] Dual-site condensation of 1,4-ketoaldehydes with 1,1,3-tris(trimethylsiloxy)-1,3-butadiene results in the oxabridged system.[11]

Conjugated thionium species are generated from allyl sulfoxides. Trapping of these species with silyl enol ethers leads to vinyl sulfides.[12]

Rearrangements. Ireland–Claisen rearrangement of allyl fluoroacetates,[13] formation of β-(N-acylamino)aldehydes from O-vinyl-N,O-acetals,[14] 1,2-group migration/trapping are some of the useful transformations initiated by TMSOTf.[15,16]

[1] H. K. Chenault and A. Castro, *TL* **35**, 9145 (1994).
[2] J.-A. Mahling and R. R. Schmidt, *S* 325 (1993).
[3] H. Kondo, S. Aoki, Y. Ichikawa, R. L. Halcomb, H. Ritzen, and C.-H. Wong, *JOC* **59**, 864 (1994).
[4] S. Kim, J. Y. Do, S. H. Kim, and D. Kim, *JCS(P1)* 2357 (1994).
[5] I. E. Marko and D. J. Bayston, *T* **50**, 7141 (1994).
[6] S. Kim, J. H. Park, and J. M. Lee, *TL* **34**, 5769 (1993).
[7] T. Ravindranathan, S. P. Chavan, J. P. Varghese, S. W. Dantale, and R. B. Tejwani, *CC* 1937 (1994).
[8] K. Kudo, Y. Hashimoto, M. Sukegawa, M. Hasegawa, and K. Saigo, *JOC* **58**, 579 (1993).
[9] Y. Kita, N. Shibata, N. Yoshida, N. Kawano, and K. Matsumoto, *JOC* **59**, 938 (1994).
[10] L. Duhamel, J. Gralak, and A. Bouyanzer, *TL* **34**, 7745 (1993).
[11] G. A. Molander and K. O. Cameron, *JACS* **115**, 830 (1993).
[12] R. Hunter, J. P. Michael, C. D. Simon, and D. S. Walter, *T* **50**, 9365 (1994).
[13] K. Arak and J. T. Welch, *TL* **34**, 2251 (1993).
[14] H. Frauenrath, T. Arenz, G. Raabe, and M. Zorn, *ACIEE* **32**, 83 (1993).
[15] Q. Liu, M. J. Simms, N. Boden, and C. M. Rayner, *JCS(P1)* 1363 (1994).
[16] I. Coldham and S. Warren, *JCS(P1)* 1637 (1993).

Trimethylsilylmethylmagnesium chloride. 15, 343

Unsaturated nitriles from 1-nitrocycloalkenes.[1] Treatment of the conjugate adducts from the Grignard reactions with PCl_3 triggers desilylative fragmentation.

[1] H.-H. Tso, B. A. Gilbert, and J. R. Hwu, *CC* 669 (1993).

Triorganoboranes. 15, 337

N-Tosylamines.[1] Reaction of R_3B with TsN=IPh in THF gives RNHTs in 60–99% yield. It is not known whether Ar_3B would undergo the same transformation.

Homolytic heteroaromatic substitutions.[2] Et_3B is often used in promoting free radical reactions, including alkylation of heteroaromatics such as pyrrole derivatives with alkyl halides.

Stereoselective radical addition to alkynes and reduction of iodoalkynes.[3]
Et_3B has been used as a catalyst for the reactions with $(Me_3Si)_3SiH$ to yield (Z)-alkenes.

Benzyl alcohols.[4] The preparation involves reaction of araldehyde tosylhydrazones with R_3B in the presence of an amine base (e.g., DBU) followed by oxidation [yielding ArCH(OH)R]. As a variation of this method, the use of Bu_4NOH as base and addition of H_2O prior to workup gives $ArCH_2R$.

Electrophilic activation of allylic alcohols.[5] Formation of lithium allyloxy-triphenylborates enables the Pd(0)-catalyzed substitution of allyl alcohols with soft nucleophiles such as malonate ester enolates.

[1] R.-Y. Yang and L.-X Dai, *S* 481 (1993).
[2] E. Baciocchi and E. Muraglia, *TL* **34**, 5015 (1993).
[3] K. Miura, K. Oshima, and K. Utimoto, *BCSJ* **66**, 2356 (1993).
[4] G.W. Kabalka, J.T. Maddox, and E. Bogas, *JOC* **59**, 5530 (1994).
[5] I. Stary, I.G. Stara, and P. Kocovsky, *TL* **34**, 179 (1993).

Triphenylphosphine.

Deoxygenation of alcohols.[1] Aldohols undergo double electrolysis in the presence of Ph_3P to give alkanes.

2,4-Alkadienones and 2,4,6-alkatrienoic esters. Both 2-alkynones and their 4-hydroxyl derivatives are converted to the dienones[2] by treatment with Ph_3P at room temperature in benzene. 4-Hydroxy-2-alkynoic esters behave analogously.

A cocatalyst for the isomerization of alkenynoic esters to the trienoic esters is phenol.[3]

Reduction–substitution of 2-alkynoic acid derivatives.[4] Reduction of the triple bond of 2-alkynoic acid derivatives with functionalization at C-4 in the umpolung sense is effected by catalytic amounts of Ph_3P, HOAc, and a nucleophile. These processes involve 2,4-dienoic intermediates.

X = OMe, NMe₂

| | Ph₃P-HOAc / DMSO | 91 | : | 9 |
| | dppp-HOAc / PhMe | 3 | : | 97 |

4-Pentenenitriles and benzo[f]indoles. Allyl iminophosphoranes are formed from the corresponding azides upon treatment with PPh_3 (Staudinger reaction). Condensation of these products with diaryl ketenes leads to 4-pentenenitriles[5] by a sigmatropic rearrangement. In the homoallyl series the Wittig reaction is followed by an intramolecular Diels–Alder reaction. Mild dehydrogenation of the products gives benzo[f]indoles.[6]

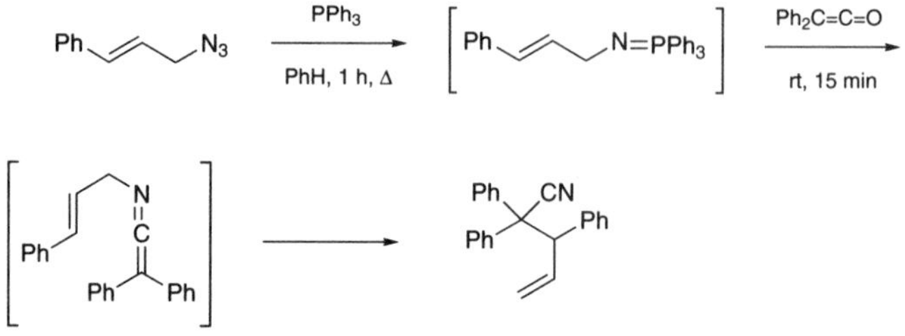

51% (one-pot reaction)

2,4-Cycloheptadienones.[7] Cyclopropylcarbene–Mo complexes react with alkynes in the presence of Ph_3P under relatively mild conditions. The products are 2,3-disubstituted 4-alkoxy-2,4-cycloheptadienones.

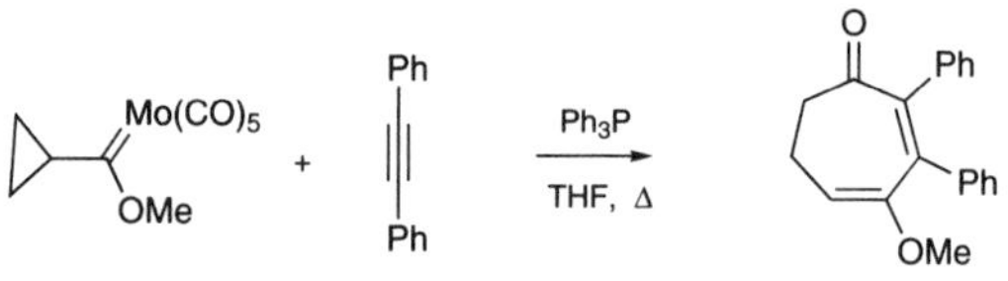

52%

[1] H. Maeda, T. Maki, K. Eguchi, T. Koide, and H. Ohmori, *TL* **35**, 4129 (1994).
[2] C. Guo and X. Lu, *JCS(PI)* 1921 (1993); *CC* 394 (1993).
[3] S. D. Rychnovsky and J. Kim, *JOC* **59**, 2659 (1994).
[4] B. M. Trost and C.-J. Li, *JACS* **35**, 3167, 10819 (1994).
[5] P. Molina, M. Alajarin, C. Lopez-Leonardo, and J. Alcantara, *T* **49**, 5153 (1993).
[6] P. Molina and C. Lopez-Leonardo, *TL* **34**, 2809 (1993).
[7] J. W. Herndon and M. Zora, *SL* 363 (1993).

Triphenylphosphine–carbon tetrahalides. 13, 331–332; 15, 352; 16, 366–368

Acid halides. The relatively mild conditions of converting acids to halides by Ph_3P–CX_4 can be exploited for a one-flask synthesis of *N*-methoxy imidoyl bromides,[1] which give rise to nitriles on photolysis. In the presence of Et_3N the Ph_3P–CX_4 combination converts amines and CF_3COOH to trifluoroacetimidoyl halides.[2]

Conjugated nitroalkenes.[3]   Dehydration of β-nitro alcohols by this reagent combination in the presence of R_3N is an improved method.

Cyclic urethanes.[4]   Carbon dioxide is activated for the heterocycle formation from amino alcohols in the presence of Ph_3P and CCl_4 or C_2Cl_6. Combinations of the halides with other P(III) reagents such as Bu_3P, $(MeO)_3P$, and $(PhO)_3P$ are also effective.

[1] Y. Kikugawa, L. H. Fu, and T. Sakamoto, *SC* **23**, 1061 (1993).
[2] K. Tamura, H. Mizukami, K. Maeda, H. Watanabe, and K. Uneyama, *JOC* **58**, 32 (1993).
[3] A. K. Saikia, N. C. Barua, R. P. Sharma, and A. C. Ghosh, *S* 685 (1994).
[4] Y. Kubota, M. Kodaka, T. Tomihiro, and H. Okuno, *JCS(PI)* 5 (1993).

Triphenylphosphine–diethyl azodicarboxylate. 13, 332; 14, 336–337; 17, 389–390

Monoesterification of saccharides.[1]   The primary hydroxyl groups of these sugars are selectively acylated under Mitsunobu reaction conditions.

Alkylations.   The reagent effects reductive *N*-alkylation of *N*-tosylamines[2] and indole derivatives[3] with alcohols. The Mitsunobu reaction of 2-(1-hydroxy-alkyl)-acrylic esters[4] and that of glycals with phenols[5] follow an S_N2' course. The reaction has been applied to the inversion of configuration at an α-cyanohydrin center,[6] whereas alkyl nitriles are prepared by this method using acetonitrile cyanohydrin as the source of nucleophile. The *C*-alkylation of *o*-nitroarylacetonitriles[7] at the benzylic position is easily controlled.

Inversion of hindered alcohols.[8]   In the Mitsunobu reaction a positive relationship between dissociation constant of the acid component (nucleophile) with reaction efficiency is indicated. Product yields are higher when using acids of lower pK_a. Among substituted benzoic acids, 4-nitro, 4-methanesulfonyl, and 4-cyano derivatives give better results.

[1] A. Bourhim, S. Czernecki, and P. Krausz, *JCC* **12**, 853 (1993).
[2] A. Garcia, L. Castedo, and D. Domingues, *SL* 271 (1993).
[3] S. S. Bhagwat and C. Gude, *TL* **35**, 1847 (1994).
[4] A. B. Charette and B. Cote, *TL* **36**, 6833 (1993).
[5] A. Sobti and G. A. Sulikowski, *TL* **35**, 3661 (1994).
[6] E. G. J. C. Warmerdam, J. Brussee, C. G. Kruse, and A. van der Gen, *T* **49**, 1063 (1993).
[7] J. E. Macor and J. M. Wehner, *H* **35**, 349 (1993).
[8] J. A. Dodge, J. I. Trujillo, and M. Presnell, *JOC* **59**, 234 (1994).

Triphenylphosphine–diisopropyl azodicarboxylate. 15, 352–353; 17, 390

Guanidines.[1]   An exceptionally mild alkylation method for guanidine is the use of a modified Mitsunobu reagent combination and an alcohol on its *N,N'*-diBoc derivative.

Closure of oxacycles.   Terminal epoxides are formed[2] selectively from acyclic 1,2,3-triols. β-Lactonization[3] proceeds with total inversion of configuration, although the yields are poor.

[1] D. S. Dodd and A. P. Kozikowski, *TL* **35**, 977 (1994).
[2] C. Gravier-Pelletier, Y. Le Merrer, and J.-C. Depezay, *SC* **24**, 2843 (1994).
[3] S. Cammas, I. Renard, K. Bautault, and P. Guerin, *TA* **4**, 1925 (1993).

Triphenylphosphine–dipyridyl disulfide. 13, 332–333

Amides.[1]   Activation of carboxylic acids by this reagent combination followed by reaction with Me_3SiNR_2 constitutes an efficient method of amide synthesis.

[1] R. Di Fabio, V. Summa, and T. Rossi, *TL* **49**, 2299 (1993).

Triphenyltin hydride–trialkylborane. 16, 369

Radical cyclization.[1]   The intramolecular free radical addition to an aldehyde in order to form a cyclohexanol is induced by $Ph_3SnH-Et_3B$ in the presence of air.

Hydrostannylation.[2]   Alkenes do not generally undergo this reaction. However, cyclopropenes are hydrostannylated with $Ph_3SnH-Bu_3B$ because of their ring strain.

[1] D. L. J. Clive and M. H. D. Postema, *CC* 429 (1993).
[2] S. Yamago, S. Ejiri, and E. Nakamura, *CL* 1889 (1994).

Triphosgene.

Anhydrides and cyclic carbonates.   Symmetrical anhydrides are formed when acids are treated with $(Cl_3CO)_2CO$ and Et_3N.[1] The transformation of the terminal glycol unit of a 1,2,3-triol to the cyclic carbonate[2] can be accomplished with triphosgene and pyridine; on the other hand, the internal diol system is protected using NaH and $(MeO)_2CO$.

Chlorides.[3]   A combination of triphosgene and Ph_3P is effective in the conversion of alcohols to chlorides. Conditions are mild (typically 0°C, 5 min., in CH_2Cl_2).

[1] R. Kocz, J. Roestamadji, and S. Mobashery, *JOC* **59**, 2913 (1994).
[2] S.-K. Kang, J.-H. Jeon, K.-S. Nam, C.-H. Park, and H.-W. Lee, *SC* **24**, 305 (1994).
[3] I. A. Rivero, R. Somanathan, and L. H. Hellberg, *SC* **23**, 711 (1993).

Tris(4-bromophenyl)aminium salts. **14**, 338; **16**, 369–370; **17**, 391

Glycosylation.[1]   The one-electron oxidant promotes activation of glycosyl sulfides toward coupling with alcohols (including sugars).

1,2-Rearrangements.   Efficient pinacol rearrangement[2] is completed at room temperature. α-Substituted arylacetaldehydes undergo rapid oxidation to give α-dicarbonyl compounds with an analogous tris(2,4-dibromophenyl)aminium salt.[3]

Desulfurization.[4]   Allyl and diallyl thiiranes undergo desulfurization at room temperature to give dienes and trienes, respectively, on exposure to the aminium salt.

[1] Y.-M. Zhang, J.-M. Mallet, and P. Sinay, *CR* **236**, 73 (1992).
[2] L. Lopez, G. Mele, and C. Mazzeo, *JCS(PI)* 779 (1994).
[3] L. Lopez, G. Mele, A. Nacci, and L. Troisi, *TL* **34**, 3897 (1993).
[4] V. Calo, L. Lopez, A. Nacci, and G. Mele, *T* **51**, 8935 (1995).

Tris(dibenzylideneacetone)dipalladium. **14**, 339; **15**, 353–355; **16**, 372; **17**, 394

Addition to unsaturated systems.   Allylsilanes are produced[1] in a Pd(0)-catalyzed reaction of conjugated dienes, organodisilanes, and acid chlorides. The acid chloride component is incorporated into the product after decarbonylation.

Addition of Bu$_3$SnH to enynes is regio- and stereoselective, thus making 2-stannyl-1,3-dienes[2] readily available. Cyclization of allyl 2-alkynoates with simultaneous carbonylation is a useful way to access functionalized γ-lactones.[3] Nonconjugated dienes undergo reactions at both ends; with aryl iodides and amines the products are allylamines bearing an ω-aryl group.[4]

Allylic substitutions.   Allylic carbonates are converted to the corresponding ethers,[5] amines,[6] sulfides[7] on exposure to various nucleophiles in the presence of Pd$_2$(dba)$_3$, and a tertiary phosphine.

Allylic compounds containing other leaving groups are also reactive toward substitution. Thus alkoxycarbonylation of allylic phosphates[8] is a method of homologation, on which an interesting β-lactam synthesis[9] is based. Using Ph$_3$SiOH as nucleophile in the displacement, 2,4-hexadiene-1,6-diols are synthesized[10] from butadienyloxiranes. Desilylation of the products is achieved with KF.

Either the S_N2 or the S_N2' pathway for the substitution of 1-tosyloxy-1-vinylcyclopropane[11] can be selected by using *N*-nucleophiles of different sizes.

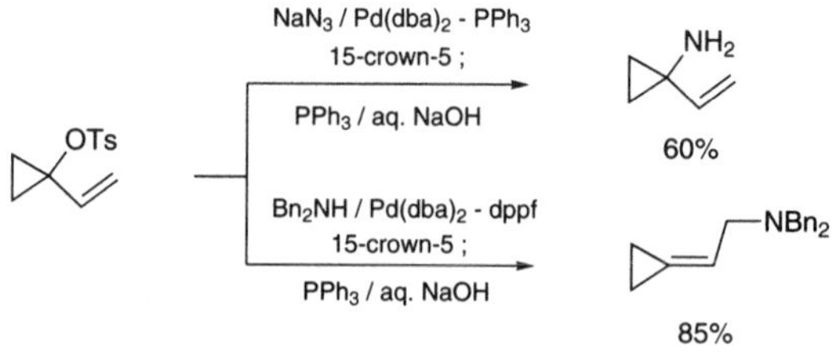

Intramolecular substitutions involving an allylsilane moiety[12] as well as following a metallo-ene reaction pattern[13] have been reported. Tandem metallo-ene cyclization and vinylstannane coupling serve to construct cycloalkanes with vicinal alkenyl chains.[14]

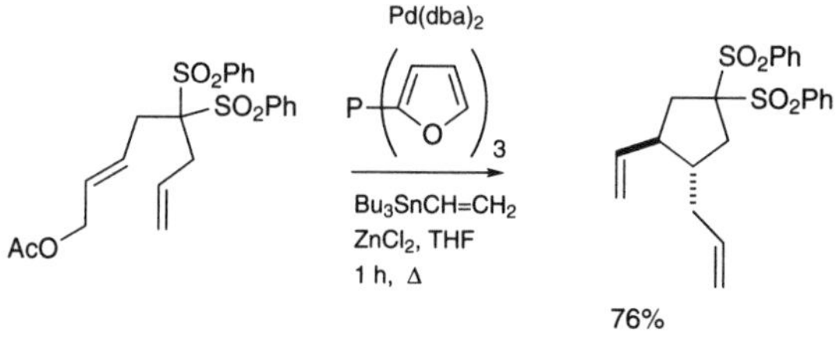

Nucleophiles for the substitution can be generated in situ. For example, a synthesis of allylic carbamates and carbonates[15] is conducted using amines and alcohols under CO_2. Since disilanes are cleaved in the presence of Pd(0) complexes, their use in the synthesis of allylsilanes[16] is feasible.

The formate anion reduces allylic acetates in the S_N2' mode. Acquisition of allylic *gem*-bimetallic compounds[17] is facilitated by this process.

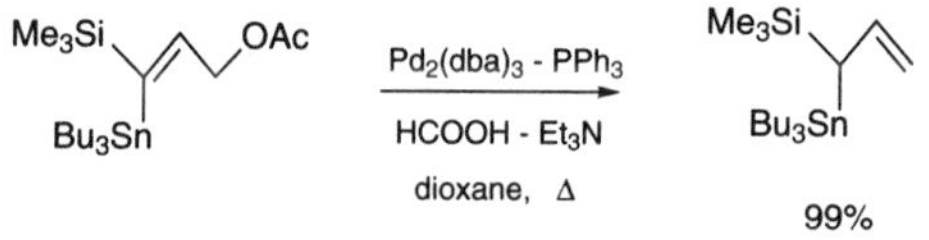

Deallylations. The facile cleavage of the allyl group from esters,[18] carbonates,[19] and carbamates by nucleophiles (amines and thiols) in the presence of a Pd(0) complex makes them useful derivatives for protection purposes.

Stille coupling. The many examples describing applications of this C–C bond-forming method attest to its versatility. A comprehensive study of the coupling of arylstannanes with sulfonates has appeared.[20] Of course, various combinations of aryl/aryl, aryl/vinyl, and vinyl/vinyl couplings are equally possible. Ph_3As is often added to the reaction media, and in other cases CuI acts as a cocatalyst.

Arylquinones,[21] arylallenes,[22] dienyl sulfoxides,[23] and dienynes[24] are some of the more unusual molecules whose assembly is greatly simplified by this method. The cross coupling with α-alkoxyalkenylstannanes[25] (and zincs) provides another approach to enol ethers of defined configuration.

A technical modification of the Stille coupling is to anchor the aryl iodide on a polymer support.[26]

Suzuki coupling. This coupling uses nontoxic boronic acids; therefore, it is preferred in the preparation of drug candidates such as 2-aryl- and 2-alkenyl-carbapenems.[27] In a synthesis of methyl 2-aryl-2,3-butadienoates the Suzuki coupling is aided by Ag_2O.[28]

Heck coupling. An intramolecular Heck coupling is a convenient way to prepare 4-methylcoumarin[29] and related compounds. Through ligand exchange with a chiral BINAP to mediate cyclization, an intermediate for vernolepin is accessible in 86% ee.[30]

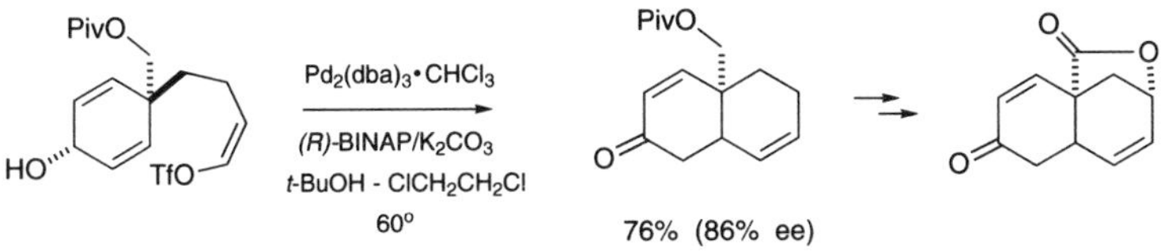

6-*endo*-*trig* Cyclization has been realized.[31]

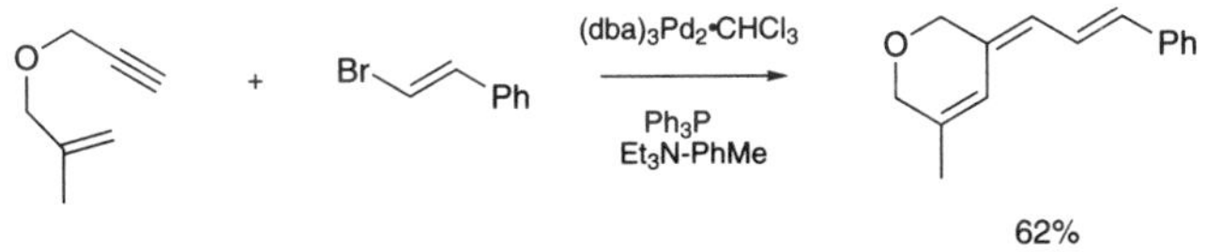

Alkylations. Molecules with active C–H and Si–H bonds add to unsaturated compounds that are activated by Pd catalysts. Substituted malononitriles are alkylated by aldehydes[32] and allenes.[33] By a three-component coupling of an acetoacetic ester, an allene, and a vinylic bromide, the construction of a diene keto ester constitutes a convenient preparation of a 2,4,5-trisubstituted benzoic ester precursor.[34] Some aryl(hydro)silanes also undergo arylation.[35]

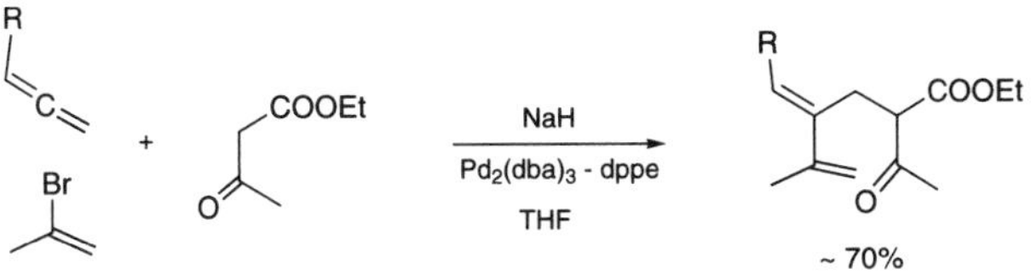

Cyclization of polyenynes.[36] Very impressive reaction sequences of ring formation from properly distanced multiple bonds are revealed.

$n = 2,3,4$

$n = 4$ 77%

Other synthetic reactions. Catalyzed elimination of allylic carbonates[37] occurs
in the absence of nucleophiles. Alkynediols undergo isomerization and dehydration,
furnishing 2,5-disubstituted furans[38] as a result. 1-Carboranyltributyltin adds to
aldehydes under the influence of the Pd catalyst to form carbinols.[39] The Pd version of
a Pauson–Khand cyclopentenone synthesis is accomplishable in the presence of CO,
and actually this version is specially suited for a one-step construction of α'-
methylenecyclopentenones.[40]

$E = CO_2Me$

43%

[1] Y. Obora, Y. Tsuji, and T. Kawamura, *JACS* **115**, 10414 (1993).

[2] B. M. Trost and C.-J. Li, *S(Spec. Issue)* 1267 (1993).

[3] J. Ji and X. Lu, *T* **50**, 9067 (1994).

[4] R. C. Larock, Y. Wang, Y.-D. Lu, and C. E. Russell, *JOC* **59**, 8107 (1994).

[5] R. Lakhmiri, P. Lhoste, B. Kryczka, and D. Sinou, *JCC* **12**, 223 (1993).

[6] H. Inami, T. Ito, H. Urabe, and F. Sato, *TL* **34**, 5919 (1993).

[7] C. Goux, P. Lhoste, and D. Sinou, *T* **50**, 10321 (1994).

[8] S.-i. Murahashi, Y. Imada, Y. Taniguchi, and S. Higashiura, *JOC* **58**, 1538 (1993).

[9] H. Tanaka, A. K. M. Abdul Hai, M. Sadakane, H. Okumoto, and S. Torii, *JOC* **59**, 3040 (1994).

[10] B. M. Trost, N. Ito, and P. D. Greenspan, *TL* **34**, 1421 (1993).

[11] P. Aufranc, J. Ollivier, A. Stolle, C. Bremer, M. Es-Sayed, A. de Meijere, and J. Salaun, *TL* **34**, 4193 (1993).

[12] M. Terakado, M. Miyazawa, and K. Yamamoto, *SL* 134 (1994).

[13] K. Hiroi and K. Hirasawa, *CPB* **42**, 786 (1994).

[14] W. Oppolzer and J. Ruiz-Montes, *HCA* **76**, 1266 (1993).

[15] W. D. McGhee, D. P. Riley, M. E. Christ, and K. M. Christ, *OM* **12**, 1429 (1993).

[16] Y. Tsuji, S. Kajita, S. Isobe, and M. Funato, *JOC* **58**, 3607 (1993).

[17] M. Lautens and P. H. M. Delanghe, *ACIEE* **33**, 2448 (1994).

[18] S. Okamoto, N. Ono, K. Tani, Y. Yoshida, and F. Sato, *CC* 279 (1994).

[19] J. P. Genet, E. Blart, M. Savignac, S. Lemeune, S. Lemeune-Audoire, and J. M. Barnard, *SL* 680 (1993).

[20] V. Farina, B. Krishnan, D. R. Marshall, and G. P. Roth, *JOC* **58**, 5434 (1993).

[21] L. S. Liebeskind and S. W. Riesinger, *JOC* **58**, 408 (1993).

[22] D. Badone, R. Cardamone, and U. Guzzi, *TL* **35**, 5477 (1994).

[23] R. S. Paley and A. de Dios, *TL* **34**, 2429 (1993).

[24] B. H. Lipshutz and A. Alami, *TL* **34**, 1433 (1993).

[25] S. Casson and P. Kocienski, *JCS(P1)* 1187 (1993).

[26] M. S. Deshpande, *TL* **35**, 5613 (1994).

[27] N. Yasuda, L. Xavier, D. L. Rieger, Y. Li, A. E. DeCamp, and U.-H. Dolling, *TL* **34**, 3211 (1993).

[28] T. Gillmann and T. Weeber, *SL* 649 (1994).

[29] M. Catellani, G. P. Chiusoli, M. C. Fagnola, and G. Solari, *TL* **35**, 5919 (1994).

[30] K. Kondo, M. Sodeoka, M. Mori, and M. Shibasaki, *S* 920 (1993).

[31] B. M. Trost and J. Dumas, *TL* **34**, 19 (1993).

[32] H. Nemoto, Y. Kubota, and Y. Yamamoto, *CC* 1665 (1994).

[33] Y. Yamamoto, M. Al-Masum, and N. Asao, *JACS* **116**, 6019 (1994).

[34] V. Gauthier, C. Grandjean, B. Cazes, and J. Gore, *BSCF* **131**, 381 (1994).

[35] Y. Uchimaru, A. M. M. El Sayed, and M. Tanaka, *OM* **12**, 2065 (1993).

[36] B. M. Trost and Y. Shi, *JACS* **115**, 9421 (1993).

[37] S.-K. Kang, D.-C. Park, C.-H. Park, and R.-K. Hong, *TL* **36**, 405 (1995).

[38] J. Ji and X. Lu, *CC* 764 (1993).

[39] H. Nakamura, N. Sadayori, M. Sekido, and Y. Yamamoto, *CC* 2581 (1994).

[40] N. C. Ihle and C. H. Heathcock, *JOC* **58**, 560 (1993).

Tris(dimethylamino)sulfur trimethylsilyldifluoride (TAS-F). 13, 336; 15, 355

β-Keto sulfoxides.[1]   TAS-F promotes the sulfinylation of silyl enol ethers with *S*-aryl arenethiosulfonate *S*-oxides.

[1] R. Caputo, C. Ferreri, L. Longobardo, G. Palumbo, and S. Pedatella, *SC* **23**, 1515 (1993).

Tris(pentafluorophenyl)borane.

Aldol and Michael reactions. $(C_6F_5)_3B$ is an air-stable and water-tolerant catalyst for these reactions. Thus the aldol condensation between a silyl enol ether and an aldehyde can be conducted at $-78°C$.[1] Aqueous HCHO can be used as electrophile.[2] The analogous condensation with imines provides a route to *β*-amino esters.[3]

Epoxide rearrangement.[4]   With the borane in nonpolar solvents, alkyl shift is favored during the rearrangement.

$Ar_3B/PhMe$, 60°	98	:	2
$SbF_5/PhMe$, -78°	15	:	85

[1] K. Ishihara, N. Hanaki, and H. Yamamoto, *SL* 577 (1993).

[2] K. Ishihara, N. Hanaki, M. Funahashi, M. Miyata, and H. Yamamoto, *BCSJ* **68**, 1721 (1995).

[3] K. Ishihara, M. Funahashi, N. Hanaki, M. Miyata, and H. Yamamoto, *SL* 963 (1994).

[4] K. Ishihara, N. Hanaki, and H. Yamamoto, *SL* 721 (1995).

Tris(trimethylsilyl)silane. 15, 358–359; 16, 374–375

Alkanes from alcohols.[1] ROH is deoxygenated via the thioxocarbamates by a free radical reaction with $(Me_3Si)_3SiH$–AIBN in refluxing benzene.

Ketone synthesis.[2] The assembly of $RCOCH_2CH_2R'$ from RI, CO, and electron-deficient alkenes $CH_2=CHR'$ is mediated by $(Me_3Si)_3SiH$–AIBN.

[1] M. Oba and K. Nishiyama, *T* **50**, 10193 (1994).

[2] I. Ryu, M. Hasegawa, A. Kurihara, A. Ogawa, S. Tsunoi, and N. Sonoda, *SL* 143 (1993).

Trityl chloride. 15, 359–360

Selective O-alkylation.[1] Polyols form monotrityl ethers with limited amount (10 mol %) of TrCl and pyridine. A primary hydroxyl is favored over a secondary alcohol, and only one of the two primary hydroxyl groups of a diol is derivatized.

[1] V. E. M. Kaats-Richters, J. W. Zwikker, E. M. D. Keegstra, and L. W. Jenneskens, *SC* **24**, 2399 (1994).

Trityl perchlorate. 13, 339–340; 14, 344–345; 15, 361–362; 16, 375–376

Glycosylation.[1] Unusual chemoselectivity is revealed using trityl ethers as glycosyl acceptors. Secondary ROTr have higher reactivity than primary ethers. In this reaction, $TrClO_4$ acts as a catalyst.

54%

[1] Y. E. Tsvetkov, P. I. Kitov, L. V. Backinowsky, and N. K. Kochetkov, *TL* **34**, 7977 (1993).

Tungsten pentacarbonyl. 14, 345–346

Pauson–Khand reaction.[1] The $W(CO)_5 \cdot THF$ complex promotes formation of bicyclic dienones from 1,6-enynes.

[1] T. R. Hoye and J. A. Suriano, *JACS* **115**, 1154 (1993).

U

Ultrasound. **15**, 363; **16**, 377–379

Reduction of esters.[1] Primary alcohols are obtained in good yields by reduction with the NaBH$_4$-polyethylene glycol-400 system under sonochemical conditions.

Reductive coupling of trimethylsilyl chloride and 1,4-disilylation of benzene.[2] (Me$_3$Si)$_2$ is formed in 89.9% yield by the Wurtz coupling. Using 4,4'-di-*t*-butylbiphenyl as an electron-transfer agent, the sonochemical reaction with benzene gives 3,6-bistrimethylsilyl-1,4-cyclohexadiene almost quantitatively.

Alkylation of alkyl cyanides.[3] Ultrasonic irradiation facilitates deprotonation of nitriles with Na in toluene and allows the alkylation with halogen compounds.

Hydrostannylation and hydroxystannylation.[4] The radical addition of an organotin hydride to a multiple bond is promoted by ultrasonic irradiation. In the presence of air, the product is a β-hydroxy stannane.

Wittig reactions.[5] Vinylic chalcogenides are formed under sonochemical conditions using K$_2$CO$_3$ as base.

Carbene formation.[6] Decomposition of diazirine gives arylchlorocarbenes, which are intercepted by alkenes.

Diels–Alder reaction of 1-azadienes.[7] Tetrahydropyridine derivatives are readily synthesized in high yields.

[1] H. Liu, X.-L. Ji, and K. Huang, *YH* **13**, 421 (1993).
[2] E. A. Mistryukov, *MC* 251 (1993).
[3] J. Berlan, H. Delmas, I. Duee, J. L. Luche, and L. Vuiglio, *SC* **24**, 1253 (1994).
[4] E. Nakamura, Y. Imanishi, and D. Machii, *JOC* **59**, 8178 (1994).
[5] C. C. Silveira, G. Perin, and A. L. Braga, *JCR(S)* 492 (1994).
[6] A. K. Bertram and M. T. H. Liu, *CC* 467 (1993).
[7] M. Villacampa, J. M. Perez, C. Avendano, and J. C. Menendez, *T* **50**, 10047 (1994).

V

Valine *t*-butyl ester. 14, 347

As chiral auxiliary.[1] Asymmetric Michael addition of β-keto esters is achievable via the chiral enamines. The stereoselectivity is dependent on solvent and additives.

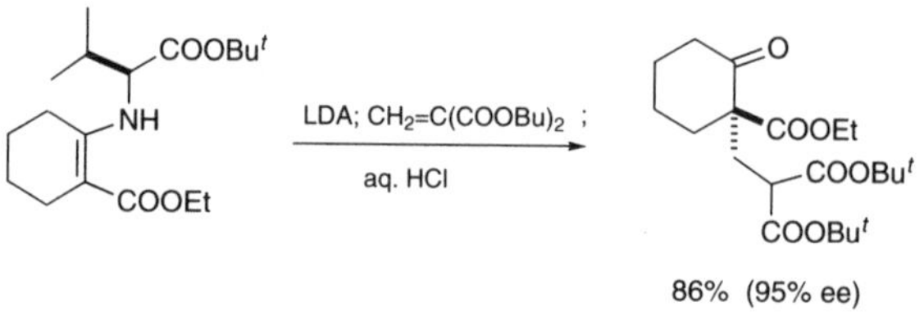

86% (95% ee)

[1] K. Ando, K. Yasuda, K. Tomioka, and K. Koga, *JCS(P1)* 277 (1994).

Vanadium(III) chloride–zinc dust.

Monodebromination of gem-dibromocyclopropanes.[1] In conjunction with diethyl phosphonate or triethyl phosphite, the reagent removes one bromine atom in a highly stereoselective manner.

Cross pinacol coupling.[2] The VCl$_3$–THF complex gives [V$_2$Cl$_3$(thf)$_6$]$_2$(Zn$_2$Cl$_6$) on treatment with zinc dust. Coupling of aldehydes furnishes *syn*-1,2-diols.

[1] T. Hirao, K. Hirano, T. Hasegawa, Y. Ohshiro, and I. Ikeda, *JOC* **58,** 6529 (1993).
[2] E. A. Kraynack and S. F. Pedersen, *JOC* **58,** 6114 (1993); A.W. Konradi, S. J. Kemp, and S. F. Pedersen, *JACS* **116,** 1316 (1994).

Vanadium(V) oxide.

Oxidation of alcohols.[1] Supported on zirconia, vanadium(V) oxide is used in the heterogeneous oxidation in an organic solvent (e.g., CH$_2$Cl$_2$) to give aldehydes and ketones.

[1] H. Nakamura, H. Matsuhashi, and K. Arata, *CL* 749 (1994).

Vanadyl 3-butylacetylacetonate.

Quinone synthesis.[1] The oxovandium complex is a catalyst for aerial oxidation of polynuclear arenes and aryl ethers.

[1] T. Takai, E. Hata, and T. Mukaiyama, *CL* 885 (1994).

Vanadyl fluoride. 16, 381–382

9,10-Phenanthrenequinones.[1]   Substituted benzils undergo oxidative cyclization on treatment with vanadyl fluoride and boron trifluoride etherate. The yields of the phenanthrenequinones range from 82% to 91% in 4 examples.

[1] B. Mohr, V. Enkelmann, and G. Wegner, *JOC* **59**, 635 (1994).

Vinylselenonium salts.

Cyclopropane synthesis.[1]   The reaction with active methylene compounds proceeds by addition, proton exchange, and intramolecular displacement to form cyclopropane derivatives.

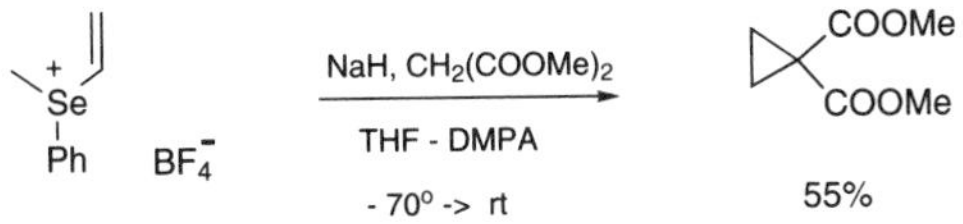

Epoxide formation.[1]   Aldehydes are attacked by the selenium ylides derived from the cation. The β-selenonium oxide intermediates decompose by selenide elimination, resulting in epoxide products.

[1] Y. Watanabe, Y. Ueno, and T. Toru, *BCSJ* **66**, 2042 (1993).

Vinylzinc reagents.

Preparation.   These reagents are accessible from vinyl halides by reaction with the metal,[1] or by transmetallation of vinyl derivatives of tin (via lithium),[2] zirconium,[3] and boron.[4]

Coupling with vinylic iodides.[1,2]   Formation of conjugated dienes is catalyzed by Pd(0) species at room temperature.

Allylic alcohols.[3]   Alkenylzirconocenes derived from alkynes are converted into the zinc reagents (with Me$_2$Zn at −65°), which react with aldehydes.

[1] R. Rossi, A. Carpita, F. Bellina, and P. Cossi, *JOMC* **451**, 33 (1993).
[2] B. H. Lipshutz, M. Alami, and R. B. Susfalk, *SL* 693 (1993).
[3] P. Wipf and W. Xu, *TL* **35**, 5197 (1994).
[4] K. A. Agrios and M. Srebnik, *JOMC* **444**, 15 (1993); idem., *JOC* **59**, 5468 (1994).

Vitamin B$_{12}$–zinc.

Ring cleavage.[1]   Cyclopropanes are isomerized to give chiral alkenes.

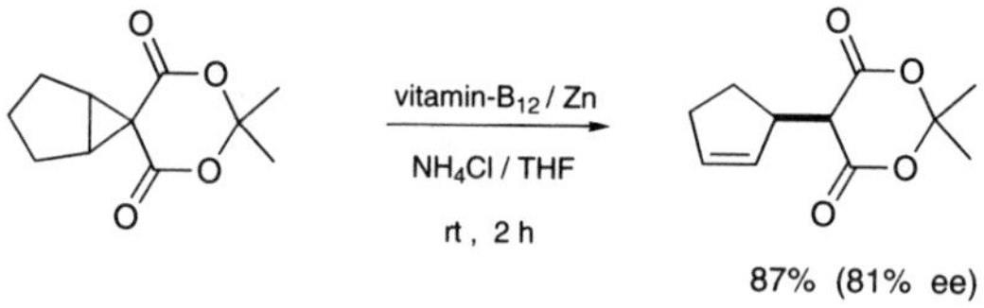

87% (81% ee)

[1] T. Troxler and R. Scheffold, *HCA* **77**, 1193 (1994).

W

Water. **16**, 383

Water has been used as a solvent, reagent, or catalyst for organic reactions at high temperatures and pressures. These reactions are typically acid- or base-catalyzed processes: hydrolysis, elimination, rearrangements, etc.[1]

[1] B. Kuhlmann, E. M. Arnett, and M. Siskin, *JOC* **59**, 3098 (1994).

X

Xenon(II) fluoride. 13, 345

Fluorination of aromatic compounds. Phenols[1] and pyrrole derivatives[2] afford fluorinated products on reaction with XeF$_2$.

Fluorodemetallation. Replacement of silyl and stannyl substituents from aryl-silanes[3] and β-trialkylstannyl enones[4] is rapid. The latter reaction is effected in the presence of AgOTf as catalyst.

Fluoroselenylation.[5] The combination of XeF$_2$ with R$_2$Se$_2$ or RSeSiR$_3$ delivers the RSe/F groups to alkynes in the *trans* fashion.

α,ω-Diphenylperfluoroalkanes.[6] The treatment of an *n*-perfluoroalkanedicar-boxylic acid and benzene with XeF$_2$ in CH$_2$Cl$_2$ at room temperature is sufficient to bring about the transformation. Yields are moderate (3 examples, 47–65%).

Fluorinated methyl aryl ethers. The facile rearrangement of oxygenated benzyl derivatives effected by XeF$_2$ to give fluorinated methyl ethers, i.e., benzyl alcohols to fluoromethyl aryl ethers[7] and benzaldehydes to difluoromethyl aryl ethers,[8] may involve the formation of fluoroxy intermediates and thence oxaspirobenzenium ions.

R = H, 2-NO$_2$, 3-F, 4-CF$_3$ 75 - 85%

67 - 86%

[1] I. Takemoto and K. Yamasaki, *BBB* **58**, 594 (1994).
[2] J. Wang and A. I. Scott, *TL* **35**, 3679 (1994).
[3] A. P. Lothian and C. A. Ramsden, *SL* 753 (1993).
[4] M. A. Tius and J. K. Kawakami, *SL* 207 (1993).
[5] H. Poleschner, M. Heydenreich, K. Spindler, and G. Haufe, *S* 1043 (1994).
[6] V. K. Brel, V. L. Uvarov, N. S. Zefirov, P. J. Stang, and R. Caple, *JOC* **58**, 6922 (1993).
[7] S. Stavber and M. Zupan, *TL* **34**, 4355 (1993).
[8] S. Stavber, Z. Koren, and M. Zupan, *SL* 265 (1994).

Y

Ytterbium. **14**, 348; **15**, 336; **16**, 384

Reduction of sulfur compounds. Formation of the metal(III) thiolates from ytterbium and disulfides is catalyzed by benzophenone.[1] The thiolates are useful in conjugate addition to enones.

Diaryl thioketones are reduced to mercaptans[2] together with minor amounts of desulfurized products.

Reductive coupling of ketones.[3] In the presence of trimethylsilyl bromide both saturated and conjugated ketones undergo coupling, at ipso and β-positions, respectively.

Silyl ketones behave differently. While aliphatic silyl ketones are reduced to α-silyl alcohols, aromatic congeners give 1,2-diarylalkynes.[4] In the presence of Me₃SiCl, the purported oxaytterbiacyclopropane intermediate can be trapped. Yet another pathway appears when the reaction is conducted in the presence of another ketone. A tandem deoxygenation–acylation reaction occurs with the silyl ketones donating the acyl moieties.[5]

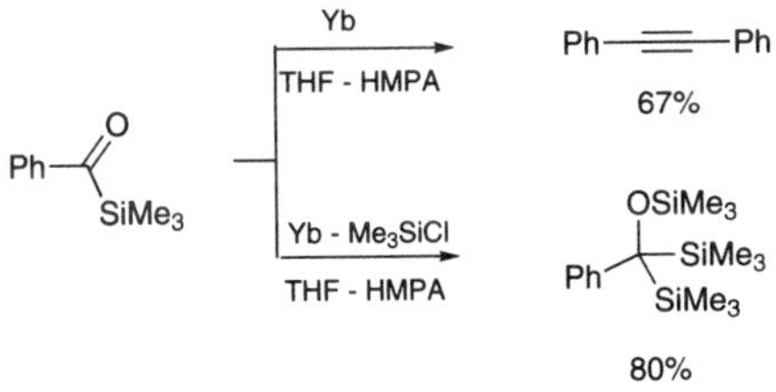

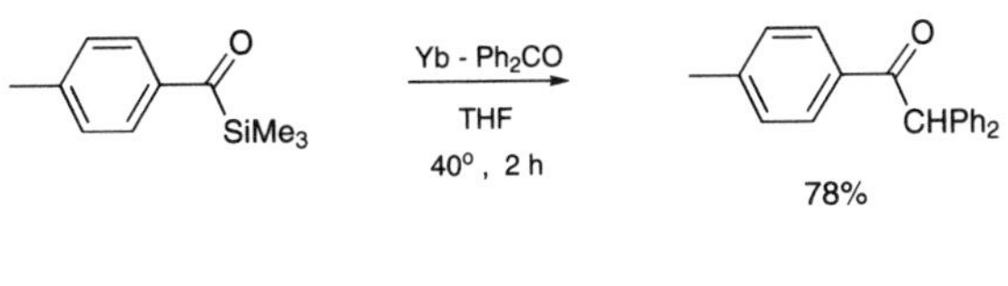

[1] Y. Taniguchi, M. Maruo, K. Takaki, and Y. Fujiwara, *TL* **35**, 7789 (1994).

[2] Y. Makioka, S.-Y. Uebori, M. Tsuno, Y. Taniguchi, K. Takaki, and Y. Fujiwara, *CL* 611 (1994).

[3] Y. Taniguchi, M. Nakahashi, T. Kuno, M. Tsuno, Y. Makioka, K. Takaki, and Y. Fujiwara, *TL* **35**, 4111 (1994).

[4] Y. Taniguchi, N. Fujii, Y. Makioka, K. Takaki, and Y. Fujiwara, *CL* 1165 (1993).

[5] Y. Taniguchi, A. Nagatiji, Y. Makioka, K. Takaki, and Y. Fujiwara, *TL* **35**, 6897 (1994).

Ytterbium(III) chloride.

Hydroperfluoroalkylation of alkenes.[1] A perfluoroalkyl group is transferred to the terminal carbon of a 1-alkene when the perfluoroalkyl iodide is treated with the YbCl$_3$–Zn system.

[1] Y. Ding, G. Zhao, and W. Huang, *TL* **34**, 1321 (1993).

Ytterbium(III) isopropoxide.

Cyanohydrin formation.[1] The transfer of a cyano group from acetone cyano-hydrin to other carbonyl compounds is rapidly accomplished with the alkoxide catalyst.

[1] H. Ohno, A. Mori, and S. Inoue, *CL* 375 (1993).

Ytterbium, tris(1,1,1,2,2,3,3-heptafluoro-7,7-dimethyl-4,6-octanedionato)-.

Ene reactions.[1] The Yb(fod)$_3$ complex promotes the ene reaction between an aldehyde and a vinyl ether at room temperature.

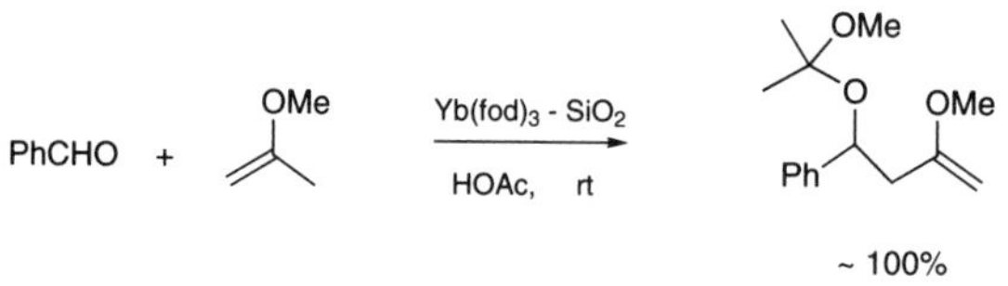

[1] M.V. Deaton and M. A. Ciufolini, *TL* **34**, 2409 (1993).

Ytterbium(III) nitrate.

Oxidation catalyst. The oxidation of alcohols to aldehydes or ketones by iodo-sylbenzene is catalyzed by this salt. Other lanthanide(III) nitrates are also effective, but LnCl$_3$ are inferior.

[1] T. Yokoo, K. Matsumoto, K. Oshima, and K. Utimoto, *CL* 571 (1993).

Ytterbium(III) triflate.

Conjugate addition. The catalyzed addition of an amine to an unsaturated ester to afford a β-amino ester (precursor of β-lactam) is subject to stereocontrol by the environment at the γ-carbon atom of the ester.[1] High pressure has a favorable effect on reactions of hindered reactants.[2]

78% (76% de)

Cross-aldol reactions. In the presence of Yb(OTf)$_3$ and Et$_3$N, ketones condense with aldehydes at room temperature.[3] As expected silyl enol ethers show better reactivities in the aldolization,[4] and it is remarkable that the reaction can be carried out in media containing water.[5]

Allylation of aldehydes.[6] Ytterbium triflate promotes the transfer of an allyl group from allyltributyltin to aldehydes at room temperature. Yields range from good to excellent (66–93%, 6 examples).

Isomerization of silyl ketene acetals.[7] Reaction of these ketone acetals proceed by migration of the silyl group from oxygen to carbon, with the formation of α-silyl-alkanoic esters, is very facile (5 min, room temperature).

Friedel–Crafts acylation.[8] The catalyst has the advantage of being readily recovered and reused without decrease in activity.

Cleavage of small heterocycles. Aziridines,[9] epoxides,[10,11] and oxetanes[12] are opened with amines when catalyzed by Yb(OTf)$_3$. The method has been compared with the pressure reaction.[11]

Glycosylation. The 1-*O*-methoxyacetoxyl group of sugar derivatives is replaced by a selected alkoxy unit in a catalyzed reaction.[13] Actually, the ester group does not need to be present, as added methoxyacetic acid can be used as cocatalyst for the glycosylation of the sugars with alcohols.[14]

[1] S. Matsubara, M. Yoshioka, and K. Utimoto, *CL* 827 (1994).

[2] G. Zenner, *TL* **36**, 233 (1995).

[3] S. Fukuzawa, T. Tsuchimoto, and T. Kanai, *BCSJ* **67**, 2227 (1994).

[4] S. Kobayashi, I. Hachiya, and T. Takahori, *S* 371 (1993).

[5] S. Kobayashi and I. Hachiya, *JOC* **59**, 3590 (1994).

[6] H. C. Aspinall, A. F. Browning, N. Greeves, and P. Ravenscroft, *TL* **35**, 4639 (1994).

[7] Y. Makioka, T. Taniguchi, K. Takaki, and Y. Fujiwara, *CL* 645 (1994).

[8] A. Kawada, S. Mitamura, and S. Kobayashi, *CC* 1157 (1993).

[9] M. Meguro, N. Asao, and Y. Yamamoto, *TL* **35**, 7395 (1994).

[10] M. Chini, P. Crotti, L. Favero, F. Macchia, and M. Pineschi, *TL* **35**, 433 (1994).

[11] M. Meguro, N. Asao, and Y. Yamamoto, *JCS(P1)* 2597 (1994).

[12] P. Crotti, L. Favero, F. Macchia, and M. Pineschi, *TL* **35**, 7089 (1994).

[13] J. Inanaga, Y. Yokoyama, and T. Hanamoto, *TL* **34**, 2791 (1993).

[14] J. Inanaga, Y. Yokoyama, and T. Hanamoto, *CC* 1090 (1993).

Yttrium(III) chloride.

Oxazolidinone formation.[1] Catalyzed epoxide opening and the subsequent formal [3+2]cycloaddition to isocyanates constitute a useful synthetic transformation.

Biscyclopentadienylyttrium chloride also promotes epoxide cleavage with acyl chlorides, giving chlorohydrin acetates.[2]

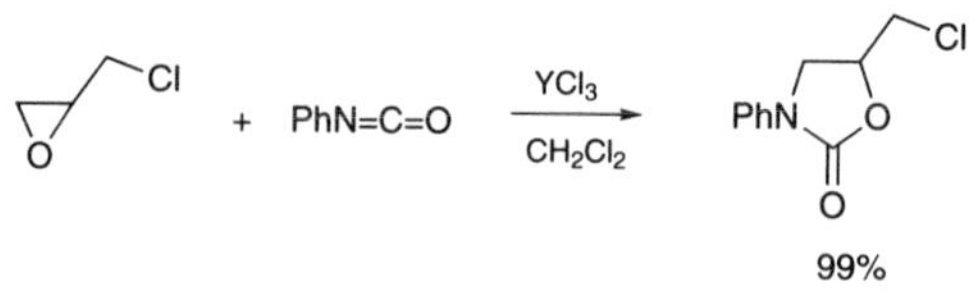

99%

[1] C. Qian and D. Zhu, *SL* 129 (1994).
[2] C. Qian and D. Zhu, *SC* **24**, 2203 (1994).

Yttrium(III) triflate.

γ-Hydroxy ketones.[1]   With Yb(OTf)$_3$ as catalyst, lithium enolates attack epoxides regioselectively.

[1] P. Crotti, V. Di Bussolo, L. Favero, F. Macchia, and M. Pineschi, *TL* **35**, 6537 (1994).

Z

Zeolites. 15, 367

Etherification.[1]   The Na–Y zeolite is an efficient catalyst for methoxymethylation of alcohols.

Trans-N-acylation.[2]   Anilides can be converted to N-(α-chloropropanoyl)-anilines on reaction with the acyl chloride over HZSM-5 zeolite.

58% (Ar=4-MeC₆H₄)

Benzoylation.[3]   The regioselective Friedel–Crafts reaction of activated arenes with α,α,α-trichlorotoluene in refluxing 1,2-dichloroethane (53–97% yield) is catalyzed by HZSM-5 zeolite.

Oxidative dimerization of arylamines.[4]   Azoxybenzenes are formed when the amines and hydrogen peroxide are heated under reflux in acetone in the presence of TS-1 zeolite.

Acetalization[5]*/dithioacetalization.*[6]   The H–Y zeolite is most commonly used as a catalyst for this reaction, which includes tetrahydropyranation of alcohols.[7]

Nitroketene S,N-acetals.[8]   In refluxing nitromethane and in the presence of a zeolite, one of the methylthio groups of bis(methylthio)methylenimines is replaced.

Cyclization.[9]   A macrocyclic lactone has been acquired (with yields up to 32%) from 15-hydroxypentadecanoic acid by exposure to dealuminated HY zeolite. Five- and six-membered heterocycles can be synthesized from diols with a modified ZSM-5 zeolite (from 18–30 mesh HZSM-5 and Cr₂O₃, and activated at 420°C for 4 h). For example, N-methylpyrrolidine is obtained in 64% yield from 1,4-butanediol and methylamine.[10]

Bromination.[11]   The regioselectivity of NBS bromination can be controlled by using the HZSM-S zeolite.

Sulfoxides.[12]   Oxidation of sulfides with NaBrO₂ · 3H₂O in aprotic solvents is assisted by zeolite.

[1] P. Kumar, S.V. N. Raju, R. S. Reddy, and B. Pandey, *TL* **35,** 1289 (1994).
[2] H. R. Sonawane, A.V. Pol, P. P. Moghe, A. Sudalai, and S. S. Biswas, *TL* **35,** 8877 (1994).
[3] V. Paul, A. Sudalai, T. Daniel, and K.V. Srinivasan, *TL* **35,** 2602 (1994).
[4] H. R. Sonawane, A.V. Pol, P. P. Moghe, S. S. Biswas, and A. Sudalai, *CC* 1215 (1994).

[5] P. Kumar, K. R. Reddy, and R. S. Reddy, *JCR(S)* 394 (1994).

[6] P. Kumar, R. S. Reddy, A. P. Singh, and B. Pandey, *S* 67 (1993).

[7] P. Kumar, C. U. Dinesh, R. S. Reddy, and B. Pandey, *S* 1069 (1993).

[8] T. I. Reddy, B. M. Bhawal, and S. Rajappa, *T* **49**, 2101 (1993).

[9] T. Tatsumi, H. Sakashita, and K. Asano, *CC* 1264 (1993).

[10] Y. V. Subba Rao, S. J. Kulkarni, M. Subrahmanyam, and A. V. Rama Rao, *JOC* **59**, 3998 (1994).

[11] V. Paul, A. Sudalai, T. Daniel, and K. V. Srinivasan, *TL* **35**, 7055 (1994).

[12] M. Hirano, H. Kubo, and T. Morimoto, *BCSJ* **67**, 1492 (1994).

Zinc. 13, 346–347; 14, 349–350; 16, 386–387; 17, 406–407

Cleavage of p-nitrobenzyl group.[1]   Protection of carboxylic acids and amines as the esters and carbamates can take advantage of the selective reduction-induced fragmentation by zinc dust, as C=C bonds, S–N bonds, benzyloxycarbonyl, and diphenylmethyl groups are not affected during the operation.

Elimination.   An improved synthesis of unsaturated pyranosides by elimination of *vic*-ditosylates is mediated by Zn–NaI with microwave heating.[2] Under such conditions higher yields and faster rates are observed. Dechlorination of β-chloro-β-(chlorodifluoromethyl)vinyl ketones to afford difluoroallyl ketones is effected with assistance of ultrasound.[3]

Glycal formation from acylated glycosyl bromides on treatment with zinc dust under aprotic conditions is favored by the addition of 1-methylimidazole.[4] A similar process based on the traditional procedure has been incorporated into a method for the hydrolysis of γ,δ-unsaturated anilides[5] and the synthesis of chiral allylic alcohols.[6]

Long-ranged defluorination of 1,4-bis(perfluoroalkyl)-1,2,3-butatrienes with zinc in DMF establishes a dienyne system.[7]

Organozinc reagents.   *o*-Phenylenedizinc(II) species are generated from the *o*-diiodo precursors with ultrasound irradiation.[8] These compounds are stabilized by complexation with TMEDA.

Rieke zinc is useful for the preparation of secondary and tertiary alkyl zinc bromides.[9] The employment of activated zinc deposited on TiO_2, which is obtained from reduction of $ZnCl_2$ by sodium dispersed on TiO_2, is another possibility (for *sec*-R and benzylic bromides).[10]

Alkylation.   Zinc-mediated reaction of cinnamyl chloride with aldehydes and ketones in aqueous THF[11] and Barbier-type reactions involving perfluoroalkyl iodides[12] have been demonstrated. Another method employs pyridinium perchlorate[13] to induce the transpositional allylation of aldehydes with allyl bromides.

Concurrent metallation and 1,4-debromination occur when 2-bromomethyl-1,4-dibromo-2-butene is treated with zinc. The resulting species is an isoprenylating agent, which may be applied to the synthesis of ipsenol and ipsdienol.[14] A more active zinc metal for this purpose can be produced by electrolysis.[15]

34 - 99%

The electrolytic method is extendable to other systems, including a direct approach to α-methylene-γ-lactones.[16]

Reformatsky and Blaise reactions. 2-Fluoro-3-hydroxyalkanoic esters are accessible[17] by a Reformatsky reaction using α-fluorinated haloesters to prepare the reagents. In some cases better results are obtained with diethylaluminum chloride as catalyst.[18]

The Reformatsky reaction with N-acyloxazolidin-2-ones and -thiazolidin-2-thiones provides a route to β-keto esters.[19] The electrolysis of α-bromoalkanoic esters in acetonitrile on a zinc anode with $ZnBr_2$ and Bu_4NBF_4 as supporting electrolytes gives acetoacetic ester derivatives.[20] The acetonitrile solvent is partially consumed.

Conjugate addition to unsaturated esters. Intramolecular[21] and intermolecular versions,[22] induced by zinc alone or in the presence of an additive such as Fe(III), Co(III),[23] have been reported.

60%

Deoxygenation of ketones. A modified Clemmensen reduction uses zinc and $AlCl_3 \cdot 6H_2O$ in aqueous THF.[24] Hydrochloric acid is generated in situ. On the other hand, ultrasound promotes the reduction with Zn-HOAc in a selective manner (of 3-oxosteroids in the presence of ketone group at C-17 or C-20).[25] Transpositional reduction of steroidal 4-en-3-ones (to 3-enes)[26] can be achieved in 15 minutes.

Reductions. The reduction of nitroarenes[27] and aldehydes[28,29] by the Zn–$CoCl_2$ combination in aqueous DMF or THF seems to offer some advantages.

[1] T. Kumagai, T. Abe, Y. Fujimoto, T. Hayashi, Y. Inoue, and Y. Nagao, *H* **36**, 1729 (1993).

[2] L. H. B. Baptistella, A. Z. Neto, H. Onaga, and E. A. M. Godoi, *TL* **34**, 8407 (1993).

[3] T. Okano, T. Shimizu, K. Sumida, and S. Eguchi, *JOC* **58**, 5163 (1993).

[4] L. Somsak and I. Nemeth, *JCC* **12**, 679 (1993).

[5] P. Metz, *T* **49**, 6367 (1993).

[6] S.-K. Kang, S.-G. Kim, D.-G. Cho, and J.-H. Jeon, *SC* **23**, 681 (1993).

[7] P. A. Morken, D. J. Burton, and D. C. Swensen, *JOC* **59**, 2119 (1994).

[8] K. Takagi, Y. Shimoshi, and K. Sasaki, *CL* 2055 (1994).

[9] M. V. Hanson, J. D. Brown, R. D. Rieke, and Q. J. Niu, *TL* **35**, 7205 (1994).

[10] H. Stadtmuller, B. Greve, K. Lennick, A. Chair, and P. Knochel, *S* 69 (1995).

[11] R. Sjoholm, R. Rairama, and M. Ahonen, *CC* 1217 (1994).

[12] Y. Shen and M. Qi, *JFC* **66**, 175 (1994).

[13] H. Maeda, J. Kawabata, and H. Ohmori, *CPB* **40**, 2834 (1992).

[14] M. Tokuda, N. Mimura, K. Yoshioka, T. Karasawa, H. Fujita, and H. Suginome, *S* 1086 (1993).

[15] M. Tokuda, N. Mimura, T. Karasawa, H. Fujita, and H. Suginome, *TL* **34**, 7607 (1993).

[16] Y. Rollin, S. Derien, E. Dunach, C. Gebehenne, and J. Perichon, *T* **49**, 7723 (1993).

[17] Y. Shen and M. Qi, *JCR(S)* 222 (1993).

[18] T. Ishihara, T. Matsuda, K. Imura, H. Matsui, and H. Yamanaka, *CL* 2167 (1994).

[19] C. Kashima, X. C. Huang, Y. Harada, and A. Hosomi, *JOC* **58**, 793 (1993).

[20] N. Zylber, J. Zylber, Y. Rollin, E. Dunach, and J. Perichon, *JOMC* **444**, 1 (1993).

[21] B. S. Bronk, S. J. Lippard, and R. L. Danheiser, *OM* **12**, 3340 (1993).

[22] P. Blanchard, A. D. Da Silva, M. S. El Kortbi, J.-L. Fourrey, and M. Robert-Gero, *JOC* **58**, 6517 (1993).

[23] J. Chen and J.-M. Hu, *JCS(PI)* 1111 (1994).

[24] P. K. Chowdhury and P. Borah, *JCR(S)* 230 (1994).

[25] J. A. R. Salvador, M. L. Sa e Melo, and A. S. Campos Neves, *TL* **34**, 361 (1993).

[26] J. A. R. Salvador, M. L. Sa e Melo, and A. S. Campos Neves, *TL* **34**, 357 (1993).

[27] R. N. Baruah, *IJC(B)* **33B**, 758 (1994).

[28] R. N. Baruah, *IJC(B)* **33B**, 182 (1994).

[29] A. Goswami and N. Borthakur, *IJC(B)* **33B**, 495 (1994).

Zinc–copper couple. 13, 348; 15, 367–368; 16, 387–388; 17, 407

Elimination. Sometimes zinc dust alone is inadequate for achieving organic reactions. Surface modification can have dramatic effects. The Zn–Cu couple is useful for the synthesis of chiral allylic alcohols[1] from epoxy tosylates and of allylamines[2] from 2-(bromomethyl)aziridines (with sonication).

Organozinc reagents. Zn–Cu couple often shows higher reactivity than zinc dust towards alkyl halides. Thus perfluoroalkylzinc iodides are rapidly formed in the presence of AIBN, and quenching such reagents in situ by DMF gives perfluoroalkyl aldehydes.[3] (Note that Zn–Ag on graphite is well suited for the preparation of arylzinc species).[4]

[1] E. Balmer, A. Germain, W. P. Jackson, and B. Lygo, *JCS(P1)* 399 (1993).
[2] N. De Kimpe, R. Jolie, and D. De Smaele, *CC* 1221 (1994).
[3] S. Benefice-Malouet, H. Blancou, and A. Commeyras, *JFC* **63**, 217 (1993).
[4] A. Fürstner, R. Singer, and P. Knochel, *TL* **35**, 1047 (1994).

Zinc bismuthate.

Deoximation.[1]   The conversion seems to be limited to allylic and benzylic oximes.

[1] H. Firouzabadi and I. Mohammadpoor-Baltork, *SC* **24**, 489 (1994).

Zinc borohydride. 14, 351; 16, 388–389

Reduction of azides.[1]   The reduction, which affords amines, proceeds at room temperature. Acyl azides give primary amides.

Reductive amination.[2]   An N-methylation method consists of $ZnCl_2$-catalyzed condensation of the amine with paraformaldehyde and treatment with $Zn(BH_4)_2$.

[1] B. C. Ranu, A. Sarkar, and R. Chakraborty, *JOC* **59**, 4114 (1994).
[2] S. Bhattacharyya, A. Chatterjee, and S. K. Duttachowdhury, *JCS(P1)* 1 (1994).

Zinc bromide. 13, 349; 15, 368; 16, 389–391

Aldol-type reactions.   The catalytic effect of $ZnBr_2$ is demonstrated in the condensation of ketene bis(trimethylsilyl)acetals with aldimines in a synthesis of β-amino acids,[1] and in the formation of (*E*)-enals from N-(*t*-butyl) bis(trimethylsilyl)-acetaldimine and aldehydes.[2]

Electrophilic substitution of β-acetoxy-β-lactams.[3]   The preparation of penem intermediates in 92–95% yields is realized by the catalyzed reaction with thiocarboxylic acid salts.

Hetero-Diels–Alder reactions.   A method of pyridine synthesis by catalyzed reaction between a 1,2,3-triazine with an aldehyde enamine[4] is useful for the elaboration of fusaric acid.

[1] M. Mladenova and M. Bellassoued, *SC* **23**, 725 (1993).
[2] M. Bellassoued and A. Majidi, *JOC* **58**, 2517 (1993).
[3] W. Cabri, I. Candiani, F. Zarini, and A. Bedeschi, *TL* **35**, 3379 (1994).
[4] J. Koyama, T. Ogura, and K. Tagahara, *H* **38**, 1595 (1994).

Zinc p-(t-butyl)benzoate.

Glycosylation.[1] The reaction between glycosyl chlorides and alcohols is catalyzed by the zinc carboxylate with fair to good β-selectivity.

[1] M. Nishizawa, D. M. Garcia, T. Shin, and H. Yamada, *CPB* **41**, 784 (1993).

Zinc chloride. 13, 349–350; 15, 368–371; 16, 391–392

Tetrahydropyranylation of alcohols.[1] $ZnCl_2$ impregnated in alumina is an effective catalyst.

Glycosylation.[2] Glycosyl phosphites are converted to ethers.

Hydrolysis of formamidines.[3] Ethanolic $ZnCl_2$ converts formamidines into formamides at room temperature. However, at reflux temperature amines are released.

Friedel–Crafts benzylation. The benzylation of arenes proceeds in good yields with benzyl chloride as reagent when $ZnCl_2$ is activated by the addition of a ketone, a primary alcohol, or water.[4]

p-Nitrobenzenesulfonyloxylation.[5] β-Keto esters undergo this reaction with bis(p-nitrobenzenesulfonyl) peroxide in the presence of $ZnCl_2$, and the products are readily converted to α-(p-nitrobenzenesulfonyloxy) ketones.

α-Phenylthioalkylation.[6,7] The standardized procedure involves transmetallation of potassium enolates to give the zinc enolates and treatment with the thioalkyl chlorides. The method is suitable for both saturated and unsaturated ketones.[8]

Michael addition. The condensation of silyl enol ethers with Michael acceptors is subject to catalysis by $ZnCl_2/Al_2O_3$.[9] The reaction is complete within 5 min at temperatures between 0° and 5°C.

The most expedient access to mono-dithioacetals of 3-cyclohexene-1,2-diones is by union of 2-acyl-1,3-dithianes with enones and subsequent aldolization, with the first step assisted by $ZnCl_2$.[10]

75%   60%

Lithium enolates of esters are inferior to the zinc enolates with respect to their reactivities towards β-amino nitroethenes.[11]

Asymmetric Michael additions involving chiral β-enamino esters[12] are catalyzed by $ZnCl_2$.

Destannylative reactions. 1-Trimethylstannyl-2,4-pentadiene reacts with aldehydes and ketones in a γ-selective manner,[13] whereas conjugated dienes are generated from 3-(tributylstannyl)propenyl pyrans.[14]

Hetero-Diels–Alder reactions. The synthesis of 4-piperidones using *N*-silylaldimines as dienophiles is promoted by ZnCl₂. Excellent stereoinduction by a chiral moiety of 2-amino-1,3-butadienes is evident.[15]

4-Amino-1-azadiene synthesis.[16] A direct approach to this class of compounds consists of reaction of the zincoenamines derived from *N*-arylketimines with nitriles.

54–95%

As radical initiator. The substitution reaction of α-bromo *N*-acylamino esters with allyltributyltin apparently occurs by a radical mechanism (forming the favorable capto-dative radical intermediates).[17] Zinc chloride plays a double role of a chelating agent and radical initiator.

Aza-Claisen rearrangement.[18] The catalyzed rearrangement of *N*-allyl-2,5-dimethoxyanilines to the 4-allyl isomers indicates participation of the *o*-methoxy substituent to direct the initial reaction at the *ipso* position.

[1] B. C. Ranu and M. Saha, *JOC* **59**, 8269 (1994).
[2] Y. Watanabe, C. Nakamoto, and S. Ozaki, *SL* 115 (1993).
[3] D. Toste, J. McNulty, and I. W. J. Still, *SC* **24**, 1617 (1994).
[4] E. Hayashi, Y. Takahashi, H. Itoh, and N. Yoneda, *BCSJ* **66**, 3520 (1993).
[5] R. V. Hoffman, H.-O. Kim, and J. C. Lee, *JOC* **59**, 1933 (1994).
[6] U. Groth, T. Huhn, and N. Richter, *LA* 49 (1993).
[7] R. Arnecke, U. Groth, and T. Kohler, *LA* 891 (1994).
[8] U. Groth, T. Kohler, and T. Taapken, *LA* 665 (1994).
[9] B. C. Ranu, M. Saha, and S. Bhar, *JCS(P1)* 2197 (1994).
[10] P. C. B. Page, A. P. Marchington, L. J. Graham, S. A. Harkin, and W. W. Wood, *T* **49**, 10369 (1993).
[11] K. Fuji, T. Kawabata, Y. Naniwa, T. Ohmori, and M. Node, *CPB* **42**, 999 (1994).
[12] A. Guingant and H. Hammami, *TA* **4**, 25 (1993).
[13] Y. Nishigaichi, M. Fujimoto, and A. Takuwa, *SL* 731 (1994).
[14] E. Kozlowska and S. Jarosz, *JCC* **13**, 889 (1994).
[15] J. Barluenga, F. Aznar, C. Valdes, A. Martin, S. Garcia-Granda, and E. Martin, *JACS* **115**, 4403 (1993).
[16] J. Barluenga, C. del Pozo Losada, and B. Olano, *TL* **34**, 5497 (1993).
[17] Y. Yamamoto, S. Onuki, M. Yumoto, and N. Asao, *JACS* **116**, 421 (1994).
[18] A. G. Mustafin, I. N. Khalilov, I. B. Abdrakhmanov, and G. A. Tolstikov, *JOCU* **28**, 1231 (1992).

Zinc fluoride.

Glycosyl fluorides.[1] Halogen exchange at the anomeric center is successful using ZnF₂ in the presence of 2,2'-bipyridine.

[1] K. D. Goggin, J. F. Lambert, and S. W. Walinsky, *SL* 162 (1994).

Zinc iodide. 13, 350–351

Addition to hindered ketones.[1]   With ZnI_2 as catalyst the formation of cyanohydrin silyl ethers using either *t*-butyldimethylsilyl or *t*-butyldiphenylsilyl cyanide is remarkably efficient.

Pummerer rearrangement.[2]   The *O*-silyl ketene acetal-induced rearrangement of sulfoxides also requires a catalyst such as ZnI_2.

Hetero-Diels–Alder reactions.   Dihydropyran synthesis by the condensation of enones and vinyl ethers in the presence of ZnI_2 is *endo*-selective.[3]

Ene reaction.[4]   3,3-Dimethyl-1-trimethylsilylmethylcyclohexene reacts with 3-butyn-2-one selectively to afford a vinylsilane product that is easily converted to γ-ionone.

[1] M. Golinski, C. P. Brock, and D. S. Watt, *JOC* **58**, 159 (1993).
[2] Y. Kita, N. Shibata, N. Yoshida, and S. Fujita, *JCS(P1)* 3335 (1994).
[3] E. Wada, H. Yasuoka, and S. Kanemasa, *CL* 145 (1994).
[4] G. Audran, H. Monti, G. Leandri, and J.-P. Monti, *TL* **34**, 3417 (1993).

Zinc triflate. 16, 392–393

Glycosylation.   Selective glycoylation is promoted by zinc triflate alone[1] or in combination with Me_3SiCl.[2]

Ene reaction.[3]

[1] K. Nakayama, K. Higashi, T. Soga, K. Uoto, and T. Kusama, *CPB* **40**, 2909 (1992).
[2] H. Susaki and K Higashi, *CPB* **41**, 201 (1993).
[3] E. Hayashi, Y. Takahashi, H. Itoh, and N. Yoneda, *BCSJ* **67**, 3040 (1994).

Zirconium(II) borohydride, polymer-supported.

Reduction.[1]   The instability of $Zr(BH_4)_2$ is greatly improved by mixing with poly(4-vinylpyridine). The consumed reagent can be regenerated.

[1] B. Tamami and N. Goudarzian, *CC* 1079 (1994).

Zirconium(IV) bromide.

Formal [3+2]cycloaddition.[1]   A one-step synthesis of γ-lactones consists of ring opening of 2,2-dialkoxycyclopropane esters to generate 1,3-dipolar species that react with carbonyl compounds. Zirconium bromide apparently activates both reactants.

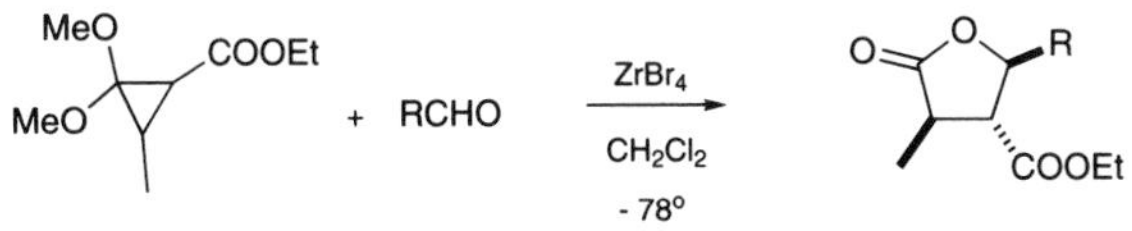

[1] S. Shimada, Y. Hashimoto, and K. Saigo, *JOC* **58**, 5226 (1993).

Zirconium(IV) chloride.

Rearrangement.[1] *vic*-Hydroxy sulfones give ketones analogous to the pinacol rearrangement.

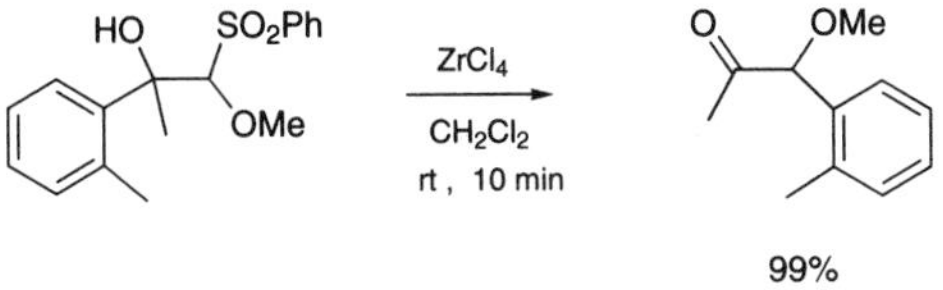

Homologation–condensation.[2] 3-Acyltetramic acids are accessible from *N*-acylacetylglycine esters. In turn the latter compounds may be assembled from α-diazo acetamide derivatives and aldehydes with elimination of dinitrogen. Zirconium chloride is a useful catalyst.

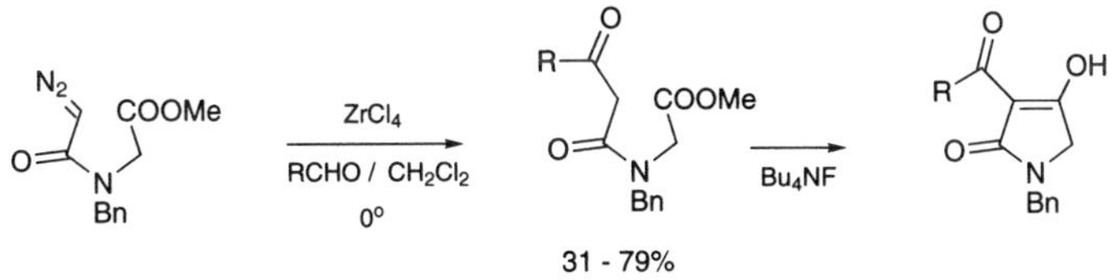

[1] J. G. Montana, N. Phillipson, and R. J. K. Taylor, *CC* 2289 (1994).
[2] T. Iida, K. Hori, K. Nomura, and E. Yoshii, *H* **38**, 1839 (1994).

Zirconium ditriflate, tetramethyl(dibenzo)tetraazaannulene.

Aldol-type reaction.[1] The low-temperature-catalyzed condensation of *O*-silyl ketene acetals with aldehydes leads to β-hydroxy esters. The same catalyst also mediates allylation of aldehydes.

[1] P. G. Cozzi, C. Floriani, A. Chiesi-Villa, and C. Rizzoli, *SL* 857 (1994).

Zirconium(IV) isopropoxide. 13, 352

Reduction.[1] Silica-supported $(i\text{-PrO})_4\text{Zr}$ mediates the reduction of carbonyl compounds to alcohols in refluxing isopropanol.

[1] K. Inada, M. Shibagaki, Y. Nakanishi, and H. Matsushita, *CL* 1795 (1993).

Zirconium(IV) oxide–trimethylsilyl chloride.

Lactonization.[1] ω-Hydroxy esters undergo cyclization with the modified ZrO_2 at room temperature.

Oxidation.[2] Aldehydes and ketones are produced using the reagent and PhCHO as hydrogen acceptor.

[1] H. Kuno, M. Shibagaki, K. Takahashi, and H. Matsushita, *BCSJ* **66**, 1305 (1993).
[2] H. Kuno, M. Shibagaki, K. Takahashi, and H. Matsushita, *BCSJ* **66**, 1699 (1993).

Zirconium phosphate.

Nef reaction.[1] The layered metal phosphate is an active catalyst for the solvent-free synthesis of β-nitro alcohols.

[1] U. Costantino, M. Curini, F. Marmottini, O. Rosati, and E. Pisani, *CL* 2215 (1994).

Zirconocene, Zr-alkylated. 15, 81

Coupling of alkenes/alkynes.[1] Cp_2ZrBu_2 is most often employed to couple unsaturated compounds to provide alkenes[1] and dienes.[2] An intramolecular version is useful for a stereoselective synthesis of *cis*-1,2-dimethyl cyclic systems.[3]

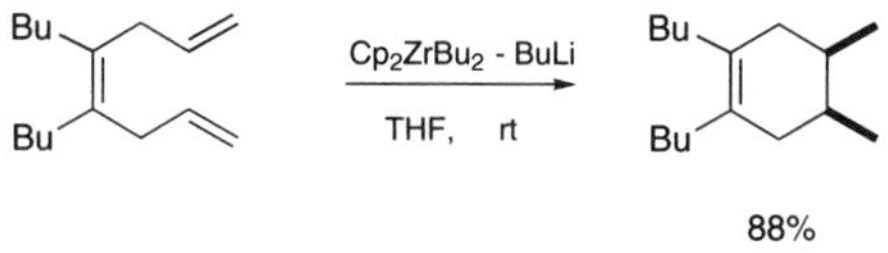

88%

Dehydrogenation and pinacol coupling. Allylamines give 1-aza-1,3-dienes,[4] whereas aldehydes furnish α-hydroxycarbanion equivalents, which condense with other carbonyl compounds.[5]

Ring contraction.[6] Vinylmorpholine derivatives are converted into *cis*-3-hydroxy-4-vinylpyrrolidines.

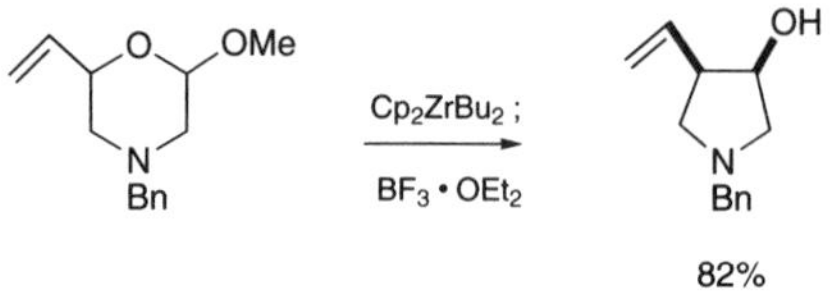

82%

[1] T. Takahashi, Z. Xi, C. J. Rousset, and N. Suzuki, *CL* 1001 (1993).
[2] T. Takahashi, D. Y. Kondakov, and N. Suzuki, *CL* 259 (1994).
[3] T. Takahashi, M. Kotora, and K. Kasai, *CC* 2693 (1994).
[4] J. Barluenga, R. Sanz, R. Gonzalez, and F. J. Fananas, *CC* 989 (1994).
[5] F. R. Askham and K. M. Carroll, *JOC* **58**, 7328 (1993).
[6] H. Ito, Y. Ikeuchi, T. Taguchi, Y. Hanzawa, and M. Shiro, *JACS* **116**, 5469 (1994).

Zirconocene π-complexes.

Reductive alkylations.[1,2] Reaction of the complexed species with electrophiles followed by protonation accomplishes reductive alkylations. However, the zirconocene–ethene complex effects cyclopropylmethylation of alkynes with 4-bromobutene.[3]

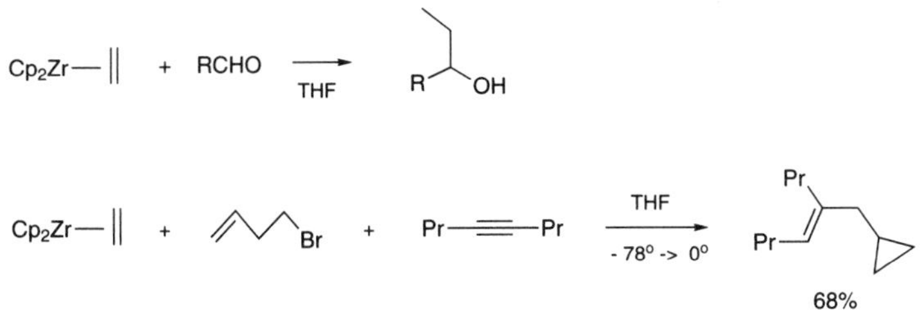

[1] N. Suzuki, C. J. Rousset, K. Aoyagi, M. Kotora, T. Takahashi, M. Hasegawa, Y. Nitto, and M. Saburi, *JOMC* **473**, 117 (1994).
[2] T. Takahashi, N. Suzuki, M. Kageyama, D. Y. Kondakov, and R. Hara, *TL* **34**, 4811 (1993).
[3] T. Takahashi, D. Y. Kondakov, and N. Suzuki, *TL* **34**, 6571 (1993).

Zirconocene dichloride. 14, 122; **15**, 120–121

Aminolysis.[1] Cp_2ZrCl_2 is a useful catalyst for aminolysis of N-acyl-2-oxazolidinones.

Deallylation.[2] The combination with BuLi forms "Cp_2Zr" which is effective for cleavage of allyl ethers and amines. The reagent shows a higher reactivity towards allyl ethers.

[1] T. Yokomatsu, A. Arakawa, and S. Shibuya, *JOC* **59**, 3506 (1994).
[2] H. Ito, T. Taguchi, and Y. Hanzawa, *JOC* **58**, 774 (1993).

Zirconocene hydride. 13, 108; **14**, 37; **15**, 52

Tishchenko reaction.[1] Aldehydes are disproportionately dimerized on reaction with Cp_2ZrH_2 at 0°C.

[1] K.-i. Morita, Y. Nishiyama, and Y. Ishii, *OM* **12**, 3748 (1993).

Zirconocene hydrochloride. **14**, 81; **15**, 80–81

Hydrozirconation.[1]   Acylsilanes containing a vinyl or ethynyl group undergo hydrozirconation instead of reaction at the carbonyl if the silyl group is hindered (e.g., *i*-Pr$_3$Si).

In the presence of Ag(I) catalyst the zirconylalkenes react with aldehydes which on hydrolysis give enals. Conjugated aldehydes with many double bonds are accessible by iterative operation of the process.[2]

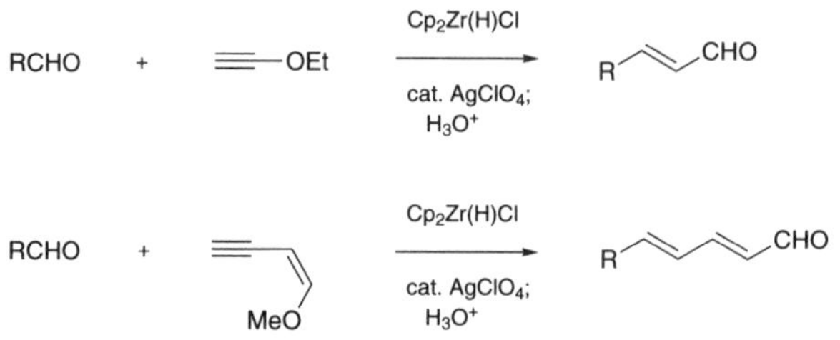

Imines.[3]   Secondary amides and large-ring lactams are converted into imines on sequential treatment with KH and Cp$_2$Zr(H)Cl.

Alkenylzinc reagents.   Hydrozirconation of 1-alkynes followed by treatment with MeLi, Me$_3$ZnLi provides alkenylzincates, which are useful for transfer of alkenyl groups to enones.[4] Prostaglandins and substances of similar skeleton are accessible by a 3-component coupling.

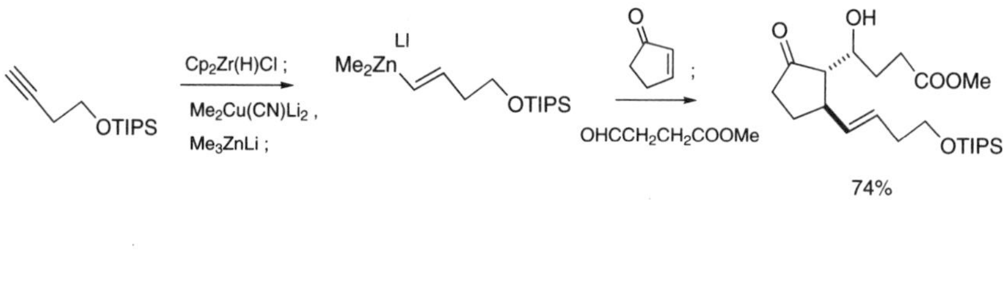

74%

Allylic alcohols.   When alkenylzirconocene chlorides derived from 1-alkynes react with epoxides in the presence of AgClO$_4$, isomerization of the epoxides to carbonyl compounds intervenes; therefore, allylic alcohols result.[5] However, epoxy carbonyl compounds are attacked at the C=O group.[6]

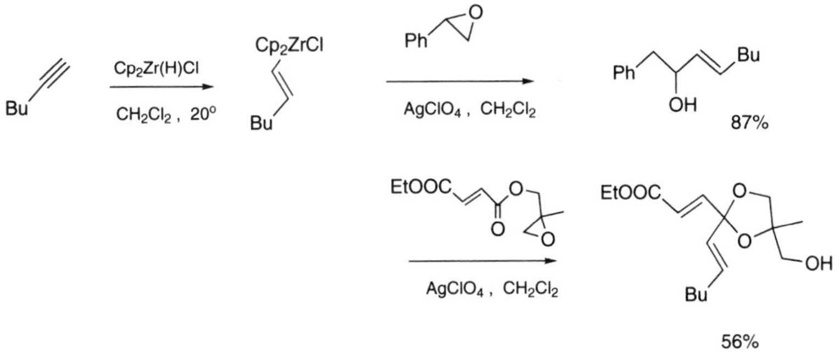

87%

56%

Allylzirconocene chlorides.   These reagents are formed from allenes. A route to 1-alken-4-ols by reaction with aldehydes has been developed.[7] 1,1-Bimetallic species derived from allenylstannanes condense with aldehydes to form 1,3-dienes.[8]

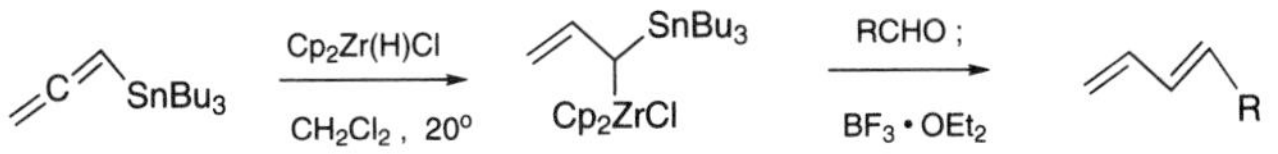

Reaction with phosphorus compounds.   Hydrozirconation of vinyl(diphenyl)-phosphine at C-1[9] is feasible, while ring cleavage of 1-phenylphospha-3-cyclopentene to give homoallylic phosphines[10] is observed. (1-Benzyl-3-pyrroline and dihydro-furan also undergo similar cleavage.)

[1] B. H. Lipshutz, C. Lindsley, and A. Bhandari, *TL* **35**, 4669 (1994).
[2] H. Maeta and K. Suzuki, *TL* **34**, 341 (1993).
[3] D. J. A. Schedler, A. G. Godfrey, and B. Ganem, *TL* **34**, 5035 (1993).
[4] B. H. Lipshutz and M. R. Wood, *JACS* **115**, 12625 (1993); *JACS* **116**, 11689 (1994).
[5] P. Wipf and W. Xu, *JOC* **58**, 825 (1993).
[6] P. Wipf and W. Xu, *JOC* **58**, 5880 (1993).
[7] M. Chino, T. Matsumoto, and K. Suzuki, *SL* 359 (1994).
[8] H. Maeta, T. Hasegawa, and K. Suzuki, *SL* 341 (1993).
[9] M. Zablocka, A. Igau, J.-P. Majoral, and K. M. Pietruziewicz, *OM* **12**, 603 (1993).
[10] N. Cenac, M. Zablocka, A. Igau, J.-P. Majoral, and K. M. Pietruziewicz, *OM* **13**, 5166 (1994).

Zirconyl chloride.

Cyclopentenones.[1]   Both aldol condensation of ketones with aldehydes and sub-sequent Nazarov cyclization of the products to yield polysubstituted cyclopentenones are catalyzed by zirconyl chloride.

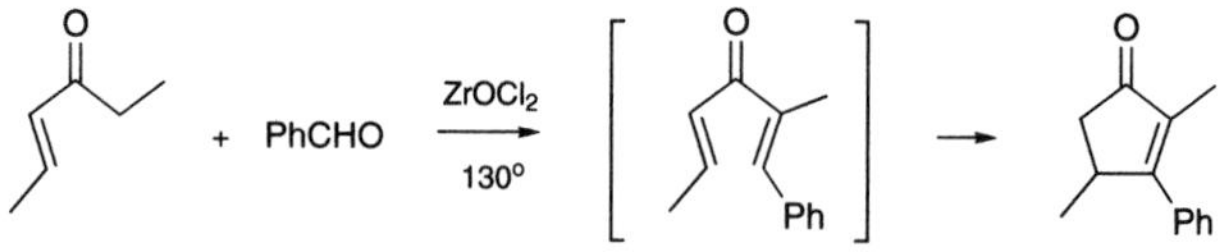

[1] T. Yuki, M. Hashimoto, Y. Nishiyama, and Y. Ishii, *JOC* **58**, 4497 (1993).

AUTHOR INDEX

Cicchi, S., 239
Cimarelli, C., 171, 221, 240, 295, 324, 340
Cimarusti, M. P., 375
Cinquini, M., 356, 362
Cintas, P., 319, 331
Cintas-Moreno, P., 331
Cipres, I., 129
Cirillo, P. F., 63
Citterio, A., 230
Ciuffreda, P., 203
Ciufolini, M. A., 84, 402
Cladingboel, D. E., 371
Claeson, G., 85, 209
Clark, D. N., 371
Clark, J. H., 298, 350
Clark, J. S., 110
Clark, R. D., 80
Clase, J. A., 125
Claudia, C., 145
Clay, R. J., 227
Clayden, J., 45, 173, 316, 349
Clerici, A., 359
Clerici, F., 356
Clery, P., 69, 291
Clive, D. L. J., 87, 272, 371, 388
Coan, P. S., 227
Cochennec, C., 214
Cochini F., 250
Cochran, B. B., 307
Coe, D. M., 15, 98
Coe, J. W., 339
Cohen, T., 111, 211, 223, 256
Colandrea, V. J., 114
Colclough, E., 204
Coldham, I., 77, 384
Cole, D. L., 153
Collazo, L., 269
Collazo, L. R., 102
Colletti, S. L., 95
Collin, J., 316
Collington, E. W., 96, 99, 249
Collington, N., 134
Collins, S., 95
Collom, T. A., 227
Collum, D. B., 327
Colomb, M., 315
Colombani, D., 288
Colombier, C., 49
Colson, P.-J., 103
Comasseto J. V., 151, 172, 262, 327

Combrink, K. D., 363
Comes-Franchini, M., 261
Cometti, G., 155
Comina, P. J., 104, 368
Comins, D. L., 170
Commenil, M.-G., 171
Commercon, M., 261
Commeyras, A., 409
Compain, P., 45
Concellon, J. M., 100
Concepcion, A. B., 21, 95, 237, 238
Congreve, M. S., 60
Connolly, C. B., 244
Consiglio, G., 97
Console, S., 276
Constantieux, T., 56
Contento, M., 141
Coogan, M. P., 179
Cooke, Jr, M. P., 170
Cooke, M. P., 65, 80, 170
Cooney, J. J. A., 173
Cooper, A. B., 102
Coote, S. J., 46, 297
Coppa, F., 24
Cordero, F. M., 239
Cordova, J. A., 298
Corey, E. J., 10, 96, 245, 371
Corley, E. G., 96
Cornell, C. L., 179
Correia, J., 289
Corriu, R. J. P., 261, 349
Cossi, P., 397
Cossy, J., 30, 107, 110, 191, 230, 269
Costa, A. L., 44
Costa, M., 283
Costantini, C., 339
Costantino, U., 414
Cote, B., 140, 387
Cotelle, P., 87
Coulter, T. S., 353
Coumbe, T., 375
Courtemanche, G., 265
Courtois, G., 16
Coustard, J.-M., 25
Coutts, S. J., 339
Couturier, D., 268
Couty, F., 166
Cox, G. G., 307
Cozzi, F., 356, 362
Cozzi, P. G., 413

Gyoung, Y. S., 340

Haaima, G., 122
Haar, J. P., 382
Habashita, H., 263
Habaue, S., 255
Habi, A., 300
Hachemi, M., 298
Hachiya, I., 95, 175, 317, 318, 403
Hackmann, C., 103
Hacksell, U., 121
Haddad, N., 106
Haemers, A., 178
Hageman, D. L., 349
Hager, L. P., 183
Hahn, J.-T., 28
Haight, D., 307
Hajicek, J., 261
Hajko, J., 208
Hakimelahi, G. H., 76
Halcomb, R. L., 384
Hale, K. J., 267
Hale, M. R., 261
Hall, D. G., 261
Hallberg, A., 276
Haller, S., 2
Hallett, D. J., 14
Halterman, R. L., 95
Hamada, M., 24
Hamada, T., 95
Hamaguchi, F., 358
Hamamoto, T., 134
Hamaue, F., 283
Hamel, P., 376
Hamelin, J., 245, 298
Hammami, H., 411
Hammock, B. D., 192
Hamon, D. P. G., 349
Hamura, S., 330
Han, G., 83, 113, 261
Han, J. L., 28
Han, S.-M., 85
Han, W., 161
Han, Y., 415
Hanaki, N., 21, 367, 394
Hanamoto, T., 130, 317, 367, 403
Hanaoka, H., 265, 310
Hanaoka, M., 361
Hanaoka, T., 129, 282
Hancock, G., 83

Handke, G., 261
Hands, N., 339
Handy, S. T., 219
Hanessian, S., 77, 96, 316
Hansen, J., 208
Hanson, M. V., 218, 408
Hansson, L., 361
Hanzawa, Y., 415
Hara, M., 25
Hara, R., 191, 415
Hara, S., 6, 181
Harada, H., 261, 316
Harada, N., 100, 129
Harada, T., 222, 265
Harada, Y., 408
Harayama, H., 45, 283, 348
Harders, J., 187
Hardinger, S. A., 71
Hardy, P. M., 179
Harimoto, T., 371
Harkin, S. A., 411
Harman, W. D., 344
Harmata, M., 261
Harms, K., 56
Harnett, J., 149
Harnett, J. J., 149
Haroutounian, S. A., 319
Harper, S., 233, 356
Harpp, D. N., 58, 145
Harring, S. R., 64
Harris, B. D., 340
Harris, C. E., 58
Harris, G. D., 280
Harrison, J., 209
Hart, D. J., 82
Hartley, R. C., 78
Harvey, D. F., 243
Harvey, I. W., 325
Harvey, J., 293
Harwood, L. M., 333
Hase, T., 153, 396, 417
Hasegawa, A., 286
Hasegawa, H., 14, 53
Hasegawa, M., 166, 361, 384, 394, 415
Hasegawa, T., 396, 417
Hashiguchi, S., 217
Hashimoto, K., 129, 245
Hashimoto, M., 24, 418
Hashimoto, T., 203
Hashimoto, Y., 263, 361, 363, 384, 413

SUBJECT INDEX